The Quark

" A Fundamental Constituent of Matter "

Edited by Paul F. Kisak

Contents

1	**Overview**	**1**
1.1	Quark	1
	1.1.1 Classification	1
	1.1.2 History	2
	1.1.3 Etymology	3
	1.1.4 Properties	4
	1.1.5 Interacting quarks	6
	1.1.6 See also	7
	1.1.7 Notes	7
	1.1.8 References	7
	1.1.9 Further reading	10
	1.1.10 External links	11
1.2	Eightfold Way	11
	1.2.1 Background	11
	1.2.2 References	12
	1.2.3 Further reading	12
1.3	Quark model	12
	1.3.1 History	13
	1.3.2 Mesons	13
	1.3.3 Baryons	14
	1.3.4 States outside the quark model	14
	1.3.5 See also	15
	1.3.6 References and external links	15
2	**Individual quarks**	**19**
2.1	Up quark	19
	2.1.1 History	19
	2.1.2 Mass	19
	2.1.3 See also	19
	2.1.4 References	20

- 2.1.5 Further reading ... 20
- 2.2 Down quark ... 20
  - 2.2.1 History ... 21
  - 2.2.2 Mass ... 21
  - 2.2.3 See also ... 21
  - 2.2.4 References ... 21
  - 2.2.5 Further reading ... 22
- 2.3 Strange quark ... 22
  - 2.3.1 History ... 22
  - 2.3.2 See also ... 23
  - 2.3.3 References ... 23
  - 2.3.4 Further reading ... 23
- 2.4 Charm quark ... 24
  - 2.4.1 Hadrons containing charm quarks ... 24
  - 2.4.2 See also ... 24
  - 2.4.3 Notes ... 24
  - 2.4.4 Further reading ... 24
- 2.5 Bottom quark ... 25
  - 2.5.1 Hadrons containing bottom quarks ... 25
  - 2.5.2 See also ... 25
  - 2.5.3 References ... 25
  - 2.5.4 Further reading ... 25
  - 2.5.5 External links ... 26
- 2.6 Top quark ... 26
  - 2.6.1 History ... 26
  - 2.6.2 Properties ... 27
  - 2.6.3 Production ... 27
  - 2.6.4 Decay ... 27
  - 2.6.5 Mass and coupling to the Higgs boson ... 28
  - 2.6.6 See also ... 29
  - 2.6.7 References ... 29
  - 2.6.8 Further reading ... 30
  - 2.6.9 External links ... 30

3 Flavour 31

- 3.1 Flavour ... 31
  - 3.1.1 Intuitive description ... 31
  - 3.1.2 Flavour symmetry ... 31
  - 3.1.3 Flavour quantum numbers ... 31

		3.1.4	Quantum chromodynamics	32
		3.1.5	Conservation laws	33
		3.1.6	History	33
		3.1.7	See also	33
		3.1.8	References	33
		3.1.9	Further reading	33
		3.1.10	External links	33
	3.2	Isospin		33
		3.2.1	Motivation for isospin	34
		3.2.2	Modern understanding of isospin	35
		3.2.3	Isospin symmetry	35
		3.2.4	Relationship to flavor	36
		3.2.5	Quark content and isospin	36
		3.2.6	Gauged isospin symmetry	37
		3.2.7	References	37
		3.2.8	Further reading	37
		3.2.9	External links	38
	3.3	Strangeness		38
		3.3.1	Conservation	38
		3.3.2	See also	38
		3.3.3	References	38
		3.3.4	Further reading	38
	3.4	Charm		38
		3.4.1	Further reading	38
	3.5	Bottomness		39
		3.5.1	Further reading	39
	3.6	Topness		39
		3.6.1	Conservation	39
		3.6.2	Further reading	39
4	**Other properties**			**40**
	4.1	Baryon number		40
		4.1.1	Baryon number vs. quark number	40
		4.1.2	Particles not formed of quarks	40
		4.1.3	Conservation	40
		4.1.4	See also	41
		4.1.5	References	41
	4.2	Color charge		41
		4.2.1	Red, green, and blue	41

- 4.2.2 Coupling constant and charge 42
- 4.2.3 Quark and gluon fields and color charges 42
- 4.2.4 See also 43
- 4.2.5 References 43
- 4.2.6 Further reading 43
- 4.3 Color confinement 44
  - 4.3.1 Origin 44
  - 4.3.2 Models exhibiting confinement 44
  - 4.3.3 Models of fully screened quarks 45
  - 4.3.4 See also 45
  - 4.3.5 References 45
  - 4.3.6 External links 45
- 4.4 Electric charge 45
  - 4.4.1 Overview 46
  - 4.4.2 Units 47
  - 4.4.3 History 47
  - 4.4.4 Static electricity and electric current 48
  - 4.4.5 Properties 49
  - 4.4.6 Conservation of electric charge 49
  - 4.4.7 See also 50
  - 4.4.8 References 50
  - 4.4.9 External links 50
- 4.5 Hypercharge 50
  - 4.5.1 Definition 50
  - 4.5.2 Relation with electric charge and isospin 50
  - 4.5.3 SU(3) model in relation to hypercharge 50
  - 4.5.4 Examples 51
  - 4.5.5 Practical obsolescence 51
  - 4.5.6 References 51
- 4.6 Spin 51
  - 4.6.1 Quantum number 51
  - 4.6.2 Magnetic moments 53
  - 4.6.3 Direction 54
  - 4.6.4 Mathematical formulation 55
  - 4.6.5 Parity 59
  - 4.6.6 Applications 59
  - 4.6.7 History 59
  - 4.6.8 See also 60

	4.6.9	References	60
	4.6.10	Further reading	61
	4.6.11	External links	61
4.7	Weak hypercharge		62
	4.7.1	Definition	62
	4.7.2	Baryon and lepton number	62
	4.7.3	See also	62
	4.7.4	Notes	62
4.8	Weak isospin		62
	4.8.1	Relation with chirality	63
	4.8.2	Weak isospin and the W bosons	63
	4.8.3	See also	63
	4.8.4	References	63
4.9	"B" − "L"		63
	4.9.1	Details	63
	4.9.2	See also	63
4.10	"X"		64
	4.10.1	See also	64
	4.10.2	Notes	64

5 Hadrons — 65

5.1	Hadron		65
	5.1.1	Etymology	65
	5.1.2	Properties	65
	5.1.3	Baryons	66
	5.1.4	Mesons	66
	5.1.5	See also	66
	5.1.6	References	66
5.2	Baryon		67
	5.2.1	Background	67
	5.2.2	Baryonic matter	67
	5.2.3	Baryogenesis	67
	5.2.4	Properties	68
	5.2.5	Nomenclature	70
	5.2.6	See also	70
	5.2.7	Notes	70
	5.2.8	References	71
	5.2.9	External links	71
5.3	List of baryons		71

- 5.3.1 Lists of baryons 72
- 5.3.2 See also 72
- 5.3.3 References 72
- 5.3.4 Further reading 73
- 5.3.5 External links 74
- 5.4 Meson 74
  - 5.4.1 History 75
  - 5.4.2 Overview 75
  - 5.4.3 Classification 78
  - 5.4.4 Exotic mesons 78
  - 5.4.5 List 78
  - 5.4.6 See also 78
  - 5.4.7 Notes 78
  - 5.4.8 References 79
  - 5.4.9 External links 80
- 5.5 List of mesons 80
  - 5.5.1 Summary table 81
  - 5.5.2 Meson properties 81
  - 5.5.3 See also 81
  - 5.5.4 References 81
  - 5.5.5 External links 83
- 5.6 Exotic hadron 83
  - 5.6.1 History 83
  - 5.6.2 Candidates 83
  - 5.6.3 See also 83
  - 5.6.4 Notes 83
- 5.7 Tetraquark 84
  - 5.7.1 History 84
  - 5.7.2 See also 84
  - 5.7.3 References 84
  - 5.7.4 External links 84
- 5.8 Pentaquark 85
  - 5.8.1 Background 85
  - 5.8.2 Structure 85
  - 5.8.3 History 86
  - 5.8.4 Applications 87
  - 5.8.5 See also 87
  - 5.8.6 Footnotes 87

		5.8.7 References	87
		5.8.8 Further reading	88
		5.8.9 External links	88

6 Appendix A – Related topics — 92

6.1 Antiparticle — 92
- 6.1.1 History — 92
- 6.1.2 Particle-antiparticle annihilation — 93
- 6.1.3 Properties of antiparticles — 94
- 6.1.4 Quantum field theory — 94
- 6.1.5 See also — 95
- 6.1.6 References — 95

6.2 Beta decay — 95
- 6.2.1 β^{*-} decay — 96
- 6.2.2 β^{*+} decay — 97
- 6.2.3 Electron capture (K-capture) — 97
- 6.2.4 Competition of beta decay types — 97
- 6.2.5 Helicity (polarization) of neutrinos, electrons and positrons emitted in beta decay — 97
- 6.2.6 Energy release — 97
- 6.2.7 Nuclear transmutation — 99
- 6.2.8 Double beta decay — 99
- 6.2.9 Bound-state β^{*-} decay — 99
- 6.2.10 Forbidden transitions — 100
- 6.2.11 Fermi transitions — 100
- 6.2.12 Gamow-Teller transitions — 100
- 6.2.13 Beta emission spectrum — 100
- 6.2.14 History — 101
- 6.2.15 See also — 102
- 6.2.16 References — 102
- 6.2.17 Bibliography — 104
- 6.2.18 External links — 104

6.3 CKM matrix — 104
- 6.3.1 The matrix — 104
- 6.3.2 Counting — 105
- 6.3.3 Observations and predictions — 105
- 6.3.4 Weak universality — 106
- 6.3.5 The unitarity triangles — 106
- 6.3.6 Parameterizations — 106
- 6.3.7 Nobel Prize — 107

	6.3.8	See also	107
	6.3.9	References	107
	6.3.10	Further reading	108
6.4	Color–flavor locking		108
	6.4.1	Color-flavor-locked Cooper pairing	108
	6.4.2	Physical properties	108
	6.4.3	References	108
6.5	Constituent quark		108
	6.5.1	Binding energy	109
	6.5.2	References	109
6.6	Current quark		109
	6.6.1	Current quark mass	109
	6.6.2	References	109
6.7	CP violation		109
	6.7.1	CP-symmetry	110
	6.7.2	Experimental status	110
	6.7.3	Strong CP problem	111
	6.7.4	CP violation and the matter–antimatter imbalance	112
	6.7.5	See also	112
	6.7.6	References	113
	6.7.7	Further reading	113
	6.7.8	External links	113
6.8	Fermion		113
	6.8.1	Elementary fermions	114
	6.8.2	Composite fermions	114
	6.8.3	See also	115
	6.8.4	Notes	115
6.9	Gell-Mann–Nishijima formula		115
	6.9.1	Formula	115
	6.9.2	References	116
	6.9.3	Further reading	116
6.10	Generation		116
	6.10.1	Overview	116
	6.10.2	Fourth generation	116
	6.10.3	See also	117
	6.10.4	References	117
6.11	Gluon		117
	6.11.1	Properties	117

| 6.11.2 Numerology of gluons 117
        6.11.3 Confinement 118
        6.11.4 Experimental observations 119
        6.11.5 See also 119
        6.11.6 References 119
        6.11.7 Further reading 120
  6.12 Hadronization 120
        6.12.1 Hadronization simulation and models 121
        6.12.2 References 121
  6.13 Lepton 121
        6.13.1 Etymology 122
        6.13.2 History 122
        6.13.3 Properties 123
        6.13.4 Universality 125
        6.13.5 Table of leptons 125
        6.13.6 See also 125
        6.13.7 Notes 125
        6.13.8 References 126
        6.13.9 External links 127
  6.14 Leptoquark 127
        6.14.1 Existence 127
        6.14.2 See also 128
        6.14.3 References 128
  6.15 November Revolution 128
        6.15.1 Background to discovery 128
        6.15.2 The name 129
        6.15.3 J/ψ melting 129
        6.15.4 Decay modes 129
        6.15.5 See also 129
        6.15.6 Notes 129
        6.15.7 References 130
  6.16 Parton 130
        6.16.1 Model 130
        6.16.2 Parton distribution functions 131
        6.16.3 References 131
        6.16.4 Further reading 132
        6.16.5 External links 132
  6.17 Preon 132

6.17.1 Background	132
6.17.2 Motivations	133
6.17.3 History	133
6.17.4 Rishon model	134
6.17.5 Criticisms	134
6.17.6 Conflicts with observed physics	134
6.17.7 Popular culture	135
6.17.8 See also	135
6.17.9 Notes	135
6.17.10 Further reading	135
6.18 QCD matter	136
6.18.1 Occurrence	136
6.18.2 Thermodynamics	136
6.18.3 Phase diagram	137
6.18.4 Theoretical challenges: calculation techniques	138
6.18.5 Experimental challenges	138
6.18.6 See also	139
6.18.7 References	139
6.18.8 Further reading	139
6.18.9 External links	139
6.19 Quark epoch	139
6.19.1 See also	139
6.19.2 References	140
6.20 Quark star	140
6.20.1 Creation	140
6.20.2 Characteristics	140
6.20.3 Other theorized quark formations	141
6.20.4 Observed overdense neutron stars	141
6.20.5 See also	141
6.20.6 References	142
6.20.7 External links	142
6.21 Quantum chromodynamics	143
6.21.1 Terminology	143
6.21.2 History	143
6.21.3 Theory	145
6.21.4 Methods	147
6.21.5 Experimental tests	148
6.21.6 Cross-relations to solid state physics	148

- 6.21.7 See also 149
- 6.21.8 References 149
- 6.21.9 Further reading 150
- 6.21.10 External links 150
- 6.22 Quarkonium 150
  - 6.22.1 Charmonium states 150
  - 6.22.2 Bottomonium states 151
  - 6.22.3 QCD and quarkonia 151
  - 6.22.4 See also 152
  - 6.22.5 References 152
- 6.23 Quark–lepton complementarity 152
  - 6.23.1 Possible evidence for QLC 152
  - 6.23.2 Open questions 153
  - 6.23.3 See also 153
  - 6.23.4 References 153
- 6.24 Standard Model 153
  - 6.24.1 Historical background 154
  - 6.24.2 Overview 154
  - 6.24.3 Particle content 154
  - 6.24.4 Theoretical aspects 156
  - 6.24.5 Fundamental forces 157
  - 6.24.6 Tests and predictions 157
  - 6.24.7 Challenges 157
  - 6.24.8 See also 158
  - 6.24.9 Notes and references 159
  - 6.24.10 References 159
  - 6.24.11 Further reading 160
  - 6.24.12 External links 161
- 6.25 Strong interaction 161
  - 6.25.1 History 162
  - 6.25.2 Details 162
  - 6.25.3 See also 163
  - 6.25.4 References 164
  - 6.25.5 Further reading 164
  - 6.25.6 External links 164
- 6.26 Weak interaction 164
  - 6.26.1 History 165
  - 6.26.2 Properties 165

- 6.26.3 Interaction types ... 166
- 6.26.4 Electroweak theory ... 167
- 6.26.5 Violation of symmetry ... 168
- 6.26.6 See also ... 168
- 6.26.7 References ... 168

6.27 List of particles ... 169
- 6.27.1 Elementary particles ... 169
- 6.27.2 Composite particles ... 171
- 6.27.3 Condensed matter ... 172
- 6.27.4 Other ... 173
- 6.27.5 Classification by speed ... 173
- 6.27.6 See also ... 173
- 6.27.7 References ... 174

6.28 Timeline of particle discoveries ... 174
- 6.28.1 See also ... 174
- 6.28.2 References ... 174

7 Appendix B – Selected biographies 177

7.1 James Bjorken ... 177
- 7.1.1 Publications ... 177
- 7.1.2 References ... 177

7.2 Nicola Cabibbo ... 177
- 7.2.1 Work ... 178
- 7.2.2 Death ... 178
- 7.2.3 References ... 178
- 7.2.4 External links ... 178

7.3 Richard Feynman ... 179
- 7.3.1 Early life ... 179
- 7.3.2 Education ... 179
- 7.3.3 Manhattan Project ... 180
- 7.3.4 Early academic career ... 181
- 7.3.5 Caltech years ... 182
- 7.3.6 Challenger disaster ... 184
- 7.3.7 Cultural identification ... 185
- 7.3.8 Personal life ... 185
- 7.3.9 Death ... 186
- 7.3.10 Popular legacy ... 186
- 7.3.11 Bibliography ... 186
- 7.3.12 See also ... 189

- 7.3.13 References 189
- 7.3.14 Bibliography 191
- 7.3.15 Further reading 192
- 7.3.16 External links 193
- 7.4 Murray Gell-Mann 193
  - 7.4.1 Early life and education 193
  - 7.4.2 Physics career 193
  - 7.4.3 Personal life 194
  - 7.4.4 Awards and honors 194
  - 7.4.5 See also 195
  - 7.4.6 References 195
  - 7.4.7 Further reading 196
  - 7.4.8 External links 196
- 7.5 Sheldon Lee Glashow 196
  - 7.5.1 Birth and education 196
  - 7.5.2 Research 196
  - 7.5.3 Superstring theory 197
  - 7.5.4 Personal life 197
  - 7.5.5 Works 197
  - 7.5.6 See also 197
  - 7.5.7 References 197
  - 7.5.8 External links 198
- 7.6 Haim Harari 198
  - 7.6.1 Biography 198
  - 7.6.2 Academic career 198
  - 7.6.3 Awards and recognition 199
  - 7.6.4 Published works 199
  - 7.6.5 References 199
  - 7.6.6 External links 199
  - 7.6.7 See also 199
- 7.7 John Iliopoulos 199
  - 7.7.1 Awards 200
  - 7.7.2 See also 200
  - 7.7.3 References 200
  - 7.7.4 External links 200
- 7.8 Makoto Kobayashi 200
  - 7.8.1 Biography 200
  - 7.8.2 Personal life 200

		7.8.3	See also	201
		7.8.4	References	201
		7.8.5	External links	201
	7.9	Leon M. Lederman	201
		7.9.1	Early life and career	201
		7.9.2	Personal life	202
		7.9.3	Publications	202
		7.9.4	Honorary degrees and awards	202
		7.9.5	See also	205
		7.9.6	References	205
		7.9.7	External links	206
	7.10	Luciano Maiani	207
		7.10.1	Academic history	207
		7.10.2	Awards	207
		7.10.3	See also	208
		7.10.4	External links	208
		7.10.5	References	208
	7.11	Toshihide Maskawa	208
		7.11.1	Biography	208
		7.11.2	Awards and honors	208
		7.11.3	Notes	208
	7.12	Yuval Ne'eman	208
		7.12.1	Biography	209
		7.12.2	Scientific career	209
		7.12.3	Awards and honours	209
		7.12.4	Political career	209
		7.12.5	Death	210
		7.12.6	See also	210
		7.12.7	References	210
		7.12.8	External links	210
	7.13	Kazuhiko Nishijima	210
		7.13.1	Life	210
		7.13.2	Books	211
		7.13.3	Awards	211
		7.13.4	References	211
		7.13.5	Further reading	211
	7.14	Burton Richter	211
		7.14.1	Life	211

- 7.14.2 See also 212
- 7.14.3 References 212
- 7.14.4 Publications 212
- 7.14.5 External links 213
- 7.15 Samuel C. C. Ting 213
  - 7.15.1 Biography 213
  - 7.15.2 Nobel Prize 213
  - 7.15.3 Alpha Magnetic Spectrometer 214
  - 7.15.4 Personal life 214
  - 7.15.5 Publications 214
  - 7.15.6 See also 214
  - 7.15.7 References 214
  - 7.15.8 External links 214
- 7.16 George Zweig 215
  - 7.16.1 Early life 215
  - 7.16.2 Career 215
  - 7.16.3 Awards and honors 215
  - 7.16.4 References 215

8 Appendix C – Selected facilities and experiments   217
- 8.1 Brookhaven National Laboratory 217
  - 8.1.1 Operation 217
  - 8.1.2 History 217
  - 8.1.3 Major programs 218
  - 8.1.4 Major facilities 218
  - 8.1.5 Plans 218
  - 8.1.6 Off-site contributions 218
  - 8.1.7 Public access 219
  - 8.1.8 Controversy 219
  - 8.1.9 Nobel Prizes 219
  - 8.1.10 See also 219
  - 8.1.11 References 220
  - 8.1.12 External links 220
- 8.2 Belle experiment 220
  - 8.2.1 See also 221
  - 8.2.2 References 221
  - 8.2.3 External links 221
- 8.3 BaBar experiment 221
  - 8.3.1 Physics 221

	8.3.2	Detector description	222
	8.3.3	Notable events	223
	8.3.4	Data record	223
	8.3.5	See also	223
	8.3.6	Notes	223
	8.3.7	External links	223
8.4	CDF experiment	223
	8.4.1	History of CDF	224
	8.4.2	Discovery of the top quark	224
	8.4.3	How CDF works	225
	8.4.4	Layer 1: the beam pipe	225
	8.4.5	Layer 2: silicon detector	225
	8.4.6	Layer 3: central outer tracker (COT)	226
	8.4.7	Layer 4: solenoid magnet	226
	8.4.8	Layers 5 and 6: electromagnetic and hadronic calorimeters	226
	8.4.9	Layer 7: muon detectors	226
	8.4.10	Conclusion	226
	8.4.11	References	227
	8.4.12	Further reading	227
	8.4.13	External links	227
8.5	DØ experiment	227
	8.5.1	Overview	228
	8.5.2	Physics research	228
	8.5.3	Detector	229
	8.5.4	Trigger and DAQ	229
	8.5.5	References	229
	8.5.6	External links	230
8.6	Fermilab	230
	8.6.1	History	230
	8.6.2	Accelerators	231
	8.6.3	Experiments	232
	8.6.4	Architecture	233
	8.6.5	Current developments	233
	8.6.6	Wildlife at Fermilab	234
	8.6.7	See also	234
	8.6.8	References	235
	8.6.9	External links	235
8.7	Large Hadron Collider	235

		8.7.1	Background	236
		8.7.2	Purpose	236
		8.7.3	Design	237
		8.7.4	Operational history	239
		8.7.5	Timeline of operations	242
		8.7.6	Findings and discoveries	242
		8.7.7	Planned "high luminosity" upgrade	243
		8.7.8	Safety of particle collisions	243
		8.7.9	Popular culture	243
		8.7.10	See also	244
		8.7.11	References	244
		8.7.12	External links	248
	8.8	SLAC National Accelerator Laboratory		248
		8.8.1	History	249
		8.8.2	Components	249
		8.8.3	Other discoveries	252
		8.8.4	See also	252
		8.8.5	References	252
		8.8.6	External links	253
	8.9	Tevatron		253
		8.9.1	History	254
		8.9.2	Mechanics	254
		8.9.3	Discoveries	255
		8.9.4	Earthquake detection	255
		8.9.5	See also	255
		8.9.6	References	256
		8.9.7	Further reading	256
		8.9.8	External links	256
9	Text and image sources, contributors, and licenses			257
	9.1	Text		257
	9.2	Images		274
	9.3	Content license		283

Chapter 1

Overview

1.1 Quark

A **quark** (/'kwɔrk/ or /'kwɑrk/) is an elementary particle and a fundamental constituent of matter. Quarks combine to form composite particles called hadrons, the most stable of which are protons and neutrons, the components of atomic nuclei.[*][1] Due to a phenomenon known as *color confinement*, quarks are never directly observed or found in isolation; they can be found only within hadrons, such as baryons (of which protons and neutrons are examples), and mesons.[*][2][*][3] For this reason, much of what is known about quarks has been drawn from observations of the hadrons themselves.

Quarks have various intrinsic properties, including electric charge, mass, color charge and spin. Quarks are the only elementary particles in the Standard Model of particle physics to experience all four fundamental interactions, also known as *fundamental forces* (electromagnetism, gravitation, strong interaction, and weak interaction), as well as the only known particles whose electric charges are not integer multiples of the elementary charge.

There are six types of quarks, known as *flavors*: up, down, strange, charm, top, and bottom.[*][4] Up and down quarks have the lowest masses of all quarks. The heavier quarks rapidly change into up and down quarks through a process of particle decay: the transformation from a higher mass state to a lower mass state. Because of this, up and down quarks are generally stable and the most common in the universe, whereas strange, charm, bottom, and top quarks can only be produced in high energy collisions (such as those involving cosmic rays and in particle accelerators). For every quark flavor there is a corresponding type of antiparticle, known as an *antiquark*, that differs from the quark only in that some of its properties have equal magnitude but opposite sign.

The quark model was independently proposed by physicists Murray Gell-Mann and George Zweig in 1964.[*][5] Quarks were introduced as parts of an ordering scheme for hadrons, and there was little evidence for their physical existence until deep inelastic scattering experiments at the Stanford Linear Accelerator Center in 1968.[*][6][*][7] Accelerator experiments have provided evidence for all six flavors. The top quark was the last to be discovered at Fermilab in 1995.[*][5]

1.1.1 Classification

See also: Standard Model
The Standard Model is the theoretical framework describ-

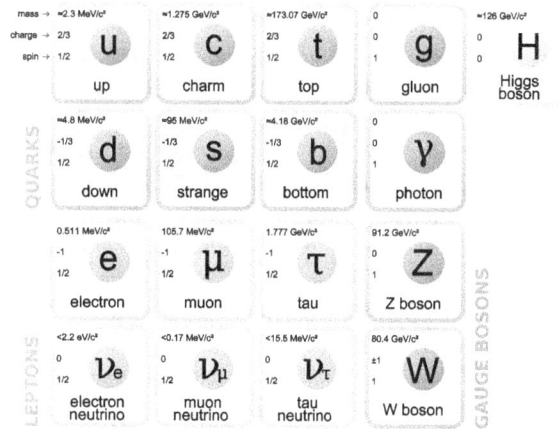

Six of the particles in the Standard Model are quarks (shown in purple). Each of the first three columns forms a generation *of matter.*

ing all the currently known elementary particles. This model contains six flavors of quarks (q), named up (u), down (d), strange (s), charm (c), bottom (b), and top (t).[*][4] Antiparticles of quarks are called *antiquarks*, and are denoted by a bar over the symbol for the corresponding quark, such as u for an up antiquark. As with antimatter in general, antiquarks have the same mass, mean lifetime, and spin as their respective quarks, but the electric charge and other charges have the opposite sign.[*][8]

Quarks are spin-$\frac{1}{2}$ particles, implying that they are fermions according to the spin-statistics theorem. They are subject to the Pauli exclusion principle, which states that no two identical fermions can simultaneously occupy the same quantum state. This is in contrast to bosons (particles with integer spin), any number of which can be in the same state.[*][9] Unlike leptons, quarks possess color charge, which causes them to engage in the strong interaction. The resulting attraction between different quarks causes the formation of composite particles known as *hadrons* (see "Strong interaction and color charge" below).

The quarks which determine the quantum numbers of hadrons are called *valence quarks*; apart from these, any hadron may contain an indefinite number of virtual (or *sea*) quarks, antiquarks, and gluons which do not influence its quantum numbers.[*][10] There are two families of hadrons: baryons, with three valence quarks, and mesons, with a valence quark and an antiquark.[*][11] The most common baryons are the proton and the neutron, the building blocks of the atomic nucleus.[*][12] A great number of hadrons are known (see list of baryons and list of mesons), most of them differentiated by their quark content and the properties these constituent quarks confer. The existence of "exotic" hadrons with more valence quarks, such as tetraquarks (qqqq) and pentaquarks (qqqqq), has been conjectured[*][13] but not proven.[*][nb 1][*][13][*][14] However, on 13 July 2015, the LHCb collaboration at CERN reported results consistent with pentaquark states.[*][15]

Elementary fermions are grouped into three generations, each comprising two leptons and two quarks. The first generation includes up and down quarks, the second strange and charm quarks, and the third bottom and top quarks. All searches for a fourth generation of quarks and other elementary fermions have failed,[*][16] and there is strong indirect evidence that no more than three generations exist.[*][nb 2][*][17] Particles in higher generations generally have greater mass and less stability, causing them to decay into lower-generation particles by means of weak interactions. Only first-generation (up and down) quarks occur commonly in nature. Heavier quarks can only be created in high-energy collisions (such as in those involving cosmic rays), and decay quickly; however, they are thought to have been present during the first fractions of a second after the Big Bang, when the universe was in an extremely hot and dense phase (the quark epoch). Studies of heavier quarks are conducted in artificially created conditions, such as in particle accelerators.[*][18]

Having electric charge, mass, color charge, and flavor, quarks are the only known elementary particles that engage in all four fundamental interactions of contemporary physics: electromagnetism, gravitation, strong interaction, and weak interaction.[*][12] Gravitation is too weak to be relevant to individual particle interactions except at extremes of energy (Planck energy) and distance scales (Planck distance). However, since no successful quantum theory of gravity exists, gravitation is not described by the Standard Model.

See the table of properties below for a more complete overview of the six quark flavors' properties.

1.1.2 History

Murray Gell-Mann at TED in 2007. Gell-Mann and George Zweig proposed the quark model in 1964.

The quark model was independently proposed by physicists Murray Gell-Mann[*][19] (pictured) and George Zweig[*][20][*][21] in 1964.[*][5] The proposal came shortly after Gell-Mann's 1961 formulation of a particle classification system known as the *Eightfold Way*—or, in more technical terms, SU(3) flavor symmetry.[*][22] Physicist Yuval Ne'eman had independently developed a scheme similar to the Eightfold Way in the same year.[*][23][*][24]

At the time of the quark theory's inception, the "particle zoo" included, amongst other particles, a multitude of hadrons. Gell-Mann and Zweig posited that they were not elementary particles, but were instead composed of combinations of quarks and antiquarks. Their model involved three flavors of quarks, up, down, and strange, to which they ascribed properties such as spin and electric charge.[*][19][*][20][*][21] The initial reaction of the physics community to the proposal was mixed. There was particular contention about whether the quark was a physical entity or a mere abstraction used to explain concepts that were not fully understood at the time.*[25]

In less than a year, extensions to the Gell-Mann–Zweig model were proposed. Sheldon Lee Glashow and James Bjorken predicted the existence of a fourth flavor of quark, which they called *charm*. The addition was proposed because it allowed for a better description of the weak interaction (the mechanism that allows quarks to decay), equalized the number of known quarks with the number of known leptons, and implied a mass formula that correctly reproduced the masses of the known mesons.*[26]

In 1968, deep inelastic scattering experiments at the Stanford Linear Accelerator Center (SLAC) showed that the proton contained much smaller, point-like objects and was therefore not an elementary particle.*[6]*[7]*[27] Physicists were reluctant to firmly identify these objects with quarks at the time, instead calling them "partons"—a term coined by Richard Feynman.*[28]*[29]*[30] The objects that were observed at SLAC would later be identified as up and down quarks as the other flavors were discovered.*[31] Nevertheless, "parton" remains in use as a collective term for the constituents of hadrons (quarks, antiquarks, and gluons).

The strange quark's existence was indirectly validated by SLAC's scattering experiments: not only was it a necessary component of Gell-Mann and Zweig's three-quark model, but it provided an explanation for the kaon (K) and pion (π) hadrons discovered in cosmic rays in 1947.*[32]

In a 1970 paper, Glashow, John Iliopoulos and Luciano Maiani presented further reasoning for the existence of the as-yet undiscovered charm quark.*[33]*[34] The number of supposed quark flavors grew to the current six in 1973, when Makoto Kobayashi and Toshihide Maskawa noted that the experimental observation of CP violation*[nb 3]*[35] could be explained if there were another pair of quarks.

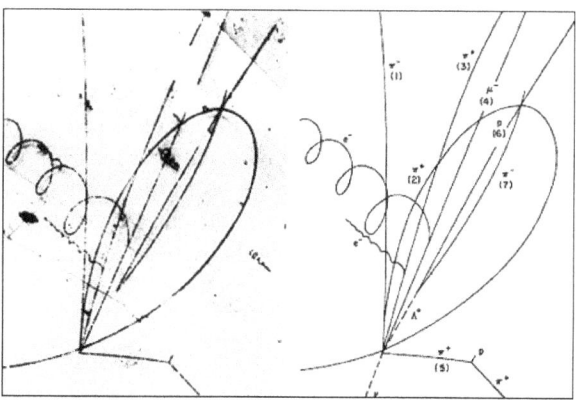

Photograph of the event that led to the discovery of the Σ++ c baryon, at the Brookhaven National Laboratory in 1974

Charm quarks were produced almost simultaneously by two teams in November 1974 (see November Revolution)—one at SLAC under Burton Richter, and one at Brookhaven National Laboratory under Samuel Ting. The charm quarks were observed bound with charm antiquarks in mesons. The two parties had assigned the discovered meson two different symbols, J and ψ; thus, it became formally known as the J/ψ meson. The discovery finally convinced the physics community of the quark model's validity.*[30]

In the following years a number of suggestions appeared for extending the quark model to six quarks. Of these, the 1975 paper by Haim Harari*[36] was the first to coin the terms *top* and *bottom* for the additional quarks.*[37]

In 1977, the bottom quark was observed by a team at Fermilab led by Leon Lederman.*[38]*[39] This was a strong indicator of the top quark's existence: without the top quark, the bottom quark would have been without a partner. However, it was not until 1995 that the top quark was finally observed, also by the CDF*[40] and DØ*[41] teams at Fermilab.*[5] It had a mass much larger than had been previously expected,*[42] almost as large as that of a gold atom.*[43]

1.1.3 Etymology

For some time, Gell-Mann was undecided on an actual spelling for the term he intended to coin, until he found the word *quark* in James Joyce's book *Finnegans Wake*:

> Three quarks for Muster Mark!
> Sure he has not got much of a bark
> And sure any he has it's all beside the mark.
> —James Joyce, *Finnegans Wake**[44]

Gell-Mann went into further detail regarding the name of the quark in his book *The Quark and the Jaguar*:*[45]

> In 1963, when I assigned the name "quark" to the fundamental constituents of the nucleon, I had the sound first, without the spelling, which could have been "kwork". Then, in one of my occasional perusals of *Finnegans Wake*, by James Joyce, I came across the word "quark" in the phrase "Three quarks for Muster Mark". Since "quark" (meaning, for one thing, the cry of the gull) was clearly intended to rhyme with "Mark", as well as "bark" and other such words, I had to find an excuse to pronounce it as "kwork". But the book represents the dream of a publican named Humphrey Chimpden Earwicker. Words in the text are typically drawn from several sources at once, like the "portmanteau" words in

"Through the Looking-Glass". From time to time, phrases occur in the book that are partially determined by calls for drinks at the bar. I argued, therefore, that perhaps one of the multiple sources of the cry "Three quarks for Muster Mark" might be "Three quarts for Mister Mark", in which case the pronunciation "kwork" would not be totally unjustified. In any case, the number three fitted perfectly the way quarks occur in nature.

Zweig preferred the name *ace* for the particle he had theorized, but Gell-Mann's terminology came to prominence once the quark model had been commonly accepted.*[46]

The quark flavors were given their names for several reasons. The up and down quarks are named after the up and down components of isospin, which they carry.*[47] Strange quarks were given their name because they were discovered to be components of the strange particles discovered in cosmic rays years before the quark model was proposed; these particles were deemed "strange" because they had unusually long lifetimes.*[48] Glashow, who co-proposed charm quark with Bjorken, is quoted as saying, "We called our construct the 'charmed quark', for we were fascinated and pleased by the symmetry it brought to the subnuclear world." *[49] The names "bottom" and "top", coined by Harari, were chosen because they are "logical partners for up and down quarks".*[36]*[37]*[48] In the past, bottom and top quarks were sometimes referred to as "beauty" and "truth" respectively, but these names have somewhat fallen out of use.*[50] While "truth" never did catch on, accelerator complexes devoted to massive production of bottom quarks are sometimes called "beauty factories".*[51]

1.1.4 Properties

Electric charge

See also: Electric charge

Quarks have fractional electric charge values – either $\frac{1}{3}$ or $\frac{2}{3}$ times the elementary charge (e), depending on flavor. Up, charm, and top quarks (collectively referred to as *up-type quarks*) have a charge of $+\frac{2}{3}$ e, while down, strange, and bottom quarks (*down-type quarks*) have $-\frac{1}{3}$ e. Antiquarks have the opposite charge to their corresponding quarks; up-type antiquarks have charges of $-\frac{2}{3}$ e and down-type antiquarks have charges of $+\frac{1}{3}$ e. Since the electric charge of a hadron is the sum of the charges of the constituent quarks, all hadrons have integer charges: the combination of three quarks (baryons), three antiquarks (antibaryons), or a quark and an antiquark (mesons) always results in integer charges.*[52] For example, the hadron constituents of atomic nuclei, neutrons and protons, have charges of 0 e and +1 e respectively; the neutron is composed of two down quarks and one up quark, and the proton of two up quarks and one down quark.*[12]

Spin

See also: Spin (physics)

Spin is an intrinsic property of elementary particles, and its direction is an important degree of freedom. It is sometimes visualized as the rotation of an object around its own axis (hence the name "spin"), though this notion is somewhat misguided at subatomic scales because elementary particles are believed to be point-like.*[53]

Spin can be represented by a vector whose length is measured in units of the reduced Planck constant $\hbar$ (pronounced "h bar"). For quarks, a measurement of the spin vector component along any axis can only yield the values $+\hbar/2$ or $-\hbar/2$; for this reason quarks are classified as spin-$\frac{1}{2}$ particles.*[54] The component of spin along a given axis – by convention the z axis – is often denoted by an up arrow ↑ for the value $+\frac{1}{2}$ and down arrow ↓ for the value $-\frac{1}{2}$, placed after the symbol for flavor. For example, an up quark with a spin of $+\frac{1}{2}$ along the z axis is denoted by u↑.*[55]

Weak interaction

Main article: Weak interaction
A quark of one flavor can transform into a quark of another flavor only through the weak interaction, one of the four fundamental interactions in particle physics. By absorbing or emitting a W boson, any up-type quark (up, charm, and top quarks) can change into any down-type quark (down, strange, and bottom quarks) and vice versa. This flavor transformation mechanism causes the radioactive process of beta decay, in which a neutron (n) "splits" into a proton (p), an electron (e−) and an electron antineutrino (ν e) (see picture). This occurs when one of the down quarks in the neutron (udd) decays into an up quark by emitting a virtual W− boson, transforming the neutron into a proton (uud). The W− boson then decays into an electron and an electron antineutrino.*[56]

Both beta decay and the inverse process of *inverse beta decay* are routinely used in medical applications such as positron emission tomography (PET) and in experiments involving neutrino detection.

While the process of flavor transformation is the same for all quarks, each quark has a preference to trans-

1.1. QUARK

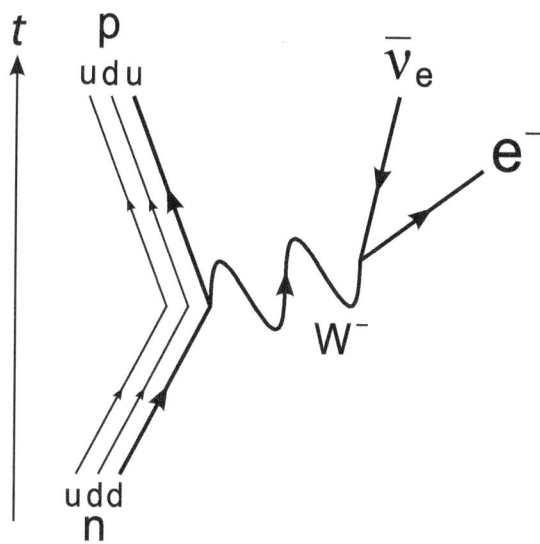

Feynman diagram of beta decay with time flowing upwards. The CKM matrix (discussed below) encodes the probability of this and other quark decays.

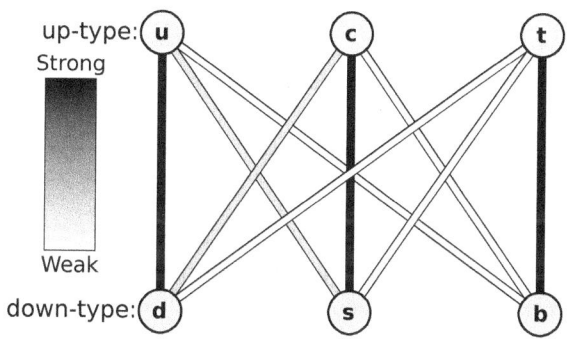

The strengths of the weak interactions between the six quarks. The "intensities" of the lines are determined by the elements of the CKM matrix.

form into the quark of its own generation. The relative tendencies of all flavor transformations are described by a mathematical table, called the Cabibbo–Kobayashi–Maskawa matrix (CKM matrix). Enforcing unitarity, the approximate magnitudes of the entries of the CKM matrix are:[57]

$$\begin{bmatrix} |V_{ud}| & |V_{us}| & |V_{ub}| \\ |V_{cd}| & |V_{cs}| & |V_{cb}| \\ |V_{td}| & |V_{ts}| & |V_{tb}| \end{bmatrix} \approx \begin{bmatrix} 0.974 & 0.225 & 0.003 \\ 0.225 & 0.973 & 0.041 \\ 0.009 & 0.040 & 0.999 \end{bmatrix},$$

where V_{ij} represents the tendency of a quark of flavor i to change into a quark of flavor j (or vice versa).*[nb 4]

There exists an equivalent weak interaction matrix for leptons (right side of the W boson on the above beta decay diagram), called the Pontecorvo–Maki–Nakagawa–Sakata matrix (PMNS matrix).*[58] Together, the CKM and PMNS matrices describe all flavor transformations, but the links between the two are not yet clear.*[59]

Strong interaction and color charge

See also: Color charge and Strong interaction

According to quantum chromodynamics (QCD), quarks possess a property called *color charge*. There are three types of color charge, arbitrarily labeled *blue*, *green*, and *red*.*[nb 5] Each of them is complemented by an anticolor – *antiblue*, *antigreen*, and *antired*. Every quark carries a color, while every antiquark carries an anticolor.*[60]

The system of attraction and repulsion between quarks charged with different combinations of the three colors is called strong interaction, which is mediated by force carrying particles known as *gluons*; this is discussed at length below. The theory that describes strong interactions is called quantum chromodynamics (QCD). A quark, which will have a single color value, can form a bound system with an antiquark carrying the corresponding anticolor. The result of two attracting quarks will be color neutrality: a quark with color charge ξ plus an antiquark with color charge $-\xi$ will result in a color charge of 0 (or "white" color) and the formation of a meson. This is analogous to the additive color model in basic optics. Similarly, the combination of three quarks, each with different color charges, or three antiquarks, each with anticolor charges, will result in the same "white" color charge and the formation of a baryon or antibaryon.*[61]

In modern particle physics, gauge symmetries – a kind of symmetry group – relate interactions between particles (see gauge theories). Color SU(3) (commonly abbreviated to SU(3)$_c$) is the gauge symmetry that relates the color charge in quarks and is the defining symmetry for quantum chromodynamics.*[62] Just as the laws of physics are independent of which directions in space are designated x, y, and z, and remain unchanged if the coordinate axes are rotated to a new orientation, the physics of quantum chromodynamics is independent of which directions in three-dimensional color space are identified as blue, red, and green. SU(3)$_c$ color transformations correspond to "rotations" in color space (which, mathematically speaking, is a complex space). Every quark flavor f, each with subtypes f_B, f_G, f_R corresponding to the quark colors,*[63] forms a triplet: a three-component quantum field which transforms under the fundamental representation of SU(3)$_c$.*[64] The requirement that SU(3)$_c$ should be local – that is, that its transformations be allowed to vary with space and time –

determines the properties of the strong interaction, in particular the existence of eight gluon types to act as its force carriers.[*][62][*][65]

Mass

See also: Invariant mass

Two terms are used in referring to a quark's mass: *current quark mass* refers to the mass of a quark by itself, while *constituent quark mass* refers to the current quark mass plus the mass of the gluon particle field surrounding the quark.[*][66] These masses typically have very different values. Most of a hadron's mass comes from the gluons that bind the constituent quarks together, rather than from the quarks themselves. While gluons are inherently massless, they possess energy – more specifically, quantum chromodynamics binding energy (QCBE) – and it is this that contributes so greatly to the overall mass of the hadron (see mass in special relativity). For example, a proton has a mass of approximately 938 MeV/c^2, of which the rest mass of its three valence quarks only contributes about 11 MeV/c^2; much of the remainder can be attributed to the gluons' QCBE.[*][67][*][68]

The Standard Model posits that elementary particles derive their masses from the Higgs mechanism, which is related to the Higgs boson. Physicists hope that further research into the reasons for the top quark's large mass of ~173 GeV/c^2, almost the mass of a gold atom,[*][67][*][69] might reveal more about the origin of the mass of quarks and other elementary particles.[*][70]

Table of properties

See also: Flavor (particle physics)

The following table summarizes the key properties of the six quarks. Flavor quantum numbers (isospin (I_3), charm (C), strangeness (S, not to be confused with spin), topness (T), and bottomness (B')) are assigned to certain quark flavors, and denote qualities of quark-based systems and hadrons. The baryon number (B) is +$\frac{1}{3}$ for all quarks, as baryons are made of three quarks. For antiquarks, the electric charge (Q) and all flavor quantum numbers (B, I_3, C, S, T, and B') are of opposite sign. Mass and total angular momentum (J; equal to spin for point particles) do not change sign for the antiquarks.

J = total angular momentum, B = baryon number, Q = electric charge, I_3 = isospin, C = charm, S = strangeness, T = topness, B' = bottomness.

* Notation such as 4190+180
−60 denotes measurement uncertainty. In the case of the top quark, the first uncertainty is statistical in nature, and the second is systematic.

1.1.5 Interacting quarks

See also: Color confinement and Gluon

As described by quantum chromodynamics, the strong interaction between quarks is mediated by gluons, massless vector gauge bosons. Each gluon carries one color charge and one anticolor charge. In the standard framework of particle interactions (part of a more general formulation known as perturbation theory), gluons are constantly exchanged between quarks through a virtual emission and absorption process. When a gluon is transferred between quarks, a color change occurs in both; for example, if a red quark emits a red–antigreen gluon, it becomes green, and if a green quark absorbs a red–antigreen gluon, it becomes red. Therefore, while each quark's color constantly changes, their strong interaction is preserved.[*][71][*][72][*][73]

Since gluons carry color charge, they themselves are able to emit and absorb other gluons. This causes *asymptotic freedom*: as quarks come closer to each other, the chromodynamic binding force between them weakens.[*][74] Conversely, as the distance between quarks increases, the binding force strengthens. The color field becomes stressed, much as an elastic band is stressed when stretched, and more gluons of appropriate color are spontaneously created to strengthen the field. Above a certain energy threshold, pairs of quarks and antiquarks are created. These pairs bind with the quarks being separated, causing new hadrons to form. This phenomenon is known as *color confinement*: quarks never appear in isolation.[*][72][*][75] This process of hadronization occurs before quarks, formed in a high energy collision, are able to interact in any other way. The only exception is the top quark, which may decay before it hadronizes.[*][76]

Sea quarks

Hadrons, along with the *valence quarks* (q
v) that contribute to their quantum numbers, contain virtual quark–antiquark (qq) pairs known as *sea quarks* (q
s). Sea quarks form when a gluon of the hadron's color field splits; this process also works in reverse in that the annihilation of two sea quarks produces a gluon. The result is a constant flux of gluon splits and creations colloquially known as "the sea".[*][77] Sea quarks are much less stable than their valence counterparts, and they typically annihilate each other within the interior of the hadron. Despite

this, sea quarks can hadronize into baryonic or mesonic particles under certain circumstances.*[78]

Other phases of quark matter

Main article: QCD matter
Under sufficiently extreme conditions, quarks may become deconfined and exist as free particles. In the course of asymptotic freedom, the strong interaction becomes weaker at higher temperatures. Eventually, color confinement would be lost and an extremely hot plasma of freely moving quarks and gluons would be formed. This theoretical phase of matter is called quark–gluon plasma.*[81] The exact conditions needed to give rise to this state are unknown and have been the subject of a great deal of speculation and experimentation. A recent estimate puts the needed temperature at $(1.90 \pm 0.02) \times 10^{12}$ kelvin.*[82] While a state of entirely free quarks and gluons has never been achieved (despite numerous attempts by CERN in the 1980s and 1990s),*[83] recent experiments at the Relativistic Heavy Ion Collider have yielded evidence for liquid-like quark matter exhibiting "nearly perfect" fluid motion.*[84]

The quark–gluon plasma would be characterized by a great increase in the number of heavier quark pairs in relation to the number of up and down quark pairs. It is believed that in the period prior to 10^{-6} seconds after the Big Bang (the quark epoch), the universe was filled with quark–gluon plasma, as the temperature was too high for hadrons to be stable.*[85]

Given sufficiently high baryon densities and relatively low temperatures – possibly comparable to those found in neutron stars – quark matter is expected to degenerate into a Fermi liquid of weakly interacting quarks. This liquid would be characterized by a condensation of colored quark Cooper pairs, thereby breaking the local $SU(3)_c$ symmetry. Because quark Cooper pairs harbor color charge, such a phase of quark matter would be color superconductive; that is, color charge would be able to pass through it with no resistance.*[86]

1.1.6 See also

- Color–flavor locking
- Neutron magnetic moment
- Leptons
- Preons – Hypothetical particles which were once postulated to be subcomponents of quarks and leptons
- Quarkonium – Mesons made of a quark and antiquark of the same flavor
- Quark star – A hypothetical degenerate neutron star with extreme density
- Quark–lepton complementarity – Possible fundamental relation between quarks and leptons

1.1.7 Notes

[1] Several research groups claimed to have proven the existence of tetraquarks and pentaquarks in the early 2000s. While the status of tetraquarks is still under debate, all known pentaquark candidates have previously been established as non-existent.

[2] The main evidence is based on the resonance width of the Z0 boson, which constrains the 4th generation neutrino to have a mass greater than ~45 GeV/c^2. This would be highly contrasting with the other three generations' neutrinos, whose masses cannot exceed 2 MeV/c^2.

[3] CP violation is a phenomenon which causes weak interactions to behave differently when left and right are swapped (P symmetry) and particles are replaced with their corresponding antiparticles (C symmetry).

[4] The actual probability of decay of one quark to another is a complicated function of (amongst other variables) the decaying quark's mass, the masses of the decay products, and the corresponding element of the CKM matrix. This probability is directly proportional (but not equal) to the magnitude squared ($|V_{ij}|^2$) of the corresponding CKM entry.

[5] Despite its name, color charge is not related to the color spectrum of visible light.

1.1.8 References

[1] "Quark (subatomic particle)". *Encyclopædia Britannica*. Retrieved 2008-06-29.

[2] R. Nave. "Confinement of Quarks". *HyperPhysics*. Georgia State University, Department of Physics and Astronomy. Retrieved 2008-06-29.

[3] R. Nave. "Bag Model of Quark Confinement". *HyperPhysics*. Georgia State University, Department of Physics and Astronomy. Retrieved 2008-06-29.

[4] R. Nave. "Quarks". *HyperPhysics*. Georgia State University, Department of Physics and Astronomy. Retrieved 2008-06-29.

[5] B. Carithers, P. Grannis (1995). "Discovery of the Top Quark" (PDF). *Beam Line* (SLAC) **25** (3): 4–16. Retrieved 2008-09-23.

[6] E.D. Bloom et al. (1969). "High-Energy Inelastic e–p Scattering at 6° and 10°". *Physical Review Letters* **23** (16): 930–934. Bibcode:1969PhRvL..23..930B. doi:10.1103/PhysRevLett.23.930.

[7] M. Breidenbach et al. (1969). "Observed Behavior of Highly Inelastic Electron–Proton Scattering". *Physical Review Letters* **23** (16): 935–939. Bibcode:1969PhRvL..23..935B. doi:10.1103/PhysRevLett.23.935.

[8] S.S.M. Wong (1998). *Introductory Nuclear Physics* (2nd ed.). Wiley Interscience. p. 30. ISBN 0-471-23973-9.

[9] K.A. Peacock (2008). *The Quantum Revolution*. Greenwood Publishing Group. p. 125. ISBN 0-313-33448-X.

[10] B. Povh, C. Scholz, K. Rith, F. Zetsche (2008). *Particles and Nuclei*. Springer. p. 98. ISBN 3-540-79367-4.

[11] Section 6.1. in P.C.W. Davies (1979). *The Forces of Nature*. Cambridge University Press. ISBN 0-521-22523-X.

[12] M. Munowitz (2005). *Knowing*. Oxford University Press. p. 35. ISBN 0-19-516737-6.

[13] W.-M. Yao (Particle Data Group) et al. (2006). "Review of Particle Physics: Pentaquark Update" (PDF). *Journal of Physics G* **33** (1): 1–1232. arXiv:astro-ph/0601168. Bibcode:2006JPhG...33....1Y. doi:10.1088/0954-3899/33/1/001.

[14] C. Amsler (Particle Data Group) et al. (2008). "Review of Particle Physics: Pentaquarks" (PDF). *Physics Letters B* **667** (1): 1–1340. Bibcode:2008PhLB..667....1P. doi:10.1016/j.physletb.2008.07.018.
C. Amsler (Particle Data Group) et al. (2008). "Review of Particle Physics: New Charmonium-Like States" (PDF). *Physics Letters B* **667** (1): 1–1340. Bibcode:2008PhLB..667....1P. doi:10.1016/j.physletb.2008.07.018.
E.V. Shuryak (2004). *The QCD Vacuum, Hadrons and Superdense Matter*. World Scientific. p. 59. ISBN 981-238-574-6.

[15] R. Aaij et al. (LHCb collaboration) (2015). "Observation of J/ψp resonances consistent with pentaquark states in Λ0 b→J/ψK−
p decays". *Physical Review Letters* **115** (7). doi:10.1103/PhysRevLett.115.072001.

[16] C. Amsler (Particle Data Group) et al. (2008). "Review of Particle Physics: b′ (4th Generation) Quarks, Searches for" (PDF). *Physics Letters B* **667** (1): 1–1340. Bibcode:2008PhLB..667....1P. doi:10.1016/j.physletb.2008.07.018.
C. Amsler (Particle Data Group) et al. (2008). "Review of Particle Physics: t′ (4th Generation) Quarks, Searches for" (PDF). *Physics Letters B* **667** (1): 1–1340. Bibcode:2008PhLB..667....1P. doi:10.1016/j.physletb.2008.07.018.

[17] D. Decamp; Deschizeaux, B.; Lees, J.-P.; Minard, M.-N.; Crespo, J.M.; Delfino, M.; Fernandez, E.; Martinez, M. et al. (1989). "Determination of the number of light neutrino species". *Physics Letters B* **231** (4): 519. Bibcode:1989PhLB..231..519D. doi:10.1016/0370-2693(89)90704-1.
A. Fisher (1991). "Searching for the Beginning of Time: Cosmic Connection". *Popular Science* **238** (4): 70.
J.D. Barrow (1997) [1994]. "The Singularity and Other Problems". *The Origin of the Universe* (Reprint ed.). Basic Books. ISBN 978-0-465-05314-8.

[18] D.H. Perkins (2003). *Particle Astrophysics*. Oxford University Press. p. 4. ISBN 0-19-850952-9.

[19] M. Gell-Mann (1964). "A Schematic Model of Baryons and Mesons". *Physics Letters* **8** (3): 214–215. Bibcode:1964PhL.....8..214G. doi:10.1016/S0031-9163(64)92001-3.

[20] G. Zweig (1964). "An SU(3) Model for Strong Interaction Symmetry and its Breaking" (PDF). *CERN Report No.8182/TH.401*.

[21] G. Zweig (1964). "An SU(3) Model for Strong Interaction Symmetry and its Breaking: II" (PDF). *CERN Report No.8419/TH.412*.

[22] M. Gell-Mann (2000) [1964]. "The Eightfold Way: A theory of strong interaction symmetry". In M. Gell-Mann, Y. Ne'eman. *The Eightfold Way*. Westview Press. p. 11. ISBN 0-7382-0299-1.
Original: M. Gell-Mann (1961). "The Eightfold Way: A theory of strong interaction symmetry". *Synchrotron Laboratory Report CTSL-20* (California Institute of Technology).

[23] Y. Ne'eman (2000) [1964]. "Derivation of strong interactions from gauge invariance". In M. Gell-Mann, Y. Ne'eman. *The Eightfold Way*. Westview Press. ISBN 0-7382-0299-1.
Original Y. Ne'eman (1961). "Derivation of strong interactions from gauge invariance". *Nuclear Physics* **26** (2): 222. Bibcode:1961NucPh..26..222N. doi:10.1016/0029-5582(61)90134-1.

[24] R.C. Olby, G.N. Cantor (1996). *Companion to the History of Modern Science*. Taylor & Francis. p. 673. ISBN 0-415-14578-3.

[25] A. Pickering (1984). *Constructing Quarks*. University of Chicago Press. pp. 114–125. ISBN 0-226-66799-5.

[26] B.J. Bjorken, S.L. Glashow; Glashow (1964). "Elementary Particles and SU(4)". *Physics Letters* **11** (3): 255–257. Bibcode:1964PhL.....11..255B. doi:10.1016/0031-9163(64)90433-0.

[27] J.I. Friedman. "The Road to the Nobel Prize". Hue University. Retrieved 2008-09-29.

[28] R.P. Feynman (1969). "Very High-Energy Collisions of Hadrons". *Physical Review Letters* **23** (24): 1415–1417. Bibcode:1969PhRvL..23.1415F. doi:10.1103/PhysRevLett.23.1415.

[29] S. Kretzer et al. (2004). "CTEQ6 Parton Distributions with Heavy Quark Mass Effects". *Physical Review D* **69** (11): 114005. arXiv:hep-ph/0307022. Bibcode:2004PhRvD..69k4005K. doi:10.1103/PhysRevD.69.114005.

[30] D.J. Griffiths (1987). *Introduction to Elementary Particles*. John Wiley & Sons. p. 42. ISBN 0-471-60386-4.

[31] M.E. Peskin, D.V. Schroeder (1995). *An introduction to quantum field theory*. Addison–Wesley. p. 556. ISBN 0-201-50397-2.

[32] V.V. Ezhela (1996). *Particle physics*. Springer. p. 2. ISBN 1-56396-642-5.

[33] S.L. Glashow, J. Iliopoulos, L. Maiani; Iliopoulos; Maiani (1970). "Weak Interactions with Lepton–Hadron Symmetry". *Physical Review D* **2** (7): 1285–1292. Bibcode:1970PhRvD...2.1285G. doi:10.1103/PhysRevD.2.1285.

[34] D.J. Griffiths (1987). *Introduction to Elementary Particles*. John Wiley & Sons. p. 44. ISBN 0-471-60386-4.

[35] M. Kobayashi, T. Maskawa; Maskawa (1973). "CP-Violation in the Renormalizable Theory of Weak Interaction". *Progress of Theoretical Physics* **49** (2): 652–657. Bibcode:1973PThPh..49..652K. doi:10.1143/PTP.49.652.

[36] H. Harari (1975). "A new quark model for hadrons". *Physics Letters B* **57B** (3): 265. Bibcode:1975PhLB...57..265H. doi:10.1016/0370-2693(75)90072-6.

[37] K.W. Staley (2004). *The Evidence for the Top Quark*. Cambridge University Press. pp. 31–33. ISBN 978-0-521-82710-2.

[38] S.W. Herb et al. (1977). "Observation of a Dimuon Resonance at 9.5 GeV in 400-GeV Proton-Nucleus Collisions". *Physical Review Letters* **39** (5): 252. Bibcode:1977PhRvL..39..252H. doi:10.1103/PhysRevLett.39.252.

[39] M. Bartusiak (1994). *A Positron named Priscilla*. National Academies Press. p. 245. ISBN 0-309-04893-1.

[40] F. Abe (CDF Collaboration) et al. (1995). "Observation of Top Quark Production in pp Collisions with the Collider Detector at Fermilab". *Physical Review Letters* **74** (14): 2626–2631. Bibcode:1995PhRvL..74.2626A. doi:10.1103/PhysRevLett.74.2626. PMID 10057978.

[41] S. Abachi (DØ Collaboration) et al. (1995). "Search for High Mass Top Quark Production in pp Collisions at $\sqrt{s}$ = 1.8 TeV". *Physical Review Letters* **74** (13): 2422–2426. Bibcode:1995PhRvL..74.2422A. doi:10.1103/PhysRevLett.74.2422.

[42] K.W. Staley (2004). *The Evidence for the Top Quark*. Cambridge University Press. p. 144. ISBN 0-521-82710-8.

[43] "New Precision Measurement of Top Quark Mass". Brookhaven National Laboratory News. 2004. Retrieved 2013-11-03.

[44] J. Joyce (1982) [1939]. *Finnegans Wake*. Penguin Books. p. 383. ISBN 0-14-006286-6.

[45] M. Gell-Mann (1995). *The Quark and the Jaguar: Adventures in the Simple and the Complex*. Henry Holt and Co. p. 180. ISBN 978-0-8050-7253-2.

[46] J. Gleick (1992). *Genius: Richard Feynman and modern physics*. Little Brown and Company. p. 390. ISBN 0-316-90316-7.

[47] J.J. Sakurai (1994). S.F Tuan, ed. *Modern Quantum Mechanics* (Revised ed.). Addison–Wesley. p. 376. ISBN 0-201-53929-2.

[48] D.H. Perkins (2000). *Introduction to high energy physics*. Cambridge University Press. p. 8. ISBN 0-521-62196-8.

[49] M. Riordan (1987). *The Hunting of the Quark: A True Story of Modern Physics*. Simon & Schuster. p. 210. ISBN 978-0-671-50466-3.

[50] F. Close (2006). *The New Cosmic Onion*. CRC Press. p. 133. ISBN 1-58488-798-2.

[51] J.T. Volk et al. (1987). "Letter of Intent for a Tevatron Beauty Factory" (PDF). Fermilab Proposal #783.

[52] G. Fraser (2006). *The New Physics for the Twenty-First Century*. Cambridge University Press. p. 91. ISBN 0-521-81600-9.

[53] "The Standard Model of Particle Physics". BBC. 2002. Retrieved 2009-04-19.

[54] F. Close (2006). *The New Cosmic Onion*. CRC Press. pp. 80–90. ISBN 1-58488-798-2.

[55] D. Lincoln (2004). *Understanding the Universe*. World Scientific. p. 116. ISBN 981-238-705-6.

[56] "Weak Interactions". *Virtual Visitor Center*. Stanford Linear Accelerator Center. 2008. Retrieved 2008-09-28.

[57] K. Nakamura et al. (2010). "Review of Particles Physics: The CKM Quark-Mixing Matrix" (PDF). *J. Phys. G* **37** (75021): 150.

[58] Z. Maki, M. Nakagawa, S. Sakata (1962). "Remarks on the Unified Model of Elementary Particles". *Progress of Theoretical Physics* **28** (5): 870. Bibcode:1962PThPh..28..870M. doi:10.1143/PTP.28.870.

[59] B.C. Chauhan, M. Picariello, J. Pulido, E. Torrente-Lujan (2007). "Quark–lepton complementarity, neutrino and standard model data predict θ_{PMNS} 13 = $9°^{+1°}_{-2°}$". *European Physical Journal* **C50** (3): 573–578. arXiv:hep-ph/0605032. Bibcode:2007EPJC...50..573C. doi:10.1140/epjc/s10052-007-0212-z.

[60] R. Nave. "The Color Force". *HyperPhysics*. Georgia State University, Department of Physics and Astronomy. Retrieved 2009-04-26.

[61] B.A. Schumm (2004). *Deep Down Things*. Johns Hopkins University Press. pp. 131–132. ISBN 0-8018-7971-X. OCLC 55229065.

[62] Part III of M.E. Peskin, D.V. Schroeder (1995). *An Introduction to Quantum Field Theory*. Addison–Wesley. ISBN 0-201-50397-2.

[63] V. Icke (1995). *The force of symmetry*. Cambridge University Press. p. 216. ISBN 0-521-45591-X.

[64] M.Y. Han (2004). *A story of light*. World Scientific. p. 78. ISBN 981-256-034-3.

[65] C. Sutton. "Quantum chromodynamics (physics)". *Encyclopædia Britannica Online*. Retrieved 2009-05-12.

[66] A. Watson (2004). *The Quantum Quark*. Cambridge University Press. pp. 285–286. ISBN 0-521-82907-0.

[67] K.A. Olive *et al.* (Particle Data Group), Chin. Phys. **C38**, 090001 (2014) (URL: http://pdg.lbl.gov)

[68] W. Weise, A.M. Green (1984). *Quarks and Nuclei*. World Scientific. pp. 65–66. ISBN 9971-966-61-1.

[69] D. McMahon (2008). *Quantum Field Theory Demystified*. McGraw–Hill. p. 17. ISBN 0-07-154382-1.

[70] S.G. Roth (2007). *Precision electroweak physics at electron–positron colliders*. Springer. p. VI. ISBN 3-540-35164-7.

[71] R.P. Feynman (1985). *QED: The Strange Theory of Light and Matter* (1st ed.). Princeton University Press. pp. 136–137. ISBN 0-691-08388-6.

[72] M. Veltman (2003). *Facts and Mysteries in Elementary Particle Physics*. World Scientific. pp. 45–47. ISBN 981-238-149-X.

[73] F. Wilczek, B. Devine (2006). *Fantastic Realities*. World Scientific. p. 85. ISBN 981-256-649-X.

[74] F. Wilczek, B. Devine (2006). *Fantastic Realities*. World Scientific. pp. 400ff. ISBN 981-256-649-X.

[75] T. Yulsman (2002). *Origin*. CRC Press. p. 55. ISBN 0-7503-0765-X.

[76] F. Garberson (2008). "Top Quark Mass and Cross Section Results from the Tevatron". arXiv:0808.0273 [hep-ex].

[77] J. Steinberger (2005). *Learning about Particles*. Springer. p. 130. ISBN 3-540-21329-5.

[78] C.-Y. Wong (1994). *Introduction to High-energy Heavy-ion Collisions*. World Scientific. p. 149. ISBN 981-02-0263-6.

[79] S.B. Rüester, V. Werth, M. Buballa, I.A. Shovkovy, D.H. Rischke; Werth; Buballa; Shovkovy; Rischke (2005). "The phase diagram of neutral quark matter: Self-consistent treatment of quark masses". *Physical Review D* **72** (3): 034003. arXiv:hep-ph/0503184. Bibcode:2005PhRvD..72c4004R. doi:10.1103/PhysRevD.72.034004.

[80] M.G. Alford, K. Rajagopal, T. Schaefer, A. Schmitt; Schmitt; Rajagopal; Schäfer (2008). "Color superconductivity in dense quark matter". *Reviews of Modern Physics* **80** (4): 1455–1515. arXiv:0709.4635. Bibcode:2008RvMP...80.1455A. doi:10.1103/RevModPhys.80.1455.

[81] S. Mrowczynski (1998). "Quark–Gluon Plasma". *Acta Physica Polonica B* **29**: 3711. arXiv:nucl-th/9905005. Bibcode:1998AcPPB..29.3711M.

[82] Z. Fodor, S.D. Katz; Katz (2004). "Critical point of QCD at finite T and μ, lattice results for physical quark masses". *Journal of High Energy Physics* **2004** (4): 50. arXiv:hep-lat/0402006. Bibcode:2004JHEP...04..050F. doi:10.1088/1126-6708/2004/04/050.

[83] U. Heinz, M. Jacob (2000). "Evidence for a New State of Matter: An Assessment of the Results from the CERN Lead Beam Programme". arXiv:nucl-th/0002042.

[84] "RHIC Scientists Serve Up "Perfect" Liquid". Brookhaven National Laboratory News. 2005. Retrieved 2009-05-22.

[85] T. Yulsman (2002). *Origins: The Quest for Our Cosmic Roots*. CRC Press. p. 75. ISBN 0-7503-0765-X.

[86] A. Sedrakian, J.W. Clark, M.G. Alford (2007). *Pairing in fermionic systems*. World Scientific. pp. 2–3. ISBN 981-256-907-3.

1.1.9 Further reading

- A. Ali, G. Kramer; Kramer (2011). "JETS and QCD: A historical review of the discovery of the quark and gluon jets and its impact on QCD". *European Physical Journal H* **36** (2): 245. arXiv:1012.2288. Bibcode:2011EPJH...36..245A. doi:10.1140/epjh/e2011-10047-1.

- D.J. Griffiths (2008). *Introduction to Elementary Particles* (2nd ed.). Wiley–VCH. ISBN 3-527-40601-8.

- I.S. Hughes (1985). *Elementary particles* (2nd ed.). Cambridge University Press. ISBN 0-521-26092-2.

- R. Oerter (2005). *The Theory of Almost Everything: The Standard Model, the Unsung Triumph of Modern Physics*. Pi Press. ISBN 0-13-236678-9.

- A. Pickering (1984). *Constructing Quarks: A Sociological History of Particle Physics*. The University of Chicago Press. ISBN 0-226-66799-5.

- B. Povh (1995). *Particles and Nuclei: An Introduction to the Physical Concepts*. Springer–Verlag. ISBN 0-387-59439-6.

- M. Riordan (1987). *The Hunting of the Quark: A true story of modern physics*. Simon & Schuster. ISBN 0-671-64884-5.

- B.A. Schumm (2004). *Deep Down Things: The Breathtaking Beauty of Particle Physics*. Johns Hopkins University Press. ISBN 0-8018-7971-X.

1.1.10 External links

- 1969 Physics Nobel Prize lecture by Murray Gell-Mann
- 1976 Physics Nobel Prize lecture by Burton Richter
- 1976 Physics Nobel Prize lecture by Samuel C.C. Ting
- 2008 Physics Nobel Prize lecture by Makoto Kobayashi
- 2008 Physics Nobel Prize lecture by Toshihide Maskawa
- The Top Quark And The Higgs Particle by T.A. Heppenheimer – A description of CERN's experiment to count the families of quarks.
- Bowley, Roger; Copeland, Ed. "Quarks". *Sixty Symbols*. Brady Haran for the University of Nottingham.

1.2 Eightfold Way

In physics, the **Eightfold Way** is a term coined by American physicist Murray Gell-Mann for a theory organizing subatomic baryons and mesons into octets (alluding to the Noble Eightfold Path of Buddhism). The theory was independently proposed by Israeli physicist Yuval Ne'eman and led to the subsequent development of the quark model.

In addition to organizing the mesons and spin−1/2 baryons into an octet, the principles of the Eightfold Way also applied to the spin-3/2 baryons, forming a decuplet. However, one of the particles of this decuplet had never been previously observed. Gell-Mann called this particle the $\Omega-$ and predicted in 1962 that it would have a strangeness −3, electric charge −1 and a mass near 1680 MeV/c^2. In 1964, a particle closely matching these predictions was discovered[1] by a particle accelerator group at Brookhaven. Gell-Mann received the 1969 Nobel Prize in Physics for his work on the theory of elementary particles.

The Eightfold Way may be understood in modern terms as a consequence of flavor symmetries between various kinds of quarks. Since the strong nuclear force affects quarks the same way regardless of their flavor, replacing one flavor of quark with another in a hadron should not alter its mass very much. Mathematically, this replacement may be described by elements of the SU(3) group. The octets and other arrangements are representations of this group.

1.2.1 Background

Flavour symmetry

Main article: Flavour (particle physics)

There is an abstract three-dimensional vector space:

$$\text{quark up} \to \begin{pmatrix} 1 \\ 0 \\ 0 \end{pmatrix}, \quad \text{quark down} \to \begin{pmatrix} 0 \\ 1 \\ 0 \end{pmatrix}, \quad \text{quark strange} \to \begin{pmatrix} 0 \\ 0 \\ 1 \end{pmatrix}$$

and the laws of physics are *approximately* invariant under applying a determinant-1 unitary transformation to this space (sometimes called a *flavour rotation*):

$$\begin{pmatrix} x \\ y \\ z \end{pmatrix} \mapsto A \begin{pmatrix} x \\ y \\ z \end{pmatrix}, \quad \text{where } A \text{ in is } SU(3)$$

Here, SU(3) refers to the Lie group of **3×3 unitary matrices with determinant 1** (Special unitary group). For example, the flavour rotation

$$A = \begin{pmatrix} 0 & 1 & 0 \\ -1 & 0 & 0 \\ 0 & 0 & 1 \end{pmatrix}$$

is a transformation that simultaneously turns all the up quarks in the universe into down quarks and vice versa. More specifically, these flavour rotations are exact symmetries if you *only* look at strong force interactions, but they are not truly exact symmetries of the universe because the three quarks have different masses and different electroweak interactions.

This approximate symmetry is called *flavour symmetry*, or more specifically *flavour SU(3) symmetry*.

(This is a slightly over-simplified description of flavour rotations, ignoring anti-quarks etc.)

Connection to representation theory

Main article: Particle physics and representation theory

Assume we have a certain particle—for example, a proton—in a quantum state $|\psi\rangle$. If we apply one of the flavour rotations A to our particle, it enters a new quantum state which we can call $A|\psi\rangle$. Depending on A, this new state might be a proton, or a neutron, or a superposition of a proton and a neutron, or various other possibilities. The set of all possible quantum states spans a vector space.

Representation theory is a mathematical theory that describes the situation where elements of a group (here, the flavour rotations A in the group SU(3)) are automorphisms of a vector space (here, the set of all possible quantum states that you get from flavour-rotating a proton). Therefore, by studying the representation theory of SU(3), we can learn the possibilities for what the vector space is and how it is affected by flavour symmetry.

Since the flavour rotations A are approximate, not exact, symmetries, each orthogonal state in the vector space corresponds to a different particle species. In the example above, when a proton is transformed by every possible flavour rotation A, it turns out that it moves around an 8-dimensional vector space. Those 8 dimensions correspond to the 8 particles in the so-called "baryon octet" (proton, neutron, Σ+, Σ0, Σ−, Ξ−, Ξ0, Λ). This corresponds to an 8-dimensional ("octet") representation of the group SU(3). Since A is an approximate symmetry, all the particles in this octet have similar mass.

Incidentally, every Lie group has a corresponding Lie algebra, and each group representation of the Lie group can be mapped to a corresponding Lie algebra representation on the same vector space. The Lie algebra SU(3) can be written as the set of 3×3 traceless Hermitian matrices. Physicists generally discuss the representation theory of the Lie algebra SU(3) instead of the Lie group SU(3), since the former is simpler and the two are ultimately equivalent.

Development

In the text above, flavour symmetry was defined and motivated using our modern understanding of quarks. But historically, it was the reverse: Quarks were motivated by our understanding of flavour symmetry. Specifically: First it was noticed that groups of particles were related to each other in a way that matched the representation theory of SU(3). From that, it was inferred that there is an approximate symmetry of the universe which is parametrized by the group SU(3). Finally, this helped lead to the discovery of quarks, three of which are interchanged by these SU(3) transformations (the three lightest: up, down, and strange).

1.2.2 References

[1] V. E. Barnes et al. (1964). "Observation of a Hyperon with Strangeness Minus Three" (PDF). *Physical Review Letters* **12** (8): 204. Bibcode:1964PhRvL..12..204B. doi:10.1103/PhysRevLett.12.204.

- D. Griffiths (2008). *Introduction to Elementary Particles 2nd.Ed.* Wiley-VCH. ISBN 3527406018.

1.2.3 Further reading

The following book contains most (if not all) historical papers on the Eightfold Way and related topics, including the Gell-Mann–Okubo mass formula.

- M. Gell-Mann; Y. Ne'eman, eds. (1964). *The Eightfold Way*. W. A. Benjamin. LCCN 65013009.

1.3 Quark model

In particle physics, the **quark model** is a classification scheme for hadrons in terms of their valence quarks—the quarks and antiquarks which give rise to the quantum numbers of the hadrons. The quark model underlies "flavor SU(3)", or the *Eightfold Way*, the successful classification scheme organizing the large number of lighter hadrons that were being discovered starting in the 1950s and continuing through the 1960s. It received experimental verification beginning in the late 1960s and is a valid effective classification of them to date. The quark model was independently proposed by physicists Murray Gell-Mann,[*][1] and George Zweig[*][2][*][3] (also see [*][4]) in 1964. Today, the model has essentially been absorbed as a component of the established quantum field theory of strong and electroweak particle interactions, dubbed the Standard Model.

Hadrons are not really "elementary", and can be regarded as bound states of their "valence quarks" and antiquarks, which give rise to the quantum numbers of the hadrons. These quantum numbers are labels identifying the hadrons, and are of two kinds. One set comes from the Poincaré symmetry—$J^{P}PC$, where J, P and C stand for the total angular momentum, P-symmetry, and C-symmetry, respectively.

The remaining are flavour quantum numbers such as the isospin, strangeness, charm, and so on. The strong interactions binding the quarks together are insensitive to these quantum numbers, so variation of them leads to systematic mass and coupling relationships among the hadrons in the same flavour multiplet.

All quarks are assigned a baryon number of ⅓. Up, charm and top quarks have an electric charge of +⅔, while the down, strange, and bottom quarks have an electric charge of $-\frac{1}{3}$. Antiquarks have the opposite quantum numbers. Quarks are spin-½ particles, and thus fermions. Each quark or antiquark obeys the Gell-Mann–Nishijima formula individually, so any additive assembly of them will as well.

Mesons are made of a valence quark–antiquark pair (thus have a baryon number of 0), while baryons are made of three quarks (thus have a baryon number of 1). This article discusses the quark model for the up, down, and strange flavors of quark (which form an approximate flavor SU(3) symmetry). There are generalizations to larger number of flavors.

1.3.1 History

Developing classification schemes for hadrons became a timely question after new experimental techniques uncovered so many of them, that it became clear that they could not all be elementary. These discoveries led Wolfgang Pauli to exclaim "Had I foreseen that, I would have gone into botany," and Enrico Fermi to advise his student Leon Lederman: "Young man, if I could remember the names of these particles, I would have been a botanist." These new schemes earned Nobel prizes for experimental particle physicists, including Luis Alvarez, who was at the forefront of many of these developments. Constructing hadrons as bound states of fewer constituents would thus organize the "zoo" at hand. Several early proposals, such as the ones by Enrico Fermi and Chen-Ning Yang (1949), and by Shoichi Sakata (1956), ended up satisfactorily covering the mesons, but failed with baryons, and so were unable to explain all the data.

The Gell-Mann–Nishijima formula, developed by Murray Gell-Mann and Kazuhiko Nishijima, led to the Eightfold way classification, invented by Gell-Mann, with important independent contributions from Yuval Ne'eman, in 1961. The hadrons were organized into SU(3) representation multiplets, octets and decuplets, of roughly the same mass, due to the strong interactions; and smaller mass differences linked to the flavor quantum numbers, invisible to the strong interactions. The Gell-Mann–Okubo mass formula systematized the quantification of these small mass differences among members of a hadronic multiplet, controlled by the explicit symmetry breaking of SU(3).

The spin-$\frac{3}{2}$ Ω^- baryon, a member of the ground-state decuplet, was a crucial prediction of that classification. After it was discovered in an experiment at Brookhaven National Laboratory, Gell-Mann received a Nobel prize in physics for his work on the Eightfold Way, in 1969.

Finally, in 1964, Gell-Mann, and, independently, George Zweig, discerned what the Eightfold Way picture encodes.

They posited elementary fermionic constituents, unobserved, and possibly unobservable in a free form, underlying and elegantly encoding the Eightfold Way classification, in an economical, tight structure, resulting in further simplicity. Hadronic mass differences were now linked to the different masses of the constituent quarks.

It would take about a decade for the unexpected nature —and physical reality—of these quarks to be appreciated more fully (See Quarks). Counter-intuitively, they cannot ever be observed in isolation (color confinement), but instead always combine with other quarks to form full hadrons, which then furnish ample indirect information on the trapped quarks themselves. Conversely, the quarks serve in the definition of Quantum chromodynamics, the fundamental theory fully describing the strong interactions; and the Eightfold Way is now understood to be a consequence of the flavor symmetry structure of the lightest three of them. To date, no Nobel prize has been awarded to Gell-Mann and Zweig for this discovery.

1.3.2 Mesons

See also: Meson and List of mesons

The Eightfold Way classification is named after the following fact. If we take three flavors of quarks, then the quarks lie in the fundamental representation, **3** (called the triplet) of flavor SU(3). The antiquarks lie in the complex conjugate representation $\bar{\mathbf{3}}$. The nine states (nonet) made out of a pair can be decomposed into the trivial representation, **1** (called the singlet), and the adjoint representation, **8** (called the octet). The notation for this decomposition is

$$\mathbf{3} \otimes \bar{\mathbf{3}} = \mathbf{8} \oplus \mathbf{1}$$

Figure 1 shows the application of this decomposition to the mesons. If the flavor symmetry were exact (as in the limit that only the strong interactions operate, but the electroweak interactions are notionally switched off), then all nine mesons would have the same mass. However, the physical content of the full theory includes consideration of the symmetry breaking induced by the quark mass differences, and considerations of mixing between various multiplets (such as the octet and the singlet).

N.B. Nevertheless, the mass splitting between the η and the η' is larger than the quark model can accommodate, and this "η–η' puzzle" has its origin in topological peculiarities of the strong interaction vacuum, such as instanton configurations.

Mesons are hadrons with zero baryon number. If the quark–antiquark pair are in an orbital angular momentum L state, and have spin S, then

- $|L - S| \leq J \leq L + S$, where $S = 0$ or 1,

- $P = (-1)^L + 1$, where the 1 in the exponent arises from the intrinsic parity of the quark–antiquark pair.

- $C = (-1)^L + S$ for mesons which have no flavor. Flavored mesons have indefinite value of C.

- For isospin $I = 1$ and 0 states, one can define a new multiplicative quantum number called the *G-parity* such that $G = (-1)^I + L + S$.

If $P = (-1)^J$, then it follows that $S = 1$, thus $PC = 1$. States with these quantum numbers are called *natural parity states*; while all other quantum numbers are thus called *exotic* (for example the state $J^P PC = 0^{*}--$).

1.3.3 Baryons

Main article: Baryon
See also: List of baryons

Since quarks are fermions, the spin-statistics theorem implies that the wavefunction of a baryon must be antisymmetric under exchange of any two quarks. This antisymmetric wavefunction is obtained by making it fully antisymmetric in color, discussed below, and symmetric in flavor, spin and space put together. With three flavors, the decomposition in flavor is

$$3 \otimes 3 \otimes 3 = 10_S \oplus 8_M \oplus 8_M \oplus 1_A$$

The decuplet is symmetric in flavor, the singlet antisymmetric and the two octets have mixed symmetry. The space and spin parts of the states are thereby fixed once the orbital angular momentum is given.

It is sometimes useful to think of the basis states of quarks as the six states of three flavors and two spins per flavor. This approximate symmetry is called spin-flavor SU(6). In terms of this, the decomposition is

$$6 \otimes 6 \otimes 6 = 56_S \oplus 70_M \oplus 70_M \oplus 20_A \ .$$

The 56 states with symmetric combination of spin and flavour decompose under flavor SU(3) into

$$56 = 10^{\frac{3}{2}} \oplus 8^{\frac{1}{2}} ,$$

where the superscript denotes the spin, S, of the baryon. Since these states are symmetric in spin and flavor, they should also be symmetric in space—a condition that is easily satisfied by making the orbital angular momentum $L = 0$. These are the ground state baryons.

The $S = \frac{1}{2}$ octet baryons are the two nucleons (p+, n0), the three Sigmas ($\Sigma+$, $\Sigma 0$, $\Sigma-$), the two Xis ($\Xi 0$, $\Xi-$), and the Lambda ($\Lambda 0$). The $S = \frac{3}{2}$ decuplet baryons are the four Deltas ($\Delta++$, $\Delta+$, $\Delta 0$, $\Delta-$), three Sigmas ($\Sigma*+$, $\Sigma*0$, $\Sigma*-$), two Xis ($\Xi*0$, $\Xi*-$), and the Omega ($\Omega-$).

Mixing of baryons, mass splittings within and between multiplets, and magnetic moments are some of the other questions that the model predicts successfully.

The discovery of color

Main article: Color charge

Color quantum numbers are the characteristic charges of the strong force, and are completely uninvolved in electroweak interactions. They were discovered as a consequence of the quark model classification, when it was appreciated that the spin $S = \frac{3}{2}$ baryon, the $\Delta++$, required three up quarks with parallel spins and vanishing orbital angular momentum. Therefore, it could not have an antisymmetric wave function, (due to the Pauli exclusion principle), *unless there were a hidden quantum number*. Oscar Greenberg noted this problem in 1964, suggesting that quarks should be para-fermions.[5]

Instead, six months later, Moo-Young Han and Yoichiro Nambu suggested the existence of three triplets of quarks to solve this problem, but flavor and color intertwined in that model— they did not commute.[6]

The modern concept of color completely commuting with all other charges and providing the strong force charge was articulated in 1973, by William Bardeen, Harald Fritzsch, and Murray Gell-Mann.[7][8]

1.3.4 States outside the quark model

While the quark model is derivable from the theory of quantum chromodynamics, the structure of hadrons is more complicated than this model allows. The full quantum mechanical wave function of any hadron must include virtual quark pairs as well as virtual gluons, and allows for a variety of mixings. There may be hadrons which lie outside the quark model. Among these are the *glueballs* (which contain only valence gluons), *hybrids* (which contain valence quarks as well as gluons) and "exotic hadrons" (such as tetraquarks or pentaquarks).

1.3.5 See also

- Subatomic particles
- Hadrons, baryons, mesons and quarks
- Exotic hadrons: exotic mesons and exotic baryons
- Quantum chromodynamics, flavor, the QCD vacuum

1.3.6 References and external links

[1] M. Gell-Mann (1964). "A Schematic Model of Baryons and Mesons". *Physics Letters* **8** (3): 214–215. Bibcode:1964PhL.....8..214G. doi:10.1016/S0031-9163(64)92001-3.

[2] G. Zweig (1964). "An SU(3) Model for Strong Interaction Symmetry and its Breaking" (PDF). *CERN Report No.8182/TH.401*.

[3] G. Zweig (1964). "An SU(3) Model for Strong Interaction Symmetry and its Breaking: II" (PDF). *CERN Report No.8419/TH.412*.

[4] Petermann, A. (1965). "Propriétés de l'étrangeté et une formule de masse pour les mésons vectoriels". *Nuclear Physics* **63** (2): 349. doi:10.1016/0029-5582(65)90348-2. which gingerly touched upon the central ideas, without quantitative substantiation;

[5] O.W. Greenberg (1964). "Spin and Unitary-Spin Independence in a Paraquark Model of Baryons and Mesons". *Physical Review Letters* **13** (20): 598. Bibcode:1964PhRvL..13..598G. doi:10.1103/PhysRevLett.13.598.

[6] M.Y. Han, Y. Nambu (1965). "Three-Triplet Model with Double SU(3) Symmetry". *Physical Review* **139** (4B): B1006. Bibcode:1965PhRv..139.1006H. doi:10.1103/PhysRev.139.B1006.

[7] W. Bardeen, H. Fritzsch, M. Gell-Mann (1973). "Light cone current algebra, π^0 decay, and $e^* + e^*$ − annihilation". In R. Gatto. *Scale and conformal symmetry in hadron physics*. John Wiley & Sons. p. 139. arXiv:hep-ph/0211388. ISBN 0-471-29292-3.

[8] Fritzsch, H.; Gell-Mann, M.; Leutwyler, H. (1973). "Advantages of the color octet gluon picture". *Physics Letters B* **47** (4): 365. doi:10.1016/0370-2693(73)90625-4.

- S. Eidelman *et al.* Particle Data Group (2004). "Review of Particle Physics" (PDF). *Physics Letters B* **592**: 1. arXiv:astro-ph/0406663. Bibcode:2004PhLB..592....1P. doi:10.1016/j.physletb.2004.06.001.

- Lichtenberg, D B (1970). *Unitary Symmetry and Elementary Particles*. Academic Press. ISBN 978-1483242729.

- Thomson, M A (2011), Lecture notes

- J.J.J. Kokkedee (1969). *The quark model*. W. A. Benjamin. ASIN B001RAVDIA.

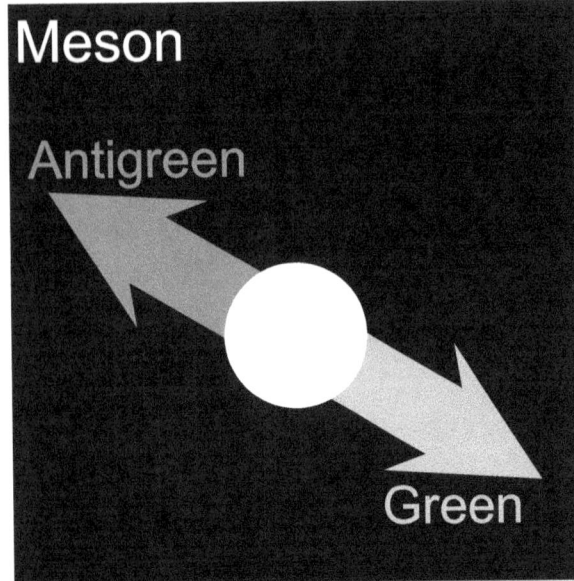

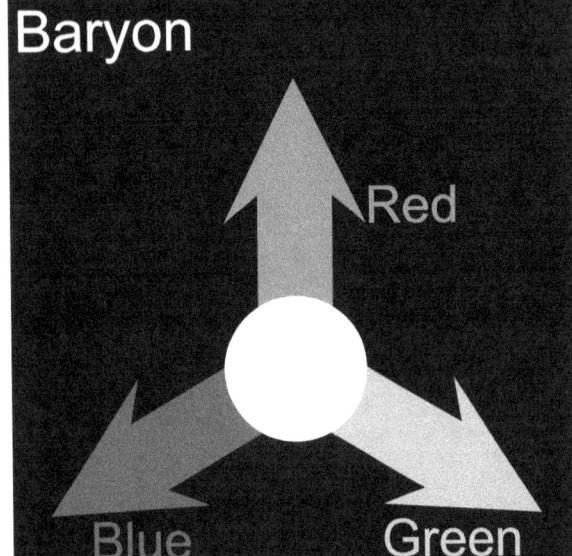

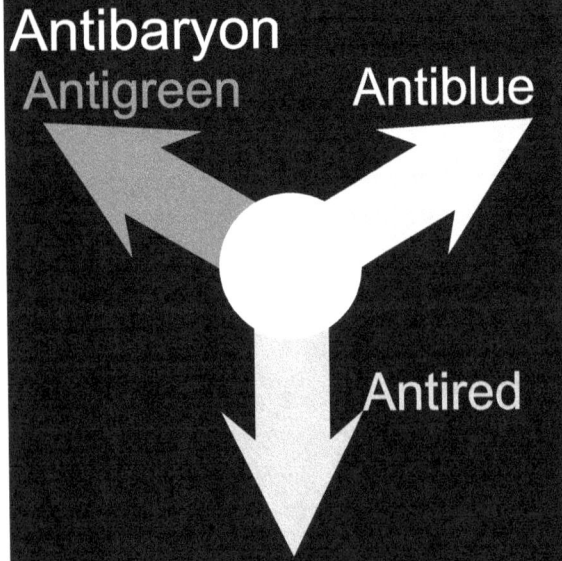

All types of hadrons have zero total color charge.

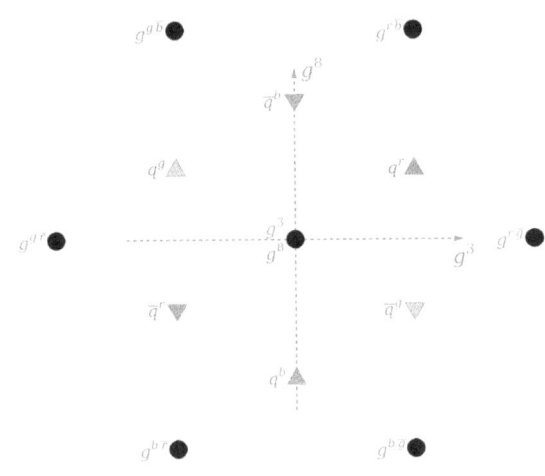

The pattern of strong charges for the three colors of quark, three antiquarks, and eight gluons (with two of zero charge overlapping).

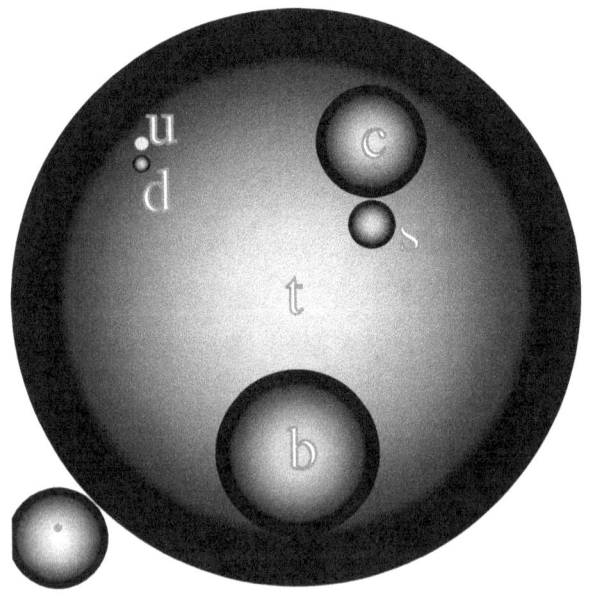

Current quark masses for all six flavors in comparison, as balls of proportional volumes. Proton and electron (red) are shown in bottom left corner for scale

1.3. QUARK MODEL

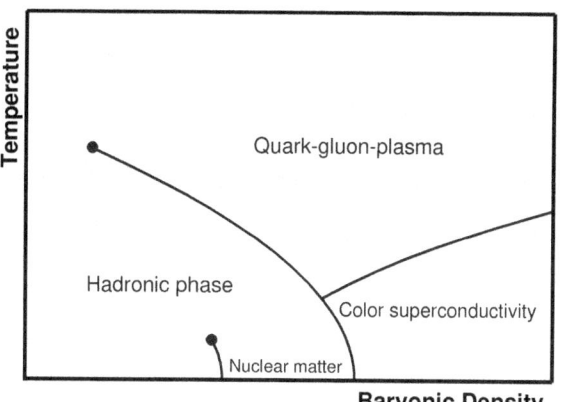

A qualitative rendering of the phase diagram of quark matter. The precise details of the diagram are the subject of ongoing research.[79]*[80]*

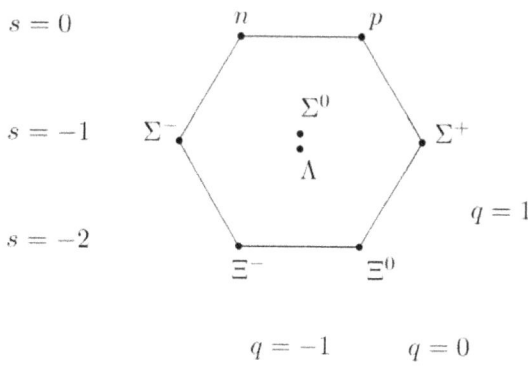

The baryon octet

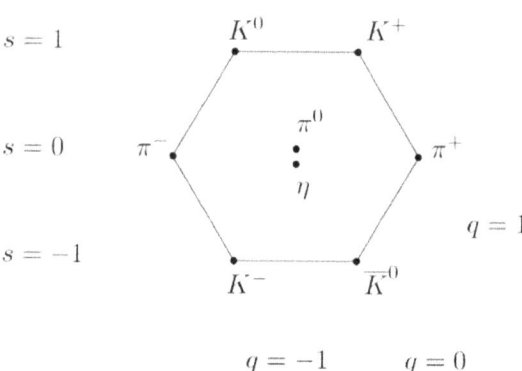

The meson octet. Particles along the same horizontal line share the same strangeness, s, while those on the same diagonals share the same charge, q.

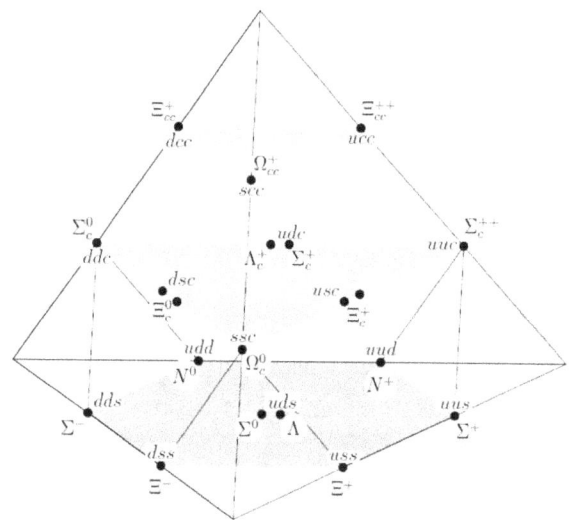

Baryon Supermultiplet using four-quark models and half spin

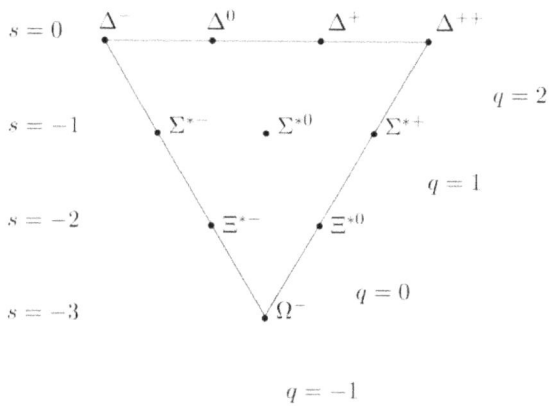

The baryon decuplet

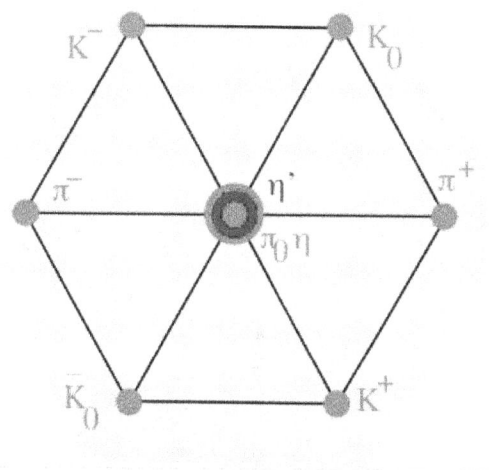

Figure 1: *The pseudoscalar meson nonet. Members of the octet are shown in green, the singlet in magenta. The name of the* Eightfold Way *derives from this classification.*

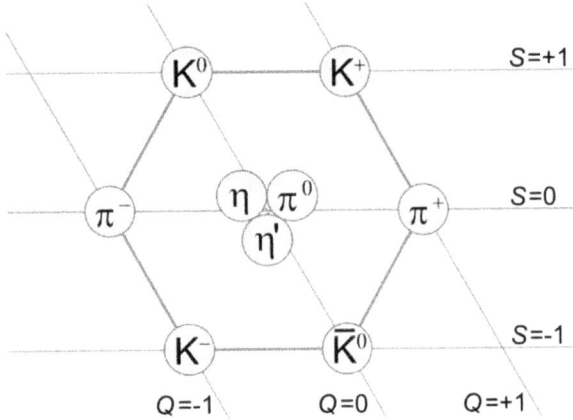

Figure 2: Pseudoscalar mesons of spin 0 form a nonet

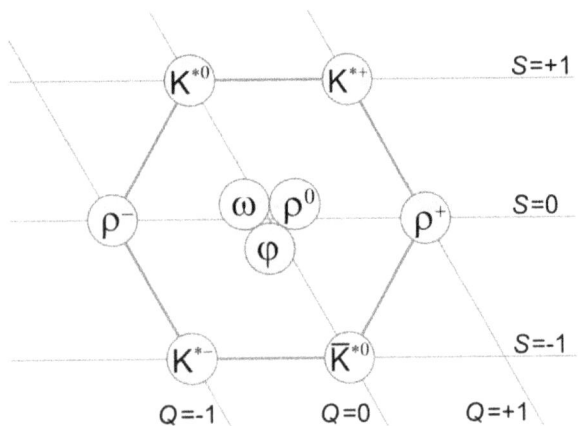

Figure 3: Mesons of spin 1 form a nonet

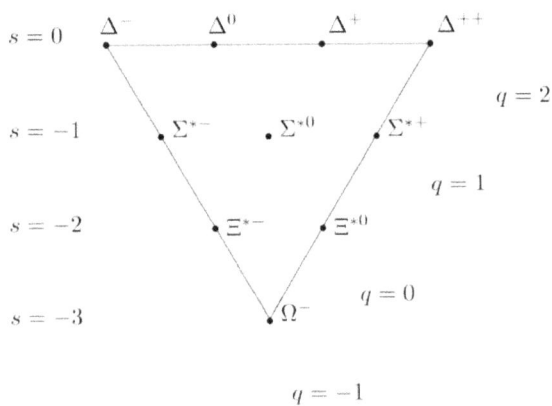

Figure 5. The S = $^3/_2$ baryon decuplet

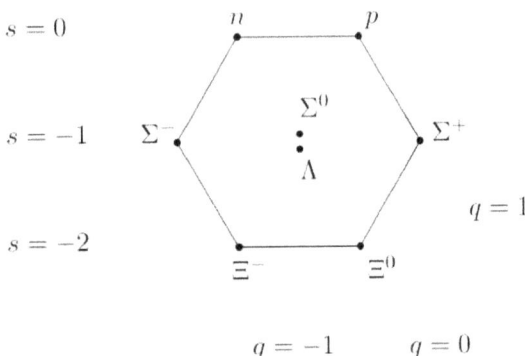

Figure 4. The S = $^1/_2$ ground state baryon octet

Chapter 2

Individual quarks

2.1 Up quark

The **up quark** or **u quark** (symbol: *u*) is the lightest of all quarks, a type of elementary particle, and a major constituent of matter. It, along with the down quark, forms the neutrons (one up quark, two down quarks) and protons (two up quarks, one down quark) of atomic nuclei. It is part of the first generation of matter, has an electric charge of $+^2/_3\,e$ and a bare mass of 1.8–3.0 MeV/c^2. Like all quarks, the up quark is an elementary fermion with spin-$^1/_2$, and experiences all four fundamental interactions: gravitation, electromagnetism, weak interactions, and strong interactions. The antiparticle of the up quark is the **up antiquark** (sometimes called *antiup quark* or simply *antiup*), which differs from it only in that some of its properties have equal magnitude but opposite sign.

Its existence (along with that of the down and strange quarks) was postulated in 1964 by Murray Gell-Mann and George Zweig to explain the *Eightfold Way* classification scheme of hadrons. The up quark was first observed by experiments at the Stanford Linear Accelerator Center in 1968.

2.1.1 History

In the beginnings of particle physics (first half of the 20th century), hadrons such as protons, neutrons and pions were thought to be elementary particles. However, as new hadrons were discovered, the 'particle zoo' grew from a few particles in the early 1930s and 1940s to several dozens of them in the 1950s. The relationships between each of them were unclear until 1961, when Murray Gell-Mann[*][2] and Yuval Ne'eman[*][3] (independently of each other) proposed a hadron classification scheme called the *Eightfold Way*, or in more technical terms, SU(3) flavor symmetry.

This classification scheme organized the hadrons into isospin multiplets, but the physical basis behind it was still unclear. In 1964, Gell-Mann[*][4] and George Zweig[*][5][*][6] (independently of each other) proposed the quark model, then consisting only of up, down, and strange quarks.[*][7] However, while the quark model explained the Eightfold Way, no direct evidence of the existence of quarks was found until 1968 at the Stanford Linear Accelerator Center.[*][8][*][9] Deep inelastic scattering experiments indicated that protons had substructure, and that protons made of three more-fundamental particles explained the data (thus confirming the quark model).[*][10]

At first people were reluctant to describe the three bodies as quarks, instead preferring Richard Feynman's parton description,[*][11][*][12][*][13] but over time the quark theory became accepted (see *November Revolution*).[*][14]

2.1.2 Mass

Despite being extremely common, the bare mass of the up quark is not well determined, but probably lies between 1.8 and 3.0 MeV/c^2.[*][1] Lattice QCD calculations give a more precise value: 2.01±0.14 MeV/c^2.[*][15]

When found in mesons (particles made of one quark and one antiquark) or baryons (particles made of three quarks), the 'effective mass' (or 'dressed' mass) of quarks becomes greater because of the binding energy caused by the gluon field between each quark (see mass–energy equivalence).The bare mass of up quarks is so light, it cannot be straightforwardly calculated because relativistic effects have to be taken into account.

2.1.3 See also

- Down quark

- Isospin

- Quark model

- Quantum Mechanics

2.1.4 References

[1] J. Beringer *et al.* (Particle Data Group) (2012). "PDGLive Particle Summary 'Quarks (u, d, s, c, b, t, b', t', Free)'" (PDF). Particle Data Group. Retrieved 2013-02-21.

[2] M. Gell-Mann (2000) [1964]. "The Eightfold Way: A theory of strong interaction symmetry". In M. Gell-Mann, Y. Ne'eman. *The Eightfold Way*. Westview Press. p. 11. ISBN 0-7382-0299-1.
Original: M. Gell-Mann (1961). "The Eightfold Way: A theory of strong interaction symmetry". *Synchrotron Laboratory Report CTSL-20* (California Institute of Technology)

[3] Y. Ne'eman (2000) [1964]. "Derivation of strong interactions from gauge invariance". In M. Gell-Mann, Y. Ne'eman. *The Eightfold Way*. Westview Press. ISBN 0-7382-0299-1.
Original Y. Ne'eman (1961). "Derivation of strong interactions from gauge invariance". *Nuclear Physics* **26** (2): 222. Bibcode:1961NucPh..26..222N. doi:10.1016/0029-5582(61)90134-1.

[4] M. Gell-Mann (1964). "A Schematic Model of Baryons and Mesons". *Physics Letters* **8** (3): 214–215. Bibcode:1964PhL.....8..214G. doi:10.1016/S0031-9163(64)92001-3.

[5] G. Zweig (1964). "An SU(3) Model for Strong Interaction Symmetry and its Breaking". *CERN Report No.8181/Th 8419*.

[6] G. Zweig (1964). "An SU(3) Model for Strong Interaction Symmetry and its Breaking: II". *CERN Report No.8419/Th 8412*.

[7] B. Carithers, P. Grannis (1995). "Discovery of the Top Quark" (PDF). *Beam Line* (SLAC) **25** (3): 4–16. Retrieved 2008-09-23.

[8] E. D. Bloom; Coward, D.; Destaebler, H.; Drees, J.; Miller, G.; Mo, L.; Taylor, R.; Breidenbach, M. et al. (1969). "High-Energy Inelastic $e-p$ Scattering at 6° and 10°". *Physical Review Letters* **23** (16): 930–934. Bibcode:1969PhRvL..23..930B. doi:10.1103/PhysRevLett.23.930.

[9] M. Breidenbach; Friedman, J.; Kendall, H.; Bloom, E.; Coward, D.; Destaebler, H.; Drees, J.; Mo, L.; Taylor, R. et al. (1969). "Observed Behavior of Highly Inelastic Electron–Proton Scattering". *Physical Review Letters* **23** (16): 935–939. Bibcode:1969PhRvL..23..935B. doi:10.1103/PhysRevLett.23.935.

[10] J. I. Friedman. "The Road to the Nobel Prize". Hue University. Retrieved 2008-09-29.

[11] R. P. Feynman (1969). "Very High-Energy Collisions of Hadrons". *Physical Review Letters* **23** (24): 1415–1417. Bibcode:1969PhRvL..23.1415F. doi:10.1103/PhysRevLett.23.1415.

[12] S. Kretzer; Lai, H.; Olness, Fredrick; Tung, W. et al. (2004). "CTEQ6 Parton Distributions with Heavy Quark Mass Effects". *Physical Review D* **69** (11): 114005. arXiv:hep-ph/0307022. Bibcode:2004PhRvD..69k4005K. doi:10.1103/PhysRevD.69.114005.

[13] D. J. Griffiths (1987). *Introduction to Elementary Particles*. John Wiley & Sons. p. 42. ISBN 0-471-60386-4.

[14] M. E. Peskin, D. V. Schroeder (1995). *An introduction to quantum field theory*. Addison–Wesley. p. 556. ISBN 0-201-50397-2.

[15] Cho, Adrian (April 2010). "Mass of the Common Quark Finally Nailed Down". Science Magazine.

2.1.5 Further reading

- A. Ali, G. Kramer; Kramer (2011). "JETS and QCD: A historical review of the discovery of the quark and gluon jets and its impact on QCD". *European Physical Journal H* **36** (2): 245. arXiv:1012.2288. Bibcode:2011EPJH...36..245A. doi:10.1140/epjh/e2011-10047-1.

- R. Nave. "Quarks". *HyperPhysics*. Georgia State University, Department of Physics and Astronomy. Retrieved 2008-06-29.

- A. Pickering (1984). *Constructing Quarks*. University of Chicago Press. pp. 114–125. ISBN 0-226-66799-5.

2.2 Down quark

The **down quark** or **d quark** (symbol: d) is the second-lightest of all quarks, a type of elementary particle, and a major constituent of matter. Together with the up quark, it forms the neutrons (one up quark, two down quarks) and protons (two up quarks, one down quark) of atomic nuclei. It is part of the first generation of matter, has an electric charge of $-\frac{1}{3}$ e and a bare mass of 4.8+0.5 −0.3 MeV/c^2.[1] Like all quarks, the down quark is an elementary fermion with spin-$\frac{1}{2}$, and experiences all four fundamental interactions: gravitation, electromagnetism, weak interactions, and strong interactions. The antiparticle of the down quark is the **down antiquark** (sometimes called *antidown quark* or simply *antidown*), which differs from it only in that some of its properties have equal magnitude but opposite sign.

Its existence (along with that of the up and strange quarks) was postulated in 1964 by Murray Gell-Mann and George Zweig to explain the *Eightfold Way* classification scheme of hadrons. The down quark was first observed by experiments at the Stanford Linear Accelerator Center in 1968.

2.2.1 History

In the beginnings of particle physics (first half of the 20th century), hadrons such as protons, neutrons, and pions were thought to be elementary particles. However, as new hadrons were discovered, the 'particle zoo' grew from a few particles in the early 1930s and 1940s to several dozens of them in the 1950s. The relationships between each of them was unclear until 1961, when Murray Gell-Mann*[2] and Yuval Ne'eman*[3] (independently of each other) proposed a hadron classification scheme called the *Eightfold Way*, or in more technical terms, SU(3) flavor symmetry.

This classification scheme organized the hadrons into isospin multiplets, but the physical basis behind it was still unclear. In 1964, Gell-Mann*[4] and George Zweig*[5]*[6] (independently of each other) proposed the quark model, then consisting only of up, down, and strange quarks.*[7] However, while the quark model explained the Eightfold Way, no direct evidence of the existence of quarks was found until 1968 at the Stanford Linear Accelerator Center.*[8]*[9] Deep inelastic scattering experiments indicated that protons had substructure, and that protons made of three more-fundamental particles explained the data (thus confirming the quark model).*[10]

At first people were reluctant to identify the three-bodies as quarks, instead preferring Richard Feynman's parton description,*[11]*[12]*[13] but over time the quark theory became accepted (see *November Revolution*).*[14]

2.2.2 Mass

Despite being extremely common, the bare mass of the down quark is not well determined, but probably lies between 4.5 and $5.3 10^0$ MeV/c^2.*[1] Lattice QCD calculations give a more precise value: 4.79 ± 0.16 MeV/c^2.*[15]

When found in mesons (particles made of one quark and one antiquark) or baryons (particles made of three quarks), the 'effective mass' (or 'dressed' mass) of quarks becomes greater because of the binding energy caused by the gluon field between quarks (see mass–energy equivalence). For example, the effective mass of down quarks in a proton is around 330 MeV/c^2. Because the bare mass of down quarks is so small, it cannot be straightforwardly calculated because relativistic effects have to be taken into account.

2.2.3 See also

- Up quark
- Isospin
- Quark model

2.2.4 References

[1] J. Beringer (Particle Data Group) et al. (2013). "PDGLive Particle Summary 'Quarks (u, d, s, c, b, t, b′, t′, Free)'" (PDF). Particle Data Group. Retrieved 2013-07-23.

[2] M. Gell-Mann (2000) [1964]. "The Eightfold Way: A theory of strong interaction symmetry". In M. Gell-Mann, Y. Ne'eman. *The Eightfold Way*. Westview Press. p. 11. ISBN 0-7382-0299-1.
Original: M. Gell-Mann (1961). "The Eightfold Way: A theory of strong interaction symmetry". *Synchrotron Laboratory Report CTSL-20* (California Institute of Technology).

[3] Y. Ne'eman (2000) [1964]. "Derivation of strong interactions from gauge invariance". In M. Gell-Mann, Y. Ne'eman. *The Eightfold Way*. Westview Press. ISBN 0-7382-0299-1.
Original Y. Ne'eman (1961). "Derivation of strong interactions from gauge invariance". *Nuclear Physics* **26** (2): 222. Bibcode:1961NucPh..26..222N. doi:10.1016/0029-5582(61)90134-1.

[4] M. Gell-Mann (1964). "A Schematic Model of Baryons and Mesons". *Physics Letters* **8** (3): 214–215. Bibcode:1964PhL......8..214G. doi:10.1016/S0031-9163(64)92001-3.

[5] G. Zweig (1964). "An SU(3) Model for Strong Interaction Symmetry and its Breaking". *CERN Report No.8181/Th 8419*.

[6] G. Zweig (1964). "An SU(3) Model for Strong Interaction Symmetry and its Breaking: II". *CERN Report No.8419/Th 8412*.

[7] B. Carithers, P. Grannis (1995). "Discovery of the Top Quark" (PDF). *Beam Line* (SLAC) **25** (3): 4–16. Retrieved 2008-09-23.

[8] E. D. Bloom; Coward, D.; Destaebler, H.; Drees, J.; Miller, G.; Mo, L.; Taylor, R.; Breidenbach, M. et al. (1969). "High-Energy Inelastic e–p Scattering at 6° and 10°". *Physical Review Letters* **23** (16): 930–934. Bibcode:1969PhRvL..23..930B. doi:10.1103/PhysRevLett.23.930.

[9] M. Breidenbach; Friedman, J.; Kendall, H.; Bloom, E.; Coward, D.; Destaebler, H.; Drees, J.; Mo, L.; Taylor, R. et al. (1969). "Observed Behavior of Highly Inelastic Electron–Proton Scattering". *Physical Review Letters* **23** (16): 935–939. Bibcode:1969PhRvL..23..935B. doi:10.1103/PhysRevLett.23.935.

[10] J. I. Friedman. "The Road to the Nobel Prize". Hue University. Retrieved 2008-09-29.

[11] R. P. Feynman (1969). "Very High-Energy Collisions of Hadrons". *Physical Review Letters* **23** (24): 1415–1417. Bibcode:1969PhRvL..23.1415F. doi:10.1103/PhysRevLett.23.1415.

[12] S. Kretzer; Lai, H.; Olness, Fredrick; Tung, W. et al. (2004). "CTEQ6 Parton Distributions with Heavy Quark Mass Effects". *Physical Review D* **69** (11): 114005. arXiv:hep-ph/0307022. Bibcode:2004PhRvD..69k4005K. doi:10.1103/PhysRevD.69.114005.

[13] D. J. Griffiths (1987). *Introduction to Elementary Particles*. John Wiley & Sons. p. 42. ISBN 0-471-60386-4.

[14] M. E. Peskin, D. V. Schroeder (1995). *An introduction to quantum field theory*. Addison–Wesley. p. 556. ISBN 0-201-50397-2.

[15] Cho, Adrian (April 2010). "Mass of the Common Quark Finally Nailed Down". Science Magazine.

2.2.5 Further reading

- A. Ali, G. Kramer; Kramer (2011). "JETS and QCD: A historical review of the discovery of the quark and gluon jets and its impact on QCD". *European Physical Journal H* **36** (2): 245. arXiv:1012.2288. Bibcode:2011EPJH...36..245A. doi:10.1140/epjh/e2011-10047-1.

- R. Nave. "Quarks". *HyperPhysics*. Georgia State University, Department of Physics and Astronomy. Retrieved 2008-06-29.

- A. Pickering (1984). *Constructing Quarks*. University of Chicago Press. pp. 114–125. ISBN 0-226-66799-5.

2.3 Strange quark

The **strange quark** or **s quark** (from its symbol, s) is the third-lightest of all quarks, a type of elementary particle. Strange quarks are found in subatomic particles called hadrons. Example of hadrons containing strange quarks include kaons (K), strange D mesons (D
s), Sigma baryons (Σ), and other strange particles.

Along with the charm quark, it is part of the second generation of matter, and has an electric charge of $-\frac{1}{3}\,e$ and a bare mass of 95+5
−5 MeV/c^2.[1] Like all quarks, the strange quark is an elementary fermion with spin-$\frac{1}{2}$, and experiences all four fundamental interactions: gravitation, electromagnetism, weak interactions, and strong interactions. The antiparticle of the strange quark is the **strange antiquark** (sometimes called *antistrange quark* or simply *antistrange*), which differs from it only in that some of its properties have equal magnitude but opposite sign.

The first strange particle (a particle containing a strange quark) was discovered in 1947 (kaons), but the existence of the strange quark itself (and that of the up and down quarks) was only postulated in 1964 by Murray Gell-Mann and George Zweig to explain the *Eightfold Way* classification scheme of hadrons. The first evidence for the existence of quarks came in 1968, in deep inelastic scattering experiments at the Stanford Linear Accelerator Center. These experiments confirmed the existence of up and down quarks, and by extension, strange quarks, as they were required to explain the Eightfold Way.

2.3.1 History

In the beginnings of particle physics (first half of the 20th century), hadrons such as protons, neutron and pions were thought to be elementary particles. However, new hadrons were discovered, the 'particle zoo' grew from a few particles in the early 1930s and 1940s to several dozens of them in the 1950s. However some particles were much longer lived than others; most particles decayed through the strong interaction and had lifetimes of around 10^{-23} seconds. But when they decayed through the weak interactions, they had lifetimes of around 10^{-10} seconds to decay. While studying these decays Murray Gell-Mann (in 1953)[2][3] and Kazuhiko Nishijima (in 1955)[4] developed the concept of *strangeness* (which Nishijima called *eta-charge*, after the eta meson (η)) which explained the 'strangeness' of the longer-lived particles. The Gell-Mann–Nishijima formula is the result of these efforts to understand strange decays.

However, the relationships between each particles and the physical basis behind the strangeness property was still unclear. In 1961, Gell-Mann[5] and Yuval Ne'eman[6] (independently of each other) proposed a hadron classification scheme called the *Eightfold Way*, or in more technical terms, SU(3) flavor symmetry. This ordered hadrons into isospin multiplets. The physical basis behind both isospin and strangeness was only explained in 1964, when Gell-Mann[7] and George Zweig[8][9] (independently of each other) proposed the quark model, then consisting only of up, down, and strange quarks.[10] Up and down quarks were the carriers of isospin, while the strange quark carried strangeness. While the quark model explained the Eightfold Way, no direct evidence of the existence of quarks was found until 1968 at the Stanford Linear Accelerator Center.[11][12] Deep inelastic scattering experiments indicated that protons had substructure, and that protons made of three more-fundamental particles explained the data (thus confirming the quark model).[13]

At first people were reluctant to identify the three-bodies as quarks, instead preferring Richard Feynman's parton description,[14][15][16] but over time the quark theory became accepted (see *November Revolution*).[17]

2.3.2 See also

- Quark model
- Strange matter
- Strangeness production
- Strangelet
- Strange star

2.3.3 References

[1] J. Beringer *et al.* (Particle Data Group) (2012). "PDGLive Particle Summary 'Quarks (u, d, s, c, b, t, b′, t′, Free)'" (PDF). Particle Data Group. Retrieved 2012-11-30.

[2] M. Gell-Mann (1953). "Isotopic Spin and New Unstable Particles". *Physical Review* **92** (3): 833. Bibcode:1953PhRv...92..833G. doi:10.1103/PhysRev.92.833.

[3] G. Johnson (2000). *Strange Beauty: Murray Gell-Mann and the Revolution in Twentieth-Century Physics*. Random House. p. 119. ISBN 0-679-43764-9. By the end of the summer... [Gell-Mann] completed his first paper, "Isotopic Spin and Curious Particles" and send it of to *Physical Review*. The editors hated the title, so he amended it to "Strange Particles". They wouldn't go for that either—never mind that almost everybody used the term—suggesting instead "Isotopic Spin and New Unstable Particles".

[4] K. Nishijima, Kazuhiko (1955). "Charge Independence Theory of V Particles". *Progress of Theoretical Physics* **13** (3): 285. Bibcode:1955PThPh..13..285N. doi:10.1143/PTP.13.285.

[5] M. Gell-Mann (2000) [1964]. "The Eightfold Way: A theory of strong interaction symmetry". In M. Gell-Mann, Y. Ne'eman. *The Eightfold Way*. Westview Press. p. 11. ISBN 0-7382-0299-1.
Original: M. Gell-Mann (1961). "The Eightfold Way: A theory of strong interaction symmetry". *Synchrotron Laboratory Report CTSL-20* (California Institute of Technology)

[6] Y. Ne'eman (2000) [1964]. "Derivation of strong interactions from gauge invariance". In M. Gell-Mann, Y. Ne'eman. *The Eightfold Way*. Westview Press. ISBN 0-7382-0299-1.
Original Y. Ne'eman (1961). "Derivation of strong interactions from gauge invariance". *Nuclear Physics* **26** (2): 222. Bibcode:1961NucPh..26..222N. doi:10.1016/0029-5582(61)90134-1.

[7] M. Gell-Mann (1964). "A Schematic Model of Baryons and Mesons". *Physics Letters* **8** (3): 214–215. Bibcode:1964PhL......8..214G. doi:10.1016/S0031-9163(64)92001-3.

[8] G. Zweig (1964). "An SU(3) Model for Strong Interaction Symmetry and its Breaking". *CERN Report No.8181/Th 8419*.

[9] G. Zweig (1964). "An SU(3) Model for Strong Interaction Symmetry and its Breaking: II". *CERN Report No.8419/Th 8412*.

[10] B. Carithers, P. Grannis (1995). "Discovery of the Top Quark" (PDF). *Beam Line* (SLAC) **25** (3): 4–16. Retrieved 2008-09-23.

[11] E. D. Bloom; Coward, D.; Destaebler, H.; Drees, J.; Miller, G.; Mo, L.; Taylor, R.; Breidenbach, M. et al. (1969). "High-Energy Inelastic e–p Scattering at 6° and 10°". *Physical Review Letters* **23** (16): 930–934. Bibcode:1969PhRvL..23..930B. doi:10.1103/PhysRevLett.23.930.

[12] M. Breidenbach; Friedman, J.; Kendall, H.; Bloom, E.; Coward, D.; Destaebler, H.; Drees, J.; Mo, L.; Taylor, R. et al. (1969). "Observed Behavior of Highly Inelastic Electron–Proton Scattering". *Physical Review Letters* **23** (16): 935–939. Bibcode:1969PhRvL..23..935B. doi:10.1103/PhysRevLett.23.935.

[13] J. I. Friedman. "The Road to the Nobel Prize". Hue University. Retrieved 2008-09-29.

[14] R. P. Feynman (1969). "Very High-Energy Collisions of Hadrons". *Physical Review Letters* **23** (24): 1415–1417. Bibcode:1969PhRvL..23.1415F. doi:10.1103/PhysRevLett.23.1415.

[15] S. Kretzer; Lai, H.; Olness, Fredrick; Tung, W. et al. (2004). "CTEQ6 Parton Distributions with Heavy Quark Mass Effects". *Physical Review D* **69** (11): 114005. arXiv:hep-th/0307022. Bibcode:2004PhRvD..69k4005K. doi:10.1103/PhysRevD.69.114005.

[16] D. J. Griffiths (1987). *Introduction to Elementary Particles*. John Wiley & Sons. p. 42. ISBN 0-471-60386-4.

[17] M. E. Peskin, D. V. Schroeder (1995). *An introduction to quantum field theory*. Addison–Wesley. p. 556. ISBN 0-201-50397-2.

2.3.4 Further reading

- R. Nave. "Quarks". *HyperPhysics*. Georgia State University, Department of Physics and Astronomy. Retrieved 2008-06-29.

- A. Pickering (1984). *Constructing Quarks*. University of Chicago Press. pp. 114–125. ISBN 0-226-66799-5.

2.4 Charm quark

The **charm quark** or **c quark** (from its symbol, c) is the third most massive of all quarks, a type of elementary particle. Charm quarks are found in hadrons, which are subatomic particles made of quarks. Example of hadrons containing charm quarks include the J/ψ meson (J/ψ), D mesons (D), charmed Sigma baryons (Σ c), and other charmed particles.

It, along with the strange quark is part of the second generation of matter, and has an electric charge of $+\frac{2}{3}$ e and a bare mass of $1.29+0.05-0.11$ GeV/c^2.[1] Like all quarks, the charm quark is an elementary fermion with spin-$\frac{1}{2}$, and experiences all four fundamental interactions: gravitation, electromagnetism, weak interactions, and strong interactions. The antiparticle of the charm quark is the **charm antiquark** (sometimes called *antischarm quark* or simply *anticharm*), which differs from it only in that some of its properties have equal magnitude but opposite sign.

The existence of a fourth quark had been speculated by a number of authors around 1964 (for instance by James Bjorken and Sheldon Glashow[4]), but its prediction is usually credited to Sheldon Glashow, John Iliopoulos and Luciano Maiani in 1970 (see GIM mechanism).[5] The first charmed particle (a particle containing a charm quark) to be discovered was the J/ψ meson. It was discovered by a team at the Stanford Linear Accelerator Center (SLAC), led by Burton Richter,[6] and one at the Brookhaven National Laboratory (BNL), led by Samuel Ting.[7]

The 1974 discovery of the J/ψ (and thus the charm quark) ushered in a series of breakthroughs which are collectively known as the *November Revolution*.

2.4.1 Hadrons containing charm quarks

Main articles: List of baryons and list of mesons

Some of the hadrons containing charm quarks include:

- D mesons contain a charm quark (or its antiparticle) and an up or down quark.
- D
  s mesons contain a charm quark and a strange quark.
- There are many charmonium states, for example the J/ψ particle. These consist of a charm quark and its antiparticle.
- Charmed baryons have been observed, and are named in analogy with strange baryons (e.g. Λ+
  c).

2.4.2 See also

- Quark model

2.4.3 Notes

[1] K. Nakamura *et al.* (Particle Data Group) (2011). "PDGLive Particle Summary 'Quarks (u, d, s, c, b, t, b′, t′, Free)'" (PDF). Particle Data Group. Retrieved 2011-08-08.

[2] Carl Rod Nave. "Transformation of Quark Flavors by the Weak Interaction". Retrieved 2010-12-06. The c quark has about 5% probability of decaying into a d quark instead of an s quark.

[3] K. Nakamura et al. (2010). "Review of Particles Physics: The CKM Quark-Mixing Matrix" (PDF). *J. Phys. G* **37** (75021): 150.

[4] B.J. Bjorken, S.L. Glashow; Glashow (1964). "Elementary particles and SU(4)". *Physics Letters* **11** (3): 255–257. Bibcode:1964PhL....11..255B. doi:10.1016/0031-9163(64)90433-0.

[5] S.L. Glashow, J. Iliopoulos, L. Maiani; Iliopoulos; Maiani (1970). "Weak Interactions with Lepton–Hadron Symmetry". *Physical Review D* **2** (7): 1285–1292. Bibcode:1970PhRvD...2.1285G. doi:10.1103/PhysRevD.2.1285.

[6] J.-E. Augustin; Boyarski, A.; Breidenbach, M.; Bulos, F.; Dakin, J.; Feldman, G.; Fischer, G.; Fryberger, D.; Hanson, G.; Jean-Marie, B.; Larsen, R.; Lüth, V.; Lynch, H.; Lyon, D.; Morehouse, C.; Paterson, J.; Perl, M.; Richter, B.; Rapidis, P.; Schwitters, R.; Tanenbaum, W.; Vannucci, F.; Abrams, G.; Briggs, D.; Chinowsky, W.; Friedberg, C.; Goldhaber, G.; Hollebeek, R.; Kadyk, J.; Lulu, B. (1974). "Discovery of a Narrow Resonance in e^*+e^*- Annihilation". *Physical Review Letters* **33** (23): 1406. Bibcode:1974PhRvL..33.1406A. doi:10.1103/PhysRevLett.33.1406.

[7] J.J. Aubert et al. (1974). "Experimental Observation of a Heavy Particle J". *Physical Review Letters* **33** (23): 1404. Bibcode:1974PhRvL..33.1404A. doi:10.1103/PhysRevLett.33.1404.

2.4.4 Further reading

- R. Nave. "Quarks". *HyperPhysics*. Georgia State University, Department of Physics and Astronomy. Retrieved 2008-06-29.

- A. Pickering (1984). *Constructing Quarks*. University of Chicago Press. pp. 114–125. ISBN 0-226-66799-5.

2.5 Bottom quark

The **bottom quark** or **b quark**, also known as the **beauty quark**, is a third-generation quark with a charge of $-\frac{1}{3}$ e. Although all quarks are described in a similar way by quantum chromodynamics, the bottom quark's large bare mass (around 4.2 GeV/c^2,[3] a bit more than four times the mass of a proton), combined with low values of the CKM matrix elements V_{ub} and V_{cb}, gives it a distinctive signature that makes it relatively easy to identify experimentally (using a technique called B-tagging). Because three generations of quark are required for CP violation (see CKM matrix), mesons containing the bottom quark are the easiest particles to use to investigate the phenomenon; such experiments are being performed at the BaBar, Belle and LHCb experiments. The bottom quark is also notable because it is a product in almost all top quark decays, and is a frequent decay product for the Higgs boson.

The bottom quark was theorized in 1973 by physicists Makoto Kobayashi and Toshihide Maskawa to explain CP violation.[1] The name "bottom" was introduced in 1975 by Haim Harari.[4][5] The bottom quark was discovered in 1977 by the Fermilab E288 experiment team led by Leon M. Lederman, when collisions produced bottomonium.[2][6][7] Kobayashi and Maskawa won the 2008 Nobel Prize in Physics for their explanation of CP-violation.[8][9] On its discovery, there were efforts to name the bottom quark "beauty", but "bottom" became the predominant usage.

The bottom quark can decay into either an up quark or charm quark via the weak interaction. Both these decays are suppressed by the CKM matrix, making lifetimes of most bottom particles (~10^{-12} s) somewhat higher than those of charmed particles (~10^{-13} s), but lower than those of strange particles (from ~10^{-10} to ~10^{-8} s).

2.5.1 Hadrons containing bottom quarks

Main articles: list of baryons and list of mesons

Some of the hadrons containing bottom quarks include:

- B mesons contain a bottom quark (or its antiparticle) and an up or down quark.
- B
  c and B
  s mesons contain a bottom quark along with a charm quark or strange quark respectively.
- There are many bottomonium states, for example the ϒ meson and χ_b(3P), the first particle discovered in LHC. These consist of a bottom quark and its antiparticle.
- Bottom baryons have been observed, and are named in analogy with strange baryons (e.g. Λ0
  b).

2.5.2 See also

- Quark model

2.5.3 References

[1] M. Kobayashi; T. Maskawa (1973). "CP-Violation in the Renormalizable Theory of Weak Interaction". *Progress of Theoretical Physics* **49** (2): 652–657. Bibcode:1973PThPh..49..652K. doi:10.1143/PTP.49.652.

[2] "Discoveries at Fermilab – Discovery of the Bottom Quark" (Press release). Fermilab. 7 August 1977. Retrieved 2009-07-24.

[3] J. Beringer (Particle Data Group) et al. (2012). "PDGLive Particle Summary 'Quarks (u, d, s, c, b, t, b′, t′, Free)'" (PDF). Particle Data Group. Retrieved 2012-12-18.

[4] H. Harari (1975). "A new quark model for hadrons". *Physics Letters B* **57** (3): 265. Bibcode:1975PhLB...57..265H. doi:10.1016/0370-2693(75)90072-6.

[5] K.W. Staley (2004). *The Evidence for the Top Quark*. Cambridge University Press. pp. 31–33. ISBN 978-0-521-82710-2.

[6] L.M. Lederman (2005). "Logbook: Bottom Quark". *Symmetry Magazine* **2** (8).

[7] S.W. Herb; Hom, D.; Lederman, L.; Sens, J.; Snyder, H.; Yoh, J.; Appel, J.; Brown, B.; Brown, C.; Innes, W.; Ueno, K.; Yamanouchi, T.; Ito, A.; Jöstlein, H.; Kaplan, D.; Kephart, R. et al. (1977). "Observation of a Dimuon Resonance at 9.5 GeV in 400-GeV Proton-Nucleus Collisions". *Physical Review Letters* **39** (5): 252. Bibcode:1977PhRvL..39..252H. doi:10.1103/PhysRevLett.39.252.

[8] 2008 Physics Nobel Prize lecture by Makoto Kobayashi

[9] 2008 Physics Nobel Prize lecture by Toshihide Maskawa

2.5.4 Further reading

- L. Lederman (1978). "The Upsilon Particle". *Scientific American* **239** (4): 72. doi:10.1038/scientificamerican1078-72.

- R. Nave. "Quarks". *HyperPhysics*. Georgia State University, Department of Physics and Astronomy. Retrieved 2008-06-29.

- A. Pickering (1984). *Constructing Quarks*. University of Chicago Press. pp. 114–125. ISBN 0-226-66799-5.

- J. Yoh (1997). "The Discovery of the b Quark at Fermilab in 1977: The Experiment Coordinator's Story" (PDF). *Proceedings of Twenty Beautiful Years of Bottom Physics*. Fermilab. Retrieved 2009-07-24.

2.5.5 External links

- History of the discovery of the bottom quark / Upsilon meson

2.6 Top quark

The **top quark**, also known as the **t quark** (symbol: t) or **truth quark**, is an elementary particle and a fundamental constituent of matter. Like all quarks, the top quark is an elementary fermion with spin-$\frac{1}{2}$, and experiences all four fundamental interactions: gravitation, electromagnetism, weak interactions, and strong interactions. It has an electric charge of $+\frac{2}{3} e$,[2] and is the most massive of all observed elementary particles. It has a mass of 173.34 ± 0.27 (stat) ± 0.71 (syst)10^0 GeV/c^2,[1] which is about the same mass as an atom of tungsten. The antiparticle of the top quark is the **top antiquark** (symbol: t, sometimes called *antitop quark* or simply *antitop*), which differs from it only in that some of its properties have equal magnitude but opposite sign.

The top quark interacts primarily by the strong interaction, but can only decay through the weak force. It decays to a W boson and either a bottom quark (most frequently), a strange quark, or, on the rarest of occasions, a down quark. The Standard Model predicts its mean lifetime to be roughly $5 \times 10^*-25$ s.[3] This is about a twentieth of the timescale for strong interactions, and therefore it does not form hadrons, giving physicists a unique opportunity to study a "bare" quark (all other quarks hadronize, meaning that they combine with other quarks to form hadrons, and can only be observed as such). Because it is so massive, the properties of the top quark allow predictions to be made of the mass of the Higgs boson under certain extensions of the Standard Model (see Mass and coupling to the Higgs boson below). As such, it is extensively studied as a means to discriminate between competing theories.

Its existence (and that of the bottom quark) was postulated in 1973 by Makoto Kobayashi and Toshihide Maskawa to explain the observed CP violations in kaon decay,[4] and was discovered in 1995 by the CDF[5] and DØ[6] experiments at Fermilab. Kobayashi and Maskawa won the 2008 Nobel Prize in Physics for the prediction of the top and bottom quark, which together form the third generation of quarks.[7]

2.6.1 History

In 1973, Makoto Kobayashi and Toshihide Maskawa predicted the existence of a third generation of quarks to explain observed CP violations in kaon decay.[4] The names top and bottom were introduced by Haim Harari in 1975,[8][9] to match the names of the first generation of quarks (up and down) reflecting the fact that the two were the 'up' and 'down' component of a weak isospin doublet.[10] The top quark was sometimes called *truth quark* in the past, but over time *top quark* became the predominant use.[11]

The proposal of Kobayashi and Maskawa heavily relied on the GIM mechanism put forward by Sheldon Lee Glashow, John Iliopoulos and Luciano Maiani,[12] which predicted the existence of the then still unobserved charm quark. When in November 1974 teams at Brookhaven National Laboratory (BNL) and the Stanford Linear Accelerator Center (SLAC) simultaneously announced the discovery of the J/ψ meson, it was soon after identified as a bound state of the missing charm quark with its antiquark. This discovery allowed the GIM mechanism to become part of the Standard Model.[13] With the acceptance of the GIM mechanism, Kobayashi and Maskawa's prediction also gained in credibility. Their case was further strengthened by the discovery of the tau by Martin Lewis Perl's team at SLAC between 1974 and 1978.[14] This announced a third generation of leptons, breaking the new symmetry between leptons and quarks introduced by the GIM mechanism. Restoration of the symmetry implied the existence of a fifth and sixth quark.

It was in fact not long until a fifth quark, the bottom, was discovered by the E288 experiment team, led by Leon Lederman at Fermilab in 1977.[15][16][17] This strongly suggested that there must also be a sixth quark, the top, to complete the pair. It was known that this quark would be heavier than the bottom, requiring more energy to create in particle collisions, but the general expectation was that the sixth quark would soon be found. However, it took another 18 years before the existence of the top was confirmed.[18]

Early searches for the top quark at SLAC and DESY (in Hamburg) came up empty-handed. When, in the early eighties, the Super Proton Synchrotron (SPS) at CERN discovered the W boson and the Z boson, it was again felt that the discovery of the top was imminent. As the SPS gained competition from the Tevatron at Fermilab there was still no sign of the missing particle, and it was announced by the group at CERN that the top mass must be at least 41 GeV/c^2. After a race between CERN and Fermilab to discover the top, the accelerator at CERN reached its limits without creating a single top, pushing the lower bound on its mass up to 77 GeV/c^2.[*][18]

The Tevatron was (until the start of LHC operation at CERN in 2009) the only hadron collider powerful enough to produce top quarks. In order to be able to confirm a future discovery, a second detector, the DØ detector, was added to the complex (in addition to the Collider Detector at Fermilab (CDF) already present). In October 1992, the two groups found their first hint of the top, with a single creation event that appeared to contain the top. In the following years, more evidence was collected and on April 22, 1994, the CDF group submitted their paper presenting tentative evidence for the existence of a top quark with a mass of about 175 GeV/c^2. In the meantime, DØ had found no more evidence than the suggestive event in 1992. A year later, on March 2, 1995, after having gathered more evidence and a reanalysis of the DØ data (who had been searching for a much lighter top), the two groups jointly reported the discovery of the top with a certainty of 99.9998% at a mass of 176±18 GeV/c^2.[*][5][*][6][*][18]

In the years leading up to the top quark discovery, it was realized that certain precision measurements of the electroweak vector boson masses and couplings are very sensitive to the value of the top quark mass. These effects become much larger for higher values of the top mass and therefore could indirectly see the top quark even if it could not be directly produced in any experiment at the time. The largest effect from the top quark mass was on the T parameter and by 1994 the precision of these indirect measurements had led to a prediction of the top quark mass to be between 145 GeV/c^2 and 185 GeV/c^2. It is the development of techniques that ultimately allowed such precision calculations that led to Gerardus 't Hooft and Martinus Veltman winning the Nobel Prize in physics in 1999.[*][19][*][20]

2.6.2 Properties

- At the final Tevatron energy of 1.96 TeV, top–antitop pairs were produced with a cross section of about 7 picobarns (pb).[*][21] The Standard Model prediction (at next-to-leading order with m_t = 175 GeV/c^2) is 6.7–7.5 pb.

- The W bosons from top quark decays carry polarization from the parent particle, hence pose themselves as a unique probe to top polarization.

- In the Standard Model, the top quark is predicted to have a spin quantum number of $1/2$ and electric charge $+2/3$. A first measurement of the top quark charge has been published, resulting in approximately 90% confidence limit that the top quark charge is indeed $+2/3$.[*][22]

2.6.3 Production

Because top quarks are very massive, large amounts of energy are needed to create one. The only way to achieve such high energies is through high energy collisions. These occur naturally in the Earth's upper atmosphere as cosmic rays collide with particles in the air, or can be created in a particle accelerator. In 2011, after the Tevatron ceased operations, the Large Hadron Collider at CERN became the only accelerator that generates a beam of sufficient energy to produce top quarks, with a center-of-mass energy of 7 TeV.

There are multiple processes that can lead to the production of a top quark. The most common is production of a top–antitop pair via strong interactions. In a collision, a highly energetic gluon is created, which subsequently decays into a top and antitop. This process was responsible for the majority of the top events at Tevatron and was the process observed when the top was first discovered in 1995.[*][23] It is also possible to produce pairs of top–antitop through the decay of an intermediate photon or Z-boson. However, these processes are predicted to be much rarer and have a virtually identical experimental signature in a hadron collider like Tevatron.

A distinctly different process is the production of single tops via weak interaction. This can happen in two ways (called channels): either an intermediate W-boson decays into a top and antibottom quark ("s-channel") or a bottom quark (probably created in a pair through the decay of a gluon) transforms to top quark by exchanging a W-boson with an up or down quark ("t-channel"). The first evidence for these processes was published by the DØ collaboration in December 2006,[*][24] and in March 2009 the CDF[*][25] and DØ[*][23] collaborations released twin papers with the definitive observation of these processes. The main significance of measuring these production processes is that their frequency is directly proportional to the | V_{tb} |^2 component of the CKM matrix.

2.6.4 Decay

The only known way that a top quark can decay is through the weak interaction producing a W-boson and a down-type quark (down, strange, or bottom). Because of its enormous mass, the top quark is extremely short-lived with a predicted lifetime of only 5×10*–25 s.[*][3] As a result top quarks do not have time to form hadrons before they decay, as other quarks do. This provides physicists with the unique opportunity to study the behavior of a "bare" quark.

In particular, it is possible to directly determine the branching ratio $\Gamma(W^*+b) / \Gamma(W^*+q$ $(q = $ b,s,d)). The best current determination of this ratio is 0.91±0.04.[26] Since this ratio is equal to $\mid V_{tb} \mid^2$ according to the Standard Model, this gives another way of determining the CKM element $\mid V_{tb} \mid$, or in combination with the determination of $\mid V_{tb} \mid$ from single top production provides tests for the assumption that the CKM matrix is unitary.[27]

The Standard Model also allows more exotic decays, but only at one loop level, meaning that they are extremely suppressed. In particular, it is possible for a top quark to decay into another up-type quark (an up or a charm) by emitting a photon or a Z-boson.[28] Searches for these exotic decay modes have provided no evidence for their existence in accordance with expectations from the Standard Model. The branching ratios for these decays have been determined to be less than 5.9 in 1,000 for photonic decay and less than 2.1 in 1,000 for Z-boson decay at 95% confidence.[26]

2.6.5 Mass and coupling to the Higgs boson

The Standard Model describes fermion masses through the Higgs mechanism. The Higgs boson has a Yukawa coupling to the left- and right-handed top quarks. After electroweak symmetry breaking (when the Higgs acquires a vacuum expectation value), the left- and right-handed components mix, becoming a mass term.

$$\mathcal{L} = y_t h q u^c \to \frac{y_t v}{\sqrt{2}}(1 + h^0/v)uu^c$$

The top quark Yukawa coupling has a value of

$$y_t = \sqrt{2}m_t/v \simeq 1$$

where v = 246 GeV is the value of the Higgs vacuum expectation value.

Yukawa couplings

See also: Beta function (physics)

In the Standard Model, all of the quark and lepton Yukawa couplings are small compared to the top quark Yukawa coupling. Understanding this hierarchy in the fermion masses is an open problem in theoretical physics. Yukawa couplings are not constants and their values change depending on the energy scale (distance scale) at which they are measured. The dynamics of Yukawa couplings are determined by the renormalization group equation.

One of the prevailing views in particle physics is that the size of the top quark Yukawa coupling is determined by the renormalization group, leading to the "quasi-infrared fixed point."

The Yukawa couplings of the up, down, charm, strange and bottom quarks, are hypothesized to have small values at the extremely high energy scale of grand unification, 10^{15} GeV. They increase in value at lower energy scales, at which the quark masses are generated by the Higgs. The slight growth is due to corrections from the QCD coupling. The corrections from the Yukawa couplings are negligible for the lower mass quarks.

If, however, a quark Yukawa coupling has a large value at very high energies, its Yukawa corrections will evolve and cancel against the QCD corrections. This is known as a (quasi-) infrared fixed point. No matter what the initial starting value of the coupling is, if it is sufficiently large it will reach this fixed point value. The corresponding quark mass is then predicted.

The top quark Yukawa coupling lies very near the infrared fixed point of the Standard Model. The renormalization group equation is:

$$\mu\frac{\partial}{\partial\mu}y_t \approx \frac{y_t}{16\pi^2}\left(\frac{9}{2}y_t^2 - 8g_3^2 - \frac{9}{4}g_2^2 - \frac{17}{20}g_1^2\right),$$

where g_3 is the color gauge coupling, g_2 is the weak isospin gauge coupling, and g_1 is the weak hypercharge gauge coupling. This equation describes how the Yukawa coupling changes with energy scale μ. Solutions to this equation for large initial values y_t cause the right-hand side of the equation to quickly approach zero, locking y_t to the QCD coupling g_3. The value of the fixed point is fairly precisely determined in the Standard Model, leading to a top quark mass of 230 GeV. However, if there is more than one Higgs doublet, the mass value will be reduced by Higgs mixing angle effects in an unpredicted way.

In the minimal supersymmetric extension of the Standard Model (MSSM), there are two Higgs doublets and the renormalization group equation for the top quark Yukawa coupling is slightly modified:

$$\mu\frac{\partial}{\partial\mu}y_t \approx \frac{y_t}{16\pi^2}\left(6y_t^2 + y_b^2 - \frac{16}{3}g_3^2 - 3g_2^2 - \frac{13}{15}g_1^2\right),$$

where y_b is the bottom quark Yukawa coupling. This leads to a fixed point where the top mass is smaller, 170–200 GeV. The uncertainty in this prediction arises because the bottom quark Yukawa coupling can be amplified in the MSSM. Some theorists believe this is supporting evidence for the MSSM.

The quasi-infrared fixed point has subsequently formed the basis of top quark condensation theories of electroweak symmetry breaking in which the Higgs boson is composite at *extremely* short distance scales, composed of a pair of top and antitop quarks.

2.6.6 See also

- CDF experiment
- Topness
- Top quark condensate
- Topcolor
- Quark model

2.6.7 References

[1] The ATLAS, CDF, CMS, D0 Collaborations (2014). "First combination of Tevatron and LHC measurements of the top-quark mass". Retrieved 2014-03-19.

[2] S. Willenbrock (2003). "The Standard Model and the Top Quark". In H.B Prosper and B. Danilov (eds.). *Techniques and Concepts of High-Energy Physics XII*. NATO Science Series **123**. Kluwer Academic. pp. 1–41. arXiv:hep-ph/0211067v3. ISBN 1-4020-1590-9.

[3] A. Quadt (2006). "Top quark physics at hadron colliders". *European Physical Journal C* **48** (3): 835–1000. Bibcode:2006EPJC...48..835Q. doi:10.1140/epjc/s2006-02631-6.

[4] M. Kobayashi, T. Maskawa (1973). "*CP*-Violation in the Renormalizable Theory of Weak Interaction". *Progress of Theoretical Physics* **49** (2): 652. Bibcode:1973PThPh..49..652K. doi:10.1143/PTP.49.652.

[5] F. Abe *et al.* (CDF Collaboration) (1995). "Observation of Top Quark Production in pp Collisions with the Collider Detector at Fermilab". *Physical Review Letters* **74** (14): 2626–2631. Bibcode:1995PhRvL..74.2626A. doi:10.1103/PhysRevLett.74.2626. PMID 10057978.

[6] S. Abachi *et al.* (DØ Collaboration) (1995). "Search for High Mass Top Quark Production in pp Collisions at $\sqrt{s}$ = 1.8 TeV". *Physical Review Letters* **74** (13): 2422–2426. Bibcode:1995PhRvL..74.2422A. doi:10.1103/PhysRevLett.74.2422.

[7] "2008 Nobel Prize in Physics". The Nobel Foundation. 2008. Retrieved 2009-09-11.

[8] H. Harari (1975). "A new quark model for hadrons". *Physics Letters B* **57** (3): 265. Bibcode:1975PhLB...57..265H. doi:10.1016/0370-2693(75)90072-6.

[9] K.W. Staley (2004). *The Evidence for the Top Quark*. Cambridge University Press. pp. 31–33. ISBN 978-0-521-82710-2.

[10] D.H. Perkins (2000). *Introduction to high energy physics*. Cambridge University Press. p. 8. ISBN 0-521-62196-8.

[11] F. Close (2006). *The New Cosmic Onion*. CRC Press. p. 133. ISBN 1-58488-798-2.

[12] S.L. Glashow, J. Iliopoulous, L. Maiani (1970). "Weak Interactions with Lepton–Hadron Symmetry". *Physical Review D* **2** (7): 1285–1292. Bibcode:1970PhRvD...2.1285G. doi:10.1103/PhysRevD.2.1285.

[13] A. Pickering (1999). *Constructing Quarks: A Sociological History of Particle Physics*. University of Chicago Press. pp. 253–254. ISBN 978-0-226-66799-7.

[14] M.L. Perl et al. (1975). "Evidence for Anomalous Lepton Production in e+e− Annihilation". *Physical Review Letters* **35** (22): 1489. Bibcode:1975PhRvL..35.1489P. doi:10.1103/PhysRevLett.35.1489.

[15] "Discoveries at Fermilab – Discovery of the Bottom Quark" (Press release). Fermilab. 7 August 1977. Retrieved 2009-07-24.

[16] L.M. Lederman (2005). "Logbook: Bottom Quark". *Symmetry Magazine* **2** (8).

[17] S.W. Herb et al. (1977). "Observation of a Dimuon Resonance at 9.5 GeV in 400-GeV Proton-Nucleus Collisions". *Physical Review Letters* **39** (5): 252. Bibcode:1977PhRvL..39..252H. doi:10.1103/PhysRevLett.39.252.

[18] T.M. Liss, P.L. Tipton (1997). "The Discovery of the Top Quark" (PDF). *Scientific American*: 54–59.

[19] "The Nobel Prize in Physics 1999". The Nobel Foundation. Retrieved 2009-09-10.

[20] "The Nobel Prize in Physics 1999, Press Release" (Press release). The Nobel Foundation. 12 October 1999. Retrieved 2009-09-10.

[21] D. Chakraborty (DØ and CDF collaborations) (2002). *Top quark and W/Z results from the Tevatron* (PDF). Rencontres de Moriond. p. 26.

[22] V.M. Abazov *et al.* (DØ Collaboration) (2007). "Experimental discrimination between charge 2*e*/3 top quark and charge 4*e*/3 exotic quark production scenarios". *Physical Review Letters* **98** (4): 041801. arXiv:hep-ex/0608044. Bibcode:2007PhRvL..98d1801A. doi:10.1103/PhysRevLett.98.041801. PMID 17358756.

[23] V.M. Abazov et al. (DØ Collaboration) (2009). "Observation of Single Top Quark Production". *Physical Review Letters* **103** (9). arXiv:0903.0850. Bibcode:2009PhRvL.103i2001A. doi:10.1103/PhysRevLett.103.092001.

[24] V.M. Abazov et al. (DØ Collaboration) (2007). "Evidence for production of single top quarks and first direct measurement of |V$_{tb}$|". *Physical Review Letters* **98** (18): 181802. arXiv:hep-ex/0612052. Bibcode:2007PhRvL..98r1802A. doi:10.1103/PhysRevLett.98.181802. PMID 17501561.

[25] T. Aaltonen et al. (CDF Collaboration) (2009). "First Observation of Electroweak Single Top Quark Production". *Physical Review Letters* **103** (9). arXiv:0903.0885. Bibcode:2009PhRvL.103i2002A. doi:10.1103/PhysRevLett.103.092002.

[26] J. Beringer et al. (Particle Data Group) (2012). "PDGLive Particle Summary 'Quarks (u, d, s, c, b, t, b', t', Free)'" (PDF). Particle Data Group. Retrieved 2013-07-23.

[27] V.M. Abazov et al. (DØ Collaboration) (2008). "Simultaneous measurement of the ratio B(t→Wb)/B(t→Wq) and the top-quark pair production cross section with the DØ detector at $\sqrt{s}$ = 1.96 TeV". *Physical Review Letters* **100** (19): 192003. arXiv:0801.1326. Bibcode:2008PhRvL.100s2003A. doi:10.1103/PhysRevLett.100.192003.

[28] S. Chekanov et al. (ZEUS Collaboration) (2003). "Search for single-top production in ep collisions at HERA". *Physics Letters B* **559** (3–4): 153. arXiv:hep-ex/0302010. Bibcode:2003PhLB..559..153Z. doi:10.1016/S0370-2693(03)00333-2.

2.6.8 Further reading

- Frank Fiedler; for the D0; CDF Collaborations (June 2005). "Top Quark Production and Properties at the Tevatron". arXiv:hep-ex/0506005 [hep-ex].

- R. Nave. "Quarks". *HyperPhysics*. Georgia State University, Department of Physics and Astronomy. Retrieved 2008-06-29.

- A. Pickering (1984). *Constructing Quarks*. University of Chicago Press. pp. 114–125. ISBN 0-226-66799-5.

2.6.9 External links

- Top quark on arxiv.org
- Tevatron Electroweak Working Group
- Top quark information on Fermilab website
- Logbook pages from CDF and DZero collaborations' top quark discovery
- Scientific American article on the discovery of the top quark
- Public Homepage of Top Quark Analysis Results from DØ Collaboration at Fermilab
- Public Homepage of Top Quark Analysis Results from CDF Collaboration at Fermilab
- Harvard Magazine article about the 1994 top quark discovery
- 1999 Nobel Prize in Physics

Chapter 3

Flavour

3.1 Flavour

In particle physics, **flavour** or **flavor** refers to a species of an elementary particle. The Standard Model counts six flavours of quarks and six flavours of leptons. They are conventionally parameterized with *flavour quantum numbers* that are assigned to all subatomic particles, including composite ones. For hadrons, these quantum numbers depend on the numbers of constituent quarks of each particular flavour.

3.1.1 Intuitive description

Elementary particles are not eternal and indestructible. Unlike in classical mechanics, where forces only change a particle's momentum, the weak force can alter the essence of a particle, even an elementary particle. This means that it can convert one quark to another quark with different mass and electric charge, and the same for leptons. From the point of view of quantum mechanics, changing the flavour of a particle by the weak force is no different in principle from changing its spin by electromagnetic interaction, and should be described with quantum numbers as well. In particular, flavour states may undergo quantum superposition.

In atomic physics the principal quantum number of an electron specifies the electron shell in which it resides, which determines the energy level of the whole atom. In an analogous way, the five flavour quantum numbers of a quark specify which of six flavours (u, d, s, c, b, t) it has, and when these quarks are combined this results in different types of baryons and mesons with different masses, electric charges, and decay modes.

3.1.2 Flavour symmetry

If there are two or more particles which have identical interactions, then they may be interchanged without affecting the physics. Any (complex) linear combination of these two particles give the same physics, as long as they are orthogonal or perpendicular to each other. In other words, the theory possesses symmetry transformations such as $M \begin{pmatrix} u \\ d \end{pmatrix}$, where u and d are the two fields, and M is any 2×2 unitary matrix with a unit determinant. Such matrices form a Lie group called SU(2) (see special unitary group). This is an example of flavour symmetry.

In quantum chromodynamics, flavour is a global symmetry. In the electroweak theory, on the other hand, this symmetry is broken, and flavour changing processes exist, such as quark decay or neutrino oscillations.

3.1.3 Flavour quantum numbers

Leptons

All leptons carry a lepton number $L = 1$. In addition, leptons carry weak isospin, T_3, which is $-1/2$ for the three charged leptons (i.e. electron, muon and tau) and $+1/2$ for the three associated neutrinos. Each doublet of a charged lepton and a neutrino consisting of opposite T_3 are said to constitute one generation of leptons. In addition, one defines a quantum number called weak hypercharge, Y_W, which is -1 for all left-handed leptons.*[1] Weak isospin and weak hypercharge are gauged in the Standard Model.

Leptons may be assigned the six flavour quantum numbers: electron number, muon number, tau number, and corresponding numbers for the neutrinos. These are conserved in strong and electromagnetic interactions, but violated by weak interactions. Therefore, such flavour quantum numbers are not of great use. A separate quantum number for each generation is more useful: electronic lepton number (+1 for electrons and electron neutrinos), muonic lepton number (+1 for muons and muon neutrinos), and tauonic lepton number (+1 for tau leptons and tau neutrinos). However, even these numbers are not absolutely conserved, as neutrinos of different generations can mix; that is, a neutrino of one flavour can transform into another flavour. The strength of such mixings is specified by a matrix called the Pontecorvo–Maki–Nakagawa–Sakata matrix (PMNS matrix).

Quarks

All quarks carry a baryon number $B = 1/3$. They also all carry weak isospin, $T_3 = \pm 1/2$. The positive-T_3 quarks (up, charm, and top quarks) are called *up-type quarks* and negative-T_3 quarks (down, strange, and bottom quarks) are called *down-type quarks*. Each doublet of up and down type quarks constitutes one generation of quarks.

For all the quark flavour quantum numbers (strangeness, charm, topness and bottomness) the convention is that the flavour charge and the electric charge of a quark have the same sign. Thus any flavour carried by a charged meson has the same sign as its charge. Quarks have the following flavour quantum numbers:

- Isospin, less ambiguously known as "isobaric spin", which has value $I_3 = 1/2$ for the up quark and $I_3 = -1/2$ for the down quark.

- Strangeness (S): Defined as $S = -(n_s - n_{\bar{s}})$, where n_s represents the number of strange quarks (s) and $n_{\bar{s}}$ represents the number of strange antiquarks (s). This quantum number was introduced by Murray Gell-Mann. This definition gives the strange quark a strangeness of -1 for the above-mentioned reason.

- Charm (C): Defined as $C = (n_c - n_{\bar{c}})$, where n_c represents the number of charm quarks (c) and $n_{\bar{c}}$ represents the number of charm antiquarks. Is +1 for the charm quark.

- Bottomness (B'): Also called 'beauty'. Defined as $B' = -(n_b - n_{\bar{b}})$, where n_b represents the number of bottom quarks (b) and $n_{\bar{b}}$ represents the number of bottom antiquarks.

- Topness (T): Also called 'truth'. Defined as $T = (n_t - n_{\bar{t}})$, where n_t represents the number of top quarks (t) and $n_{\bar{t}}$ represents the number of top antiquarks. However, because of the extremely short half-life of the top quark, by the time it can interact strongly it has already decayed to another flavour of quark (usually to a bottom quark). For that reason the top quark doesn't hadronize, that is it never forms any meson or baryon.

These five quantum numbers, together with baryon number (which is not a flavour quantum number) completely specify numbers of all 6 quark flavours separately (as $n_q - n_{\bar{q}}$, i.e. an antiquark is counted with the minus sign). They are conserved by both the electromagnetic and strong interactions (but not the weak interaction). From them can be built the derived quantum numbers:

- Hypercharge (Y): $Y = B + S + C + B' + T$
- Electric charge: $Q = I_3 + 1/2 Y$ (see Gell-Mann–Nishijima formula)

The terms "strange" and "strangeness" predate the discovery of the quark, but continued to be used after its discovery for the sake of continuity (i.e. the strangeness of each type of hadron remained the same); strangeness of anti-particles being referred to as +1, and particles as −1 as per the original definition. Strangeness was introduced to explain the rate of decay of newly discovered particles, such as the kaon, and was used in the Eightfold Way classification of hadrons and in subsequent quark models. These quantum numbers are preserved under strong and electromagnetic interactions, but not under weak interactions.

For first-order weak decays, that is processes involving only one quark decay, these quantum numbers (e.g. charm) can only vary by 1 ($|C| = \pm 1$); $\Delta B' = \pm 1$. Since first-order processes are more common than second-order processes (involving two quark decays), this can be used as an approximate "selection rule" for weak decays.

A quark of a given flavour is an eigenstate of the weak interaction part of the Hamiltonian: it will interact in a definite way with the W and Z bosons. On the other hand, a fermion of a fixed mass (an eigenstate of the kinetic and strong interaction parts of the Hamiltonian) is normally a superposition of various flavours. As a result, the flavour content of a quantum state may change as it propagates freely. The transformation from flavour to mass basis for quarks is given by the Cabibbo–Kobayashi–Maskawa matrix (CKM matrix). This matrix is analogous to the PMNS matrix for neutrinos, and defines the strength of flavour changes under weak interactions of quarks.

The CKM matrix allows for CP violation if there are at least three generations.

Antiparticles and hadrons

Flavour quantum numbers are additive. Hence antiparticles have flavour equal in magnitude to the particle but opposite in sign. Hadrons inherit their flavour quantum number from their valence quarks: this is the basis of the classification in the quark model. The relations between the hypercharge, electric charge and other flavour quantum numbers hold for hadrons as well as quarks.

3.1.4 Quantum chromodynamics

Flavour symmetry is closely related to chiral symmetry. This part of the article is best read along with the one on chirality.

Quantum chromodynamics (QCD) contains six flavours of quarks. However, their masses differ and as a result they are not strictly interchangeable with each other. The up and down flavours are close to having equal masses, and the theory of these two quarks possesses an approximate SU(2) symmetry (isospin symmetry).

Under some circumstances, the masses of the quarks can be neglected entirely. One can then make flavour transformations independently on the left- and right-handed parts of each quark field. The flavour group is then a chiral group $SU_L(N_f) \times SU_R(N_f)$.

If all quarks had non-zero but equal masses, then this chiral symmetry is broken to the *vector symmetry* of the "diagonal flavour group" $SU(N_f)$, which applies the same transformation to both helicities of the quarks. Such a reduction of the symmetry is called *explicit symmetry breaking*. The amount of explicit symmetry breaking is controlled by the current quark masses in QCD.

Even if quarks are massless, chiral flavour symmetry can be spontaneously broken if the vacuum of the theory contains a chiral condensate (as it does in low-energy QCD). This gives rise to an effective mass for the quarks, often identified with the valence quark mass in QCD.

Symmetries of QCD

Analysis of experiments indicate that the current quark masses of the lighter flavours of quarks are much smaller than the QCD scale, Λ_{QCD}, hence chiral flavour symmetry is a good approximation to QCD for the up, down and strange quarks. The success of chiral perturbation theory and the even more naive chiral models spring from this fact. The valence quark masses extracted from the quark model are much larger than the current quark mass. This indicates that QCD has spontaneous chiral symmetry breaking with the formation of a chiral condensate. Other phases of QCD may break the chiral flavour symmetries in other ways.

3.1.5 Conservation laws

All of the various charges discussed above are conserved by the fact that the charge operator is best understood as the generator of a symmetry that commutes with the Hamiltonian. Thus, the eigenvalues of the various charge operators are conserved.

Absolutely conserved flavour quantum numbers are: (including the baryon number for completeness)

- electric charge (Q)
- weak isospin (I_3)
- baryon number (B)
- lepton number (L)

In some theories, the individual baryon and lepton number conservation can be violated, if the difference between them ($B - L$) is conserved (see chiral anomaly). All other flavour quantum numbers are violated by the electroweak interactions. Strong interactions conserve all flavours.

3.1.6 History

Some of the historical events that lead to the development of flavour symmetry are discussed in the article on isospin.

3.1.7 See also

- Standard Model (mathematical formulation)
- Cabibbo–Kobayashi–Maskawa matrix
- Strong CP problem and chirality (physics)
- Chiral symmetry breaking and quark matter
- Quark flavour tagging, such as B-tagging, is an example of particle identification in experimental particle physics.

3.1.8 References

[1] See table in S. Raby, R. Slanky (1997). "Neutrino Masses: How to add them to the Standard Model" (PDF). *Los Alamos Science* (25): 64.

3.1.9 Further reading

- Lessons in Particle Physics Luis Anchordoqui and Francis Halzen, University of Wisconsin, 18th Dec. 2009

3.1.10 External links

- The particle data group.

3.2 Isospin

In nuclear physics and particle physics, **isospin** (*isotopic spin*, *isobaric spin*) is a quantum number related to the strong interaction. Particles that are affected equally by the strong force but have different charges (e.g. protons and neutrons) can be treated as being different states of the same particle with isospin values related to the number of charge states.*[1]

Although it does not have the units of angular momentum and is not a type of spin, the formalism that describes it is mathematically similar to that of angular momentum in quantum mechanics, which means it can be coupled in the same manner. For example, a proton-neutron pair can be coupled in a state of total isospin 1 or 0.*[2] It is a dimensionless quantity and the name derives from the fact that the mathematical structures used to describe it are very similar to those used to describe the intrinsic angular momentum (spin).

This term was derived from *isotopic spin*, a confusing term to which nuclear physicists prefer *isobaric spin*, which is more precise in meaning. Isospin symmetry is a subset of the flavour symmetry seen more broadly in the interactions of baryons and mesons. Isospin symmetry remains an important concept in particle physics, and a close examination of this symmetry historically led directly to the discovery and understanding of quarks and of the development of Yang–Mills theory.

3.2.1 Motivation for isospin

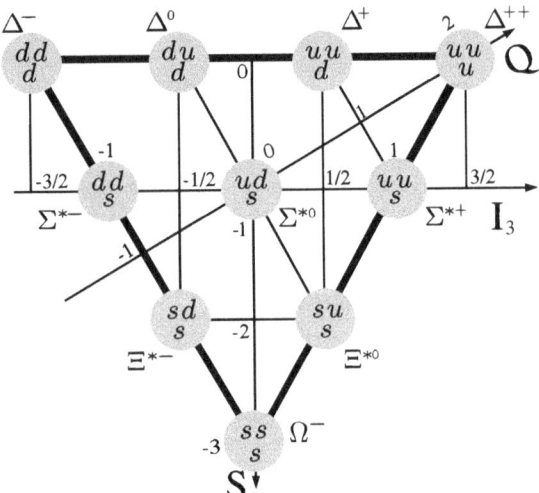

Combinations of three u, d or s-quarks forming baryons with spin-$\frac{3}{2}$ form the baryon decuplet.

Isospin was introduced by Werner Heisenberg in 1932*[3] to explain symmetries of the then newly discovered neutron:

- The mass of the neutron and the proton are almost identical: they are nearly degenerate, and both are thus often called nucleons. Although the proton has a positive charge, and the neutron is neutral, they are almost

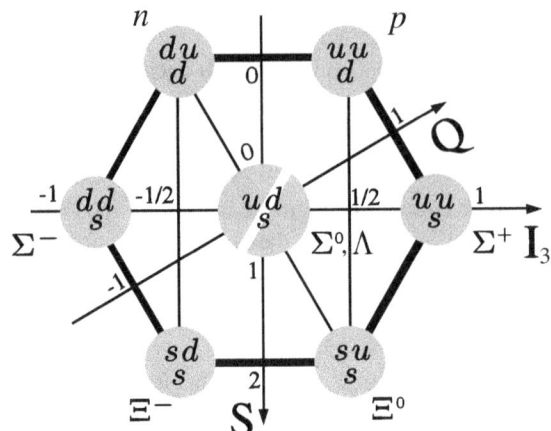

Combinations of three u, d or s-quarks forming baryons with spin-$\frac{1}{2}$ form the baryon octet

identical in all other respects.

- The strength of the strong interaction between any pair of nucleons is the same, independent of whether they are interacting as protons or as neutrons.

Thus, isospin was introduced as a concept well before the development in the 1960s of the quark model which provides our modern understanding. The name *isospin* however, was introduced by Eugene Wigner in 1937.*[4]

Protons and neutrons, baryons of spin $\frac{1}{2}$, were grouped together as nucleons because they both have nearly the same mass and interact in nearly the same way. Thus, it was convenient to treat them as being different states of the same particle. Since a spin $\frac{1}{2}$ particle has two states, the two were said to be of isospin $\frac{1}{2}$. The proton and neutron were then associated with different isospin projections $I_3 = +\frac{1}{2}$ and $-\frac{1}{2}$ respectively. When constructing a physical theory of nuclear forces, one could then simply assume that it does not depend on isospin.

These considerations would also prove useful in the analysis of meson-nucleon interactions after the discovery of the pions in 1947. The three pions (π+, π0, π–) could be assigned to an isospin triplet with $I = 1$ and $I_3 = +1$, 0 or -1. By assuming that isospin was conserved by nuclear interactions, the new mesons were more easily accommodated by nuclear theory.

As further particles were discovered, they were assigned into isospin multiplets according to the number of different charge states seen: 2 doublets, $I = -\frac{1}{2}$ and $I = \frac{1}{2}$ of K mesons (K–, K0),(K+, K0), a triplet $I = 1$ of Sigma baryons (Σ+, Σ0, Σ–) a singlet $I = 0$ Lambda baryon (Λ0), a quartet $I = \frac{3}{2}$ Delta baryons (Δ++, Δ+, Δ0, Δ–), and so on. This multiplet structure was combined with strangeness in

3.2. ISOSPIN

Murray Gell-Mann's eightfold way, ultimately leading to the quark model and quantum chromodynamics.

3.2.2 Modern understanding of isospin

Observation of the light baryons (those made of up, down and strange quarks) lead us to believe that some of these particles are so similar in terms of their strong interactions that they can be treated as different states of the same particle. In the modern understanding of quantum chromodynamics, this is because up and down quarks are very similar in mass, and have the same strong interactions. Particles made of the same numbers of up and down quarks have similar masses and are grouped together. For examples, the particles known as the Delta baryons—baryons of spin $\frac{3}{2}$ made of a mix of three up and down quarks—are grouped together because they all have nearly the same mass (approximately 1232 MeV/c^2), and interact in nearly the same way.

However, because the up and down quarks have different charges ($\frac{2}{3} e$ and $-\frac{1}{3} e$ respectively), the four Deltas also have different charges (Δ++ (uuu), Δ+ (uud), Δ0 (udd), Δ− (ddd)). These Deltas could be treated as the same particle and the difference in charge being due to the particle being in different states. Isospin was devised as a parallel to spin to associate an isospin projection (denoted I_3) to each charged state. Since there were four Deltas, four projections were needed. Because isospin was modeled on spin, the isospin projections were made to vary in increments of 1 and to have four increments of 1, you needed an isospin value of $\frac{3}{2}$ (giving the projections $I_3 = \frac{3}{2}, \frac{1}{2}, -\frac{1}{2}, -\frac{3}{2}$). Thus, all the Deltas were said to have isospin $I = \frac{3}{2}$ and each individual charge had different I_3 (e.g. the Δ++ was associated with $I_3 = +\frac{3}{2}$). In the isospin picture, the four Deltas and the two nucleons were thought to be the different states of two particles. In the quark model, the Deltas can be thought of as the excited states of the nucleons.

After the quark model was elaborated, it was noted that the isospin projection was related to the up and down quark content of particles. The relation is

$$I_3 = \frac{1}{2}\left[(n_u - n_{\bar{u}}) - (n_d - n_{\bar{d}})\right]$$

where n_u and n_d are the numbers of up and down quarks respectively, and $n_{\bar{u}}$ and $n_{\bar{d}}$ are the numbers of up and down antiquarks respectively.

By this, the value of I_3 of the nucleons proton (symbol p) and neutron (symbol n) is determined by their quark composition, uud for the proton and udd for the neutron.

3.2.3 Isospin symmetry

In quantum mechanics, when a Hamiltonian has a symmetry, that symmetry manifests itself through a set of states that have the same energy; that is, the states are degenerate. In particle physics, the near mass-degeneracy of the neutron and proton points to an approximate symmetry of the Hamiltonian describing the strong interactions. The neutron does have a slightly higher mass due to isospin breaking; this is due to the difference in the masses of the up and down quarks and the effects of the electromagnetic interaction. However, the appearance of an approximate symmetry is still useful, since the small breakings can be described by a perturbation theory, which gives rise to slight differences between the near-degenerate states.

SU(2)

See also: Representation theory of SU(2)

Heisenberg's contribution was to note that the mathematical formulation of this symmetry was in certain respects similar to the mathematical formulation of spin, whence the name "isospin" derives. To be precise, the isospin symmetry is given by the invariance of the Hamiltonian of the strong interactions under the action of the Lie group SU(2). The neutron and the proton are assigned to the doublet (the spin-$\frac{1}{2}$, **2**, or fundamental representation) of SU(2). The pions are assigned to the triplet (the spin-1, **3**, or adjoint representation) of SU(2). Though, there is a difference from the theory of spin: the group action does not preserve flavor.

Like the case for regular spin, the isospin operator **I** is vector-valued: it has three components $\mathbf{I}_x, \mathbf{I}_y, \mathbf{I}_z$ which are coordinates in the same 3-dimensional vector space where the **3** representation acts. Note that it has nothing to do with the physical space, except similar mathematical formalism. Isospin is described by two quantum numbers: I, the total isospin, and I_3, an eigenvalue of the $\mathbf{I}_z$ projection for which flavor states are eigenstates, not an *arbitrary projection* as in the case of spin. In other words, each I_3 state specifies certain flavor state of a multiplet. The third coordinate (z), to which the "3" subscript refers, is chosen due to notational conventions which relate bases in **2** and **3** representation spaces. Namely, for the spin-$\frac{1}{2}$ case, components of **I** are equal to Pauli matrices divided by 2 and $\mathbf{I}_z = \frac{1}{2}\tau_3$, where

$$\tau_3 = \begin{pmatrix} 1 & 0 \\ 0 & -1 \end{pmatrix}$$

While the forms of these matrices are the isomorphic to those of spin, *these* Pauli matrices only acts within the Hilbert space of isospin, not that of spin, and therefore is common to denote them with **τ** rather than **σ** to avoid confusion.

The power of isospin symmetry and related methods such as the Eightfold Way come from the observation that families of particles with similar masses tend to correspond to the invariant subspaces associated with the irreducible representations of the Lie algebra 𝖘𝖚(2). In this context, an invariant subspace is spanned by basis vectors which correspond to particles in a family. Under the action of the Lie algebra 𝖘𝖚(2), which generates rotations in isospin space, elements corresponding to definite particle states or superpositions of states can be rotated into each other, but can never leave the space (since the subspace is in fact invariant). This is reflective of the symmetry present. The fact that unitary matrices will commute with the Hamiltonian means that the physical quantities calculated do not change even under unitary transformation. In the case of isospin, this machinery is used to reflect the fact that the strong force behaves the same under the exchange of the up and down quark (and by extension the exchange of the proton and the neutron).

3.2.4 Relationship to flavor

The discovery and subsequent analysis of additional particles, both mesons and baryons, made it clear that the concept of isospin symmetry could be broadened to an even larger symmetry group, now called flavor symmetry. Once the kaons and their property of strangeness became better understood, it started to become clear that these, too, seemed to be a part of an enlarged symmetry that contained isospin as a subgroup. The larger symmetry was named the Eightfold Way by Murray Gell-Mann, and was promptly recognized to correspond to the adjoint representation of SU(3). To better understand the origin of this symmetry, Gell-Mann proposed the existence of up, down and strange quarks which would belong to the fundamental representation of the SU(3) flavor symmetry.

Although isospin symmetry is very slightly broken, SU(3) symmetry is more badly broken, due to the much higher mass of the strange quark compared to the up and down. The discovery of charm, bottomness and topness could lead to further expansions up to SU(6) flavour symmetry, but the very large masses of these quarks makes such symmetries almost useless. In modern applications, such as lattice QCD, isospin symmetry is often treated as exact while the heavier quarks must be treated separately.

3.2.5 Quark content and isospin

Up and down quarks each have isospin $I = \frac{1}{2}$, and isospin 3-components (I_3) of $\frac{1}{2}$ and $-\frac{1}{2}$ respectively. All other quarks have $I = 0$. In general

$$I_3 = \frac{1}{2}(n_u - n_d).$$

Hadron nomenclature

Main articles: Baryon and Mesons

Hadron nomenclature is based on isospin.[*][5]

- Particles of isospin $\frac{3}{2}$ can only be made by a mix of three u and d quarks (Delta baryons).
- Particles of isospin 1 are made of a mix of two u and d quarks (Pi mesons, Rho mesons, Sigma baryons with one heavier quark, etc.).
- Particles of isospin $\frac{1}{2}$ can be made of a mix of three u and d quarks (nucleons) or from one u or d quark with heavier quarks (K mesons, D mesons, Xi baryons, etc.)
- Particles of isospin 0 can be made of one u and one d quark (Eta mesons, Omega mesons, Lambda baryons, etc.), or from no u or d quarks at all (Omega baryons, Phi mesons, etc.), with heavier quarks in all cases.

Isospin symmetry of quarks

In the framework of the Standard Model, the isospin symmetry of the proton and neutron are reinterpreted as the isospin symmetry of the up and down quarks. Technically, the nucleon doublet states are seen to be linear combinations of products of 3-particle isospin doublet states and spin doublet states. That is, the (spin-up) proton wave function, in terms of quark-flavour eigenstates, is described by

$$|p\uparrow\rangle = \tfrac{1}{3\sqrt{2}} \begin{pmatrix} |duu\rangle & |udu\rangle & |uud\rangle \end{pmatrix} \begin{pmatrix} 2 & -1 & -1 \\ -1 & 2 & -1 \\ -1 & -1 & 2 \end{pmatrix}$$

[*][6]

and the (spin-up) neutron by

$$|n\uparrow\rangle = \tfrac{1}{3\sqrt{2}} \begin{pmatrix} |udd\rangle & |dud\rangle & |ddu\rangle \end{pmatrix} \begin{pmatrix} 2 & -1 & -1 \\ -1 & 2 & -1 \\ -1 & -1 & 2 \end{pmatrix}$$

[*][6]

Here, $|u\rangle$ is the up quark flavour eigenstate, and $|d\rangle$ is the down quark flavour eigenstate, while $|\uparrow\rangle$ and $|\downarrow\rangle$ are the eigenstates of S_z. Although these superpositions are the technically correct way of denoting a proton and neutron in terms of quark flavour and spin eigenstates, for brevity, they are often simply referred to as "*uud*" and "*udd*". Note also that the derivation above assumes exact isospin symmetry and is modified by SU(2)-breaking terms.

Similarly, the isospin symmetry of the pions are given by:

$$|\pi^+\rangle = |u\bar{d}\rangle$$
$$|\pi^0\rangle = \frac{1}{\sqrt{2}}\left(|u\bar{u}\rangle - |d\bar{d}\rangle\right)$$
$$|\pi^-\rangle = -|d\bar{u}\rangle$$

Weak isospin

Main article: weak isospin

Isospin is similar to, but should not be confused with weak isospin. Briefly, weak isospin is the gauge symmetry of the weak interaction which connects quark and lepton doublets of left-handed particles in all generations; for example, up and down quarks, top and bottom quarks, electrons and electron neutrinos. By contrast (strong) isospin connects only up and down quarks, acts on both chiralities (left and right) and is a global (not a gauge) symmetry.

3.2.6 Gauged isospin symmetry

Attempts have been made to promote isospin from a global to a local symmetry. In 1954, Chen Ning Yang and Robert Mills suggested that the notion of protons and neutrons, which are continuously rotated into each other by isospin, should be allowed to vary from point to point. To describe this, the proton and neutron direction in isospin space must be defined at every point, giving local basis for isospin. A gauge connection would then describe how to transform isospin along a path between two points.

This Yang–Mills theory describes interacting vector bosons, like the photon of electromagnetism. Unlike the photon, the SU(2) gauge theory would contain self-interacting gauge bosons. The condition of gauge invariance suggests that they have zero mass, just as in electromagnetism.

Ignoring the massless problem, as Yang and Mills did, the theory makes a firm prediction: the vector particle should couple to all particles of a given isospin *universally*. The coupling to the nucleon would be the same as the coupling to the kaons. The coupling to the pions would be the same as the self-coupling of the vector bosons to themselves.

When Yang and Mills proposed the theory, there was no candidate vector boson. J. J. Sakurai in 1960 predicted that there should be a massive vector boson which is coupled to isospin, and predicted that it would show universal couplings. The rho mesons were discovered a short time later, and were quickly identified as Sakurai's vector bosons. The couplings of the rho to the nucleons and to each other were verified to be universal, as best as experiment could measure. The fact that the diagonal isospin current contains part of the electromagnetic current led to the prediction of rho-photon mixing and the concept of vector meson dominance, ideas which led to successful theoretical pictures of GeV-scale photon-nucleus scattering.

Although the discovery of the quarks led to reinterpretation of the rho meson as a vector bound state of a quark and an antiquark, it is sometimes still useful to think of it as the gauge boson of a hidden local symmetry[*][7]

3.2.7 References

[1] http://www.thefreedictionary.com/isospin

[2] Povh, Bogdan; Klaus, Rith; Scholz, Christoph; Zetsche, Frank (2008) [1993]. "2". *Particles and Nuclei*. p. 21. ISBN 978-3-540-79367-0.

[3] Heisenberg, W. (1932). "Über den Bau der Atomkerne". *Zeitschrift für Physik* (in German) **77**: 1–11. Bibcode:1932ZPhy...77....1H. doi:10.1007/BF01342433.

[4] Wigner, E. (1937). "On the Consequences of the Symmetry of the Nuclear Hamiltonian on the Spectroscopy of Nuclei". *Physical Review* **51** (2): 106–119. Bibcode:1937PhRv...51..106W. doi:10.1103/PhysRev.51.106.

[5] C. Amsler et al.; (Particle Data Group) (2008). "Review of Particle Physics: Naming scheme for hadrons" (PDF). *Physics Letters B* **667**: 1. Bibcode:2008PhLB..667....1P. doi:10.1016/j.physletb.2008.07.018.

[6] Greiner, W.; Müller, B. (1989). *Quantum Mechanics: Symmetries*. Springer-Verlag. p. 279. ISBN 3-540-58080-8.

[7] Bando, M.; Kugo, T.; Uehara, S.; Yamawaki, K.; Yanagida, T. (1985). "Is the ρ Meson a Dynamical Gauge Boson of Hidden Local Symmetry?". *Physical Review Letters* **54** (12): 1215–1218. Bibcode:1985PhRvL..54.1215B. doi:10.1103/PhysRevLett.54.1215. PMID 10030967.

3.2.8 Further reading

- Itzykson, C.; Zuber, J.-B. (1980). *Quantum Field Theory*. McGraw-Hill. ISBN 0-07-032071-3.

- Griffiths, D. (1987). *Introduction to Elementary Particles*. John Wiley & Sons. ISBN 0-471-60386-4.

3.2.9 External links

- Nuclear Structure and Decay Data - IAEA Nuclides' Isospin

3.3 Strangeness

This article is about a concept in particle physics. For the definition of "strangeness", see wikt:strangeness. For other uses, see Strange (disambiguation).

In particle physics, **strangeness** ("S") is a property of particles, expressed as a quantum number, for describing decay of particles in strong and electromagnetic reactions, which occur in a short period of time. The strangeness of a particle is defined as:

$$S = -(n_s - n_{\bar{s}})$$

where n_s represents the number of strange quarks (s) and $n_{\bar{s}}$ represents the number of strange antiquarks (s).

The terms *strange* and *strangeness* predate the discovery of the quark, and were adopted after its discovery in order to preserve the continuity of the phrase; strangeness of antiparticles being referred to as +1, and particles as −1 as per the original definition. For all the quark flavor quantum numbers (strangeness, charm, topness and bottomness) the convention is that the flavor charge and the electric charge of a quark have the same sign. With this, any flavor carried by a charged meson has the same sign as its charge.

3.3.1 Conservation

Strangeness was introduced by Murray Gell-Mann and Kazuhiko Nishijima to explain the fact that certain particles, such as the kaons or certain hyperons, were created easily in particle collisions, yet decayed much more slowly than expected for their large masses and large production cross sections. Noting that collisions seemed to always produce pairs of these particles, it was postulated that a new conserved quantity, dubbed "strangeness", was preserved during their creation, but *not* conserved in their decay.

In our modern understanding, strangeness is conserved during the strong and the electromagnetic interactions, but not during the weak interactions. Consequently, the lightest particles containing a strange quark cannot decay by the strong interaction, and must instead decay via the much slower weak interaction. In most cases these decays change the value of the strangeness by one unit. However, this doesn't necessarily hold in second-order weak reactions, where there are mixes of K0 and K0 mesons. All in all, the amount of strangeness can change in a weak interaction reaction by +1, 0 or −1 (depending on the reaction).

3.3.2 See also

- Strangeness production

3.3.3 References

- D.J. Griffiths (1987). *Introduction to Elementary Particles*. John Wiley & Sons. ISBN 0-471-60386-4.

3.3.4 Further reading

- Lessons in Particle Physics Luis Anchordoqui and Francis Halzen, University of Wisconsin, 18th Dec. 2009

3.4 Charm

Charm (symbol *C*) is a flavour quantum number representing the difference between the number of charm quarks (c) and charm antiquarks (c) that are present in a particle:

$$C = n_c - n_{\bar{c}}.$$

By convention, the sign of flavour quantum numbers agree with the sign of the electric charge carried by the quark of corresponding flavour. The charm quark, which carries an electric charge (Q) of $+^2/_3$, therefore carries a charm of +1. The charm antiquarks have the opposite charge ($Q = -^2/_3$), and flavour quantum numbers ($C = -1$).

As with any flavour-related quantum numbers, charm is preserved under strong and electromagnetic interaction, but not under weak interaction (see CKM matrix). For first-order weak decays, that is processes involving only one quark decay, charm can only vary by 1 ($\Delta C = \pm 1, 0$). Since first-order processes are more common than second-order processes (involving two quark decays), this can be used as an approximate "selection rule" for weak decays.

3.4.1 Further reading

- Lessons in Particle Physics Luis Anchordoqui and Francis Halzen, University of Wisconsin, 18th Dec. 2009

3.5 Bottomness

In physics, **bottomness** (symbol B') also called **beauty**, is a flavour quantum number reflecting the difference between the number of bottom antiquarks ($n_{\bar{b}}$) and the number of bottom quarks (n_b) that are present in a particle:

$$B' = -(n_b - n_{\bar{b}})$$

Bottom quarks have (by convention) a bottomness of −1 while bottom antiquarks have a bottomness of +1. The convention is that the flavour quantum number sign for the quark is the same as the sign of the electric charge (symbol Q) of that quark (in this case, Q = −1/3).

As with other flavour-related quantum numbers, bottomness is preserved under strong and electromagnetic interactions, but not under weak interactions. For first-order weak reactions, it holds that $\Delta B' = \pm 1$.

This term is rarely used. Most physicists simply refer to "the number of bottom quarks" and "the number of bottom antiquarks".

3.5.1 Further reading

- Anchordoqui, L.; Halzen, F. (2009). "Lessons in Particle Physics". arXiv:0906.1271 [physics.ed-ph].

3.6 Topness

Topness (also called **truth**), a flavour quantum number, represents the difference between the number of top quarks (t) and number of top antiquarks (t) that are present in a particle:

$$T = n_t - n_{\bar{t}}$$

By convention, top quarks have a topness of +1 and top antiquarks have a topness of −1. The term "topness" is rarely used; most physicists simply refer to "the number of top quarks" and "the number of top antiquarks".

3.6.1 Conservation

Like all flavour quantum numbers, topness is preserved under strong and electromagnetic interactions, but not under weak interaction. However the top quark is extremely unstable, with a half-life under 10^*-23 s, which is the required time for the strong interaction to take place. For that reason the top quark does not hadronize, that is it never forms any meson or baryon, so the topness of a meson or a baryon is every time equal at zero. By the time it can interact strongly it has already decayed to another flavour of quark (usually to a bottom quark).

3.6.2 Further reading

- Anchordoqui, L.; Halzen, F. (2009). "Lessons in Particle Physics". arXiv:0906.1271 [physics.ed-ph].

Chapter 4

Other properties

4.1 Baryon number

In particle physics, the **baryon number** is a strictly conserved additive quantum number of a system. It is defined as

$$B = \frac{1}{3}(n_q - n_{\bar{q}}),$$

where n_q is the number of quarks, and $n_{\bar{q}}$ is the number of antiquarks. Baryons (three quarks) have a baryon number of +1, mesons (one quark, one antiquark) have a baryon number of 0, and antibaryons (three antiquarks) have a baryon number of −1. Exotic hadrons like pentaquarks (four quarks, one antiquark) and tetraquarks (two quarks, two antiquarks) are also classified as baryons and mesons depending on their baryon number.

4.1.1 Baryon number vs. quark number

See also: Color charge

Quarks carry not only electric charge, but also charges such as color charge and weak isospin. Because of a phenomenon known as *color confinement*, a hadron cannot have a net color charge; that is, the total color charge of a particle has to be zero ("white"). A quark can have one of three "colors", dubbed "red", "green", and "blue".

For normal hadrons, a white color can thus be achieved in one of three ways:

- A quark of one color with an antiquark of the corresponding anticolor, giving a meson with baryon number 0,
- Three quarks of different colors, giving a baryon with baryon number +1,
- Three antiquarks into an antibaryon with baryon number −1.

The baryon number was defined long before the quark model was established, so rather than changing the definitions, particle physicists simply gave quarks one third the baryon number. Nowadays it might be more accurate to speak of the conservation of **quark number**.

In theory, exotic hadrons can be formed by adding pairs of quark and antiquark, provided that each pair has a matching color/anticolor. For example, a pentaquark (four quarks, one antiquark) could have the individual quark colors: red, green, blue, blue, and antiblue.

4.1.2 Particles not formed of quarks

Particles without any quarks have a baryon number of zero. Such particles include leptons (electron, muon, tau and their neutrinos) and gauge bosons (photon, W and Z bosons, gluons, and the Higgs boson); or the hypothetical graviton.

4.1.3 Conservation

See also: Conservation law (physics)

The baryon number is conserved in nearly all the interactions of the Standard Model. 'Conserved' means that the sum of the baryon number of all incoming particles is the same as the sum of the baryon numbers of all particles resulting from the reaction. An exception is the chiral anomaly proposed by some extensions of the standard model. However, sphalerons are not all that common. Electroweak sphalerons can only change the baryon number by 3. No experimental evidence of sphalerons has yet been observed.

The still hypothetical idea of a grand unified theory allows for the changing of a baryon into several leptons (see $B - L$), thus violating the conservation of both baryon and lepton numbers.*[1] Proton decay would be an example of such a process taking place, but has never been observed.

4.1.4 See also

- Lepton number
- Flavour (particle physics)
- Isospin
- Hypercharge
- Proton decay
- $B-L$

4.1.5 References

[1] Griffiths, David (2008). *Introduction to Elementary Particles* (2nd ed.). New York: John Wiley & Sons. p. 77. ISBN 9783527618477. In the grand unified theories new interactions are contemplated, permitting decays such as p+ → e+ + π0 or p+ → ν
μ + π+ in which baryon number and lepton number change.

4.2 Color charge

Color charge is a property of quarks and gluons that is related to the particles' strong interactions in the theory of quantum chromodynamics (QCD). The color charge of quarks and gluons is completely unrelated to visual perception of color,*[1] because it is a property that has almost no manifestation at distances above the size of an atomic nucleus. The term *color* was chosen because the charge responsible for the strong force between particles can be analogized to the three primary colors of human vision: red, green, and blue.*[2] Another color scheme is "red, yellow, and blue",*[3] using paint, rather than light as the perceptible analogy.

Particles have corresponding antiparticles. A particle with red, green, or blue charge has a corresponding antiparticle in which the color charge must be the anticolor of red, green, and blue, respectively, for the color charge to be conserved in particle-antiparticle creation and annihilation. Particle physicists call these antired, antigreen, and antiblue. All three colors mixed together, or any one of these colors and its complement (or negative), is "colorless" or "white" and has a net color charge of zero. Free particles have a color charge of zero: baryons are composed of three quarks, but the individual quarks can have red, green, or blue charges, or negatives; mesons are made from a quark and antiquark, the quark can be any color, and the antiquark will have the negative of that color. This color charge differs from electromagnetic charges since electromagnetic charges have only one kind of value. Positive and negative electrical charges are the same kind of charge as they only differ by the sign.

Shortly after the existence of quarks was first proposed in 1964, Oscar W. Greenberg introduced the notion of color charge to explain how quarks could coexist inside some hadrons in otherwise identical quantum states without violating the Pauli exclusion principle. The theory of quantum chromodynamics has been under development since the 1970s and constitutes an important component of the Standard Model of particle physics.

4.2.1 Red, green, and blue

In QCD, a quark's color can take one of three values or charges, red, green, and blue. An antiquark can take one of three anticolors, called antired, antigreen, and antiblue (represented as cyan, magenta and yellow, respectively). Gluons are mixtures of two colors, such as red and antigreen, which constitutes their color charge. QCD considers eight gluons of the possible nine color–anticolor combinations to be unique; see *eight gluon colors* for an explanation.

The following illustrates the coupling constants for color-charged particles:

- The quark colors (red, green, blue) combine to be colorless
- The quark anticolors (antired, antigreen, antiblue) also combine to be colorless
- A hadron with 3 quarks (red, green, blue) before a color change
- Blue quark emits a blue-antigreen gluon
- Green quark has absorbed the blue-antigreen gluon and is now blue; color remains conserved
- An animation of the interaction inside a neutron. The gluons are represented as circles with the color charge in the center and the anti-color charge on the outside.

Field lines from color charges

Main article: Field (physics)

Analogous to an electric field and electric charges, the strong force acting between color charges can be depicted using field lines. However, the color field lines do not arc outwards from one charge to another as much, because they are pulled together tightly by gluons (within 1 fm).*[4] This effect confines quarks within hadrons.

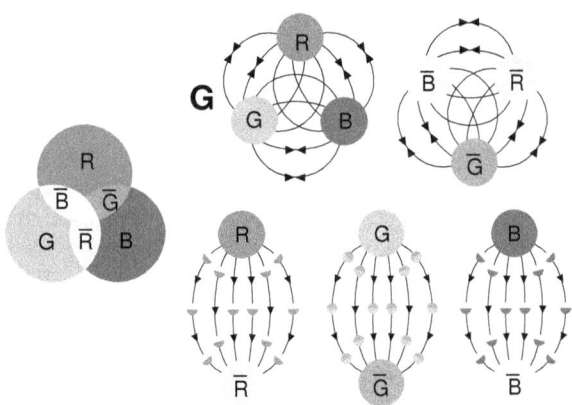

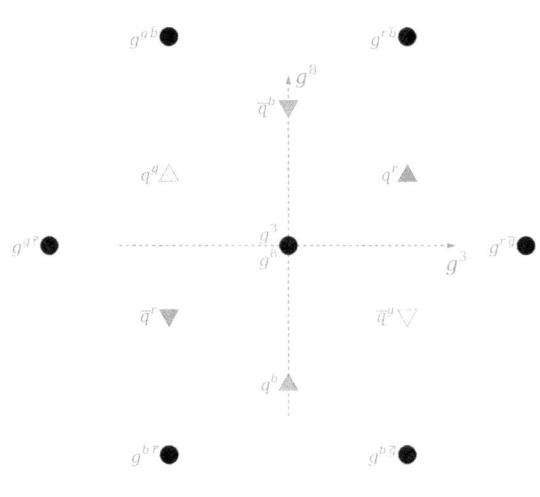

*Fields due to color charges, as in quarks (**G** is the gluon field strength tensor). These are "colorless" combinations.* **Top:** *Color charge has "ternary neutral states" as well as binary neutrality (analogous to electric charge).* **Bottom:** *Quark/antiquark combinations.*[5]*[6]

The pattern of strong charges for the three colors of quark, three antiquarks, and eight gluons (with two of zero charge overlapping).

4.2.2 Coupling constant and charge

In a quantum field theory, a coupling constant and a charge are different but related notions. The coupling constant sets the magnitude of the force of interaction; for example, in quantum electrodynamics, the fine-structure constant is a coupling constant. The charge in a gauge theory has to do with the way a particle transforms under the gauge symmetry; i.e., its representation under the gauge group. For example, the electron has charge −1 and the positron has charge +1, implying that the gauge transformation has opposite effects on them in some sense. Specifically, if a local gauge transformation $\phi(x)$ is applied in electrodynamics, then one finds (using tensor index notation):

$$\begin{aligned} A_\mu &\rightarrow A_\mu + \partial_\mu \phi(x) \;, \\ \psi &\rightarrow \exp[iQ\phi(x)]\psi \text{ and} \\ \overline{\psi} &\rightarrow \exp[-iQ\phi(x)]\overline{\psi} \end{aligned}$$

where A_μ is the photon field, and ψ is the electron field with $Q = -1$ (a bar over ψ denotes its antiparticle — the positron). Since QCD is a non-abelian theory, the representations, and hence the color charges, are more complicated. They are dealt with in the next section.

4.2.3 Quark and gluon fields and color charges

In QCD the gauge group is the non-abelian group SU(3). The *running coupling* is usually denoted by α_s. Each flavor of quark belongs to the fundamental representation (**3**) and contains a triplet of fields together denoted by ψ. The antiquark field belongs to the complex conjugate representation (**3***) and also contains a triplet of fields. We can write

$$\psi = \begin{pmatrix} \psi_1 \\ \psi_2 \\ \psi_3 \end{pmatrix} \text{ and } \overline{\psi} = \begin{pmatrix} \overline{\psi}_1^* \\ \overline{\psi}_2^* \\ \overline{\psi}_3^* \end{pmatrix}.$$

The gluon contains an octet of fields (see gluon field), and belongs to the adjoint representation (**8**), and can be written using the Gell-Mann matrices as

$$\mathbf{A}_\mu = A_\mu^a \lambda_a.$$

(there is an implied summation over $a = 1, 2, \ldots 8$). All other particles belong to the trivial representation (**1**) of color SU(3). The **color charge** of each of these fields is fully specified by the representations. Quarks have a color charge of red, green or blue and antiquarks have a color charge of antired, antigreen or antiblue. Gluons have a combination of two color charges (one of red, green or blue

4.2. COLOR CHARGE

and one of antired, antigreen and antiblue) in a superposition of states which are given by the Gell-Mann matrices. All other particles have zero color charge. Mathematically speaking, the color charge of a particle is the value of a certain quadratic Casimir operator in the representation of the particle.

In the simple language introduced previously, the three indices "1", "2" and "3" in the quark triplet above are usually identified with the three colors. The colorful language misses the following point. A gauge transformation in color SU(3) can be written as $\psi \to U\psi$, where U is a 3 × 3 matrix which belongs to the group SU(3). Thus, after gauge transformation, the new colors are linear combinations of the old colors. In short, the simplified language introduced before is not gauge invariant.

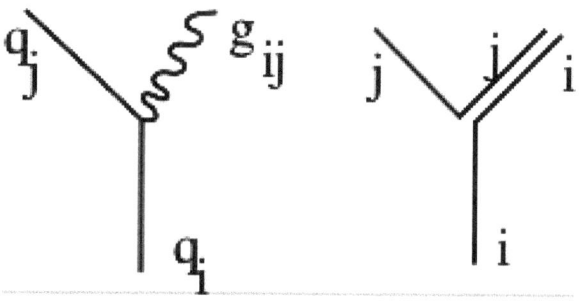

Color-line representation of QCD vertex

Color charge is conserved, but the book-keeping involved in this is more complicated than just adding up the charges, as is done in quantum electrodynamics. One simple way of doing this is to look at the interaction vertex in QCD and replace it by a color-line representation. The meaning is the following. Let ψ_i represent the i-th component of a quark field (loosely called the i-th color). The *color* of a gluon is similarly given by **A** which corresponds to the particular Gell-Mann matrix it is associated with. This matrix has indices i and j. These are the *color labels* on the gluon. At the interaction vertex one has $q_i \to g_{ij} + q_j$. The **color-line** representation tracks these indices. Color charge conservation means that the ends of these color-lines must be either in the initial or final state, equivalently, that no lines break in the middle of a diagram.

Since gluons carry color charge, two gluons can also interact. A typical interaction vertex (called the three gluon vertex) for gluons involves $g + g \to g$. This is shown here, along with its color-line representation. The color-line diagrams can be restated in terms of conservation laws of color; however, as noted before, this is not a gauge invariant language. Note that in a typical non-abelian gauge theory the gauge boson carries the charge of the theory, and hence has interactions of this kind; for example, the W boson in the electroweak theory. In the electroweak theory, the W also

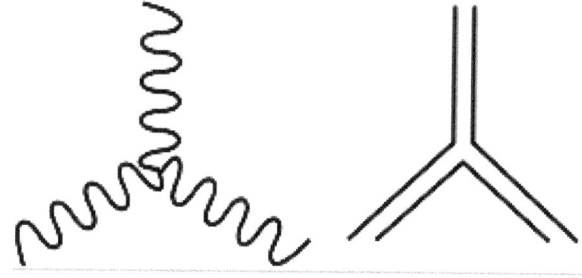

Color-line representation of 3-gluon vertex

carries electric charge, and hence interacts with a photon.

4.2.4 See also

- Color confinement
- Gluon field strength tensor

4.2.5 References

[1] Feynman, Richard (1985), *QED: The Strange Theory of Light and Matter*, Princeton University Press, p. 136, ISBN 0-691-08388-6, The idiot physicists, unable to come up with any wonderful Greek words anymore, call this type of polarization by the unfortunate name of 'color,' which has nothing to do with color in the normal sense.

[2] Close (2007)

[3] R. Penrose (2005). *The Road to Reality*. Vintage books. p. 648. ISBN 978-00994-40680.

[4] R. Resnick, R. Eisberg (1985), *Quantum Physics of Atoms, Molecules, Solids, Nuclei and Particles* (2nd ed.), John Wiley & Sons, p. 684, ISBN 978-0-471-87373-0

[5] Parker, C.B. (1994), *McGraw Hill Encyclopaedia of Physics* (2nd ed.), Mc Graw Hill, ISBN 0-07-051400-3

[6] M. Mansfield, C. O' Sullivan (2011), *Understanding Physics* (4th ed.), John Wiley & Sons, ISBN 978-0-47-0746370

4.2.6 Further reading

- Georgi, Howard (1999), *Lie algebras in particle physics*, Perseus Books Group, ISBN 0-7382-0233-9.
- Griffiths, David J. (1987), *Introduction to Elementary Particles*, New York: John Wiley & Sons, ISBN 0-471-60386-4.
- Christman, J. Richard (2001), "Colour and Charm" (PDF), *Project PHYSNET document MISN-0-283*.

- Hawking, Stephen (1998), *A Brief History of Time*, Bantam Dell Publishing Group, ISBN 978-0-553-10953-5.

- Close, Frank (2007), *The New Cosmic Onion*, Taylor & Francis, ISBN 1-58488-798-2.

4.3 Color confinement

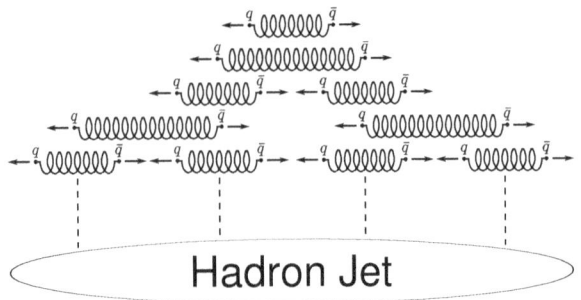

The color force favors confinement because at a certain range it is more energetically favorable to create a quark-antiquark pair than to continue to elongate the color flux tube. This is analogous to the behavior of an elongated rubber-band.

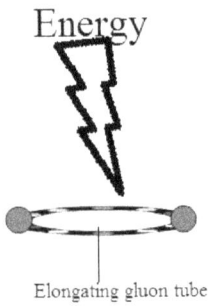

An animation of color confinement. Energy is supplied to the quarks, and the gluon tube elongates until it reaches a point where it "snaps" and forms a quark-antiquark pair.

Color confinement, often simply called **confinement**, is the phenomenon that color charged particles (such as quarks) cannot be isolated singularly, and therefore cannot be directly observed.[1] Quarks, by default, clump together to form groups, or hadrons. The two types of hadrons are the mesons (one quark, one antiquark) and the baryons (three quarks).

The constituent quarks in a group cannot be separated from their parent hadron, and this is why quarks currently cannot be studied or observed in any more direct way than at a hadron level.[2]

4.3.1 Origin

The reasons for quark confinement are somewhat complicated; no analytic proof exists that quantum chromodynamics should be confining. The current theory is that confinement is due to the force-carrying gluons having color charge. As any two electrically charged particles separate, the electric fields between them diminish quickly, allowing (for example) electrons to become unbound from atomic nuclei. However, as a quark-antiquark pair separates, the gluon field forms a narrow tube (or string) of color field between them. This is quite different from the behavior of the electric field of a pair of positive and negative electric charges, which extends into the whole surrounding space and diminishes at large distances. Because of this behavior of the gluon field, a strong force between the quark pair acts constantly—regardless of their distance[3][4]—with a force of around 10,000 newtons.[5]

When two quarks become separated, as happens in particle accelerator collisions, at some point it is more energetically favorable for a new quark–antiquark pair to spontaneously appear, than to allow the tube to extend further. As a result of this, when quarks are produced in particle accelerators, instead of seeing the individual quarks in detectors, scientists see "jets" of many color-neutral particles (mesons and baryons), clustered together. This process is called *hadronization*, *fragmentation*, or *string breaking*, and is one of the least understood processes in particle physics.

The confining phase is usually defined by the behavior of the action of the Wilson loop, which is simply the path in spacetime traced out by a quark–antiquark pair created at one point and annihilated at another point. In a non-confining theory, the action of such a loop is proportional to its perimeter. However, in a confining theory, the action of the loop is instead proportional to its area. Since the area will be proportional to the separation of the quark–antiquark pair, free quarks are suppressed. Mesons are allowed in such a picture, since a loop containing another loop in the opposite direction will have only a small area between the two loops.

4.3.2 Models exhibiting confinement

Besides QCD in four spacetime dimensions, another model which exhibits confinement is the Schwinger model.[6] Compact Abelian gauge theories also exhibit confinement in 2 and 3 spacetime dimensions.[7] Confinement has recently been found in elementary excitations of magnetic systems called spinons.[8]

4.3.3 Models of fully screened quarks

Besides the quark confinement idea, there is a potential possibility, that color charge of quarks gets fully screened by the gluonic color, surrounding the quark. Exact solutions of SU(3) classical Yang–Mills theory, which provide full screening (by gluon fields) of the color charge of a quark have been found.*[9] However, such classical solutions do not take into account non-trivial properties of QCD vacuum. Therefore, a significance of such full gluonic screening solutions for a separated quark is not clear.

4.3.4 See also

- Gluon field strength tensor
- Asymptotic freedom
- Center vortices
- Deconfining phase
- Quantum mechanics
- Particle physics
- Fundamental force
- Dual superconducting model
- Beta-function
- Infrared safety

4.3.5 References

[1] V. Barger, R. Phillips (1997). *Collider Physics*. Addison–Wesley. ISBN 0-201-14945-1.

[2] T.-Y. Wu, W.-Y. Pauchy Hwang (1991). *Relativistic quantum mechanics and quantum fields*. World Scientific. p. 321. ISBN 981-02-0608-9.

[3] T. Muta (2009). *Foundations of quantum chromodynamics: an introduction to perturbative methods in gauge theories* (3rd ed.). World Scientific. ISBN 978-981-279-353-9.

[4] A. Smilga (2001). *Lectures on quantum chromodynamics*. World Scientific. ISBN 978-981-02-4331-9.

[5] Fritzsch, op. cite, p. 164. The author states that the force between differently coloured quarks remains constant at any distance after they travel only a tiny distance from each other, and is equal to that need to raise one ton, which is 1000 kg x 9.8 m/s^2 = ~10,000 N.

[6] Wilson, Kenneth G. (1974-10-15). "Confinement of Quarks". *Physical Review D* (College Park, MD, USA: American Physical Society) **10**: 2445–2459. Bibcode:1974PhRvD..10.2445W. doi:10.1103/PhysRevD.10.2445. ISSN 1550-2368. OCLC 55589778. Retrieved 2014-04-12.

[7] Schön, Verena; Michael, Thies (2000-08-22). "2d Model Field Theories at Finite Temperature and Density (Section 2.5)". arXiv:hep-th/0008175v1 [hep-th].

[8] Lake, Bella; Tsvelik, Alexei M.; Notbohm, Susanne; Tennant, D. Alan; Perring, Toby G.; Reehuis, Manfred; Sekar, Chinnathambi; Krabbes, Gernot; Büchner, Bernd (2009-11-29). "Confinement of fractional quantum number particles in a condensed-matter system". *Nature Physics* (London, UK: Nature Publishing Group) **6** (1): 50–55. arXiv:0908.1038. Bibcode:2010NatPh...6...50L. doi:10.1038/nphys1462. ISSN 1745-2481. OCLC 150143123. Retrieved 2014-04-12. (subscription required (help)).

[9] Cahill, Kevin (1978-08-28). "Example of Color Screening". *Physical Review Letters* (American Physical Society) **41** (9): 599–601. Bibcode:1978PhRvL..41..599C. doi:10.1103/PhysRevLett.41.599. ISSN 1079-7114. OCLC 31492939. Retrieved 2014-04-12.

4.3.6 External links

- Quarks

4.4 Electric charge

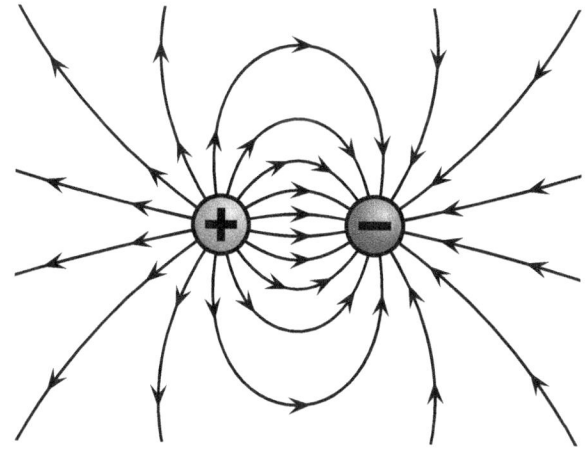

Electric field of a positive and a negative point charge.

Electric charge is the physical property of matter that causes it to experience a force when placed in an electromagnetic field. There are two types of electric charges: positive and negative. Positively charged substances are repelled from other positively charged substances, but attracted to negatively charged substances; negatively charged substances are repelled from negative and attracted to positive. An object is negatively charged if it has an excess of electrons, and is otherwise positively charged or uncharged. The SI derived unit of electric charge is the coulomb (C), although in electrical engineering it is also common to use the ampere-hour (Ah), and in chemistry it is common to use the elementary charge (e) as a unit. The symbol Q is often used to denote charge. The early knowledge of how charged substances interact is now called classical electrodynamics, and is still very accurate if quantum effects do not need to be considered.

The *electric charge* is a fundamental conserved property of some subatomic particles, which determines their electromagnetic interaction. Electrically charged matter is influenced by, and produces, electromagnetic fields. The interaction between a moving charge and an electromagnetic field is the source of the electromagnetic force, which is one of the four fundamental forces (See also: magnetic field).

Twentieth-century experiments demonstrated that electric charge is *quantized*; that is, it comes in integer multiples of individual small units called the elementary charge, e, approximately equal to 1.602×10^{-19} coulombs (except for particles called quarks, which have charges that are integer multiples of $e/3$). The proton has a charge of $+e$, and the electron has a charge of $-e$. The study of charged particles, and how their interactions are mediated by photons, is called quantum electrodynamics.

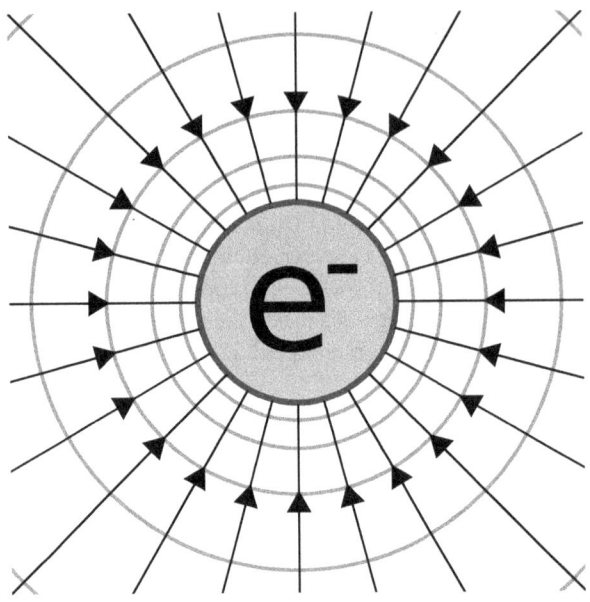

Diagram showing field lines and equipotentials around an electron, a negatively charged particle. In an electrically neutral atom, the number of electrons is equal to the number of protons (which are positively charged), resulting in a net zero overall charge

4.4.1 Overview

Charge is the fundamental property of forms of matter that exhibit electrostatic attraction or repulsion in the presence of other matter. Electric charge is a characteristic property of many subatomic particles. The charges of free-standing particles are integer multiples of the elementary charge e; we say that electric charge is *quantized*. Michael Faraday, in his electrolysis experiments, was the first to note the discrete nature of electric charge. Robert Millikan's oil-drop experiment demonstrated this fact directly, and measured the elementary charge.

By convention, the charge of an electron is -1, while that of a proton is $+1$. Charged particles whose charges have the same sign repel one another, and particles whose charges have different signs attract. Coulomb's law quantifies the electrostatic force between two particles by asserting that the force is proportional to the product of their charges, and inversely proportional to the square of the distance between them.

The charge of an antiparticle equals that of the corresponding particle, but with opposite sign. Quarks have fractional charges of either $-\frac{1}{3}$ or $+\frac{2}{3}$, but free-standing quarks have never been observed (the theoretical reason for this fact is asymptotic freedom).

The electric charge of a macroscopic object is the sum of the electric charges of the particles that make it up. This charge is often small, because matter is made of atoms, and atoms typically have equal numbers of protons and electrons, in which case their charges cancel out, yielding a net charge of zero, thus making the atom neutral.

An *ion* is an atom (or group of atoms) that has lost one or more electrons, giving it a net positive charge (cation), or that has gained one or more electrons, giving it a net negative charge (anion). *Monatomic ions* are formed from single atoms, while *polyatomic ions* are formed from two or more atoms that have been bonded together, in each case yielding an ion with a positive or negative net charge.

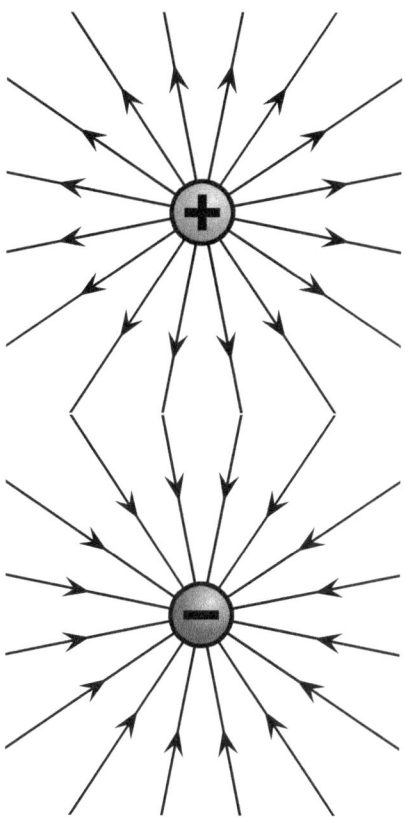

Electric field induced by a positive electric charge (left) and a field induced by a negative electric charge (right).

During formation of macroscopic objects, constituent atoms and ions usually combine to form structures composed of neutral *ionic compounds* electrically bound to neutral atoms. Thus macroscopic objects tend toward being neutral overall, but macroscopic objects are rarely perfectly net neutral.

Sometimes macroscopic objects contain ions distributed throughout the material, rigidly bound in place, giving an overall net positive or negative charge to the object. Also, macroscopic objects made of conductive elements, can more or less easily (depending on the element) take on or give off electrons, and then maintain a net negative or positive charge indefinitely. When the net electric charge of an object is non-zero and motionless, the phenomenon is known as static electricity. This can easily be produced by rubbing two dissimilar materials together, such as rubbing amber with fur or glass with silk. In this way nonconductive materials can be charged to a significant degree, either positively or negatively. Charge taken from one material is moved to the other material, leaving an opposite charge of the same magnitude behind. The law of *conservation of charge* always applies, giving the object from which a negative charge has been taken a positive charge of the same magnitude, and vice versa.

Even when an object's net charge is zero, charge can be distributed non-uniformly in the object (e.g., due to an external electromagnetic field, or bound polar molecules). In such cases the object is said to be polarized. The charge due to polarization is known as bound charge, while charge on an object produced by electrons gained or lost from outside the object is called *free charge*. The motion of electrons in conductive metals in a specific direction is known as electric current.

4.4.2 Units

The SI unit of quantity of electric charge is the coulomb, which is equivalent to about 6.242×10^{18} e (e is the charge of a proton). Hence, the charge of an electron is approximately $-1.602 \times 10^{*}-19$ C. The coulomb is defined as the quantity of charge that has passed through the cross section of an electrical conductor carrying one ampere within one second. The symbol Q is often used to denote a quantity of electricity or charge. The quantity of electric charge can be directly measured with an electrometer, or indirectly measured with a ballistic galvanometer.

After finding the quantized character of charge, in 1891 George Stoney proposed the unit 'electron' for this fundamental unit of electrical charge. This was before the discovery of the particle by J.J. Thomson in 1897. The unit is today treated as nameless, referred to as "elementary charge", "fundamental unit of charge", or simply as "e". A measure of charge should be a multiple of the elementary charge e, even if at large scales charge seems to behave as a real quantity. In some contexts it is meaningful to speak of fractions of a charge; for example in the charging of a capacitor, or in the fractional quantum Hall effect.

In systems of units other than SI such as cgs, electric charge is expressed as combination of only three fundamental quantities such as length, mass and time and not four as in SI where electric charge is a combination of length, mass, time and electric current.

4.4.3 History

As reported by the ancient Greek mathematician Thales of Miletus around 600 BC, charge (or *electricity*) could be accumulated by rubbing fur on various substances, such as amber. The Greeks noted that the charged amber buttons could attract light objects such as hair. They also noted that if they rubbed the amber for long enough, they could even get an electric spark to jump. This property derives from the triboelectric effect.

In 1600, the English scientist William Gilbert returned to the subject in *De Magnete*, and coined the New Latin word

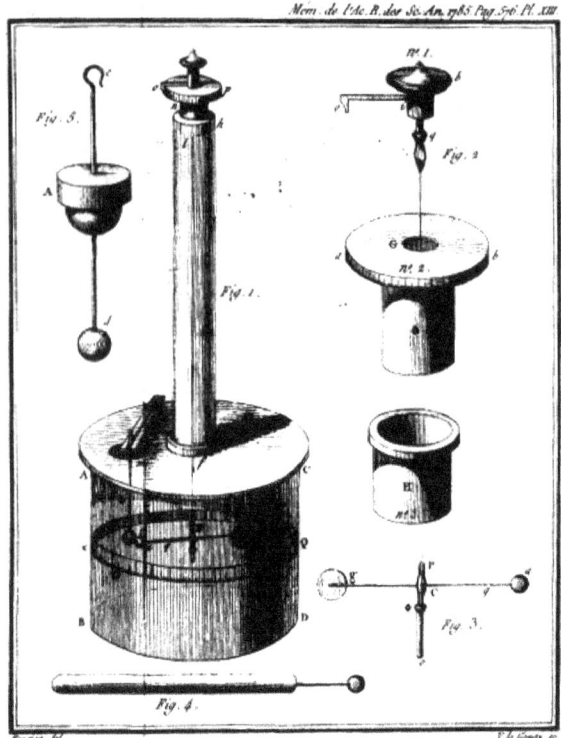

Coulomb's torsion balance

electricus from ηλεκτρον (*elektron*), the Greek word for *amber*, which soon gave rise to the English words "electric" and "electricity." He was followed in 1660 by Otto von Guericke, who invented what was probably the first electrostatic generator. Other European pioneers were Robert Boyle, who in 1675 stated that electric attraction and repulsion can act across a vacuum; Stephen Gray, who in 1729 classified materials as conductors and insulators; and C. F. du Fay, who proposed in 1733*[1] that electricity comes in two varieties that cancel each other, and expressed this in terms of a two-fluid theory. When glass was rubbed with silk, du Fay said that the glass was charged with *vitreous electricity*, and, when amber was rubbed with fur, the amber was said to be charged with *resinous electricity*. In 1839, Michael Faraday showed that the apparent division between static electricity, current electricity, and bioelectricity was incorrect, and all were a consequence of the behavior of a single kind of electricity appearing in opposite polarities. It is arbitrary which polarity is called positive and which is called negative. Positive charge can be defined as the charge left on a glass rod after being rubbed with silk.*[2]

One of the foremost experts on electricity in the 18th century was Benjamin Franklin, who argued in favour of a one-fluid theory of electricity. Franklin imagined electricity as being a type of invisible fluid present in all matter; for example, he believed that it was the glass in a Leyden jar that held the accumulated charge. He posited that rubbing insulating surfaces together caused this fluid to change location, and that a flow of this fluid constitutes an electric current. He also posited that when matter contained too little of the fluid it was "negatively" charged, and when it had an excess it was "positively" charged. For a reason that was not recorded, he identified the term "positive" with vitreous electricity and "negative" with resinous electricity. William Watson arrived at the same explanation at about the same time.

4.4.4 Static electricity and electric current

Static electricity and electric current are two separate phenomena. They both involve electric charge, and may occur simultaneously in the same object. Static electricity refers to the electric charge of an object and the related electrostatic discharge when two objects are brought together that are not at equilibrium. An electrostatic discharge creates a change in the charge of each of the two objects. In contrast, electric current is the flow of electric charge through an object, which produces no net loss or gain of electric charge.

Electrification by friction

Further information: triboelectric effect

When a piece of glass and a piece of resin—neither of which exhibit any electrical properties—are rubbed together and left with the rubbed surfaces in contact, they still exhibit no electrical properties. When separated, they attract each other.

A second piece of glass rubbed with a second piece of resin, then separated and suspended near the former pieces of glass and resin causes these phenomena:

- The two pieces of glass repel each other.
- Each piece of glass attracts each piece of resin.
- The two pieces of resin repel each other.

This attraction and repulsion is an *electrical phenomena,* and the bodies that exhibit them are said to be *electrified*, or *electrically charged*. Bodies may be electrified in many other ways, as well as by friction. The electrical properties of the two pieces of glass are similar to each other but opposite to those of the two pieces of resin: The glass attracts what the resin repels and repels what the resin attracts.

If a body electrified in any manner whatsoever behaves as the glass does, that is, if it repels the glass and attracts the resin, the body is said to be 'vitreously' electrified, and if it attracts the glass and repels the resin it is said to be 'resinously' electrified. All electrified bodies are found to be either vitreously or resinously electrified.

It is the established convention of the scientific community to define the vitreous electrification as positive, and the resinous electrification as negative. The exactly opposite properties of the two kinds of electrification justify our indicating them by opposite signs, but the application of the positive sign to one rather than to the other kind must be considered as a matter of arbitrary convention, just as it is a matter of convention in mathematical diagram to reckon positive distances towards the right hand.

No force, either of attraction or of repulsion, can be observed between an electrified body and a body not electrified.*[3]

Actually, all bodies are electrified, but may appear not to be so by the relative similar charge of neighboring objects in the environment. An object further electrified + or − creates an equivalent or opposite charge by default in neighboring objects, until those charges can equalize. The effects of attraction can be observed in high-voltage experiments, while lower voltage effects are merely weaker and therefore less obvious. The attraction and repulsion forces are codified by Coulomb's Law (attraction falls off at the square of the distance, which has a corollary for acceleration in a gravitational field, suggesting that gravitation may be merely electrostatic phenomenon between relatively weak charges in terms of scale). See also the Casimir effect.

It is now known that the Franklin/Watson model was fundamentally correct. There is only one kind of electrical charge, and only one variable is required to keep track of the amount of charge.*[4] On the other hand, just knowing the charge is not a complete description of the situation. Matter is composed of several kinds of electrically charged particles, and these particles have many properties, not just charge.

The most common charge carriers are the positively charged proton and the negatively charged electron. The movement of any of these charged particles constitutes an electric current. In many situations, it suffices to speak of the *conventional current* without regard to whether it is carried by positive charges moving in the direction of the conventional current or by negative charges moving in the opposite direction. This macroscopic viewpoint is an approximation that simplifies electromagnetic concepts and calculations.

At the opposite extreme, if one looks at the microscopic situation, one sees there are many ways of carrying an electric current, including: a flow of electrons; a flow of electron "holes" that act like positive particles; and both negative and positive particles (ions or other charged particles) flowing in opposite directions in an electrolytic solution or a plasma.

Beware that, in the common and important case of metallic wires, the direction of the conventional current is opposite to the drift velocity of the actual charge carriers, i.e., the electrons. This is a source of confusion for beginners.

4.4.5 Properties

Aside from the properties described in articles about electromagnetism, charge is a relativistic invariant. This means that any particle that has charge Q, no matter how fast it goes, always has charge Q. This property has been experimentally verified by showing that the charge of *one* helium nucleus (two protons and two neutrons bound together in a nucleus and moving around at high speeds) is the same as *two* deuterium nuclei (one proton and one neutron bound together, but moving much more slowly than they would if they were in a helium nucleus).

4.4.6 Conservation of electric charge

Main article: Charge conservation

The total electric charge of an isolated system remains constant regardless of changes within the system itself. This law is inherent to all processes known to physics and can be derived in a local form from gauge invariance of the wave function. The conservation of charge results in the charge-current continuity equation. More generally, the net change in charge density ϱ within a volume of integration V is equal to the area integral over the current density $\mathbf{J}$ through the closed surface $S = \partial V$, which is in turn equal to the net current I:

$$-\tfrac{d}{dt}\int_V \rho\, dV = \oiint_{\partial V} \mathbf{J}\cdot d\mathbf{S} = \int J\, dS \cos\theta = I.$$

Thus, the conservation of electric charge, as expressed by the continuity equation, gives the result:

$$I = \frac{dQ}{dt}.$$

The charge transferred between times t_i and t_f is obtained by integrating both sides:

$$Q = \int_{t_i}^{t_f} I\, dt$$

where *I* is the net outward current through a closed surface and *Q* is the electric charge contained within the volume defined by the surface.

4.4.7 See also

- Quantity of electricity
- SI electromagnetism units

4.4.8 References

[1] Two Kinds of Electrical Fluid: Vitreous and Resinous – 1733

[2] Electromagnetic Fields (2nd Edition), Roald K. Wangsness, Wiley, 1986. ISBN 0-471-81186-6 (intermediate level textbook)

[3] James Clerk Maxwell *A Treatise on Electricity and Magnetism*, pp. 32-33, Dover Publications Inc., 1954 ASIN: B000HFDK0K, 3rd ed. of 1891

[4] One Kind of Charge

4.4.9 External links

- How fast does a charge decay?
- Science Aid: Electrostatic charge Easy-to-understand page on electrostatic charge.
- History of the electrical units.

4.5 Hypercharge

In particle physics, the **hypercharge** (from **hyper**onic + **charge**) *Y* of a particle is related to the strong interaction, and is distinct from the similarly named weak hypercharge, which has an analogous role in the electroweak interaction. The concept of hypercharge combines and unifies isospin and flavour into a single charge operator.

4.5.1 Definition

Hypercharge in particle physics is a quantum number relating the strong interactions of the SU(3) model. Isospin is defined in the SU(2) model while the SU(3) model defines hypercharge.

SU(3) weight diagrams (see below) are 2-dimensional with the coordinates referring to two quantum numbers, I_z, which is the z-component of isospin and *Y*, which is the hypercharge (the sum of strangeness (*S*), charm (*C*), bottomness (*B'*), topness (*T*), and baryon number (*B*)). Mathematically, hypercharge is

$$Y = S + C + B' + T + B$$

and conservation of hypercharge implies a conservation of flavour. Strong interactions conserve hypercharge, but weak interactions do not.

4.5.2 Relation with electric charge and isospin

Main article: Gell-Mann–Nishijima formula

The Gell-Mann–Nishijima formula relates isospin and electric charge

$$Q = I_3 + \frac{1}{2}Y,$$

where I_3 is the third component of isospin and *Q* is the particle's charge.

Isospin creates multiplets of particles whose average charge is related to the hypercharge by:

$$Y = 2\bar{Q}.$$

since the hypercharge is the same for all members of a multiplet, and the average of the I_3 values is 0.

4.5.3 SU(3) model in relation to hypercharge

The SU(2) model has multiplets characterized by a quantum number *J*, which is the total angular momentum. Each multiplet consists of 2*J* + 1 substates with equally spaced values of J_z, forming a symmetric arrangement seen in atomic spectra and isospin. This formalises the observation that certain strong baryon decays were not observed, leading to the prediction of the mass, strangeness and charge of the $\Omega-$ baryon.

The SU(3) has *supermultiplets* containing SU(2) multiplets. SU(3) now needs 2 numbers to specify all its sub-states which are denoted by λ_1 and λ_2.

(λ_1 + 1) specifies the number of points in the topmost side of the hexagon while (λ_2 + 1) specifies the number of points on the bottom side.

4.5.4 Examples

- The nucleon group (protons with $Q = +1$ and neutrons with $Q = 0$) have an average charge of $+1/2$, so they both have hypercharge $Y = 1$ (baryon number $B = +1$, $S = C = B' = T = 0$). From the Gell-Mann–Nishijima formula we know that proton has isospin $I_3 = +1/2$, while neutron has $I_3 = -1/2$.

- This also works for quarks: for the *up* quark, with a charge of $+2/3$, and an I_3 of $+1/2$, we deduce a hypercharge of $1/3$, due to its baryon number (since you need 3 quarks to make a baryon, a quark has baryon number of $1/3$).

- For a *strange* quark, with charge $-1/3$, a baryon number of $1/3$ and strangeness of -1 we get a hypercharge $Y = -2/3$, so we deduce an $I_3 = 0$. That means that a *strange* quark makes a singlet of its own (same happens with *charm*, *bottom* and *top* quarks), while *up* and *down* constitute an isospin doublet.

4.5.5 Practical obsolescence

Hypercharge was a concept developed in the 1960s, to organize groups of particles in the *"particle zoo"* and to develop *ad hoc* conservation laws based on their observed transformations. With the advent of the quark model, it is now obvious that (if one only includes the up, down and strange quarks out of the total 6 quarks in the Standard Model), hypercharge Y is the following combination of the numbers of up (n_u), down (n_d), and strange quarks (n_s):

$$Y = \frac{1}{3}(n_u + n_d - 2n_s).$$

In modern descriptions of hadron interaction, it has become more obvious to draw Feynman diagrams that trace through individual quarks composing the interacting baryons and mesons, rather than counting hypercharge quantum numbers. Weak hypercharge, however, remains of practical use in various theories of the electroweak interaction.

4.5.6 References

- Henry Semat, John R. Albright (1984). *Introduction to atomic and nuclear physics*. Chapman and Hall. ISBN 0-412-15670-9.

4.6 Spin

This article is about spin in quantum mechanics. For rotation in classical mechanics, see angular momentum.

In quantum mechanics and particle physics, **spin** is an intrinsic form of angular momentum carried by elementary particles, composite particles (hadrons), and atomic nuclei.*[1]*[2]

Spin is one of two types of angular momentum in quantum mechanics, the other being *orbital angular momentum*. The orbital angular momentum operator is the quantum-mechanical counterpart to the classical notion of angular momentum: it arises when a particle executes a rotating or twisting trajectory (such as when an electron orbits a nucleus).*[3]*[4] The existence of spin angular momentum is inferred from experiments, such as the Stern–Gerlach experiment, in which particles are observed to possess angular momentum that cannot be accounted for by orbital angular momentum alone.*[5]

In some ways, spin is like a vector quantity; it has a definite magnitude, and it has a "direction" (but quantization makes this "direction" different from the direction of an ordinary vector). All elementary particles of a given kind have the same magnitude of spin angular momentum, which is indicated by assigning the particle a *spin quantum number*.*[2]

The SI unit of spin is the joule-second, just as with classical angular momentum. In practice, however, it is written as a multiple of the reduced Planck constant $\hbar$, usually in natural units, where the $\hbar$ is omitted, resulting in a unitless number. Spin quantum numbers are unitless numbers by definition.

When combined with the spin-statistics theorem, the spin of electrons results in the Pauli exclusion principle, which in turn underlies the periodic table of chemical elements.

Wolfgang Pauli was the first to propose the concept of spin, but he did not name it. In 1925, Ralph Kronig, George Uhlenbeck and Samuel Goudsmit at Leiden University suggested a physical interpretation of particles spinning around their own axis. The mathematical theory was worked out in depth by Pauli in 1927. When Paul Dirac derived his relativistic quantum mechanics in 1928, electron spin was an essential part of it.

4.6.1 Quantum number

Main article: Spin quantum number

As the name suggests, spin was originally conceived as the rotation of a particle around some axis. This picture is correct so far as spin obeys the same mathematical laws as quantized angular momenta do. On the other hand, spin has some peculiar properties that distinguish it from orbital angular momenta:

- Spin quantum numbers may take half-integer values.

- Although the direction of its spin can be changed, an elementary particle cannot be made to spin faster or slower.

- The spin of a charged particle is associated with a magnetic dipole moment with a g-factor differing from 1. This could only occur classically if the internal charge of the particle were distributed differently from its mass.

The conventional definition of the **spin quantum number**, s, is $s = n/2$, where n can be any non-negative integer. Hence the allowed values of s are 0, 1/2, 1, 3/2, 2, etc. The value of s for an elementary particle depends only on the type of particle, and cannot be altered in any known way (in contrast to the *spin direction* described below). The spin angular momentum, S, of any physical system is quantized. The allowed values of S are:

$$S = \frac{h}{2\pi}\sqrt{s(s+1)} = \frac{h}{4\pi}\sqrt{n(n+2)},$$

where h is the Planck constant. In contrast, orbital angular momentum can only take on integer values of s; i.e., even-numbered values of n.

Fermions and bosons

Those particles with half-integer spins, such as 1/2, 3/2, 5/2, are known as fermions, while those particles with integer spins, such as 0, 1, 2, are known as bosons. The two families of particles obey different rules and *broadly* have different roles in the world around us. A key distinction between the two families is that fermions obey the Pauli exclusion principle; that is, there cannot be two identical fermions simultaneously having the same quantum numbers (meaning, roughly, having the same position, velocity and spin direction). In contrast, bosons obey the rules of Bose–Einstein statistics and have no such restriction, so they may "bunch together" even if in identical states. Also, composite particles can have spins different from the particles which comprise them. For example, a helium atom can have spin 0 and therefore can behave like a boson even though the quarks and electrons which make it up are all fermions.

This has profound practical applications:

- Quarks and leptons (including electrons and neutrinos), which make up what is classically known as matter, are all fermions with spin 1/2. The common idea that "matter takes up space" actually comes from the Pauli exclusion principle acting on these particles to prevent the fermions that make up matter from being in the same quantum state. Further compaction would require electrons to occupy the same energy states, and therefore a kind of pressure (sometimes known as degeneracy pressure of electrons) acts to resist the fermions being overly close. It is also this pressure which prevents stars collapsing inwardly, and which, when it finally gives way under immense gravitational pressure in a dying massive star, triggers inward collapse and the dramatic explosion into a supernova.

Elementary fermions with other spins (3/2, 5/2 etc.) are not known to exist, as of 2014.

- Elementary particles which are thought of as carrying forces are all bosons with spin 1. They include the photon which carries the electromagnetic force, the gluon (strong force), and the W and Z bosons (weak force). The ability of bosons to occupy the same quantum state is used in the laser, which aligns many photons having the same quantum number (the same direction and frequency), superfluid liquid helium resulting from helium-4 atoms being bosons, and superconductivity where pairs of electrons (which individually are fermions) act as single composite bosons.

Elementary bosons with other spins (0, 2, 3 etc.) were not historically known to exist, although they have received considerable theoretical treatment and are well established within their respective mainstream theories. In particular theoreticians have proposed the graviton (predicted to exist by some quantum gravity theories) with spin 2, and the Higgs boson (explaining electroweak symmetry breaking) with spin 0. Since 2013 the Higgs boson with spin 0 has been considered proven to exist. It is the first scalar particle (spin 0) known to exist in nature.

Theoretical and experimental studies have shown that the spin possessed by elementary particles cannot be explained by postulating that they are made up of even smaller particles rotating about a common center of mass analogous to a classical electron radius; as far as can be presently determined, these elementary particles have no inner structure. The spin of an elementary particle is therefore seen as a truly intrinsic physical property, akin to the particle's electric charge and rest mass.

Spin-statistics theorem

The proof that particles with half-integer spin (fermions) obey Fermi–Dirac statistics and the Pauli Exclusion Principle, and particles with integer spin (bosons) obey Bose–Einstein statistics, occupy "symmetric states", and thus can share quantum states, is known as the spin-statistics theorem. The theorem relies on both quantum mechanics and the theory of special relativity, and this connection between spin and statistics has been called "one of the most important applications of the special relativity theory".*[6]

4.6.2 Magnetic moments

Main article: Spin magnetic moment

Particles with spin can possess a magnetic dipole moment,

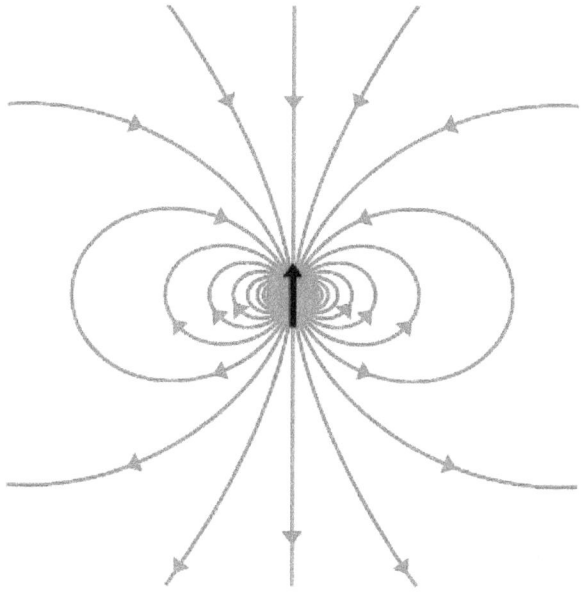

Schematic diagram depicting the spin of the neutron as the black arrow and magnetic field lines associated with the neutron magnetic moment. The neutron has a negative magnetic moment. While the spin of the neutron is upward in this diagram, the magnetic field lines at the center of the dipole are downward.

just like a rotating electrically charged body in classical electrodynamics. These magnetic moments can be experimentally observed in several ways, e.g. by the deflection of particles by inhomogeneous magnetic fields in a Stern–Gerlach experiment, or by measuring the magnetic fields generated by the particles themselves.

The intrinsic magnetic moment $\boldsymbol{\mu}$ of a spin-1/2 particle with charge q, mass m, and spin angular momentum $\mathbf{S}$, is*[7]

$$\boldsymbol{\mu} = \frac{g_s q}{2m}\mathbf{S}$$

where the dimensionless quantity g_s is called the spin g-factor. For exclusively orbital rotations it would be 1 (assuming that the mass and the charge occupy spheres of equal radius).

The electron, being a charged elementary particle, possesses a nonzero magnetic moment. One of the triumphs of the theory of quantum electrodynamics is its accurate prediction of the electron g-factor, which has been experimentally determined to have the value −2.0023193043622(15), with the digits in parentheses denoting measurement uncertainty in the last two digits at one standard deviation.*[8] The value of 2 arises from the Dirac equation, a fundamental equation connecting the electron's spin with its electromagnetic properties, and the correction of 0.002319304... arises from the electron's interaction with the surrounding electromagnetic field, including its own field.*[9] Composite particles also possess magnetic moments associated with their spin. In particular, the neutron possesses a non-zero magnetic moment despite being electrically neutral. This fact was an early indication that the neutron is not an elementary particle. In fact, it is made up of quarks, which are electrically charged particles. The magnetic moment of the neutron comes from the spins of the individual quarks and their orbital motions.

Neutrinos are both elementary and electrically neutral. The minimally extended Standard Model that takes into account non-zero neutrino masses predicts neutrino magnetic moments of:*[10]*[11]*[12]

$$\mu_\nu \approx 3 \times 10^{-19} \mu_B \frac{m_\nu}{\text{eV}}$$

where the μ_ν are the neutrino magnetic moments, m_ν are the neutrino masses, and μ_B is the Bohr magneton. New physics above the electroweak scale could, however, lead to significantly higher neutrino magnetic moments. It can be shown in a model independent way that neutrino magnetic moments larger than about 10^{-14} μ_B are unnatural, because they would also lead to large radiative contributions to the neutrino mass. Since the neutrino masses cannot exceed about 1 eV, these radiative corrections must then be assumed to be fine tuned to cancel out to a large degree.*[13]

The measurement of neutrino magnetic moments is an active area of research. As of 2001, the latest experimental results have put the neutrino magnetic moment at less than 1.2×10^{-10} times the electron's magnetic moment.

In ordinary materials, the magnetic dipole moments of individual atoms produce magnetic fields that cancel one another, because each dipole points in a random direction. Ferromagnetic materials below their Curie temperature, however, exhibit magnetic domains in which the atomic dipole moments are locally aligned, producing a macroscopic, non-zero magnetic field from the domain. These are the ordinary "magnets" with which we are all familiar.

In paramagnetic materials, the magnetic dipole moments of individual atoms spontaneously align with an externally applied magnetic field. In diamagnetic materials, on the other hand, the magnetic dipole moments of individual atoms spontaneously align oppositely to any externally applied magnetic field, even if it requires energy to do so.

The study of the behavior of such "spin models" is a thriving area of research in condensed matter physics. For instance, the Ising model describes spins (dipoles) that have only two possible states, up and down, whereas in the Heisenberg model the spin vector is allowed to point in any direction. These models have many interesting properties, which have led to interesting results in the theory of phase transitions.

4.6.3 Direction

Further information: Angular momentum operator

Spin projection quantum number and multiplicity

In classical mechanics, the angular momentum of a particle possesses not only a magnitude (how fast the body is rotating), but also a direction (either up or down on the axis of rotation of the particle). Quantum mechanical spin also contains information about direction, but in a more subtle form. Quantum mechanics states that the component of angular momentum measured along any direction can only take on the values *[14]

$$S_i = \hbar s_i, \quad s_i \in \{-s, -(s-1), \ldots, s-1, s\}$$

where S_i is the spin component along the i-axis (either x, y, or z), s_i is the spin projection quantum number along the i-axis, and s is the principal spin quantum number (discussed in the previous section). Conventionally the direction chosen is the z-axis:

$$S_z = \hbar s_z, \quad s_z \in \{-s, -(s-1), \ldots, s-1, s\}$$

where S_z is the spin component along the z-axis, s_z is the spin projection quantum number along the z-axis.

One can see that there are $2s+1$ possible values of s_z. The number "$2s + 1$" is the multiplicity of the spin system. For example, there are only two possible values for a spin-1/2 particle: $s_z = +1/2$ and $s_z = -1/2$. These correspond to quantum states in which the spin is pointing in the +z or −z directions respectively, and are often referred to as "spin up" and "spin down". For a spin-3/2 particle, like a delta baryon, the possible values are +3/2, +1/2, −1/2, −3/2.

Vector

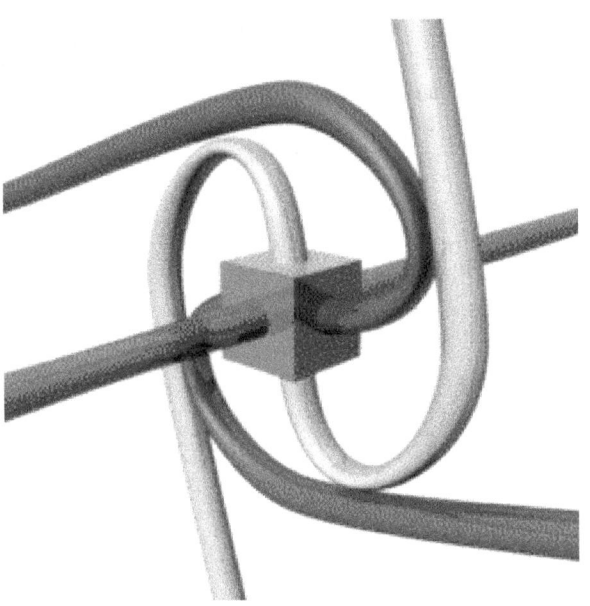

A single point in space can spin continuously without becoming tangled. Notice that after a 360 degree rotation, the spiral flips between clockwise and counterclockwise orientations. It returns to its original configuration after spinning a full 720 degrees.

For a given quantum state, one could think of a spin vector $\langle S \rangle$ whose components are the expectation values of the spin components along each axis, i.e., $\langle S \rangle = [\langle S_x \rangle, \langle S_y \rangle, \langle S_z \rangle]$. This vector then would describe the "direction" in which the spin is pointing, corresponding to the classical concept of the axis of rotation. It turns out that the spin vector is not very useful in actual quantum mechanical calculations, because it cannot be measured directly: s_x, s_y and s_z cannot possess simultaneous definite values, because of a quantum uncertainty relation between them. However, for statistically large collections of particles that have been placed in the same pure quantum state, such as through the use of a Stern–Gerlach apparatus, the spin vector does have a well-defined experimental meaning: It specifies the direction in ordinary space in which a subsequent detector must be oriented in order to achieve the maximum possible probability (100%) of detecting every particle in the collection. For spin-1/2 particles, this maximum probability drops off smoothly as the angle between the spin vector and the detector increases, until at an angle of 180 degrees—that is, for detectors oriented in the opposite direction to the spin vector—the expectation of detecting particles from the collection reaches a minimum of 0%.

4.6. SPIN

As a qualitative concept, the spin vector is often handy because it is easy to picture classically. For instance, quantum mechanical spin can exhibit phenomena analogous to classical gyroscopic effects. For example, one can exert a kind of "torque" on an electron by putting it in a magnetic field (the field acts upon the electron's intrinsic magnetic dipole moment—see the following section). The result is that the spin vector undergoes precession, just like a classical gyroscope. This phenomenon is known as electron spin resonance (ESR). The equivalent behaviour of protons in atomic nuclei is used in nuclear magnetic resonance (NMR) spectroscopy and imaging.

Mathematically, quantum mechanical spin states are described by vector-like objects known as spinors. There are subtle differences between the behavior of spinors and vectors under coordinate rotations. For example, rotating a spin-1/2 particle by 360 degrees does not bring it back to the same quantum state, but to the state with the opposite quantum phase; this is detectable, in principle, with interference experiments. To return the particle to its exact original state, one needs a 720 degree rotation. A spin-zero particle can only have a single quantum state, even after torque is applied. Rotating a spin-2 particle 180 degrees can bring it back to the same quantum state and a spin-4 particle should be rotated 90 degrees to bring it back to the same quantum state. The spin 2 particle can be analogous to a straight stick that looks the same even after it is rotated 180 degrees and a spin 0 particle can be imagined as sphere which looks the same after whatever angle it is turned through.

4.6.4 Mathematical formulation

Operator

Spin obeys commutation relations analogous to those of the orbital angular momentum:

$$[S_i, S_j] = i\hbar \epsilon_{ijk} S_k$$

where ϵ_{ijk} is the Levi-Civita symbol. It follows (as with angular momentum) that the eigenvectors of S^2 and S_z (expressed as kets in the total S basis) are:

$$S^2|s,m\rangle = \hbar^2 s(s+1)|s,m\rangle$$
$$S_z|s,m\rangle = \hbar m|s,m\rangle.$$

The spin raising and lowering operators acting on these eigenvectors give:

$$S_\pm|s,m\rangle = \hbar\sqrt{s(s+1) - m(m\pm 1)}|s, m\pm 1\rangle, \text{ where } S_\pm = S_x \pm iS_y.$$

But unlike orbital angular momentum the eigenvectors are not spherical harmonics. They are not functions of θ and φ. There is also no reason to exclude half-integer values of s and m.

In addition to their other properties, all quantum mechanical particles possess an intrinsic spin (though it may have the intrinsic spin 0, too). The spin is quantized in units of the reduced Planck constant, such that the state function of the particle is, say, not $\psi = \psi(\mathbf{r})$, but $\psi = \psi(\mathbf{r}, \sigma)$ where σ is out of the following discrete set of values:

$$\sigma \in \{-s\hbar, -(s-1)\hbar, \cdots, +(s-1)\hbar, +s\hbar\}.$$

One distinguishes bosons (integer spin) and fermions (half-integer spin). The total angular momentum conserved in interaction processes is then the *sum* of the orbital angular momentum and the spin.

Pauli matrices

The quantum mechanical operators associated with spin-$\frac{1}{2}$ observables are:

$$\hat{\mathbf{S}} = \frac{\hbar}{2}\sigma$$

where in Cartesian components:

$$S_x = \frac{\hbar}{2}\sigma_x, \quad S_y = \frac{\hbar}{2}\sigma_y, \quad S_z = \frac{\hbar}{2}\sigma_z.$$

For the special case of spin-1/2 particles, σ_x, σ_y and σ_z are the three Pauli matrices, given by:

$$\sigma_x = \begin{pmatrix} 0 & 1 \\ 1 & 0 \end{pmatrix} \quad \sigma_y = \begin{pmatrix} 0 & -i \\ i & 0 \end{pmatrix} \quad \sigma_z = \begin{pmatrix} 1 & 0 \\ 0 & -1 \end{pmatrix}.$$

Pauli exclusion principle

For systems of N identical particles this is related to the Pauli exclusion principle, which states that by interchanges of any two of the N particles one must have

$$\psi(\cdots \mathbf{r}_i, \sigma_i \cdots \mathbf{r}_j, \sigma_j \cdots) = (-1)^{2s}\psi(\cdots \mathbf{r}_j, \sigma_j \cdots \mathbf{r}_i, \sigma_i \cdots).$$

Thus, for bosons the prefactor $(-1)^*2s$ will reduce to $+1$, for fermions to -1. In quantum mechanics all particles are either bosons or fermions. In some speculative relativistic quantum field theories "supersymmetric" particles also exist, where linear combinations of bosonic and fermionic components appear. In two dimensions, the prefactor $(-1)^{*}2s$ can be replaced by any complex number of magnitude 1 such as in the Anyon.

The above permutation postulate for N-particle state functions has most-important consequences in daily life, e.g. the periodic table of the chemists or biologists.

Rotations

See also: symmetries in quantum mechanics

As described above, quantum mechanics states that components of angular momentum measured along any direction can only take a number of discrete values. The most convenient quantum mechanical description of particle's spin is therefore with a set of complex numbers corresponding to amplitudes of finding a given value of projection of its intrinsic angular momentum on a given axis. For instance, for a spin 1/2 particle, we would need two numbers $a_{\pm 1/2}$, giving amplitudes of finding it with projection of angular momentum equal to $\hbar/2$ and $-\hbar/2$, satisfying the requirement

$$\left|a_{\frac{1}{2}}\right|^2 + \left|a_{-\frac{1}{2}}\right|^2 = 1.$$

For a generic particle with spin s, we would need $2s + 1$ such parameters. Since these numbers depend on the choice of the axis, they transform into each other non-trivially when this axis is rotated. It's clear that the transformation law must be linear, so we can represent it by associating a matrix with each rotation, and the product of two transformation matrices corresponding to rotations A and B must be equal (up to phase) to the matrix representing rotation AB. Further, rotations preserve the quantum mechanical inner product, and so should our transformation matrices:

$$\sum_{m=-j}^{j} a_m^* b_m = \sum_{m=-j}^{j} \left(\sum_{n=-j}^{j} U_{nm} a_n\right)^* \left(\sum_{k=-j}^{j} U_{km} b_k\right)$$

$$\sum_{n=-j}^{j} \sum_{k=-j}^{j} U_{np}^* U_{kq} = \delta_{pq}.$$

Mathematically speaking, these matrices furnish a unitary projective representation of the rotation group SO(3). Each such representation corresponds to a representation of the covering group of SO(3), which is SU(2).[15] There is one n-dimensional irreducible representation of SU(2) for each dimension, though this representation is n-dimensional real for odd n and n-dimensional complex for even n (hence of real dimension $2n$). For a rotation by angle θ in the plane with normal vector $\hat{\boldsymbol{\theta}}$, U can be written

$$U = e^{-\frac{i}{\hbar}\boldsymbol{\theta}\cdot\mathbf{S}},$$

where $\boldsymbol{\theta} = \theta\hat{\boldsymbol{\theta}}$ is a and $\mathbf{S}$ is the vector of spin operators.

(Click "show" at right to see a proof or "hide" to hide it.)

Working in the coordinate system where $\hat{\boldsymbol{\theta}} = \hat{z}$, we would like to show that S_x and S_y are rotated into each other by the angle θ. Starting with S_x. Using units where $\hbar = 1$:

$$S_x \to U^\dagger S_x U = e^{i\theta S_z} S_x e^{-i\theta S_z}$$
$$= S_x + (i\theta)[S_z, S_x] + \left(\frac{1}{2!}\right)(i\theta)^2[S_z,[S_z,S_x]] + \ldots$$

Using the spin operator commutation relations, we see that the commutators evaluate to iS_y for the odd terms in the series, and to S_x for all of the even terms. Thus:

$$U^\dagger S_x U = S_x \left[1 - \frac{\theta^2}{2!} + \ldots\right] - S_y \left[\theta - \frac{\theta^3}{3!} \ldots\right]$$
$$= S_x \cos\theta - S_y \sin\theta$$

as expected. Note that since we only relied on the spin operator commutation relations, this proof holds for any dimension (i.e., for any principal spin quantum number s).*[16]

A generic rotation in 3-dimensional space can be built by compounding operators of this type using Euler angles:

$$\mathcal{R}(\alpha,\beta,\gamma) = e^{-i\alpha S_x} e^{-i\beta S_y} e^{-i\gamma S_z}$$

An irreducible representation of this group of operators is furnished by the Wigner D-matrix:

$$D_{m'm}^s(\alpha,\beta,\gamma) \equiv \langle sm'|\mathcal{R}(\alpha,\beta,\gamma)|sm\rangle = e^{-im'\alpha} d_{m'm}^s(\beta) e^{-im\gamma}$$

where

$$d_{m'm}^s(\beta) = \langle sm'|e^{-i\beta s_y}|sm\rangle$$

4.6. SPIN

is Wigner's small d-matrix. Note that for $\gamma = 2\pi$ and $\alpha = \beta = 0$; i.e., a full rotation about the z-axis, the Wigner D-matrix elements become

$$D^s_{m'm}(0,0,2\pi) = d^s_{m'm}(0)e^{-im2\pi} = \delta_{m'm}(-1)^{2m}.$$

Recalling that a generic spin state can be written as a superposition of states with definite m, we see that if s is an integer, the values of m are all integers, and this matrix corresponds to the identity operator. However, if s is a half-integer, the values of m are also all half-integers, giving $(-1)^*2m = -1$ for all m, and hence upon rotation by 2π the state picks up a minus sign. This fact is a crucial element of the proof of the spin-statistics theorem.

Lorentz transformations

We could try the same approach to determine the behavior of spin under general Lorentz transformations, but we would immediately discover a major obstacle. Unlike SO(3), the group of Lorentz transformations SO(3,1) is non-compact and therefore does not have any faithful, unitary, finite-dimensional representations.

In case of spin 1/2 particles, it is possible to find a construction that includes both a finite-dimensional representation and a scalar product that is preserved by this representation. We associate a 4-component Dirac spinor ψ with each particle. These spinors transform under Lorentz transformations according to the law

$$\psi' = \exp\left(\frac{1}{8}\omega_{\mu\nu}[\gamma_\mu,\gamma_\nu]\right)\psi$$

where γ_μ are gamma matrices and $\omega_{\mu\nu}$ is an antisymmetric 4×4 matrix parametrizing the transformation. It can be shown that the scalar product

$$\langle\psi|\phi\rangle = \bar{\psi}\phi = \psi^\dagger\gamma_0\phi$$

is preserved. It is not, however, positive definite, so the representation is not unitary.

Metrology along the x, y, and z axes

Each of the (Hermitian) Pauli matrices has two eigenvalues, +1 and −1. The corresponding normalized eigenvectors are:

$$\psi_{x+} = \frac{1}{\sqrt{2}}\begin{pmatrix}1\\1\end{pmatrix}, \quad \psi_{x-} = \frac{1}{\sqrt{2}}\begin{pmatrix}1\\-1\end{pmatrix},$$
$$\psi_{y+} = \frac{1}{\sqrt{2}}\begin{pmatrix}1\\i\end{pmatrix}, \quad \psi_{y-} = \frac{1}{\sqrt{2}}\begin{pmatrix}1\\-i\end{pmatrix},$$
$$\psi_{z+} = \begin{pmatrix}1\\0\end{pmatrix}, \quad \psi_{z-} = \begin{pmatrix}0\\1\end{pmatrix}.$$

By the postulates of quantum mechanics, an experiment designed to measure the electron spin on the x, y or z axis can only yield an eigenvalue of the corresponding spin operator (S_x, S_y or S_z) on that axis, i.e. $\hbar/2$ or $-\hbar/2$. The quantum state of a particle (with respect to spin), can be represented by a two component spinor:

$$\psi = \begin{pmatrix}a+bi\\c+di\end{pmatrix}.$$

When the spin of this particle is measured with respect to a given axis (in this example, the x-axis), the probability that its spin will be measured as $\hbar/2$ is just $|\langle\psi_{x+}|\psi\rangle|^2$. Correspondingly, the probability that its spin will be measured as $-\hbar/2$ is just $|\langle\psi_{x-}|\psi\rangle|^2$. Following the measurement, the spin state of the particle will collapse into the corresponding eigenstate. As a result, if the particle's spin along a given axis has been measured to have a given eigenvalue, all measurements will yield the same eigenvalue (since $|\langle\psi_{x+}|\psi_{x+}\rangle|^2 = 1$, etc), provided that no measurements of the spin are made along other axes.

Metrology along an arbitrary axis

The operator to measure spin along an arbitrary axis direction is easily obtained from the Pauli spin matrices. Let $u = (u_x, u_y, u_z)$ be an arbitrary unit vector. Then the operator for spin in this direction is simply

$$S_u = \frac{\hbar}{2}(u_x\sigma_x + u_y\sigma_y + u_z\sigma_z)$$

The operator S_u has eigenvalues of $\pm\hbar/2$, just like the usual spin matrices. This method of finding the operator for spin in an arbitrary direction generalizes to higher spin states, one takes the dot product of the direction with a vector of the three operators for the three x, y, z axis directions.

A normalized spinor for spin-1/2 in the (u_x, u_y, u_z) direction (which works for all spin states except spin down where it will give 0/0), is:

$$\frac{1}{\sqrt{2+2u_z}}\begin{pmatrix}1+u_z\\u_x+iu_y\end{pmatrix}.$$

The above spinor is obtained in the usual way by diagonalizing the σ_u matrix and finding the eigenstates corresponding to the eigenvalues. In quantum mechanics, vectors are termed "normalized" when multiplied by a normalizing factor, which results in the vector having a length of unity.

$$S_x = \frac{\hbar}{\sqrt{2}} \begin{pmatrix} 0 & 1 & 0 \\ 1 & 0 & 1 \\ 0 & 1 & 0 \end{pmatrix}$$

$$S_y = \frac{\hbar}{\sqrt{2}} \begin{pmatrix} 0 & -i & 0 \\ i & 0 & -i \\ 0 & i & 0 \end{pmatrix}$$

$$S_z = \hbar \begin{pmatrix} 1 & 0 & 0 \\ 0 & 0 & 0 \\ 0 & 0 & -1 \end{pmatrix}$$

Compatibility of metrology

Since the Pauli matrices do not commute, measurements of spin along the different axes are incompatible. This means that if, for example, we know the spin along the x-axis, and we then measure the spin along the y-axis, we have invalidated our previous knowledge of the x-axis spin. This can be seen from the property of the eigenvectors (i.e. eigenstates) of the Pauli matrices that:

for spin 3/2 they are

$$S_x = \frac{\hbar}{2} \begin{pmatrix} 0 & \sqrt{3} & 0 & 0 \\ \sqrt{3} & 0 & 2 & 0 \\ 0 & 2 & 0 & \sqrt{3} \\ 0 & 0 & \sqrt{3} & 0 \end{pmatrix}$$

$$S_y = \frac{\hbar}{2} \begin{pmatrix} 0 & -i\sqrt{3} & 0 & 0 \\ i\sqrt{3} & 0 & -2i & 0 \\ 0 & 2i & 0 & -i\sqrt{3} \\ 0 & 0 & i\sqrt{3} & 0 \end{pmatrix}$$

$$|\langle \psi_{x\pm} | \psi_{y\pm} \rangle|^2 = |\langle \psi_{x\pm} | \psi_{z\pm} \rangle|^2 = |\langle \psi_{y\pm} | \psi_{z\pm} \rangle|^2 = \frac{1}{2}.$$

$$S_z = \frac{\hbar}{2} \begin{pmatrix} 3 & 0 & 0 & 0 \\ 0 & 1 & 0 & 0 \\ 0 & 0 & -1 & 0 \\ 0 & 0 & 0 & -3 \end{pmatrix}$$

So when physicists measure the spin of a particle along the x-axis as, for example, $\hbar/2$, the particle's spin state collapses into the eigenstate $|\psi_{x+}\rangle$. When we then subsequently measure the particle's spin along the y-axis, the spin state will now collapse into either $|\psi_{y+}\rangle$ or $|\psi_{y-}\rangle$, each with probability 1/2. Let us say, in our example, that we measure $-\hbar/2$. When we now return to measure the particle's spin along the x-axis again, the probabilities that we will measure $\hbar/2$ or $-\hbar/2$ are each 1/2 (i.e. they are $|\langle \psi_{x+} | \psi_{y-} \rangle|^2$ and $|\langle \psi_{x-} | \psi_{y-} \rangle|^2$ respectively). This implies that the original measurement of the spin along the x-axis is no longer valid, since the spin along the x-axis will now be measured to have either eigenvalue with equal probability.

and for spin 5/2 they are

$$S_x = \frac{\hbar}{2} \begin{pmatrix} 0 & \sqrt{5} & 0 & 0 & 0 & 0 \\ \sqrt{5} & 0 & 2\sqrt{2} & 0 & 0 & 0 \\ 0 & 2\sqrt{2} & 0 & 3 & 0 & 0 \\ 0 & 0 & 3 & 0 & 2\sqrt{2} & 0 \\ 0 & 0 & 0 & 2\sqrt{2} & 0 & \sqrt{5} \\ 0 & 0 & 0 & 0 & \sqrt{5} & 0 \end{pmatrix}$$

$$S_y = \frac{\hbar}{2} \begin{pmatrix} 0 & -i\sqrt{5} & 0 & 0 & 0 & 0 \\ i\sqrt{5} & 0 & -2i\sqrt{2} & 0 & 0 & 0 \\ 0 & 2i\sqrt{2} & 0 & -3i & 0 & 0 \\ 0 & 0 & 3i & 0 & -2i\sqrt{2} & 0 \\ 0 & 0 & 0 & 2i\sqrt{2} & 0 & -i\sqrt{5} \\ 0 & 0 & 0 & 0 & i\sqrt{5} & 0 \end{pmatrix}$$

Higher spins

The spin-1/2 operator $\mathbf{S} = \hbar/2\boldsymbol{\sigma}$ form the fundamental representation of SU(2). By taking Kronecker products of this representation with itself repeatedly, one may construct all higher irreducible representations. That is, the resulting spin operators for higher spin systems in three spatial dimensions, for arbitrarily large s, can be calculated using this spin operator and ladder operators.

$$S_z = \frac{\hbar}{2} \begin{pmatrix} 5 & 0 & 0 & 0 & 0 & 0 \\ 0 & 3 & 0 & 0 & 0 & 0 \\ 0 & 0 & 1 & 0 & 0 & 0 \\ 0 & 0 & 0 & -1 & 0 & 0 \\ 0 & 0 & 0 & 0 & -3 & 0 \\ 0 & 0 & 0 & 0 & 0 & -5 \end{pmatrix}.$$

The resulting spin matrices for spin 1 are:

The generalization of these matrices for arbitrary s is

$$(S_x)_{ab} = \frac{\hbar}{2}(\delta_{a,b+1} + \delta_{a+1,b})\sqrt{(s+1)(a+b-1) - ab}$$
$$(S_y)_{ab} = \frac{\hbar}{2i}(\delta_{a,b+1} - \delta_{a+1,b})\sqrt{(s+1)(a+b-1) - ab}$$
$$(S_z)_{ab} = \hbar(s+1-a)\delta_{a,b} = \hbar(s+1-b)\delta_{a,b}.$$

Also useful in the quantum mechanics of multiparticle systems, the general Pauli group G_n is defined to consist of all n-fold tensor products of Pauli matrices.

The analog formula of Euler's formula in terms of the Pauli matrices:

$$e^{i\theta(\hat{\mathbf{n}} \cdot \boldsymbol{\sigma})} = I \cos\theta + i(\hat{\mathbf{n}} \cdot \boldsymbol{\sigma})\sin\theta$$

for higher spins is tractable, but less simple.*[17]

4.6.5 Parity

In tables of the spin quantum number s for nuclei or particles, the spin is often followed by a "+" or "−". This refers to the parity with "+" for even parity (wave function unchanged by spatial inversion) and "−" for odd parity (wave function negated by spatial inversion). For example, see the isotopes of bismuth.

4.6.6 Applications

Spin has important theoretical implications and practical applications. Well-established *direct* applications of spin include:

- Nuclear magnetic resonance (NMR) spectroscopy in chemistry;
- Electron spin resonance spectroscopy in chemistry and physics;
- Magnetic resonance imaging (MRI) in medicine, a type of applied NMR, which relies on proton spin density;
- Giant magnetoresistive (GMR) drive head technology in modern hard disks.

Electron spin plays an important role in magnetism, with applications for instance in computer memories. The manipulation of *nuclear spin* by radiofrequency waves (nuclear magnetic resonance) is important in chemical spectroscopy and medical imaging.

Spin-orbit coupling leads to the fine structure of atomic spectra, which is used in atomic clocks and in the modern definition of the second. Precise measurements of the g-factor of the electron have played an important role in the development and verification of quantum electrodynamics. *Photon spin* is associated with the polarization of light.

An emerging application of spin is as a binary information carrier in spin transistors. The original concept, proposed in 1990, is known as Datta-Das spin transistor.*[18] Electronics based on spin transistors are referred to as spintronics. The manipulation of spin in dilute magnetic semiconductor materials, such as metal-doped ZnO or TiO_2 imparts a further degree of freedom and has the potential to facilitate the fabrication of more efficient electronics.*[19]

There are many *indirect* applications and manifestations of spin and the associated Pauli exclusion principle, starting with the periodic table of chemistry.

4.6.7 History

Spin was first discovered in the context of the emission spectrum of alkali metals. In 1924 Wolfgang Pauli introduced what he called a "two-valued quantum degree of freedom" associated with the electron in the outermost shell. This allowed him to formulate the Pauli exclusion principle, stating that no two electrons can share the same quantum state at the same time.

Wolfgang Pauli

The physical interpretation of Pauli's "degree of freedom" was initially unknown. Ralph Kronig, one of Landé's assistants, suggested in early 1925 that it was produced by the self-rotation of the electron. When Pauli heard about the idea, he criticized it severely, noting that the electron's hypothetical surface would have to be moving faster than the speed of light in order for it to rotate quickly enough to produce the necessary angular momentum. This would violate the theory of relativity. Largely due to Pauli's criticism, Kronig decided not to publish his idea.

In the autumn of 1925, the same thought came to two Dutch physicists, George Uhlenbeck and Samuel Goudsmit at Leiden University. Under the advice of Paul Ehrenfest, they published their results. It met a favorable response, especially after Llewellyn Thomas managed to resolve a factor-of-two discrepancy between experimental results and Uhlenbeck and Goudsmit's calculations (and Kronig's unpublished results). This discrepancy was due to the orientation of the electron's tangent frame, in addition to its position.

Mathematically speaking, a fiber bundle description is needed. The tangent bundle effect is additive and relativistic; that is, it vanishes if c goes to infinity. It is one half of the value obtained without regard for the tangent space orientation, but with opposite sign. Thus the combined effect differs from the latter by a factor two (Thomas precession).

Despite his initial objections, Pauli formalized the theory of spin in 1927, using the modern theory of quantum mechanics invented by Schrödinger and Heisenberg. He pioneered the use of Pauli matrices as a representation of the spin operators, and introduced a two-component spinor wavefunction.

Pauli's theory of spin was non-relativistic. However, in 1928, Paul Dirac published the Dirac equation, which described the relativistic electron. In the Dirac equation, a four-component spinor (known as a "Dirac spinor") was used for the electron wave-function. In 1940, Pauli proved the *spin-statistics theorem*, which states that fermions have half-integer spin and bosons integer spin.

In retrospect, the first direct experimental evidence of the electron spin was the Stern–Gerlach experiment of 1922. However, the correct explanation of this experiment was only given in 1927.*[20]

4.6.8 See also

- Einstein–de Haas effect
- Spin-orbital
- Chirality (physics)
- Dynamic nuclear polarisation
- Helicity (particle physics)
- Pauli equation
- Pauli–Lubanski pseudovector
- Rarita–Schwinger equation
- Representation theory of SU(2)
- Spin-½
- Spin-flip
- Spin isomers of hydrogen
- Spin tensor
- Spin wave
- Spin engineering
- Yrast
- Zitterbewegung

4.6.9 References

[1] Merzbacher, Eugen (1998). *Quantum Mechanics* (3rd ed.). pp. 372–3.

[2] Griffiths, David (2005). *Introduction to Quantum Mechanics* (2nd ed.). pp. 183–4.

[3] "Angular Momentum Operator Algebra", class notes by Michael Fowler

[4] *A modern approach to quantum mechanics*, by Townsend, p. 31 and p. 80

[5] Eisberg, Robert; Resnick, Robert (1985). *Quantum Physics of Atoms, Molecules, Solids, Nuclei, and Particles* (2nd ed.). pp. 272–3.

[6] Pauli, Wolfgang (1940). "The Connection Between Spin and Statistics" (PDF). *Phys. Rev* **58** (8): 716–722. Bibcode:1940PhRv...58..716P. doi:10.1103/PhysRev.58.716.

[7] Physics of Atoms and Molecules, B.H. Bransden, C.J.Joachain, Longman, 1983, ISBN 0-582-44401-2

[8] "CODATA Value: electron g factor". *The NIST Reference on Constants, Units, and Uncertainty*. NIST. 2006. Retrieved 2013-11-15.

[9] R.P. Feynman (1985). "Electrons and Their Interactions". *QED: The Strange Theory of Light and Matter*. Princeton, New Jersey: Princeton University Press. p. 115. ISBN 0-691-08388-6.

"After some years, it was discovered that this value [−$g/2$] was not exactly 1, but slightly more—something like 1.00116. This correction was worked out for the first time in 1948 by Schwinger as $j*j$ divided by 2 pi [sic] [where j is the square root of the fine-structure constant], and was due to an alternative way the electron can go from place to place: instead of going directly from one point to another, the electron goes along for a while and suddenly emits a photon; then (horrors!) it absorbs its own photon."

[10] W.J. Marciano, A.I. Sanda (1977). "Exotic decays of the muon and heavy leptons in gauge theories". *Physics Letters* **B67** (3): 303–305. Bibcode:1977PhLB...67..303M. doi:10.1016/0370-2693(77)90377-X.

[11] B.W. Lee, R.E. Shrock (1977). "Natural suppression of symmetry violation in gauge theories: Muon- and electron-lepton-number nonconservation". *Physical Review* **D16** (5): 1444–1473. Bibcode:1977PhRvD..16.1444L. doi:10.1103/PhysRevD.16.1444.

[12] K. Fujikawa, R. E. Shrock (1980). "Magnetic Moment of a Massive Neutrino and Neutrino-Spin Rotation". *Physical Review Letters* **45** (12): 963–966. Bibcode:1980PhRvL..45..963F. doi:10.1103/PhysRevLett.45.963.

[13] N.F. Bell; Cirigliano, V.; Ramsey-Musolf, M.; Vogel, P.; Wise, Mark et al. (2005). "How Magnetic is the Dirac Neutrino?". *Physical Review Letters* **95** (15): 151802. arXiv:hep-ph/0504134. Bibcode:2005PhRvL..95o1802B. doi:10.1103/PhysRevLett.95.151802. PMID 16241715.

[14] Quanta: A handbook of concepts, P.W. Atkins, Oxford University Press, 1974, ISBN 0-19-855493-1

[15] B.C. Hall (2013). *Quantum Theory for Mathematicians*. Springer. pp. 354–358.

[16] *Modern Quantum Mechanics*, by J. J. Sakurai, p159

[17] Curtright, T L; Fairlie, D B; Zachos, C K (2014). "A compact formula for rotations as spin matrix polynomials". *SIGMA* **10**: 084. doi:10.3842/SIGMA.2014.084.

[18] Datta. S and B. Das (1990). "Electronic analog of the electrooptic modulator". *Applied Physics Letters* **56** (7): 665–667. Bibcode:1990ApPhL..56..665D. doi:10.1063/1.102730.

[19] Assadi, M.H.N; Hanaor, D.A.H (2013). "Theoretical study on copper's energetics and magnetism in TiO_2 polymorphs" (PDF). *Journal of Applied Physics* **113** (23): 233913. doi:10.1063/1.4811539.

[20] B. Friedrich, D. Herschbach (2003). "Stern and Gerlach: How a Bad Cigar Helped Reorient Atomic Physics". *Physics Today* **56** (12): 53. Bibcode:2003PhT....56l..53F. doi:10.1063/1.1650229.

4.6.10 Further reading

- Cohen-Tannoudji, Claude; Diu, Bernard; Laloë, Franck (2006). *Quantum Mechanics* (2 volume set ed.). John Wiley & Sons. ISBN 978-0-471-56952-7.

- Condon, E. U.; Shortley, G. H. (1935). "Especially Chapter 3". *The Theory of Atomic Spectra*. Cambridge University Press. ISBN 0-521-09209-4.

- Hipple, J. A.; Sommer, H.; Thomas, H.A. (1949). *A precise method of determining the faraday by magnetic resonance*. doi:10.1103/PhysRev.76.1877.2. https://www.academia.edu/6483539/John_A._Hipple_1911-1985_technology_as_knowledge

- Edmonds, A. R. (1957). *Angular Momentum in Quantum Mechanics*. Princeton University Press. ISBN 0-691-07912-9.

- Jackson, John David (1998). *Classical Electrodynamics* (3rd ed.). John Wiley & Sons. ISBN 978-0-471-30932-1.

- Serway, Raymond A.; Jewett, John W. (2004). *Physics for Scientists and Engineers* (6th ed.). Brooks/Cole. ISBN 0-534-40842-7.

- Thompson, William J. (1994). *Angular Momentum: An Illustrated Guide to Rotational Symmetries for Physical Systems*. Wiley. ISBN 0-471-55264-X.

- Tipler, Paul (2004). *Physics for Scientists and Engineers: Mechanics, Oscillations and Waves, Thermodynamics* (5th ed.). W. H. Freeman. ISBN 0-7167-0809-4.

- Sin-Itiro Tomonaga, The Story of Spin, 1997

4.6.11 External links

- "Spintronics. Feature Article" in *Scientific American*, June 2002.

- Goudsmit on the discovery of electron spin.

- *Nature*: "Milestones in 'spin' since 1896."

- ECE 495N Lecture 36: Spin Online lecture by S. Datta

4.7 Weak hypercharge

The **weak hypercharge** in particle physics is a quantum number relating the electric charge and the third component of weak isospin. It is conserved (only terms that are overall weak-hypercharge neutral are allowed in the Lagrangian) and is similar to the Gell-Mann–Nishijima formula for the hypercharge of strong interactions (which is not conserved in weak interactions). It is frequently denoted Y_W and corresponds to the gauge symmetry U(1).[1]

4.7.1 Definition

Weak hypercharge is the generator of the U(1) component of the electroweak gauge group, SU(2)×U(1) and its associated quantum field B mixes with the W^3 electroweak quantum field to produce the observed Z gauge boson and the photon of quantum electrodynamics.

Weak hypercharge, usually written as Y_W, satisfies the equality:

$$Q = T_3 + \frac{Y_W}{2}$$

where Q is the electrical charge (in elementary charge units) and T_3 is the third component of weak isospin. Rearranging, the weak hypercharge can be explicitly defined as:

$$Y_W = 2(Q - T_3)$$

Note: sometimes weak hypercharge is scaled so that

$$Y_W = Q - T_3$$

although this is a minority usage.[2]

Hypercharge assignments in the Standard Model are determined up to a twofold ambiguity by demanding cancellation of all anomalies.

4.7.2 Baryon and lepton number

Weak hypercharge is related to baryon number minus lepton number via:

$$X + 2Y_W = 5(B - L)$$

where X is a GUT-associated conserved quantum number. Since weak hypercharge is always conserved this implies that baryon number minus lepton number is also always conserved, within the Standard Model and most extensions.

Neutron decay

$$n \to p + e- + \nu_e$$

Hence neutron decay conserves baryon number B and lepton number L separately, so also the difference $B - L$ is conserved.

Proton decay

Proton decay is a prediction of many grand unification theories.

$$p+ \to e+ + \pi 0 \to e+ + 2\gamma$$

Hence proton decay conserves $B - L$, even though it violates both lepton number and baryon number conservation.

4.7.3 See also

- Standard Model (mathematical formulation)

4.7.4 Notes

[1] J. F. Donoghue, E. Golowich, B. R. Holstein (1994). *Dynamics of the standard model.* Cambridge University Press. p. 52. ISBN 0-521-47652-6.

[2] M. R. Anderson (2003). *The mathematical theory of cosmic strings.* CRC Press. p. 12. ISBN 0-7503-0160-0.

4.8 Weak isospin

In particle physics, **weak isospin** is a quantum number relating to the weak interaction, and parallels the idea of isospin under the strong interaction. Weak isospin is usually given the symbol T or I with the third component written as T_z, T_3, I_z or I_3.[1] Weak isospin is a complement of the weak hypercharge, which unifies weak interactions with electromagnetic interactions. It can be understood as the eigenvalue of a charge operator.

The **weak isospin conservation law** relates the conservation of T_3; all weak interactions must preserve T_3. It is also conserved by the other interactions and is therefore a conserved quantity in general. For this reason T_3 is more important than T and often the term "weak isospin" refers to the "3rd component of weak isospin".

4.8.1 Relation with chirality

Fermions with negative chirality (also called left-handed fermions) have $T = \frac{1}{2}$ and can be grouped into doublets with $T_3 = \pm\frac{1}{2}$ that behave the same way under the weak interaction. For example, up-type quarks (u, c, t) have $T_3 = +\frac{1}{2}$ and always transform into down-type quarks (d, s, b), which have $T_3 = -\frac{1}{2}$, and vice versa. On the other hand, a quark never decays weakly into a quark of the same T_3. Something similar happens with left-handed leptons, which exist as doublets containing a charged lepton (e−, μ−, τ−) with $T_3 = -\frac{1}{2}$ and a neutrino (ν

e, ν

μ, ν

τ) with $T_3 = \frac{1}{2}$.

Fermions with positive chirality (also called right-handed fermions) have $T = 0$ and form singlets that do not undergo weak interactions.

Electric charge, Q, is related to weak isospin, T_3, and weak hypercharge, Y_W, by

$$Q = T_3 + \frac{Y_W}{2}.$$

4.8.2 Weak isospin and the W bosons

The symmetry associated with spin is SU(2). This requires gauge bosons to transform between weak isospin charges: bosons W+, W− and W0. This implies that W bosons have a $T = 1$, with three different values of T_3.

- W+ boson ($T_3 = +1$) is emitted in transitions $\{(T_3 = +\frac{1}{2}) \to (T_3 = -\frac{1}{2})\}$,

- W− boson ($T_3 = -1$) is emitted in transitions $\{(T_3 = -\frac{1}{2}) \to (T_3 = +\frac{1}{2})\}$.

- W0 boson ($T_3 = 0$) would be emitted in reactions where T_3 does not change. However, under electroweak unification, the W0 boson mixes with the weak hypercharge gauge boson B, resulting in the observed Z0 boson and the photon of Quantum Electrodynamics.

4.8.3 See also

- Field theoretical formulation of standard model
- Weak hypercharge

4.8.4 References

[1] Ambiguities: I is also used as sign for the 'normal' isospin, same for the third component I_3 aka I_z. T is also used as the sign for Topness. This article uses T and T_3.

4.9 "B" − "L"

"B-L" redirects here. For other uses, see BL (disambiguation).

In high energy physics, **B − L** (pronounced "bee minus ell") is the difference between the baryon number (B) and the lepton number (L).

4.9.1 Details

This quantum number is the charge of a global/gauge U(1) symmetry in some Grand Unified Theory models, called $U(1)_{B-L}$. Unlike baryon number alone or lepton number alone, this hypothetical symmetry would not be broken by chiral anomalies or gravitational anomalies, as long as this symmetry is global, which is why this symmetry is often invoked. If $B − L$ exists as a symmetry, it has to be spontaneously broken to give the neutrinos a nonzero mass if we assume the seesaw mechanism. The gauge bosons associated to this symmetry are commonly called X and Y bosons.

The anomalies that would break baryon number conservation and lepton number conservation individually cancel in such a way that $B − L$ is always conserved. One hypothetical example is proton decay where a proton ($B = 1; L = 0$) would decay into a pion ($B = 0, L = 0$) and positron ($B = 0; L = −1$).

Weak hypercharge Y
W is related to $B − L$ via:

X + 2Y
W = 5(B − L)

where X is the U(1) symmetry Grand Unified Theory-associated conserved quantum number.

4.9.2 See also

- Baryogenesis
- Leptogenesis
- Majoron

- Proton decay
- X and Y bosons
- X (charge)
- Leptoquark

4.10 "X"

In particle physics, the **X-charge** (or simply X) is a conserved quantum number associated with the SO(10) grand unification theory.

X is related to the difference between the baryon number B and the lepton number L (that is $B - L$), and the weak hypercharge Y_W via the relation:

$$X = 5(B - L) - 2Y_W$$

4.10.1 See also

- Standard Model (mathematical formulation)
- Noether's theorem
- X and Y bosons

4.10.2 Notes

Chapter 5

Hadrons

5.1 Hadron

In particle physics, a **hadron** 🔊*i/ˈhædrɒn/ (Greek: ἁδρός, *hadrós*, "stout, thick") is a composite particle made of quarks held together by the strong force (in a similar way as molecules are held together by the electromagnetic force).

Hadrons are categorized into two families: baryons, made of three quarks, and mesons, made of one quark and one antiquark. Protons and neutrons are examples of baryons; pions are an example of a meson. Hadrons containing more than three valence quarks (exotic hadrons) have been discovered in recent years. A tetraquark state (an exotic meson), named the Z(4430)*−, was discovered in 2007 by the Belle Collaboration *[1] and confirmed as a resonance in 2014 by the LHCb collaboration.*[2] Two pentaquark states (exotic baryons), named P+c(4380) and P+c(4450), were discovered in 2015 by the LHCb collaboration.*[3] There are several more exotic hadron candidates, and other colour-singlet quark combinations may also exist.

Of the hadrons, protons are stable, and neutrons bound within atomic nuclei are stable. Other hadrons are unstable under ordinary conditions; free neutrons decay with a half-life of about 611 seconds. Experimentally, hadron physics is studied by colliding protons or nuclei of heavy elements such as lead, and detecting the debris in the produced particle showers.

5.1.1 Etymology

The term "hadron" was introduced by Lev B. Okun in a plenary talk at the 1962 International Conference on High Energy Physics.*[4] In this talk he said:

> Not withstanding the fact that this report deals with weak interactions, we shall frequently have to speak of strongly interacting particles. These particles pose not only numerous scientific problems, but also a terminological problem.

The point is that "strongly interacting particles" is a very clumsy term which does not yield itself to the formation of an adjective. For this reason, to take but one instance, decays into strongly interacting particles are called non-leptonic. This definition is not exact because "non-leptonic" may also signify "photonic". In this report I shall call strongly interacting particles "hadrons", and the corresponding decays "hadronic" (the Greek ἁδρός signifies "large" , "massive" , in contrast to λεπτός which means "small" , "light"). I hope that this terminology will prove to be convenient. —Lev B. Okun, 1962

5.1.2 Properties

According to the quark model,*[5] the properties of hadrons are primarily determined by their so-called *valence quarks*. For example, a proton is composed of two up quarks (each with electric charge $+2/3$, for a total of $+4/3$ together) and one down quark (with electric charge $-1/3$). Adding these together yields the proton charge of +1. Although quarks also carry color charge, hadrons must have zero total color charge because of a phenomenon called color confinement. That is, hadrons must be "colorless" or "white" . These are the simplest of the two ways: three quarks of different colors, or a quark of one color and an antiquark carrying the corresponding anticolor. Hadrons with the first arrangement are called baryons, and those with the second arrangement are mesons.

Hadrons, however, are not composed of just three or two quarks, because of the strength of the strong force. More accurately, strong force gluons have enough energy (E) to have resonances composed of massive (m) quarks ($E > mc^2$) . Thus, virtual quarks and antiquarks, in a 1:1 ratio, form the majority of massive particles inside a hadron. The two or three quarks are the excess of quarks vs. antiquarks in hadrons, and vice versa in anti-hadrons. Because the virtual quarks are not stable wave packets (quanta), but irregular and transient phenomena, it is not meaningful to ask which quark is real and which virtual; only the excess is apparent from the outside. Massless virtual gluons compose the numerical majority of particles inside hadrons.

Like all subatomic particles, hadrons are assigned quantum numbers corresponding to the representations of the Poincaré group: $J^P C(m)$, where J is the spin quantum number, P the intrinsic parity (or P-parity), and C, the charge conjugation (or C-parity), and the particle's mass, m. Note that the mass of a hadron has very little to do with the mass of its valence quarks; rather, due to mass–energy equivalence, most of the mass comes from the large amount of energy associated with the strong interaction. Hadrons may also carry flavor quantum numbers such as isospin (or G parity), and strangeness. All quarks carry an additive, conserved quantum number called a baryon number (B), which is $+1/3$ for quarks and $-1/3$ for antiquarks. This means that baryons (groups of three quarks) have $B = 1$ whereas mesons have $B = 0$.

Hadrons have excited states known as resonances. Each ground state hadron may have several excited states; several hundreds of resonances have been observed in particle physics experiments. Resonances decay extremely quickly (within about 10^*-24 seconds) via the strong nuclear force.

In other phases of matter the hadrons may disappear. For example, at very high temperature and high pressure, unless there are sufficiently many flavors of quarks, the theory of quantum chromodynamics (QCD) predicts that quarks and gluons will no longer be confined within hadrons, "because the strength of the strong interaction diminishes with energy". This property, which is known as asymptotic freedom, has been experimentally confirmed in the energy range between 1 GeV (gigaelectronvolt) and 1 TeV (teraelectronvolt).[*][6]

All free hadrons except the proton (and antiproton) are unstable.

5.1.3 Baryons

Main article: Baryon

All known baryons are made of three valence quarks, so they are fermions, *i.e.*, they have odd half-integral spin, because they have an odd number of quarks. As quarks possess baryon number $B = 1/3$, baryons have baryon number $B = 1$. The best-known baryons are the proton and the neutron.

One can hypothesise baryons with further quark-antiquark pairs in addition to their three quarks. Hypothetical baryons with one extra quark-antiquark pair (5 quarks in all) are called pentaquarks. As of August 2015, there are two known pentaquarks, P+c(4380) and P+c(4450), both discovered in 2015 by the LHCb collaboration.[*][3]

Each type of baryon has a corresponding antiparticle (antibaryon) in which quarks are replaced by their corresponding antiquarks. For example, just as a proton is made of two up-quarks and one down-quark, its corresponding antiparticle, the antiproton, is made of two up-antiquarks and one down-antiquark.

5.1.4 Mesons

Main article: Meson

Mesons are hadrons composed of a quark-antiquark pair. They are bosons, meaning they have integral spin, *i.e.*, 0, 1, or -1, as they have an even number of quarks. They have baryon number $B = 0$. Examples of mesons commonly produced in particle physics experiments include pions and kaons. Pions also play a role in holding atomic nuclei together via the residual strong force.

In principle, mesons with more than one quark-antiquark pair may exist; a hypothetical meson with two pairs is called a tetraquark. Several tetraquark candidates were found in the 2000s, but their status is under debate.[*][7] Several other hypothetical "exotic" mesons lie outside the quark model of classification. These include glueballs and hybrid mesons (mesons bound by excited gluons).

5.1.5 See also

- Hadronization, the formation of hadrons out of quarks and gluons
- Large Hadron Collider (LHC)
- List of particles
- Standard model
- Subatomic particles
- Hadron therapy, a.k.a. particle therapy
- Exotic hadrons

5.1.6 References

[1] Choi, S.-K.; Belle Collaboration et al. (2007). "Observation of a resonance-like structure in the π±Ψ′ mass distribution in exclusive B→Kπ±Ψ′ decays". *Physical Review Letters* **100** (14). arXiv:0708.1790. Bibcode:2008PhRvL.100n2001C. doi:10.1103/PhysRevLett.100.142001.

[2] LHCb collaboration (2014): Observation of the resonant character of the Z(4430)* – state

[3] R. Aaij et al. (LHCb collaboration) (2015). "Observation of J/ψp resonances consistent with pentaquark states in Λ0
b→J/ψK−
p decays". *Physical Review Letters* **115** (7). doi:10.1103/PhysRevLett.115.072001.

[4] Lev B. Okun (1962). "The Theory of Weak Interaction". *Proceedings of 1962 International Conference on High-Energy Physics at CERN*. Geneva. p. 845. Bibcode:1962hep..conf..845O.

[5] C. Amsler *et al.* (Particle Data Group) (2008). "Review of Particle Physics – Quark Model" (PDF). *Physics Letters B* **667**: 1. Bibcode:2008PhLB..667....1P. doi:10.1016/j.physletb.2008.07.018.

[6] S. Bethke (2007). "Experimental tests of asymptotic freedom". *Progress in Particle and Nuclear Physics* **58** (2): 351. arXiv:hep-ex/0606035. Bibcode:2007PrPNP..58..351B. doi:10.1016/j.ppnp.2006.06.001.

[7] Mysterious Subatomic Particle May Represent Exotic New Form of Matter

5.2 Baryon

Not to be confused with Baryonyx.

A **baryon** is a composite subatomic particle made up of three quarks (as distinct from mesons, which are composed of one quark and one antiquark). Baryons and mesons belong to the hadron family of particles, which are the quark-based particles. The name "baryon" comes from the Greek word for "heavy" (βαρύς, *barys*), because, at the time of their naming, most known elementary particles had lower masses than the baryons.

As quark-based particles, baryons participate in the strong interaction, whereas leptons, which are not quark-based, do not. The most familiar baryons are the protons and neutrons that make up most of the mass of the visible matter in the universe. Electrons (the other major component of the atom) are leptons.

Each baryon has a corresponding antiparticle (antibaryon) where quarks are replaced by their corresponding antiquarks. For example, a proton is made of two up quarks and one down quark; and its corresponding antiparticle, the antiproton, is made of two up antiquarks and one down antiquark.

5.2.1 Background

Baryons are strongly interacting fermions that is, they experience the strong nuclear force and are described by Fermi–Dirac statistics, which apply to all particles obeying the Pauli exclusion principle. This is in contrast to the bosons, which do not obey the exclusion principle.

Baryons, along with mesons, are hadrons, meaning they are particles composed of quarks. Quarks have baryon numbers of $B = 1/3$ and antiquarks have baryon number of $B = -1/3$. The term "baryon" usually refers to *triquarks*— baryons made of three quarks ($B = 1/3 + 1/3 + 1/3 = 1$).

Other exotic baryons have been proposed, such as pentaquarks—baryons made of four quarks and one antiquark ($B = 1/3 + 1/3 + 1/3 + 1/3 - 1/3 = 1$), but their existence is not generally accepted. In theory, heptaquarks (5 quarks, 2 antiquarks), nonaquarks (6 quarks, 3 antiquarks), etc. could also exist. Until recently, it was believed that some experiments showed the existence of pentaquarks— baryons made of four quarks and one antiquark.*[1]*[2] The particle physics community as a whole did not view their existence as likely in 2006,*[3] and in 2008, considered evidence to be overwhelmingly against the existence of the reported pentaquarks.*[4] However, in July 2015, the LHCb experiment observed two resonances consistent with pentaquark states in the Λ0
b → J/ψK−
p decay, with a combined statistical significance of 15σ.*[5]*[6]

5.2.2 Baryonic matter

Nearly all matter that may be encountered or experienced in everyday life is baryonic matter, which includes atoms of any sort, and provides those with the quality of mass. Non-baryonic matter, as implied by the name, is any sort of matter that is not composed primarily of baryons. Those might include neutrinos or free electrons, dark matter, such as supersymmetric particles, axions, or black holes.

The very existence of baryons is also a significant issue in cosmology because it is assumed that the Big Bang produced a state with equal amounts of baryons and antibaryons. The process by which baryons came to outnumber their antiparticles is called baryogenesis.

5.2.3 Baryogenesis

Main article: Baryogenesis

Experiments are consistent with the number of quarks in the universe being a constant and, to be more specific, the number of baryons being a constant ; in technical language, the total baryon number appears to be *conserved*. Within the prevailing Standard Model of particle physics, the number of baryons may change in multiples of three due to the action of sphalerons, although this is rare and has not been observed under experiment. Some grand unified theories of particle physics also predict that a single proton can decay, changing the baryon number by one; however, this has not yet been observed under experiment. The excess of baryons over antibaryons in the present universe is thought to be due to non-conservation of baryon number in the very early universe, though this is not well understood.

5.2.4 Properties

Isospin and charge

Main article: Isospin

The concept of isospin was first proposed by Werner Heisenberg in 1932 to explain the similarities between protons and neutrons under the strong interaction.[7] Although they had different electric charges, their masses were so similar that physicists believed they were actually the same particle. The different electric charges were explained as being the result of some unknown excitation similar to spin. This unknown excitation was later dubbed *isospin* by Eugene Wigner in 1937.[8]

This belief lasted until Murray Gell-Mann proposed the quark model in 1964 (containing originally only the u, d, and s quarks).[9] The success of the isospin model is now understood to be the result of the similar masses of the u and d quarks. Since the u and d quarks have similar masses, particles made of the same number then also have similar masses. The exact specific u and d quark composition determines the charge, as u quarks carry charge $+2/3$ while d quarks carry charge $-1/3$. For example the four Deltas all have different charges (Δ++ (uuu), Δ+ (uud), Δ0 (udd), Δ- (ddd)), but have similar masses (~1,232 MeV/c^2) as they are each made of a combination of three u and d quarks. Under the isospin model, they were considered to be a single particle in different charged states.

The mathematics of isospin was modeled after that of spin. Isospin projections varied in increments of 1 just like those of spin, and to each projection was associated a "charged state". Since the "Delta particle" had four "charged states", it was said to be of isospin $I = 3/2$. Its "charged states" Δ++, Δ+, Δ0, and Δ-, corresponded to the isospin projections $I_3 = +3/2$, $I_3 = +1/2$, $I_3 = -1/2$, and $I_3 = -3/2$, respectively. Another example is the "nucleon particle". As there were two nucleon "charged states", it was said to be of isospin $1/2$. The positive nucleon N+ (proton) was identified with $I_3 = +1/2$ and the neutral nucleon N0 (neutron) with $I_3 = -1/2$.[10] It was later noted that the isospin projections were related to the up and down quark content of particles by the relation:

$$I_3 = \frac{1}{2}[(n_u - n_{\bar{u}}) - (n_d - n_{\bar{d}})],$$

where the *n*'s are the number of up and down quarks and antiquarks.

In the "isospin picture", the four Deltas and the two nucleons were thought to be the different states of two particles. However in the quark model, Deltas are different states of nucleons (the N*++ or N*- are forbidden by Pauli's exclusion principle). Isospin, although conveying an inaccurate picture of things, is still used to classify baryons, leading to unnatural and often confusing nomenclature.

Flavour quantum numbers

Main article: Flavour (particle physics) § Flavour quantum numbers

The strangeness flavour quantum number S (not to be confused with spin) was noticed to go up and down along with particle mass. The higher the mass, the lower the strangeness (the more s quarks). Particles could be described with isospin projections (related to charge) and strangeness (mass) (see the uds octet and decuplet figures on the right). As other quarks were discovered, new quantum numbers were made to have similar description of udc and udb octets and decuplets. Since only the u and d mass are similar, this description of particle mass and charge in terms of isospin and flavour quantum numbers works well only for octet and decuplet made of one u, one d, and one other quark, and breaks down for the other octets and decuplets (for example, ucb octet and decuplet). If the quarks all had the same mass, their behaviour would be called *symmetric*, as they would all behave in exactly the same way with respect to the strong interaction. Since quarks do not have the same mass, they do not interact in the same way (exactly like an electron placed in an electric field will accelerate more than a proton placed in the same field because of its lighter mass), and the symmetry is said to be broken.

It was noted that charge (Q) was related to the isospin projection (I_3), the baryon number (B) and flavour quantum numbers (S, C, B', T) by the Gell-Mann–Nishijima formula:[10]

$$Q = I_3 + \frac{1}{2}(B + S + C + B' + T),$$

where S, C, B', and T represent the strangeness, charm, bottomness and topness flavour quantum numbers, respectively. They are related to the number of strange, charm, bottom, and top quarks and antiquark according to the relations:

$S = -(n_s - n_{\bar{s}})$,

$C = +(n_c - n_{\bar{c}})$,

$B' = -(n_b - n_{\bar{b}})$,

$T = +(n_t - n_{\bar{t}})$,

meaning that the Gell-Mann–Nishijima formula is equivalent to the expression of charge in terms of quark content:

$$Q = \frac{2}{3}[(n_u - n_{\bar{u}}) + (n_c - n_{\bar{c}}) + (n_t - n_{\bar{t}})] - \frac{1}{3}[(n_d - n_{\bar{d}}) + (n_s - n_{\bar{s}}) + (n_b - n_{\bar{b}})]$$

Spin, orbital angular momentum, and total angular momentum

Main articles: Spin (physics), Angular momentum operator, Quantum numbers and Clebsch–Gordan coefficients

Spin (quantum number S) is a vector quantity that represents the "intrinsic" angular momentum of a particle. It comes in increments of $\frac{1}{2}\hbar$ (pronounced "h-bar"). The $\hbar$ is often dropped because it is the "fundamental" unit of spin, and it is implied that "spin 1" means "spin 1 $\hbar$". In some systems of natural units, $\hbar$ is chosen to be 1, and therefore does not appear anywhere.

Quarks are fermionic particles of spin $\frac{1}{2}$ ($S = \frac{1}{2}$). Because spin projections vary in increments of 1 (that is 1 $\hbar$), a single quark has a spin vector of length $\frac{1}{2}$, and has two spin projections ($S_z = +\frac{1}{2}$ and $S_z = -\frac{1}{2}$). Two quarks can have their spins aligned, in which case the two spin vectors add to make a vector of length $S = 1$ and three spin projections ($S_z = +1$, $S_z = 0$, and $S_z = -1$). If two quarks have unaligned spins, the spin vectors add up to make a vector of length $S = 0$ and has only one spin projection ($S_z = 0$), etc. Since baryons are made of three quarks, their spin vectors can add to make a vector of length $S = \frac{3}{2}$, which has four spin projections ($S_z = +\frac{3}{2}$, $S_z = +\frac{1}{2}$, $S_z = -\frac{1}{2}$, and $S_z = -\frac{3}{2}$), or a vector of length $S = \frac{1}{2}$ with two spin projections ($S_z = +\frac{1}{2}$, and $S_z = -\frac{1}{2}$).[11]

There is another quantity of angular momentum, called the orbital angular momentum, (azimuthal quantum number L), that comes in increments of 1 $\hbar$, which represent the angular moment due to quarks orbiting around each other. The total angular momentum (total angular momentum quantum number J) of a particle is therefore the combination of intrinsic angular momentum (spin) and orbital angular momentum. It can take any value from $J = |L - S|$ to $J = |L + S|$, in increments of 1.

Particle physicists are most interested in baryons with no orbital angular momentum ($L = 0$), as they correspond to ground states—states of minimal energy. Therefore the two groups of baryons most studied are the $S = \frac{1}{2}$; $L = 0$ and $S = \frac{3}{2}$; $L = 0$, which corresponds to $J = \frac{1}{2}^+$ and $J = \frac{3}{2}^+$, respectively, although they are not the only ones. It is also possible to obtain $J = \frac{3}{2}^+$ particles from $S = \frac{1}{2}$ and $L = 2$, as well as $S = \frac{3}{2}$ and $L = 2$. This phenomenon of having multiple particles in the same total angular momentum configuration is called *degeneracy*. How to distinguish between these degenerate baryons is an active area of research in baryon spectroscopy.[12][13]

Parity

Main article: Parity (physics)

If the universe were reflected in a mirror, most of the laws of physics would be identical—things would behave the same way regardless of what we call "left" and what we call "right". This concept of mirror reflection is called *intrinsic parity* or *parity* (*P*). Gravity, the electromagnetic force, and the strong interaction all behave in the same way regardless of whether or not the universe is reflected in a mirror, and thus are said to conserve parity (P-symmetry). However, the weak interaction *does* distinguish "left" from "right", a phenomenon called parity violation (P-violation).

Based on this, one might think that, if the wavefunction for each particle (in more precise terms, the quantum field for each particle type) were simultaneously mirror-reversed, then the new set of wavefunctions would perfectly satisfy the laws of physics (apart from the weak interaction). It turns out that this is not quite true: In order for the equations to be satisfied, the wavefunctions of certain types of particles have to be multiplied by −1, in addition to being mirror-reversed. Such particle types are said to have *negative* or *odd* parity ($P = -1$, or alternatively $P = -$), while the other particles are said to have *positive* or *even* parity ($P = +1$, or alternatively $P = +$).

For baryons, the parity is related to the orbital angular momentum by the relation:[14]

$$P = (-1)^L.$$

As a consequence, baryons with no orbital angular momentum ($L = 0$) all have even parity ($P = +$).

5.2.5 Nomenclature

Baryons are classified into groups according to their isospin (I) values and quark (q) content. There are six groups of baryons—nucleon (N), Delta (Δ), Lambda (Λ), Sigma (Σ), Xi (Ξ), and Omega (Ω). The rules for classification are defined by the Particle Data Group. These rules consider the up (u), down (d) and strange (s) quarks to be *light* and the charm (c), bottom (b), and top (t) quarks to be *heavy*. The rules cover all the particles that can be made from three of each of the six quarks, even though baryons made of t quarks are not expected to exist because of the t quark's short lifetime. The rules do not cover pentaquarks.*[15]

- Baryons with three u and/or d quarks are N's ($I = \frac{1}{2}$) or Δ's ($I = \frac{3}{2}$).
- Baryons with two u and/or d quarks are Λ's ($I = 0$) or Σ's ($I = 1$). If the third quark is heavy, its identity is given by a subscript.
- Baryons with one u or d quark are Ξ's ($I = \frac{1}{2}$). One or two subscripts are used if one or both of the remaining quarks are heavy.
- Baryons with no u or d quarks are Ω's ($I = 0$), and subscripts indicate any heavy quark content.
- Baryons that decay strongly have their masses as part of their names. For example, Σ^0 does not decay strongly, but Δ^*++(1232) does.

It is also a widespread (but not universal) practice to follow some additional rules when distinguishing between some states that would otherwise have the same symbol.*[10]

- Baryons in total angular momentum $J = \frac{3}{2}$ configuration that have the same symbols as their $J = \frac{1}{2}$ counterparts are denoted by an asterisk (*).
- Two baryons can be made of three different quarks in $J = \frac{1}{2}$ configuration. In this case, a prime (′) is used to distinguish between them.
  - *Exception*: When two of the three quarks are one up and one down quark, one baryon is dubbed Λ while the other is dubbed Σ.

Quarks carry charge, so knowing the charge of a particle indirectly gives the quark content. For example, the rules above say that a Λ^+_c contains a c quark and some combination of two u and/or d quarks. The c quark has a charge of ($Q = +\frac{2}{3}$), therefore the other two must be a u quark ($Q = +\frac{2}{3}$), and a d quark ($Q = -\frac{1}{3}$) to have the correct total charge ($Q = +1$).

5.2.6 See also

- Eightfold way
- List of baryons
- List of particles
- Meson
- Timeline of particle discoveries

5.2.7 Notes

[1] H. Muir (2003)

[2] K. Carter (2003)

[3] W.-M. Yao *et al.* (2006): Particle listings – Θ^*+

[4] C. Amsler *et al.* (2008): Pentaquarks

[5] LHCb (14 July 2015). "Observation of particles composed of five quarks, pentaquark-charmonium states, seen in $\Lambda_b^0 \to J/\psi p K^*$- decays." . *CERN website*. Retrieved 2015-07-14.

[6] R. Aaij et al. (LHCb collaboration) (2015). "Observation of J/ψp resonances consistent with pentaquark states in $\Lambda0_b \to J/\psi K^-p$ decays". *Physical Review Letters* **115** (7). Bibcode:2015PhRvL.115g2001A. doi:10.1103/PhysRevLett.115.072001.

[7] W. Heisenberg (1932)

[8] E. Wigner (1937)

[9] M. Gell-Mann (1964)

[10] S.S.M. Wong (1998a)

[11] R. Shankar (1994)

[12] H. Garcilazo *et al.* (2007)

[13] D.M. Manley (2005)

[14] S.S.M. Wong (1998b)

[15] C. Amsler *et al.* (2008): Naming scheme for hadrons

5.2.8 References

- C. Amsler *et al.* (Particle Data Group) (2008). "Review of Particle Physics". *Physics Letters B* **667** (1): 1–1340. Bibcode:2008PhLB..667....1P. doi:10.1016/j.physletb.2008.07.018.

- H. Garcilazo, J. Vijande, and A. Valcarce (2007). "Faddeev study of heavy-baryon spectroscopy". *Journal of Physics G* **34** (5): 961–976. doi:10.1088/0954-3899/34/5/014.

- K. Carter (2006). "The rise and fall of the pentaquark". Fermilab and SLAC. Retrieved 2008-05-27.

- W.-M. Yao *et al.*(Particle Data Group) (2006). "Review of Particle Physics". *Journal of Physics G* **33**: 1–1232. arXiv:astro-ph/0601168. Bibcode:2006JPhG...33....1Y. doi:10.1088/0954-3899/33/1/001.

- D.M. Manley (2005). "Status of baryon spectroscopy". *Journal of Physics: Conference Series* **5**: 230–237. Bibcode:2005JPhCS...9..230M. doi:10.1088/1742-6596/9/1/043.

- H. Muir (2003). "Pentaquark discovery confounds sceptics". New Scientist. Retrieved 2008-05-27.

- S.S.M. Wong (1998a). "Chapter 2—Nucleon Structure". *Introductory Nuclear Physics* (2nd ed.). New York (NY): John Wiley & Sons. pp. 21–56. ISBN 0-471-23973-9.

- S.S.M. Wong (1998b). "Chapter 3—The Deuteron". *Introductory Nuclear Physics* (2nd ed.). New York (NY): John Wiley & Sons. pp. 57–104. ISBN 0-471-23973-9.

- R. Shankar (1994). *Principles of Quantum Mechanics* (2nd ed.). New York (NY): Plenum Press. ISBN 0-306-44790-8.

- E. Wigner (1937). "On the Consequences of the Symmetry of the Nuclear Hamiltonian on the Spectroscopy of Nuclei". *Physical Review* **51** (2): 106–119. Bibcode:1937PhRv...51..106W. doi:10.1103/PhysRev.51.106.

- M. Gell-Mann (1964). "A Schematic of Baryons and Mesons". *Physics Letters* **8** (3): 214–215. Bibcode:1964PhL.....8..214G. doi:10.1016/S0031-9163(64)92001-3.

- W. Heisenberg (1932). "Über den Bau der Atomkerne I". *Zeitschrift für Physik* (in German) **77**: 1–11. Bibcode:1932ZPhy...77....1H. doi:10.1007/BF01342433.

- W. Heisenberg (1932). "Über den Bau der Atomkerne II". *Zeitschrift für Physik* (in German) **78** (3–4): 156–164. Bibcode:1932ZPhy...78..156H. doi:10.1007/BF01337585.

- W. Heisenberg (1932). "Über den Bau der Atomkerne III". *Zeitschrift für Physik* (in German) **80** (9–10): 587–596. Bibcode:1933ZPhy...80..587H. doi:10.1007/BF01335696.

5.2.9 External links

- Particle Data Group—Review of Particle Physics (2008).

- Georgia State University—HyperPhysics

- Baryons made thinkable, an interactive visualisation allowing physical properties to be compared

5.3 List of baryons

Baryons are composite particles made of three quarks, as opposed to mesons, which are composite particles made of one quark and one antiquark. Baryons and mesons are both hadrons, which are particles composed solely of quarks or both quarks and antiquarks. The term *baryon* is derived from the Greek "βαρύς" (*barys*), meaning "heavy", because, at the time of their naming, it was believed that baryons were characterized by having greater masses than other particles that were classed as matter.

Until a few years ago, it was believed that some experiments showed the existence of pentaquarks – baryons made of four quarks and one antiquark.[1][2] The particle physics community as a whole did not view their existence as likely by 2006.[3] On 13 July 2015, the LHCb collaboration at CERN reported results consistent with pentaquark states in the decay of bottom Lambda baryons (Λ0b).[4]

Since baryons are composed of quarks, they participate in the strong interaction. Leptons, on the other hand, are not composed of quarks and as such do not participate in the strong interaction. The most famous baryons are the protons and neutrons that make up most of the mass of the visible matter in the universe, whereas electrons, the other major component of atoms, are leptons. Each baryon has a corresponding antiparticle known as an antibaryon in which quarks are replaced by their corresponding antiquarks. For example, a proton is made of two up quarks and one down quark, while its corresponding antiparticle, the antiproton, is made of two up antiquarks and one down antiquark.

5.3.1 Lists of baryons

These lists detail all known and predicted baryons in total angular momentum $J = \frac{1}{2}$ and $J = \frac{3}{2}$ configurations with positive parity.

- Baryons composed of one type of quark (uuu, ddd, ...) can exist in $J = \frac{3}{2}$ configuration, but $J = \frac{1}{2}$ is forbidden by the Pauli exclusion principle.

- Baryons composed of two types of quarks (uud, uus, ...) can exist in both $J = \frac{1}{2}$ and $J = \frac{3}{2}$ configurations

- Baryons composed of three types of quarks (uds, udc, ...) can exist in both $J = \frac{1}{2}$ and $J = \frac{3}{2}$ configurations. Two $J = \frac{1}{2}$ configurations are possible for these baryons.

The symbols encountered in these lists are: *I* (isospin), *J* (total angular momentum), *P* (parity), u (up quark), d (down quark), s (strange quark), c (charm quark), b (bottom quark), *Q* (charge), *B* (baryon number), *S* (strangeness), *C* (charm), *B′* (bottomness), as well as a wide array of subatomic particles (hover for name). (See the *baryon* article for a detailed explanation of these symbols.)

Antibaryons are not listed in the tables; however, they simply would have all quarks changed to antiquarks, and *Q*, *B*, *S*, *C*, *B′*, would be of opposite signs. Particles with *† next to their names have been predicted by the Standard Model but not yet observed. Values in red have not been firmly established by experiments, but are predicted by the quark model and are consistent with the measurements.*[5]*[6]

$J^*P = \frac{1}{2}^*$+ baryons

*†^ Particle has not yet been observed.

*[a] ^ The masses of the proton and neutron are known with much better precision in atomic mass units (u) than in MeV/c^2, due to the relatively poorly known value of the elementary charge. In atomic mass unit, the mass of the proton is 1.007276466812(90) u whereas that of the neutron is 1.00866491600(43) u.

*[b] ^ At least 10^{35} years. See proton decay.

*[c] ^ For free neutrons; in most common nuclei, neutrons are stable.

*[d] ^ PDG reports the resonance width (Γ). Here the conversion $\tau = $ *ℏ/Γ is given instead.

[e] ^ Some controversy exists about this data.[23]

$J^*P = \frac{3}{2}^*$+ baryons

*†^ Particle has not yet been observed.

*[h] ^ PDG reports the resonance width (Γ). Here the conversion $\tau = $ *ℏ/Γ is given instead.

Baryon resonance particles

This table gives the name, quantum numbers (where known), and experimental status of baryons resonances confirmed by the PDG.*[36] Baryon resonance particles are excited baryon states with short half lives and higher masses. Despite significant research, the fundamental degrees of freedom behind baryon excitation spectra are still poorly understood.*[37] The spin-parity J*P (when known) is given with each particle. For the strongly decaying particles, the J*P values are considered to be part of the names, as is the mass for all resonances.

5.3.2 See also

- Eightfold way (physics)
- List of mesons
- List of particles
- Timeline of particle discoveries

5.3.3 References

[1] H. Muir (2003)

[2] K. Carter (2003)

[3] W.-M. Yao *et al.* (2006): Particle listings – Positive Theta

[4] R. Aaij et al. (LHCb collaboration) (2015). "Observation of J/ψp resonances consistent with pentaquark states in Λ0 b→J/ψK− p decays". *Physical Review Letters* **115** (7). doi:10.1103/PhysRevLett.115.072001.

[5] J. Beringer *et al.* (2012) and 2013 partial update for the 2014 edition: Particle summary tables – Baryons

[6] J.G. Körner *et al.* (1994)

[7] J. Beringer *et al.* (2012): Particle listings – p+

[8] J. Beringer *et al.* (2012): Particle listings – n0

[9] J. Beringer *et al.* (2012): Particle listings – Λ

[10] J. Beringer *et al.* (2012): Particle listings – Λ c

[11] J. Beringer *et al.* (2012): Particle listings – Λ b

[12] J. Beringer *et al.* (2012): Particle listings – Σ+

[13] J. Beringer *et al.* (2012): Particle listings – Σ0

[14] J. Beringer *et al.* (2012): Particle listings – Σ−

5.3. LIST OF BARYONS

[15] J. Beringer *et al.* (2012): Particle listings – Σ_c

[16] J. Beringer *et al.* (2012): Particle listings – Σ_b

[17] J. Beringer *et al.* (2012): Particle listings – Ξ^0

[18] J. Beringer *et al.* (2012): Particle listings – Ξ^-

[19] J. Beringer *et al.* (2012): Particle listings – Ξ_c^+

[20] J. Beringer *et al.* (2012): Particle listings – Ξ_c^0

[21] J. Beringer *et al.* (2012): Particle listings – Ξ'^+_c

[22] J. Beringer *et al.* (2012): Particle listings – Ξ'^0_c

[23] J. Beringer *et al.* (2012): Particle listings – Ξ^+_{cc}

[24] J. Beringer *et al.* (2012): Particle listings – Ξ_b

[25] J. Beringer *et al.* (2012): Particle listings – Ω_c^0

[26] J. Beringer *et al.* (2012): Particle listings – Ω_b^-

[27] J. Beringer *et al.* (2012): Particle listings – $\Delta(1232)$

[28] J. Beringer *et al.* (2012): Particle listings – $\Sigma(1385)$

[29] J. Beringer *et al.* (2012): Particle listings – $\Sigma_c(2520)$

[30] J. Beringer *et al.* (2012): Particle listings – Σ^*_b

[31] J. Beringer *et al.* (2012): Particle listings – $\Xi(1530)$

[32] J. Beringer *et al.* (2012): Particle listings – $\Xi_c(2645)$

[33] J. Beringer *et al.* (2012): Particle listings – $\Xi^0_b(5945)$

[34] J. Beringer *et al.* (2012): Particle listings – Ω^-

[35] J. Beringer *et al.* (2012): Particle listings – $\Omega^0_c(2770)$

[36] http://pdg.lbl.gov/2014/tables/rpp2014-qtab-baryons.pdf

[37] Crede, Volker; Roberts, Winston (2013). "Progress Toward Understanding Baryon Resonances". *Rep. Prog. Phys.* **76**. doi:10.1088/0034-4885/76/7/076301. Retrieved 23 July 2015.

Bibliography

- R. Aaij et al. (LHCb collaboration) (2015). "Observation of J/ψp resonances consistent with pentaquark states in $\Lambda^0_b \rightarrow$ J/ψK$^-$p decays". arXiv:1507.03414 [hep-ex].

- J. Beringer *et al.* (Particle Data Group) (2012). "Review of Particle Physics". *Physical Review D* **86** (01): 010001. Bibcode:2012PhRvD...86a0001B. doi:10.1103/PhysRevD.86.010001.

- K. Nakamura *et al.* (Particle Data Group) (2010). "Review of Particle Physics". *Journal of Physics G* **37** (7A): 075021. Bibcode:2010JPhG...37g5021N. doi:10.1088/0954-3899/37/7A/075021.

- C. Amsler *et al.* (Particle Data Group) (2008). "Review of Particle Physics". *Physics Letters B* **667** (1): 1–1340. Bibcode:2008PhLB..667....1P. doi:10.1016/j.physletb.2008.07.018.

- V.M. Abazov (DØ Collaboration) (2008). "Observation of the doubly strange b baryon Ω^-_b" (PDF). Fermilab-Pub-08/335-E.

- K. Carter (2006). "The rise and fall of the pentaquark". *Symmetry Magazine*. Fermilab/SLAC. Retrieved 2008-05-27.

- W.-M. Yao *et al.* (Particle Data Group) (2006). "Review of Particle Physics". *Journal of Physics G* **33**: 1–1232. arXiv:astro-ph/0601168. Bibcode:2006JPhG...33....1Y. doi:10.1088/0954-3899/33/1/001.

- H. Muir (2003). "Pentaquark discovery confounds sceptics". *New Scientist*. Retrieved 2008-05-27.

- J.G. Körner, M. Krämer, and D. Pirjol (1994). "Heavy Baryons". *Progress in Particle and Nuclear Physics* **33**: 787–868. arXiv:hep-ph/9406359. Bibcode:1994PrPNP..33..787K. doi:10.1016/0146-6410(94)90053-1.

5.3.4 Further reading

- H. Garcilazo, J. Vijande, and A. Valcarce (2007). "Faddeev study of heavy-baryon spectroscopy". *Journal of Physics G* **34** (5): 961–976. doi:10.1088/0954-3899/34/5/014.

- S. Robbins (2006). "Physics Particle Overview – Baryons". *Journey Through the Galaxy*. Retrieved 2008-04-20.

- D.M. Manley (2005). "Status of baryon spectroscopy". *Journal of Physics: Conference Series* **5**: 230–237. Bibcode:2005JPhCS...9..230M. doi:10.1088/1742-6596/9/1/043.

- S.S.M. Wong (1998). *Introductory Nuclear Physics* (2nd ed.). New York (NY): John Wiley & Sons. ISBN 0-471-23973-9.

- R. Shankar (1994). *Principles of Quantum Mechanics* (2nd ed.). New York (NY): Plenum Press. ISBN 0-306-44790-8.

- E. Wigner (1937). "On the Consequences of the Symmetry of the Nuclear Hamiltonian on the Spectroscopy of Nuclei". *Physical Review* **51** (2): 106–119. Bibcode:1937PhRv...51..106W. doi:10.1103/PhysRev.51.106.

- M. Gell-Mann (1964). "A Schematic of Baryons and Mesons". *Physics Letters* **8** (3): 214–215. Bibcode:1964PhL.....8..214G. doi:10.1016/S0031-9163(64)92001-3.

- W. Heisenberg (1932). "Über den Bau der Atomkerne I". *Zeitschrift für Physik* (in German) **77**: 1–11. Bibcode:1932ZPhy...77....1H. doi:10.1007/BF01342433.

- W. Heisenberg (1932). "Über den Bau der Atomkerne II". *Zeitschrift für Physik* (in German) **78** (3–4): 156–164. Bibcode:1932ZPhy...78..156H. doi:10.1007/BF01337585.

- W. Heisenberg (1932). "Über den Bau der Atomkerne III". *Zeitschrift für Physik* (in German) **80** (9–10): 587–596. Bibcode:1933ZPhy...80..587H. doi:10.1007/BF01335696.

5.3.5 External links

- Particle Data Group – Review of Particle Physics (2008).

- Georgia State University – HyperPhysics

- Baryons made thinkable, an interactive visualisation allowing physical properties to be compared

5.4 Meson

In particle physics, **mesons** (/ˈmiːzɒnz/ or /ˈmɛzɒnz/) are hadronic subatomic particles composed of one quark and one antiquark, bound together by the strong interaction. Because mesons are composed of sub-particles, they have a physical size, with a diameter of roughly one fermi, which is about $2/3$ the size of a proton or neutron. All mesons are unstable, with the longest-lived lasting for only a few hundredths of a microsecond. Charged mesons decay (sometimes through intermediate particles) to form electrons and neutrinos. Uncharged mesons may decay to photons.

Mesons are not produced by radioactive decay, but appear in nature only as short-lived products of very high-energy interactions in matter, between particles made of quarks. In cosmic ray interactions, for example, such particles are ordinary protons and neutrons. Mesons are also frequently produced artificially in high-energy particle accelerators that collide protons, anti-protons, or other particles.

In nature, the importance of lighter mesons is that they are the associated quantum-field particles that transmit the nuclear force, in the same way that photons are the particles that transmit the electromagnetic force. The higher energy (more massive) mesons were created momentarily in the Big Bang, but are not thought to play a role in nature today. However, such particles are regularly created in experiments, in order to understand the nature of the heavier types of quark that compose the heavier mesons.

Mesons are part of the hadron particle family, defined simply as particles composed of two quarks. The other members of the hadron family are the baryons: subatomic particles composed of three quarks rather than two. Some experiments show evidence of exotic mesons, which don't have the conventional valence quark content of one quark and one antiquark.

Because quarks have a spin of $1/2$, the difference in quark-number between mesons and baryons results in conventional two-quark mesons being bosons, whereas baryons are fermions.

Each type of meson has a corresponding antiparticle (antimeson) in which quarks are replaced by their corresponding antiquarks and vice versa. For example, a positive pion (π+) is made of one up quark and one down antiquark; and its corresponding antiparticle, the negative pion (π−), is made of one up antiquark and one down quark.

Because mesons are composed of quarks, they participate in both the weak and strong interactions. Mesons with net electric charge also participate in the electromagnetic interaction. They are classified according to their quark content, total angular momentum, parity and various other properties, such as C-parity and G-parity. Although no meson is stable, those of lower mass are nonetheless more stable than the most massive mesons, and are easier to observe and study in particle accelerators or in cosmic ray experiments. They are also typically less massive than baryons, meaning that they are more easily produced in experiments, and thus exhibit certain higher energy phenomena more readily than baryons composed of the same quarks would. For example, the charm quark was first seen in the J/Psi meson (J/ψ) in 1974,[*][1][*][2] and the bottom quark in the upsilon meson (ϒ) in 1977.[*][3]

5.4.1 History

From theoretical considerations, in 1934 Hideki Yukawa[*][4][*][5] predicted the existence and the approximate mass of the "meson" as the carrier of the nuclear force that holds atomic nuclei together. If there were no nuclear force, all nuclei with two or more protons would fly apart because of the electromagnetic repulsion. Yukawa called his carrier particle the meson, from μέσος *mesos*, the Greek word for "intermediate," because its predicted mass was between that of the electron and that of the proton, which has about 1,836 times the mass of the electron. Yukawa had originally named his particle the "mesotron", but he was corrected by the physicist Werner Heisenberg (whose father was a professor of Greek at the University of Munich). Heisenberg pointed out that there is no "tr" in the Greek word "mesos".[*][6]

The first candidate for Yukawa's meson, now known in modern terminology as the muon, was discovered in 1936 by Carl David Anderson and others in the decay products of cosmic ray interactions. The mu meson had about the right mass to be Yukawa's carrier of the strong nuclear force, but over the course of the next decade, it became evident that it was not the right particle. It was eventually found that the "mu meson" did not participate in the strong nuclear interaction at all, but rather behaved like a heavy version of the electron, and was eventually classed as a lepton like the electron, rather than a meson. Physicists in making this choice decided that properties other than particle mass should control their classification.

There were years of delays in the subatomic particle research during World War II in 1939–45, with most physicists working in applied projects for wartime necessities. When the war ended in August 1945, many physicists gradually returned to peacetime research. The first true meson to be discovered was what would later be called the "pi meson" (or pion). This discovery was made in 1947, by Cecil Powell, César Lattes, and Giuseppe Occhialini, who were investigating cosmic ray products at the University of Bristol in England, based on photographic films placed in the Andes mountains. Some mesons in these films had about the same mass as the already-known meson, yet seemed to decay into it, leading physicist Robert Marshak to hypothesize in 1947 that it was actually a new and different meson. Over the next few years, more experiments showed that the pion was indeed involved in strong interactions. The pion (as a virtual particle) is the primary force carrier for the nuclear force in atomic nuclei. Other mesons, such as the rho mesons are involved in mediating this force as well, but to lesser extents. Following the discovery of the pion, Yukawa was awarded the 1949 Nobel Prize in Physics for his predictions.

The word *meson* has at times been used to mean *any* force carrier, such as the "Z^0 meson", which is involved in mediating the weak interaction.[*][7] However, this spurious usage has fallen out of favor. Mesons are now defined as particles composed of pairs of quarks and antiquarks.

5.4.2 Overview

Spin, orbital angular momentum, and total angular momentum

Main articles: Spin (physics), angular momentum operator, Total angular momentum and Quantum numbers

Spin (quantum number S) is a vector quantity that represents the "intrinsic" angular momentum of a particle. It comes in increments of $\frac{1}{2}$ ℏ. The ℏ is often dropped because it is the "fundamental" unit of spin, and it is implied that "spin 1" means "spin 1 ℏ". (In some systems of natural units, ℏ is chosen to be 1, and therefore does not appear in equations).

Quarks are fermions—specifically in this case, particles having spin $\frac{1}{2}$ ($S = \frac{1}{2}$). Because spin projections vary in increments of 1 (that is 1 ℏ), a single quark has a spin vector of length $\frac{1}{2}$, and has two spin projections ($S_z = +\frac{1}{2}$ and $S_z = -\frac{1}{2}$). Two quarks can have their spins aligned, in which case the two spin vectors add to make a vector of length $S = 1$ and three spin projections ($S_z = +1$, $S_z = 0$, and $S_z = -1$), called the spin-1 triplet. If two quarks have unaligned spins, the spin vectors add up to make a vector of length $S = 0$ and only one spin projection ($S_z = 0$), called the spin-0 singlet. Because mesons are made of one quark and one antiquark, they can be found in triplet and singlet spin states.

There is another quantity of quantized angular momentum, called the orbital angular momentum (quantum number L), that comes in increments of 1 ℏ, which represent the angular momentum due to quarks orbiting around each other. The total angular momentum (quantum number J) of a particle is therefore the combination of intrinsic angular momentum (spin) and orbital angular momentum. It can take any value from $J = |L - S|$ to $J = |L + S|$, in increments of 1.

Particle physicists are most interested in mesons with no orbital angular momentum ($L = 0$), therefore the two groups of mesons most studied are the $S = 1; L = 0$ and $S = 0; L = 0$, which corresponds to $J = 1$ and $J = 0$, although they are not the only ones. It is also possible to obtain $J = 1$ particles from $S = 0$ and $L = 1$. How to distinguish between the $S = 1$, $L = 0$ and $S = 0$, $L = 1$ mesons is an active area of research in meson spectroscopy.

Parity

Main article: Parity (physics)

If the universe were reflected in a mirror, most of the laws of physics would be identical—things would behave the same way regardless of what we call "left" and what we call "right". This concept of mirror reflection is called parity (P). Gravity, the electromagnetic force, and the strong interaction all behave in the same way regardless of whether or not the universe is reflected in a mirror, and thus are said to conserve parity (P-symmetry). However, the weak interaction does distinguish "left" from "right", a phenomenon called parity violation (P-violation).

Based on this, one might think that, if the wavefunction for each particle (more precisely, the quantum field for each particle type) were simultaneously mirror-reversed, then the new set of wavefunctions would perfectly satisfy the laws of physics (apart from the weak interaction). It turns out that this is not quite true: In order for the equations to be satisfied, the wavefunctions of certain types of particles have to be multiplied by −1, in addition to being mirror-reversed. Such particle types are said to have *negative* or *odd* parity ($P = -1$, or alternatively $P = -$), whereas the other particles are said to have *positive* or *even* parity ($P = +1$, or alternatively $P = +$).

For mesons, the parity is related to the orbital angular momentum by the relation:*[8]

$$P = (-1)^{L+1}$$

where the L is a result of the parity of the corresponding spherical harmonic of the wavefunction. The '+1' in the exponent comes from the fact that, according to the Dirac equation, a quark and an antiquark have opposite intrinsic parities. Therefore, the intrinsic parity of a meson is the product of the intrinsic parities of the quark (+1) and antiquark (−1). As these are different, their product is −1, and so it contributes a +1 in the exponent.

As a consequence, mesons with no orbital angular momentum ($L = 0$) all have odd parity ($P = -1$).

C-parity

Main article: C-parity

C-parity is only defined for mesons that are their own antiparticle (i.e. neutral mesons). It represents whether or not the wavefunction of the meson remains the same under the interchange of their quark with their antiquark.*[9] If

$$|q\bar{q}\rangle = |\bar{q}q\rangle$$

then, the meson is "C even" (C = +1). On the other hand, if

$$|q\bar{q}\rangle = -|\bar{q}q\rangle$$

then the meson is "C odd" (C = −1).

C-parity rarely is studied on its own, but more commonly in combination with P-parity into CP-parity. CP-parity was thought to be conserved, but was later found to be violated in weak interactions.*[10]*[11]*[12]

G-parity

Main article: G-parity

G parity is a generalization of the C-parity. Instead of simply comparing the wavefunction after exchanging quarks and antiquarks, it compares the wavefunction after exchanging the meson for the corresponding antimeson, regardless of quark content.*[13] In the case of neutral meson, G-parity is equivalent to C-parity because neutral mesons are their own antiparticles.

If

$$|q_1\bar{q_2}\rangle = |\bar{q_1}q_2\rangle$$

then, the meson is "G even" (G = +1). On the other hand, if

$$|q_1\bar{q_2}\rangle = -|\bar{q_1}q_2\rangle$$

then the meson is "G odd" (G = −1).

Isospin and charge

Main article: Isospin

The concept of isospin was first proposed by Werner

Heisenberg in 1932 to explain the similarities between protons and neutrons under the strong interaction.*[14] Although they had different electric charges, their masses were so similar that physicists believed that they were actually the same particle. The different electric charges were explained as being the result of some unknown excitation similar to spin. This unknown excitation was later dubbed *isospin* by Eugene Wigner in 1937.*[15] When the first mesons were discovered, they too were seen through the eyes of isospin and so the three pions were believed to be the same particle, but in different isospin states.

This belief lasted until Murray Gell-Mann proposed the quark model in 1964 (containing originally only the u, d, and s quarks).*[16] The success of the isospin model is now understood to be the result of the similar masses of the u and d quarks. Because the u and d quarks have similar masses, particles made of the same number of them also have similar masses. The exact specific u and d quark composition determines the charge, because u quarks carry charge $+^2/_3$ whereas d quarks carry charge $-^1/_3$. For example the three pions all have different charges (π+ (ud), π0 (a quantum superposition of uu and dd states), π− (du)), but have similar masses (~140 MeV/c^2) as they are each made of a same number of total of up and down quarks and antiquarks. Under the isospin model, they were considered to be a single particle in different charged states.

The mathematics of isospin was modeled after that of spin. Isospin projections varied in increments of 1 just like those of spin, and to each projection was associated a "charged state". Because the "pion particle" had three "charged states", it was said to be of isospin $I = 1$. Its "charged states" π+, π0, and π−, corresponded to the isospin projections $I_3 = +1$, $I_3 = 0$, and $I_3 = -1$ respectively. Another example is the "rho particle", also with three charged states. Its "charged states" ρ+, ρ0, and ρ−, corresponded to the isospin projections $I_3 = +1$, $I_3 = 0$, and $I_3 = -1$ respectively. It was later noted that the isospin projections were related to the up and down quark content of particles by the relation

$$I_3 = \frac{1}{2}[(n_u - n_{\bar{u}}) - (n_d - n_{\bar{d}})],$$

where the *n*'s are the number of up and down quarks and antiquarks.

In the "isospin picture", the three pions and three rhos were thought to be the different states of two particles. However, in the quark model, the rhos are excited states of pions. Isospin, although conveying an inaccurate picture of things, is still used to classify hadrons, leading to unnatural and often confusing nomenclature. Because mesons are hadrons, the isospin classification is also used, with $I_3 = +^1/_2$ for up quarks and down antiquarks, and $I_3 = -^1/_2$ for up antiquarks and down quarks.

Flavour quantum numbers

Main article: Flavour (particle physics) § Flavour quantum numbers

The strangeness quantum number S (not to be confused with spin) was noticed to go up and down along with particle mass. The higher the mass, the lower the strangeness (the more s quarks). Particles could be described with isospin projections (related to charge) and strangeness (mass) (see the uds nonet figures). As other quarks were discovered, new quantum numbers were made to have similar description of udc and udb nonets. Because only the u and d mass are similar, this description of particle mass and charge in terms of isospin and flavour quantum numbers only works well for the nonets made of one u, one d and one other quark and breaks down for the other nonets (for example ucb nonet). If the quarks all had the same mass, their behaviour would be called *symmetric*, because they would all behave in exactly the same way with respect to the strong interaction. However, as quarks do not have the same mass, they do not interact in the same way (exactly like an electron placed in an electric field will accelerate more than a proton placed in the same field because of its lighter mass), and the symmetry is said to be broken.

It was noted that charge (Q) was related to the isospin projection (I_3), the baryon number (B) and flavour quantum numbers (S, C, B', T) by the Gell-Mann–Nishijima formula:*[17]

$$Q = I_3 + \frac{1}{2}(B + S + C + B' + T),$$

where S, C, B', and T represent the strangeness, charm, bottomness and topness flavour quantum numbers respectively. They are related to the number of strange, charm, bottom, and top quarks and antiquark according to the relations:

$$S = -(n_s - n_{\bar{s}})$$
$$C = +(n_c - n_{\bar{c}})$$
$$B' = -(n_b - n_{\bar{b}})$$
$$T = +(n_t - n_{\bar{t}}),$$

meaning that the Gell-Mann–Nishijima formula is equivalent to the expression of charge in terms of quark content:

$$Q = \frac{2}{3}[(n_u - n_{\bar{u}}) + (n_c - n_{\bar{c}}) + (n_t - n_{\bar{t}})] - \frac{1}{3}[(n_d - n_{\bar{d}}) + (n_s - n_{\bar{s}}) + (n_b - n_{\bar{b}})]$$

5.4.3 Classification

Mesons are classified into groups according to their isospin (I), total angular momentum (J), parity (P), G-parity (G) or C-parity (C) when applicable, and quark (q) content. The rules for classification are defined by the Particle Data Group, and are rather convoluted.*[18] The rules are presented below, in table form for simplicity.

Types of meson

Mesons are classified into types according to their spin configurations. Some specific configurations are given special names based on the mathematical properties of their spin configuration.

Nomenclature

Flavourless mesons Flavourless mesons are mesons made of pair of quark and antiquarks of the same flavour (all their flavour quantum numbers are zero: $S = 0$, $C = 0$, $B' = 0$, $T = 0$).*[20] The rules for flavourless mesons are:*[18]

> *† ^ The C parity is only relevant to neutral mesons.
> *†† ^ For $J^{PC}=1^{--}$, the ψ is called the J/ψ

In addition:

- When the spectroscopic state of the meson is known, it is added in parentheses.
- When the spectroscopic state is unknown, mass (in MeV/c^2) is added in parentheses.
- When the meson is in its ground state, nothing is added in parentheses.

Flavoured mesons Flavoured mesons are mesons made of pair of quark and antiquarks of different flavours. The rules are simpler in this case: the main symbol depends on the heavier quark, the superscript depends on the charge, and the subscript (if any) depends on the lighter quark. In table form, they are:*[18]

In addition:

- If J^*P is in the "normal series" (i.e., $J^*P = 0^*+$, 1^*-, 2^*+, 3^*-, ...), a superscript $*$ is added.

- If the meson is not pseudoscalar ($J^*P = 0^*-$) or vector ($J^*P = 1^*-$), J is added as a subscript.
- When the spectroscopic state of the meson is known, it is added in parentheses.
- When the spectroscopic state is unknown, mass (in MeV/c^2) is added in parentheses.
- When the meson is in its ground state, nothing is added in parentheses.

5.4.4 Exotic mesons

Main article: Exotic meson

There is experimental evidence for particles that are hadrons (i.e., are composed of quarks) and are color-neutral with zero baryon number, and thus by conventional definition are mesons. Yet, these particles do not consist of a single quark-antiquark pair, as all the other conventional mesons discussed above do. A tentative category for these particles is exotic mesons.

There are at least five exotic meson resonances that have been experimentally confirmed to exist by two or more independent experiments. The most statistically significant of these is the Z(4430), discovered by the Belle experiment in 2007 and confirmed by LHCb in 2014. It is a candidate for being a tetraquark: a particle composed of two quarks and two antiquarks.*[21] See the main article above for other particle resonances that are candidates for being exotic mesons.

5.4.5 List

Main article: List of mesons

5.4.6 See also

- Standard Model

5.4.7 Notes

[1] J.J. Aubert *et al.* (1974)

[2] J.E. Augustin *et al.* (1974)

[3] S.W. Herb *et al.* (1977)

[4] The Noble Foundation (1949) Nobel Prize in Physics 1949 – Presentation Speech

[5] H. Yukawa (1935)

[6] G. Gamow (1961)

[7] J. Steinberger (1998)

[8] C. Amsler *et al.* (2008): Quark Model

[9] M.S. Sozzi (2008b)

[10] J.W. Cronin (1980)

[11] V.L. Fitch (1980)

[12] M.S. Sozzi (2008c)

[13] K. Gottfried, V.F. Weisskopf (1986)

[14] W. Heisenberg (1932)

[15] E. Wigner (1937)

[16] M. Gell-Mann (1964)

[17] S.S.M Wong (1998)

[18] C. Amsler *et al.* (2008): Naming scheme for hadrons

[19] W.E. Burcham, M. Jobes (1995)

[20] For the purpose of nomenclature, the isospin projection I_3 isn't considered a flavour quantum number. This means that the charged pion-like mesons ($\pi^*\pm$, $a^*\pm$, $b^*\pm$, and $\rho^*\pm$ mesons) follow the rules of flavourless mesons, even if they aren't truly "flavourless".

[21] LHCb collaborators (2014): Observation of the resonant character of the Z(4430)− state

5.4.8 References

- M.S. Sozzi (2008a). "Parity". *Discrete Symmetries and CP Violation: From Experiment to Theory*. Oxford University Press. pp. 15–87. ISBN 0-19-929666-9.

- M.S. Sozzi (2008b). "Charge Conjugation". *Discrete Symmetries and CP Violation: From Experiment to Theory*. Oxford University Press. pp. 88–120. ISBN 0-19-929666-9.

- M.S. Sozzi (2008c). "CP-Symmetry". *Discrete Symmetries and CP Violation: From Experiment to Theory*. Oxford University Press. pp. 231–275. ISBN 0-19-929666-9.

- C. Amsler *et al.* (Particle Data Group) (2008). "Review of Particle Physics". *Physics Letters B* **667** (1): 1–1340. Bibcode:2008PhLB..667....1P. doi:10.1016/j.physletb.2008.07.018.

- S.S.M. Wong (1998). "Nucleon Structure". *Introductory Nuclear Physics* (2nd ed.). New York (NY): John Wiley & Sons. pp. 21–56. ISBN 0-471-23973-9.

- W.E. Burcham, M. Jobes (1995). *Nuclear and Particle Physics* (2nd ed.). Longman Publishing. ISBN 0-582-45088-8.

- R. Shankar (1994). *Principles of Quantum Mechanics* (2nd ed.). New York (NY): Plenum Press. ISBN 0-306-44790-8.

- J. Steinberger (1989). "Experiments with high-energy neutrino beams". *Reviews of Modern Physics* **61** (3): 533–545. Bibcode:1989RvMP...61..533S. doi:10.1103/RevModPhys.61.533.

- K. Gottfried, V.F. Weisskopf (1986). "Hadronic Spectroscopy: G-parity". *Concepts of Particle Physics* **2**. Oxford University Press. pp. 303–311. ISBN 0-19-503393-0.

- J.W. Cronin (1980). "CP Symmetry Violation—The Search for its origin" (PDF). The Nobel Foundation.

- V.L. Fitch (1980). "The Discovery of Charge—Conjugation Parity Asymmetry" (PDF). The Nobel Foundation.

- S.W. Herb; Hom, D.; Lederman, L.; Sens, J.; Snyder, H.; Yoh, J.; Appel, J.; Brown, B. et al. (1977). "Observation of a Dimuon Resonance at 9.5 Gev in 400-GeV Proton-Nucleus Collisions". *Physical Review Letters* **39** (5): 252–255. Bibcode:1977PhRvL..39..252H. doi:10.1103/PhysRevLett.39.252.

- J.J. Aubert; Becker, U.; Biggs, P.; Burger, J.; Chen, M.; Everhart, G.; Goldhagen, P.; Leong, J. et al. (1974). "Experimental Observation of a Heavy Particle J". *Physical Review Letters* **33** (23): 1404–1406. Bibcode:1974PhRvL..33.1404A. doi:10.1103/PhysRevLett.33.1404.

- J.E. Augustin; Boyarski, A.; Breidenbach, M.; Bulos, F.; Dakin, J.; Feldman, G.; Fischer, G.; Fryberger, D. et al. (1974). "Discovery of a Narrow Resonance in e^*+e^*- Annihilation". *Physical Review Letters* **33** (23): 1406–1408. Bibcode:1974PhRvL..33.1406A. doi:10.1103/PhysRevLett.33.1406.

- M. Gell-Mann (1964). "A Schematic of Baryons and Mesons". *Physics Letters* **8** (3): 214–215. Bibcode:1964PhL......8..214G. doi:10.1016/S0031-9163(64)92001-3.

- Ishfaq Ahmad (1965). "the Interactions of 200 MeV $\pi\pm$ -Mesons with Complex Nuclei Proposal to Study the Interactions of 200 MeV $\pi\pm$ -Mesons with Complex Nuclei" (PDF). *CERN documents* **3** (5).

- G. Gamow (1988) [1961]. *The Great Physicists from Galileo to Einstein* (Reprint ed.). Dover Publications. p. 315. ISBN 978-0-486-25767-9.

- E. Wigner (1937). "On the Consequences of the Symmetry of the Nuclear Hamiltonian on the Spectroscopy of Nuclei". *Physical Review* **51** (2): 106–119. Bibcode:1937PhRv...51..106W. doi:10.1103/PhysRev.51.106.

- H. Yukawa (1935). "On the Interaction of Elementary Particles" (PDF). *Proc. Phys. Math. Soc. Jap.* **17** (48).

- W. Heisenberg (1932). "Über den Bau der Atomkerne I". *Zeitschrift für Physik* (in German) **77**: 1–11. Bibcode:1932ZPhy...77....1H. doi:10.1007/BF01342433.

- W. Heisenberg (1932). "Über den Bau der Atomkerne II". *Zeitschrift für Physik* (in German) **78** (3–4): 156–164. Bibcode:1932ZPhy...78..156H. doi:10.1007/BF01337585.

- W. Heisenberg (1932). "Über den Bau der Atomkerne III". *Zeitschrift für Physik* (in German) **80** (9–10): 587–596. Bibcode:1933ZPhy...80..587H. doi:10.1007/BF01335696.

5.4.9 External links

- A table of some mesons and their properties

- *Particle Data Group*—Compiles authoritative information on particle properties

- hep-ph/0211411: The light scalar mesons within quark models

- Naming scheme for hadrons (a PDF file)

- Mesons made thinkable, an interactive visualisation allowing physical properties to be compared

Recent findings

- What Happened to the Antimatter? Fermilab's DZero Experiment Finds Clues in Quick-Change Meson

- CDF experiment's definitive observation of matter-antimatter oscillations in the Bs meson

5.5 List of mesons

This list is of all known and predicted scalar, pseudoscalar and vector mesons. See list of particles for a more detailed list of particles found in particle physics.

Mesons are unstable subatomic particles composed of one quark and one antiquark. They are part of the hadron particle family – particles made of quarks. The other members of the hadron family are the baryons – subatomic particles composed of three quarks. The main difference between mesons and baryons is that mesons have integer spin (thus are bosons) while baryons are fermions (half-integer spin). Because mesons are bosons, the Pauli exclusion principle does not apply to them. Because of this, they can act as force mediating particles on short distances, and thus play a part in processes such as the nuclear interaction.

Since mesons are composed of quarks, they participate in both the weak and strong interactions. Mesons with net electric charge also participate in the electromagnetic interaction. They are classified according to their quark content, total angular momentum, parity, and various other properties such as C-parity and G-parity. While no meson is stable, those of lower mass are nonetheless more stable than the most massive mesons, and are easier to observe and study in particle accelerators or in cosmic ray experiments. They are also typically less massive than baryons, meaning that they are more easily produced in experiments, and will exhibit higher-energy phenomena sooner than baryons would. For example, the charm quark was first seen in the J/Psi meson (J/ψ) in 1974,[1][2] and the bottom quark in the upsilon meson (ϒ) in 1977.[3]

Each meson has a corresponding antiparticle (antimeson) where quarks are replaced by their corresponding antiquarks and vice versa. For example, a positive pion (π+) is made of one up quark and one down antiquark; and its corresponding antiparticle, the negative pion (π−), is made of one up antiquark and one down quark. Some experiments show the evidence of *tetraquarks* – "exotic" mesons made of two quarks and two antiquarks, but the particle physics community as a whole does not view their existence as likely, although still possible.[4]

The symbols encountered in these lists are: I (*isospin*), J (*total angular momentum*), P (*parity*), C (*C-parity*), G (*G-parity*), u (*up quark*), d (*down quark*), s (*strange quark*), c (*charm quark*), b (*bottom quark*), Q (*charge*), B (*baryon number*), S (*strangeness*), C (*charm*), and B′ (*bottomness*), as well as a wide array of subatomic particles (hover for name).

5.5.1 Summary table

Because this table was initially derived from published results and many of those results were preliminary, as many as 64 of the mesons in the following table may not exist or have the wrong mass or quantum numbers.

5.5.2 Meson properties

The following lists detail all known and predicted pseudoscalar ($J^P = 0^-$) and vector ($J^P = 1^-$) mesons.

The properties and quark content of the particles are tabulated below; for the corresponding antiparticles, simply change quarks into antiquarks (and vice versa) and flip the sign of Q, B, S, C, and B′. Particles with *† next to their names have been predicted by the standard model but not yet observed. Values in red have not been firmly established by experiments, but are predicted by the quark model and are consistent with the measurements.

Pseudoscalar mesons

*[a] ^ Makeup inexact due to non-zero quark masses.

*[b] ^ PDG reports the resonance width (Γ). Here the conversion $\tau = *\hbar/\Gamma$ is given instead.

*[c] ^ Strong eigenstate. No definite lifetime (see kaon notes below)

*[d] ^ The mass of the K0
L and K0
S are given as that of the K0. However, it is known that a difference between the masses of the K0
L and K0
S on the order of $2.2 \times 10^{*-11}$ MeV/c^2 exists.*[15]

*[e] ^ Weak eigenstate. Makeup is missing small CP–violating term (see notes on neutral kaons below).

Vector mesons

*[f] ^ PDG reports the resonance width (Γ). Here the conversion $\tau = *\hbar/\Gamma$ is given instead.

*[g] ^ The exact value depends on the method used. See the given reference for detail.

Notes on neutral kaons

There are two complications with neutral kaons:*[34]

- Due to neutral kaon mixing, the K0
  S and K0
  L are not eigenstates of strangeness. However, they *are* eigenstates of the weak force, which determines how they decay, so these are the particles with definite lifetime.

- The linear combinations given in the table for the K0
  S and K0
  L are not exactly correct, since there is a small correction due to CP violation. See CP violation in kaons.

Note that these issues also exist in principle for other neutral flavored mesons; however, the weak eigenstates are considered separate particles only for kaons because of their dramatically different lifetimes.*[34]

5.5.3 See also

- List of baryons
- List of particles
- Timeline of particle discoveries

5.5.4 References

[1] J.J. Aubert *et al.* (1974)

[2] J.E. Augustin *et al.* (1974)

[3] S.W. Herb *et al.* (1977)

[4] C. Amsler *et al.* (2008): Charmonium States

[5] K.A. Olive *et al.* (2014): Meson Summary Table

[6] K.A. Olive *et al.* (2014): Particle listings – $\pi\pm$

[7] K.A. Olive *et al.* (2014): Particle listings – $\pi 0$

[8] K.A. Olive *et al.* (2014): Particle listings – η

[9] K.A. Olive *et al.* (2014): Particle listings – η'

[10] K.A. Olive *et al.* (2014): Particle listings – η
c

[11] K.A. Olive *et al.* (2014): Particle listings – η
b

[12] K.A. Olive *et al.* (2014): Particle listings – K$\pm$

[13] K.A. Olive *et al.* (2014): Particle listings – K0

[14] K.A. Olive *et al.* (2014): Particle listings – K0
S

[15] K.A. Olive *et al.* (2014): Particle listings – K0
L

[16] K.A. Olive *et al.* (2014): Particle listings – D$\pm$

[17] K.A. Olive *et al.* (2014): Particle listings – D0

[18] K.A. Olive *et al.* (2014): Particle listings – D±s

[19] K.A. Olive *et al.* (2014): Particle listings – B±

[20] K.A. Olive *et al.* (2014): Particle listings – B0

[21] K.A. Olive *et al.* (2014): Particle listings – B0s

[22] K.A. Olive *et al.* (2014): Particle listings – B±c

[23] K.A. Olive *et al.* (2014): Particle listings – ρ

[24] K.A. Olive *et al.* (2014): Particle listings – ω(782)

[25] K.A. Olive *et al.* (2014): Particle listings – φ

[26] K.A. Olive *et al.* (2014): Particle listings – J/Ψ

[27] K.A. Olive *et al.* (2014): Particle listings – Υ(1S)

[28] K.A. Olive *et al.* (2014): Particle listings – K∗(892)

[29] K.A. Olive *et al.* (2014): Particle listings – D∗±(2010)

[30] K.A. Olive *et al.* (2014): Particle listings – D∗0(2007)

[31] K.A. Olive *et al.* (2014): Particle listings – D∗±s

[32] K.A. Olive *et al.* (2014): Particle listings – B∗

[33] K.A. Olive *et al.* (2014): Particle listings – B∗s

[34] J.W. Cronin (1980)

Bibliography

- K.A. Olive *et al.* (Particle Data Group) (2014). "Review of Particle Physics". *Chinese Physics C* **38** (9): 090001.

- M.S. Sozzi (2008a). "Parity". *Discrete Symmetries and CP Violation: From Experiment to Theory*. Oxford University Press. pp. 15–87. ISBN 0-19-929666-9.

- M.S. Sozzi (2008a). "Charge Conjugation". *Discrete Symmetries and CP Violation: From Experiment to Theory*. Oxford University Press. pp. 88–120. ISBN 0-19-929666-9.

- M.S. Sozzi (2008c). "CP-Symmetry". *Discrete Symmetries and CP Violation: From Experiment to Theory*. Oxford University Press. pp. 231–275. ISBN 0-19-929666-9.

- C. Amsler *et al.* (Particle Data Group); Amsler; Doser; Antonelli; Asner; Babu; Baer; Band; Barnett; Bergren; Beringer; Bernardi; Bertl; Bichsel; Biebel; Bloch; Blucher; Blusk; Cahn; Carena; Caso; Ceccucci; Chakraborty; Chen; Chivukula; Cowan; Dahl; d'Ambrosio; Damour et al. (2008). "Review of Particle Physics". *Physics Letters B* **667** (1): 1–1340. Bibcode:2008PhLB..667....1P. doi:10.1016/j.physletb.2008.07.018.

- S.S.M. Wong (1998). "Nucleon Structure". *Introductory Nuclear Physics* (2nd ed.). John Wiley & Sons. pp. 21–56. ISBN 0-471-23973-9.

- R. Shankar (1994). *Principles of Quantum Mechanics* (2nd ed.). Plenum Press. ISBN 0-306-44790-8.

- K. Gottfried, V.F. Weisskopf (1986). "Hadronic Spectroscopy: G-parity". *Concepts of Particle Physics* **2**. Oxford University Press. pp. 303–311. ISBN 0-19-503393-0.

- J.W. Cronin (1980). "CP Symmetry Violation – The Search for its origin" (PDF). *Nobel Lecture*. The Nobel Foundation.

- V.L. Fitch (1980). "The Discovery of Charge – Conjugation Parity Asymmetry" (PDF). *Nobel Lecture*. The Nobel Foundation.

- S.W. Herb; Hom, D.; Lederman, L.; Sens, J.; Snyder, H.; Yoh, J.; Appel, J.; Brown, B. et al. (1977). "Observation of a Dimuon Resonance at 9.5 Gev in 400-GeV Proton-Nucleus Collisions". *Physical Review Letters* **39** (5): 252–255. Bibcode:1977PhRvL..39..252H. doi:10.1103/PhysRevLett.39.252.

- J.J. Aubert; Becker, U.; Biggs, P.; Burger, J.; Chen, M.; Everhart, G.; Goldhagen, P.; Leong, J. et al. (1974). "Experimental Observation of a Heavy Particle J". *Physical Review Letters* **33** (23): 1404–1406. Bibcode:1974PhRvL..33.1404A. doi:10.1103/PhysRevLett.33.1404.

- J.E. Augustin; Boyarski, A.; Breidenbach, M.; Bulos, F.; Dakin, J.; Feldman, G.; Fischer, G.; Fryberger, D. et al. (1974). "Discovery of a Narrow Resonance in e^+e^- Annihilation". *Physical Review Letters* **33** (23): 1406–1408. Bibcode:1974PhRvL..33.1406A. doi:10.1103/PhysRevLett.33.1406.

- M. Gell-Mann (1964). "A Schematic of Baryons and Mesons". *Physics Letters* **8** (3): 214–215. Bibcode:1964PhL.....8..214G. doi:10.1016/S0031-9163(64)92001-3.

- E. Wigner (1937). "On the Consequences of the Symmetry of the Nuclear Hamiltonian on the Spectroscopy of Nuclei". *Physical Review* **51** (2): 106–119. Bibcode:1937PhRv...51..106W. doi:10.1103/PhysRev.51.106.

- W. Heisenberg (1932). "Über den Bau der Atomkerne I". *Zeitschrift für Physik* (in German) **77** (1–2): 1–11. Bibcode:1932ZPhy...77....1H. doi:10.1007/BF01342433.

- W. Heisenberg (1932). "Über den Bau der Atomkerne II". *Zeitschrift für Physik* (in German) **78** (3–4): 156–164. Bibcode:1932ZPhy...78..156H. doi:10.1007/BF01337585.

- W. Heisenberg (1932). "Über den Bau der Atomkerne III". *Zeitschrift für Physik* (in German) **80** (9–10): 587–596. Bibcode:1933ZPhy...80..587H. doi:10.1007/BF01335696.

5.5.5 External links

- Particle Data Group – The Review of Particle Physics (2008)

- Mesons made thinkable, an interactive visualisation allowing physical properties to be compared

5.6 Exotic hadron

Exotic hadrons are subatomic particles composed of quarks and gluons, but which do not fit into the usual scheme of hadrons. While bound by the strong interaction they are not predicted by the simple quark model. That is, exotic hadrons do not have the same quark content as ordinary hadrons: **exotic baryons** have more than just the three quarks of ordinary baryons and **exotic mesons** do not have one quark and one antiquark like ordinary mesons. Exotic hadrons can be searched for by looking for S-matrix poles with quantum numbers forbidden to ordinary hadrons. Experimental signatures for such exotic hadrons have been seen recently*[1] but remain a topic of controversy in particle physics.

Jaffe and Low *[2] suggested that the exotic hadrons manifest themselves as poles of the P matrix, and not of the S matrix. Experimental P-matrix poles are determined reliably in both the meson-meson channels and nucleon-nucleon channels.

5.6.1 History

When the quark model was first postulated by Murray Gell-Mann and others in the 1960s, it was to organize the states known then to be in existence in a meaningful way. As Quantum Chromodynamics (QCD) developed over the next decade, it became apparent that there was no reason why only 3-quark and quark-antiquark combinations could exist. In addition, it seemed that gluons, the mediator particles of the strong interaction, could also form bound states by themselves (glueballs) and with quarks (hybrid hadrons). Several decades have passed without conclusive evidence of an exotic hadron that could be associated with the S-matrix pole.

In April 2014, The LHCb collaboration confirmed the existence of the Z(4430)*−. Examinations of the character of the particle suggest that it may be exotic.*[3]

5.6.2 Candidates

There are several exotic hadron candidates:

- X(3872) – Discovered by the Belle detector at KEK in Japan, this particle has been variously hypothesized to be diquark or a mesonic molecule.

- Y(3940) – This particle fails to fit into the Charmonium spectrum predicted by theorists.

- Y(4140) – Discovered at Fermilab in March 2009.

- Y(4260) – Discovered by the BaBar detector at SLAC in Menlo Park, California this particle is hypothesized to be made up of a gluon bound to a quark and antiquark.

- Zc(3900) – Discovered by Belle and BES III

- Z(4430) – Discovered by Belle and later confirmed by LHCb with 13.9σ significance

5.6.3 See also

- Pentaquark
- Tetraquark

5.6.4 Notes

[1] See "note on non-q qbar mesons" in PDG 2006, Journal of Physics, G 33 (2006) 1.

[2] R. L. Jaffe and F. E. Low, Phys. Rev. D 19, 2105 (1979). doi:10.1103/PhysRevD.19.2105

[3] LHCb collaboration (7 April 2014). "Observation of the resonant character of the Z(4430)*− state". arXiv:1404.1903.

5.7 Tetraquark

A **tetraquark**, in particle physics, is an exotic meson composed of four valence quarks. In principle, a tetraquark state may be allowed in quantum chromodynamics, the modern theory of strong interactions. Any established tetraquark state would be an example of an exotic hadron which lies outside the quark model classification.

5.7.1 History

In 2003 a particle temporarily called X(3872), by the Belle experiment in Japan, was proposed to be a tetraquark candidate,*[2] as originally theorized.*[3] The name X is a temporary name, indicating that there are still some questions about its properties to be tested. The number following is the mass of the particle in 10^0 MeV/c^2.

In 2004, the D_{sJ}(2632) state seen in Fermilab's SELEX was suggested as a possible tetraquark candidate.

In 2007, Belle announced the observation of the Z(4430) state, a ccdu tetraquark candidate. In 2014, the Large Hadron Collider experiment LHCb confirmed this resonance with a significance of over 13.9σ.*[4]*[5] There are also indications that the Y(4660), also discovered by Belle in 2007, could be a tetraquark state.*[6]

In 2009, Fermilab announced that they have discovered a particle temporarily called Y(4140), which may also be a tetraquark.*[7]

In 2010, two physicists from DESY and a physicist from Quaid-i-Azam University re-analyzed former experimental data and announced that, in connection with the ϒ(5S) meson (a form of bottomonium), a well-defined tetraquark resonance exists.*[8]*[9]

In June 2013, two independent groups reported on Z_c(3900).*[10] *[11]

5.7.2 See also

- Color confinement

- Hadron

- Pentaquark

5.7.3 References

[1] N. Cardoso, M. Cardoso, and P. Bicudo (2011). "Colour Fields Computed in SU(3) Lattice QCD for the Static Tetraquark System". *Physical Review D* **84** (5): 054508. arXiv:1107.1355. doi:10.1103/PhysRevD.84.054508.

[2] D. Harris (13 April 2008). "The charming case of X(3872)". *Symmetry Magazine*. Retrieved 2009-12-17.

[3] L. Maiani, F. Piccinini, V. Riquer and A.D. Polosa (2005). "Diquark-antidiquarks with hidden or open charm and the nature of X(3872)". *Physical Review D* **71**: 014028. arXiv:hep-ph/0412098. Bibcode:2005PhRvD..71a4028M. doi:10.1103/PhysRevD.71.014028.

[4] "LHCb confirms existence of exotic hadrons".

[5] LHCb collaboration, LHCb; Aaij, R.; Adeva, B.; Adinolfi, M.; Affolder, A.; Ajaltouni, Z.; Albrecht, J.; Alessio, F.; Alexander, M.; Ali, S.; Alkhazov, G.; Alvarez Cartelle, P.; Alves Jr, A. A.; Amato, S.; Amerio, S.; Amhis, Y.; An, L.; Anderlini, L.; Anderson, J.; Andreassen, R.; Andreotti, M.; Andrews, J. E.; Appleby, R. B.; Aquines Gutierrez, O.; Archilli, F.; Artamonov, A.; Artuso, M.; Aslanides, E.; Auriemma, G. et al. (2014). "Observation of the resonant character of the Z(4430)− state". arXiv:1404.1903v1 [hep-ex].

[6] G. Cotugno, R. Faccini, A.D. Polosa and C. Sabelli (2010). "Charmed Baryonium". *Physical Review Letters* **104** (13): 132005. arXiv:0911.2178. Bibcode:2010PhRvL.104m2005C. doi:10.1103/PhysRevLett.104.132005.

[7] Anne Minard (2009-03-18). "New Particle Throws Monkeywrench in Particle Physics". Universetoday.com. Retrieved 2014-04-12.

[8] "Evidence grows for tetraquarks". physicsworld.com. Retrieved 2014-04-12.

[9] A. Ali, C. Hambrock, M.J. Aslam; Hambrock; Aslam (2010). "Tetraquark Interpretation of the BELLE Data on the Anomalous ϒ(1S)π+π- and ϒ(2S)π+π- Production near the ϒ(5S) Resonance". *Physical Review Letters* **104** (16): 162001. arXiv:0912.5016. Bibcode:2010PhRvL.104p2001A. doi:10.1103/PhysRevLett.104.162001.

[10] "Physics - New Particle Hints at Four-Quark Matter". Physics.aps.org. 2013-06-17. Retrieved 2014-04-12.

[11] Eric Swanson (2013). "Viewpoint: New Particle Hints at Four-Quark Matter". *Physics* **69** (6). Bibcode:2013PhyOJ...6...69S. doi:10.1103/Physics.6.69.

5.7.4 External links

- The Belle experiment (press release)

5.8. PENTAQUARK

- O'Luanaigh, Cian. "LHCb confirms existence of exotic hadrons". *cern.ch*. Geneva, Switzerland: CERN. Retrieved 2014-04-12.

5.8 Pentaquark

Two models of a generic pentaquark

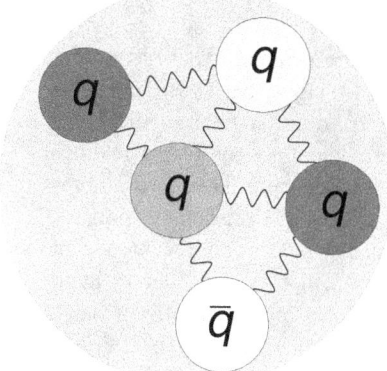

A five-quark "bag"

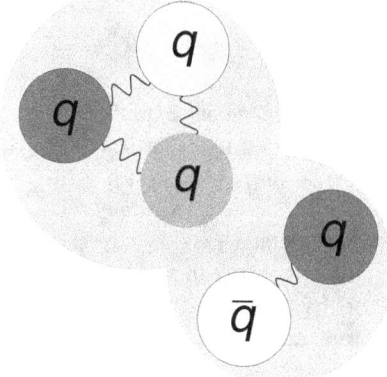

A meson-baryon molecule

q indicates a quark, whereas $\bar{q}$ indicates an antiquark. The wavy lines are gluons, which mediate the strong interaction between the quarks. The colours correspond to the various colour charges of quarks. The colours red, green, and blue must each be present and the remaining quark and antiquark must share corresponding colour and anticolour, here chosen to be blue and antiblue (shown as yellow).

A **pentaquark** is a subatomic particle consisting of four quarks and one antiquark bound together.

As quarks have a baryon number of $+\frac{1}{3}$, and antiquarks of $-\frac{1}{3}$, the pentaquark would have a total baryon number of 1, and thus would be a baryon. Further, because it has five quarks instead of the usual three found in regular baryons (aka 'triquarks'), it would be classified as an exotic baryon. The name pentaquark was coined by Harry J. Lipkin in 1987,[1] however, the possibility of five-quark particles was identified as early as 1964 when Murray Gell-Mann first postulated the existence of quarks.[2] Although predicted for decades, pentaquarks have proved surprisingly difficult to discover and some physicists were beginning to suspect that an unknown law of nature prevented their production.[3]

The first claim of pentaquark discovery was recorded at LEPS in Japan in 2003, and several experiments in the mid-2000s also reported discoveries of other pentaquark states.[4] Others were not able to replicate the LEPS results, however, and the other pentaquark discoveries were not accepted because of poor data and statistical analysis.[5] On 13 July 2015, the LHCb collaboration at CERN reported results consistent with pentaquark states in the decay of bottom Lambda baryons (Λ^0_b).[6]

Outside of particle physics laboratories pentaquarks also could be produced naturally by supernovae as part of the process of forming a neutron star.[7] The scientific study of pentaquarks might offer insights into how these stars form, as well as, allowing more thorough study of particle interactions and the strong force.

5.8.1 Background

Main article: Quark

A quark is a type of elementary particle that has mass, electric charge, and colour charge, as well as an additional property called flavour, which describes what type of quark it is (up, down, strange, charm, top, or bottom). Due to an effect known as colour confinement, quarks are never seen on their own. Instead, they form composite particles known as hadrons so that their colour charges cancel out. Hadrons made of one quark and one antiquark are known as mesons, while those made of three quarks are known as baryons. These 'regular' hadrons are well documented and characterized, however, there is nothing in theory to prevent quarks from forming 'exotic' hadrons such as tetraquarks with two quarks and two antiquarks, or pentaquarks with four quarks and one antiquark.[3]

5.8.2 Structure

A wide variety of pentaquarks are possible, with different quark combinations producing different particles. To identify which quarks compose a given pentaquark, physicists use the notation $qqqq\bar{q}$, where q and $\bar{q}$ respectively refer to any of the six flavours of quarks and antiquarks. The symbols u, d, s, c, b, and t stand for the up, down, strange, charm, bottom, and top quarks respectively, with the symbols of u, d, s, c, b, t corresponding to the respective antiquarks. For instance a pentaquark made of two up quarks, one down quark, one charm quark, and one charm antiquark would be denoted uudcc.

The quarks are bound together by the strong force, which acts in such a way as to cancel the colour charges within the particle. In a meson, this means a quark is partnered with an antiquark with an opposite colour charge – blue and antiblue, for example – while in a baryon, the three quarks have between them all three colour charges – red, blue, and green.*[nb 1] In a pentaquark, the colours also need to cancel out, and the only feasible combination is to have one quark with one colour (e.g. red), one quark with a second colour (e.g. green), two quarks with the third colour (e.g. blue), and one antiquark to counteract the surplus colour (e.g. antiblue).*[8]

The binding mechanism for pentaquarks is not yet clear. They may consist of five quarks tightly bound together, but it is also possible that they are more loosely bound and consist of a three-quark baryon and a two-quark meson interacting relatively weakly with each other via pion exchange (the same force that binds atomic nuclei) in a "meson-baryon molecule" .*[9]*[2]*[10]

5.8.3 History

Mid-2000s

The requirement to include an antiquark means that many classes of pentaquark are hard to identify experimentally – if the flavour of the antiquark matches the flavour of any other quark in the quintuplet, it will cancel out and the particle will resemble its three-quark hadron cousin. For this reason, early pentaquark searches looked for particles where the antiquark did not cancel.*[8] In the mid-2000s, several experiments claimed to reveal pentaquark states. In particular, a resonance with a mass of 1540 MeV/c^2 (4.6 σ) was reported by LEPS in 2003, the Θ+.*[11] This coincided with a pentaquark state with a mass of 1530 MeV/c^2 predicted in 1997.*[12]

The proposed state was composed of two up quarks, two down quarks, and one strange antiquark (uudds). Following this announcement, nine other independent experiments reported seeing narrow peaks from nK+ and pK0, with masses between 1522 MeV/c^2 and 1555 MeV/c^2, all above 4 σ.*[11] While concerns existed about the validity of these states, the Particle Data Group gave the Θ+ a 3-star rating (out of 4) in the 2004 *Review of Particle Physics*.*[11] Two other pentaquark states were reported albeit with low statistical significance—the Φ−− (ddssu), with a mass of 1860 MeV/c^2 and the Θ0
c (uuddc), with a mass of 3099 MeV/c^2. Both were later found to be statistical effects rather than true resonances.*[11]

Ten experiments then looked for the Θ+, but came out empty-handed.*[11] Two in particular (one at BELLE, and the other at CLAS) had nearly the same conditions as other experiments which claimed to have detected the Θ+ (DIANA and SAPHIR respectively).*[11] The 2006 *Review of Particle Physics* concluded:*[11]

> [T]here has not been a high-statistics confirmation of any of the original experiments that claimed to see the Θ+; there have been two high-statistics repeats from Jefferson Lab that have clearly shown the original positive claims in those two cases to be wrong; there have been a number of other high-statistics experiments, none of which have found any evidence for the Θ+; and all attempts to confirm the two other claimed pentaquark states have led to negative results. The conclusion that pentaquarks in general, and the Θ+, in particular, do not exist, appears compelling.

The 2008 *Review of Particle Physics* went even further:*[5]

> There are two or three recent experiments that find weak evidence for signals near the nominal masses, but there is simply no point in tabulating them in view of the overwhelming evidence that the claimed pentaquarks do not exist... The whole story—the discoveries themselves, the tidal wave of papers by theorists and phenomenologists that followed, and the eventual "undiscovery"—is a curious episode in the history of science.

Despite these null results, LEPS results as of 2009 continue to show the existence of a narrow state with a mass of 1524±4 MeV/c^2, with a statistical significance of 5.1 σ.*[13] Experiments continue to study this controversy.

2015 LHCb results

In July 2015, the LHCb collaboration identified pentaquarks in the Λ0
b→J/ψK−
p channel, which represents the decay of the bottom lambda baryon (Λ0
b) into a J/ψ meson (J/ψ), a kaon (K−
) and a proton (p). The results showed that sometimes, instead of decaying directly into mesons and baryons, the Λ0
b decayed via intermediate pentaquark states. The two

states, named P+
c(4380) and P+
c(4450), had individual statistical significances of 9 σ and 12 σ, respectively, and a combined significance of 15 σ — enough to claim a formal discovery. The analysis ruled out the possibility that the effect was caused by conventional particles."[2] The two pentaquark states were both observed decaying strongly to J/ψp, hence must have a valence quark content of two up quarks, a down quark, a charm quark, and an anti-charm quark (uudcc̄), making them charmonium-pentaquarks.*[6]*[7]*[14]

The search for pentaquarks was not an objective of the LHCb experiment (which is primarily designed to investigate matter-antimatter asymmetry)*[15] and the apparent discovery of pentaquarks was described as an "accident" and "something we've stumbled across" by a CERN spokesperson.*[9]

5.8.4 Applications

The discovery of pentaquarks will allow physicists to study the strong force in greater detail and aid understanding of quantum chromodynamics. In addition, current theories suggest that some very large stars produce pentaquarks as they collapse. The study of pentaquarks might help shed light on the physics of neutron stars.*[7]

5.8.5 See also

- Exotic matter
- List of particles
- Quark model
- Tetraquark
- Triquark

5.8.6 Footnotes

[1] The colour charges do not correspond to physical visible colours. They are arbitrary labels used to help scientists describe and visualise the charges of quarks.

5.8.7 References

[1] H. J. Lipkin (1987). "New possibilities for exotic hadrons —anticharmed strange baryons". *Physics Letters B* **195** (3): 484–488. Bibcode:1987PhLB..195..484L. doi:10.1016/0370-2693(87)90055-4.

[2] "Observation of particles composed of five quarks, pentaquark-charmonium states, seen in Λ0 b→J/ψpK− decays". CERN/LHCb. 14 July 2015. Retrieved 2015-07-14.

[3] H. Muir (2 July 2003). "Pentaquark discovery confounds sceptics". *New Scientist*. Retrieved 2010-01-08.

[4] K. Hicks (23 July 2003). "Physicists find evidence for an exotic baryon". Ohio University. Retrieved 2010-01-08.

[5] See p. 1124 in C. Amsler et al. (Particle Data Group) (2008). "Review of particle physics" (PDF). *Physics Letters B* **667** (1-5): 1. Bibcode:2008PhLB..667....1A. doi:10.1016/j.physletb.2008.07.018.

[6] R. Aaij et al. (LHCb collaboration) (2015). "Observation of J/ψp resonances consistent with pentaquark states in Λ0 b→J/ψK−p decays". *Physical Review Letters* **115** (7). doi:10.1103/PhysRevLett.115.072001.

[7] I. Sample (14 July 2015). "Large Hadron Collider scientists discover new particles: pentaquarks". *The Guardian*. Retrieved 2015-07-14.

[8] J. Pochodzalla (2005). "Duets of strange quarks". *Hadron Physics*. p. 268. ISBN 161499014X.

[9] G. Amit (14 July 2015). "Pentaquark discovery at LHC shows long-sought new form of matter". *New Scientist*. Retrieved 2015-07-14.

[10] T. D. Cohen, P. M. Hohler, R. F. Lebed (2005). "On the Existence of Heavy Pentaquarks: The large N_c and Heavy Quark Limits and Beyond". *Physical Review D* **72** (7): 074010. arXiv:hep-ph/0508199. Bibcode:2005PhRvD..72g4010C. doi:10.1103/PhysRevD.72.074010.

[11] W.-M. Yao et al. (Particle Data Group) (2006). "Review of particle physics: Θ+" (PDF). *Journal of Physics G* **33**: 1. arXiv:astro-ph/0601168. Bibcode:2006JPhG...33....1Y. doi:10.1088/0954-3899/33/1/001.

[12] D. Diakonov, V. Petrov, and M. Polyakov (1997). "Exotic anti-decuplet of baryons: prediction from chiral solitons". *Zeitschrift für Physik A* **359** (3): 305. arXiv:hep-ph/9703373. Bibcode:1997ZPhyA.359..305D. doi:10.1007/s002180050406.

[13] T. Nakano et al. (LEPS Collaboration) (2009). "Evidence of the Θ*+ in the γd→K*+K−pn reaction". *Physical Review C* **79** (2): 025210. arXiv:0812.1035. Bibcode:2009PhRvC..79b5210N. doi:10.1103/PhysRevC.79.025210.

[14] P. Rincon (14 July 2015). "Large Hadron Collider discovers new pentaquark particle". *BBC News*. Retrieved 2015-07-14.

[15] "Where has all the antimatter gone?". CERN/LHCb. 2008. Retrieved 2015-07-15.

[16] N. Cardoso, M. Cardoso, and P. Bicudo (2013). "Color fields of the static pentaquark system computed in SU(3) lattice QCD". *Physical Review D* **87** (3): 034504. arXiv:1209.1532. doi:10.1103/PhysRevD.87.034504.

5.8.8 Further reading

- David Whitehouse (1 July 2003). "Behold the Pentaquark (BBC News)". BBC News. Retrieved 2010-01-08.

- Thomas E. Browder, Igor R. Klebanov, Daniel R. Marlow (2004). "Prospects for Pentaquark Production at Meson Factories". *Physics Letters B* **587**: 62. arXiv:hep-ph/0401115. Bibcode:2004PhLB..587...62B. doi:10.1016/j.physletb.2004.03.003.

- Akio Sugamoto (2004). "An Attempt to Study Pentaquark Baryons in String Theory". arXiv:hep-ph/0404019 [hep-ph].

- Kenneth Hicks (2005). "An Experimental Review of the Θ+ Pentaquark". *Journal of Physics: Conference Series* **9**: 183. arXiv:hep-ex/0412048. Bibcode:2005JPhCS...9..183H. doi:10.1088/1742-6596/9/1/035.

- Mark Peplow (18 April 2005). "Doubt is Cast on Pentaquarks". *Nature*. doi:10.1038/news050418-1.

- Maggie McKie (20 April 2005). "Pentaquark hunt draws blanks". *New Scientist*. Retrieved 2010-01-08.

- Thomas Jefferson National Accelerator Facility (21 April 2005). "Is It Or Isn't It? Pentaquark Debate Heats Up". *Space Daily*. Retrieved 2010-01-08.

- Dmitri Diakonov (2005). "Relativistic Mean Field Approximation to Baryons". *European Physical Journal A* **24**: 3. Bibcode:2005EPJAS..24a...3D. doi:10.1140/epjad/s2005-05-001-3.

- Schumacher, R. A. (2006). "The Rise and Fall of Pentaquarks in Experiments". *AIP Conference Proceedings* **842**: 409. arXiv:nucl-ex/0512042. doi:10.1063/1.2220285.

- Kandice Carter (2006). "The Rise and Fall of the Pentaquark". *Symmetry Magazine* **3** (7): 16.

5.8.9 External links

- "Pentaquark on arxiv.org".

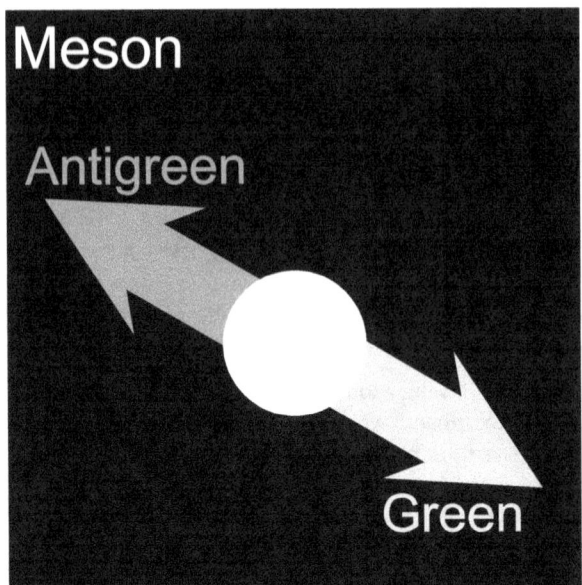
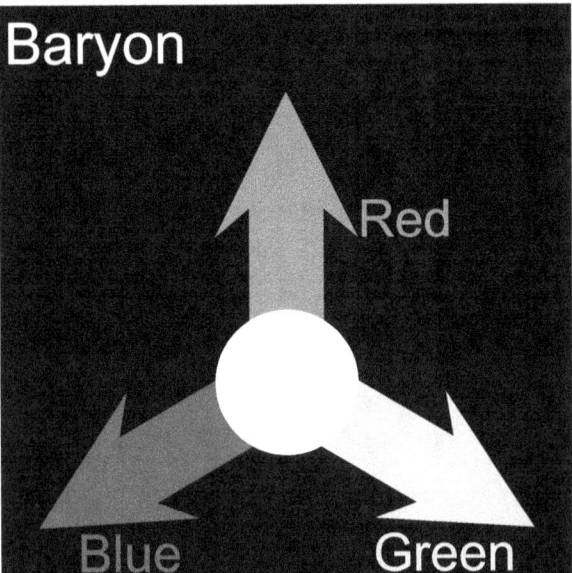
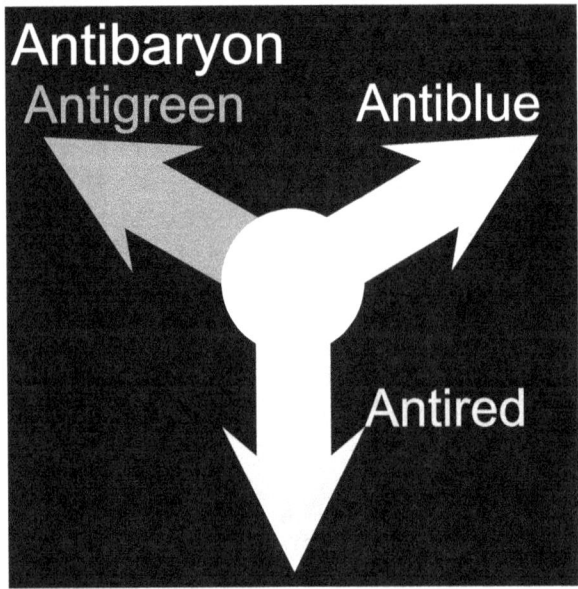

All types of hadrons have zero total color charge. (three examples shown)

5.8. PENTAQUARK

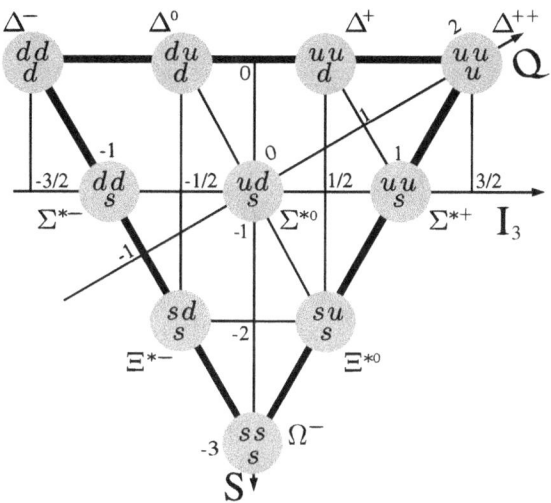

*Combinations of three **u**, **d** or s quarks forming baryons with a spin-$\frac{3}{2}$ form the* uds baryon decuplet

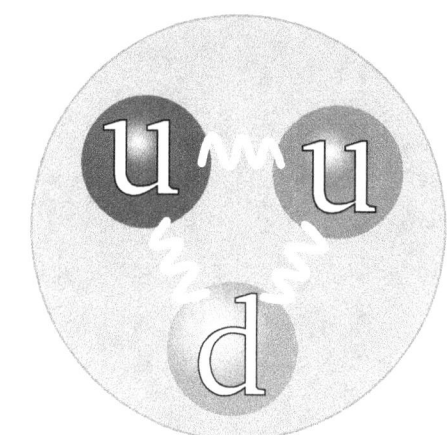

A diagram of a proton, one of the most famous baryons, containing two up quarks and one down quark

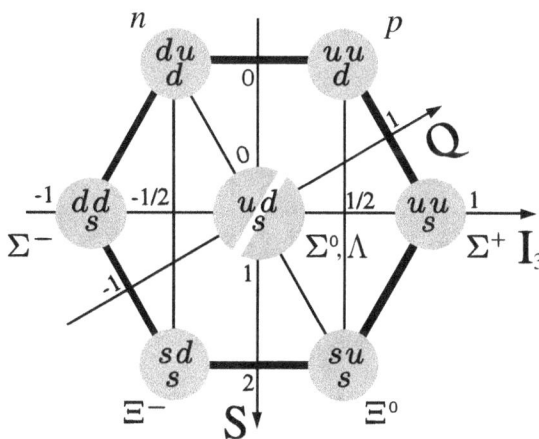

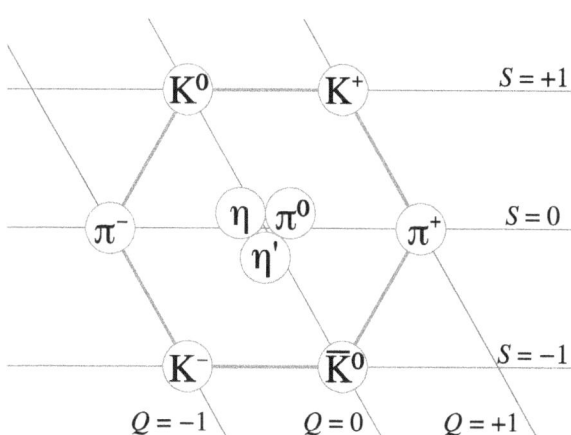

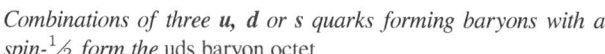

*Combinations of three **u**, **d** or s quarks forming baryons with a spin-$\frac{1}{2}$ form the* uds baryon octet

Combinations of one u, d or s quarks and one u, d, or s antiquark in $J^P = 0^-$ configuration form a nonet.

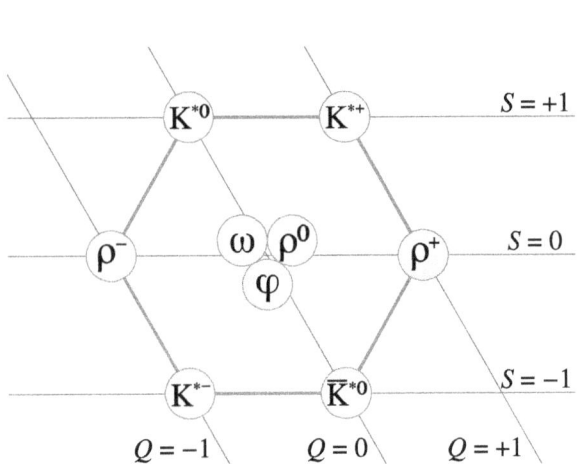

Combinations of one u, d or s quarks and one u, d, or s antiquark in $J^*P = 1^*-$ configuration also form a nonet.

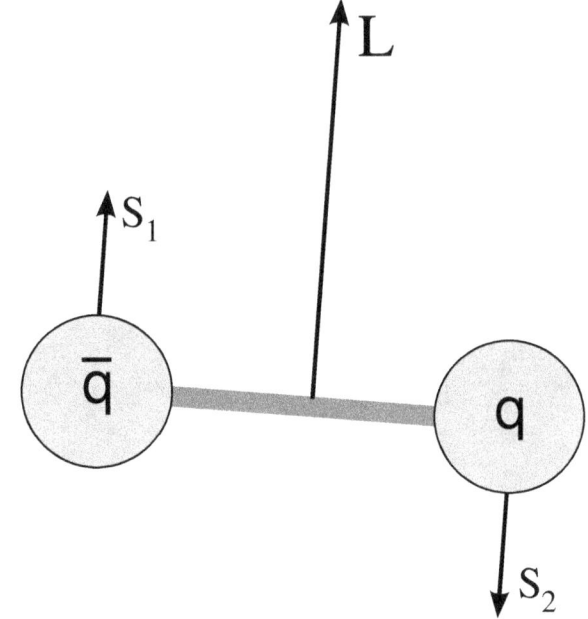

A regular meson made from a quark (q) and an antiquark (q) with spins s_2 and s_1 respectively and having an overall angular momentum L

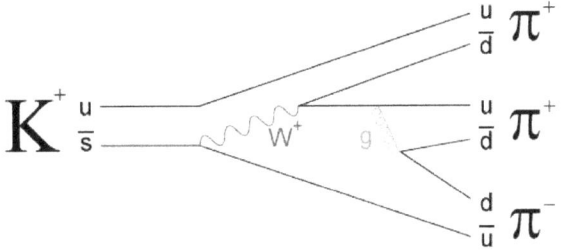

The decay of a kaon (K+) into three pions (2 π+, 1 π−) is a process that involves both weak and strong interactions.
Weak interactions: The strange antiquark (s) of the kaon transmutes into an up antiquark (u) by the emission of a W+ boson; the W+ boson subsequently decays into a down antiquark (d) and an up quark (u).
Strong interactions: An up quark (u) emits a gluon (g) which decays into a down quark (d) and a down antiquark (d).

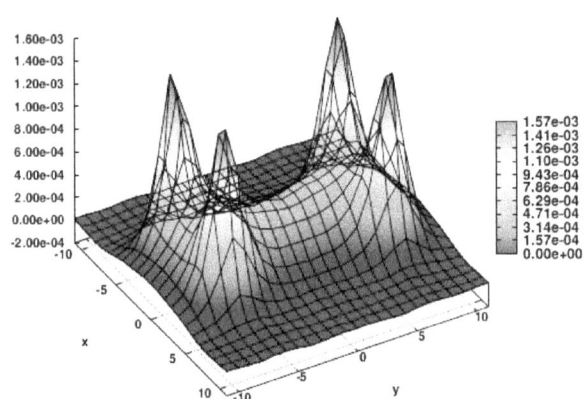

Colour flux tubes produced by four static quark and antiquark charges, computed in lattice QCD.*[1] Confinement in Quantum Chromo Dynamics leads to the production of flux tubes connecting colour charges. The flux tubes act as attractive QCD string-like potentials.

5.8. PENTAQUARK

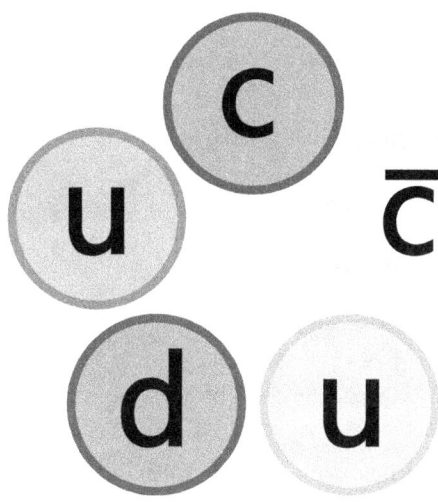

*A diagram of the P+
c type pentaquark possibly discovered in July 2015, showing the flavours of each quark and one possible colour configuration.*

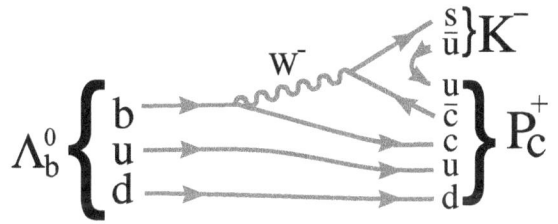

*Feynman diagram representing the decay of a lambda baryon Λ0
b into a kaon K−
and a pentaquark P+
c.*

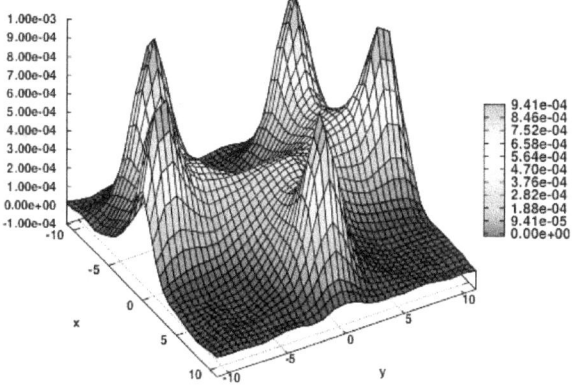

Colour flux tubes produced by five static quark and antiquark charges, computed in lattice QCD.[16] *Confinement in Quantum Chromo Dynamics leads to the production of flux tubes connecting colour charges. The flux tubes act as attractive QCD string-like potentials.*

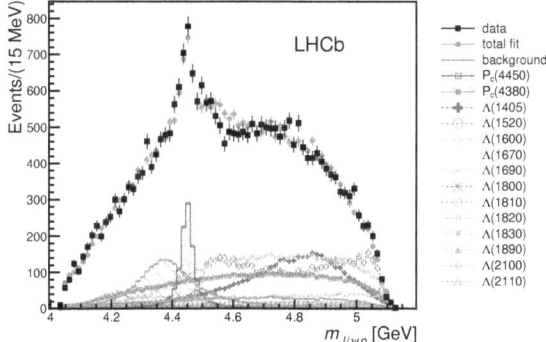

*A fit to the J/ψp invariant mass spectrum for the Λ0
b→J/ψK−
p decay, with each fit component shown individually. The contribution of the pentaquarks are shown by hatched histograms.*

Chapter 6

Appendix A – Related topics

6.1 Antiparticle

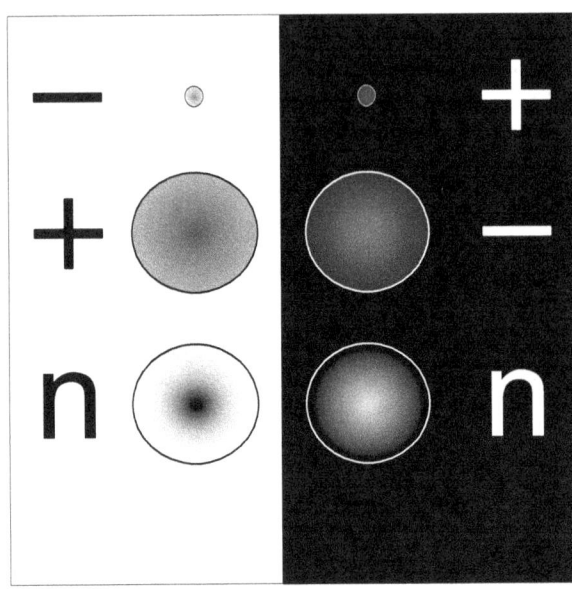

Illustration of electric charge of particles (left) and antiparticles (right). From top to bottom; electron/positron, proton/antiproton, neutron/antineutron.

Corresponding to most kinds of particles, there is an associated antimatter **antiparticle** with the same mass and opposite charge (including electric charge). For example, the antiparticle of the electron is the positively charged positron, which is produced naturally in certain types of radioactive decay.

The laws of nature are very nearly symmetrical with respect to particles and antiparticles. For example, an antiproton and a positron can form an antihydrogen atom, which is believed to have the same properties as a hydrogen atom. This leads to the question of why the formation of matter after the Big Bang resulted in a universe consisting almost entirely of matter, rather than being a half-and-half mixture of matter and antimatter. The discovery of Charge Parity violation helped to shed light on this problem by showing that this symmetry, originally thought to be perfect, was only approximate.

Particle-antiparticle pairs can annihilate each other, producing photons; since the charges of the particle and antiparticle are opposite, total charge is conserved. For example, the positrons produced in natural radioactive decay quickly annihilate themselves with electrons, producing pairs of gamma rays, a process exploited in positron emission tomography.

Antiparticles are produced naturally in beta decay, and in the interaction of cosmic rays in the Earth's atmosphere. Because charge is conserved, it is not possible to create an antiparticle without either destroying a particle of the same charge (as in beta decay) or creating a particle of the opposite charge. The latter is seen in many processes in which both a particle and its antiparticle are created simultaneously, as in particle accelerators. This is the inverse of the particle-antiparticle annihilation process.

Although particles and their antiparticles have opposite charges, electrically neutral particles need not be identical to their antiparticles. The neutron, for example, is made out of quarks, the antineutron from antiquarks, and they are distinguishable from one another because neutrons and antineutrons annihilate each other upon contact. However, other neutral particles are their own antiparticles, such as photons, hypothetical gravitons, and some WIMPs.

6.1.1 History

Experiment

In 1932, soon after the prediction of positrons by Paul Dirac, Carl D. Anderson found that cosmic-ray collisions produced these particles in a cloud chamber—a particle detector in which moving electrons (or positrons) leave behind trails as they move through the gas. The electric charge-to-mass ratio of a particle can be measured by observing the radius of curling of its cloud-chamber track in a magnetic field. Positrons, because of the direction that their paths curled, were at first mistaken for electrons travelling in the opposite direction. Positron paths in a cloud-chamber trace the same helical path as an electron but rotate in the opposite direction with respect to the magnetic field direction due to their having the same magnitude of charge-to-mass ratio but with opposite charge and, therefore, opposite signed charge-to-mass ratios.

The antiproton and antineutron were found by Emilio Segrè and Owen Chamberlain in 1955 at the University of California, Berkeley. Since then, the antiparticles of many other subatomic particles have been created in particle accelerator experiments. In recent years, complete atoms of antimatter have been assembled out of antiprotons and positrons, collected in electromagnetic traps.*[1]

Dirac's Hole theory

... the development of quantum field theory made the interpretation of antiparticles as holes unnecessary, even though it lingers on in many textbooks.

Steven Weinberg*[2]

Solutions of the Dirac equation contained negative energy quantum states. As a result, an electron could always radiate energy and fall into a negative energy state. Even worse, it could keep radiating infinite amounts of energy because there were infinitely many negative energy states available. To prevent this unphysical situation from happening, Dirac proposed that a "sea" of negative-energy electrons fills the universe, already occupying all of the lower-energy states so that, due to the Pauli exclusion principle, no other electron could fall into them. Sometimes, however, one of these negative-energy particles could be lifted out of this Dirac sea to become a positive-energy particle. But, when lifted out, it would leave behind a *hole* in the sea that would act exactly like a positive-energy electron with a reversed charge. These he interpreted as "negative-energy electrons" and attempted to identify them with protons in his 1930 paper *A Theory of Electrons and Protons**[3] However, these "negative-energy electrons" turned out to be positrons, and not protons.

This picture implied an infinite negative charge for the universe--a problem of which Dirac was aware. Dirac tried to argue that we would perceive this as the normal state of zero charge. Another difficulty was the difference in masses of the electron and the proton. Dirac tried to argue that this was due to the electromagnetic interactions with the sea, until Hermann Weyl proved that hole theory was completely symmetric between negative and positive charges. Dirac also predicted a reaction $e- + p+ \rightarrow \gamma + \gamma$, where an electron and a proton annihilate to give two photons. Robert Oppenheimer and Igor Tamm proved that this would cause ordinary matter to disappear too fast. A year later, in 1931, Dirac modified his theory and postulated the positron, a new particle of the same mass as the electron. The discovery of this particle the next year removed the last two objections to his theory.

However, the problem of infinite charge of the universe remains. Also, as we now know, bosons also have antiparticles, but since bosons do not obey the Pauli exclusion principle (only fermions do), hole theory does not work for them. A unified interpretation of antiparticles is now available in quantum field theory, which solves both these problems.

6.1.2 Particle-antiparticle annihilation

Main article: Annihilation

If a particle and antiparticle are in the appropriate quan-

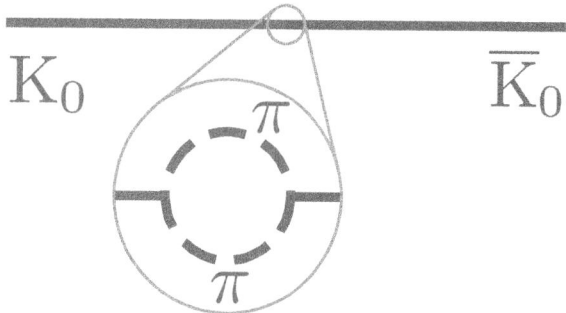

An example of a virtual pion pair that influences the propagation of a kaon, causing a neutral kaon to mix *with the antikaon. This is an example of renormalization in quantum field theory—the field theory being necessary because of the change in particle number.*

tum states, then they can annihilate each other and produce other particles. Reactions such as $e- + e+ \rightarrow \gamma + \gamma$ (the two-photon annihilation of an electron-positron pair) are an example. The single-photon annihilation of an electron-positron pair, $e- + e+ \rightarrow \gamma$, cannot occur in free space because it is impossible to conserve energy and momentum together in this process. However, in the Coulomb field of a nucleus the translational invariance is broken and single-photon annihilation may occur.*[4] The reverse reaction (in free space, without an atomic nucleus) is also impossible for this reason. In quantum field theory, this process is allowed only as an intermediate quantum state for times short enough that the violation of energy conservation can be accommodated by the uncertainty principle. This opens the way for virtual pair production or annihilation in which a one particle quantum state may *fluctuate* into a two particle state and back. These processes are important in the vacuum state and renormalization of a quantum field theory. It also opens the way for neutral particle mixing through processes such as the one pictured here, which is a complicated example of mass renormalization.

6.1.3 Properties of antiparticles

Quantum states of a particle and an antiparticle can be interchanged by applying the charge conjugation (**C**), parity (**P**), and time reversal (**T**) operators. If $|p, \sigma, n\rangle$ denotes the quantum state of a particle (**n**) with momentum **p**, spin **J** whose component in the z-direction is σ, then one has

$$CPT\,|p,\sigma,n\rangle = (-1)^{J-\sigma}\,|p,-\sigma,n^c\rangle,$$

where **n*c** denotes the charge conjugate state, *i.e.*, the antiparticle. This behaviour under **CPT** is the same as the statement that the particle and its antiparticle lie in the same irreducible representation of the Poincaré group. Properties of antiparticles can be related to those of particles through this. If **T** is a good symmetry of the dynamics, then

$$T\,|p,\sigma,n\rangle \propto |-p,-\sigma,n\rangle,$$
$$CP\,|p,\sigma,n\rangle \propto |-p,\sigma,n^c\rangle,$$
$$C\,|p,\sigma,n\rangle \propto |p,\sigma,n^c\rangle,$$

where the proportionality sign indicates that there might be a phase on the right hand side. In other words, particle and antiparticle must have

- the same mass **m**
- the same spin state **J**
- opposite electric charges **q** and **-q**.

6.1.4 Quantum field theory

This section draws upon the ideas, language and notation of canonical quantization of a quantum field theory.

One may try to quantize an electron field without mixing the annihilation and creation operators by writing

$$\psi(x) = \sum_k u_k(x) a_k e^{-iE(k)t},$$

where we use the symbol k to denote the quantum numbers p and σ of the previous section and the sign of the energy, $E(k)$, and a_k denotes the corresponding annihilation operators. Of course, since we are dealing with fermions, we have to have the operators satisfy canonical anti-commutation relations. However, if one now writes down the Hamiltonian

$$H = \sum_k E(k) a_k^\dagger a_k,$$

then one sees immediately that the expectation value of H need not be positive. This is because $E(k)$ can have any sign whatsoever, and the combination of creation and annihilation operators has expectation value 1 or 0.

So one has to introduce the charge conjugate *antiparticle* field, with its own creation and annihilation operators satisfying the relations

$$b_{k'} = a_k^\dagger \text{ and } b_{k'}^\dagger = a_k,$$

where k has the same p, and opposite σ and sign of the energy. Then one can rewrite the field in the form

$$\psi(x) = \sum_{k_+} u_k(x) a_k e^{-iE(k)t} + \sum_{k_-} u_k(x) b_k^\dagger e^{-iE(k)t},$$

where the first sum is over positive energy states and the second over those of negative energy. The energy becomes

$$H = \sum_{k_+} E_k a_k^\dagger a_k + \sum_{k_-} |E(k)| b_k^\dagger b_k + E_0,$$

where E_0 is an infinite negative constant. The vacuum state is defined as the state with no particle or antiparticle, *i.e.*, $a_k|0\rangle = 0$ and $b_k|0\rangle = 0$. Then the energy of the vacuum is exactly E_0. Since all energies are measured relative to the vacuum, **H** is positive definite. Analysis of the properties of a_k and b_k shows that one is the annihilation operator for particles and the other for antiparticles. This is the case of a fermion.

This approach is due to Vladimir Fock, Wendell Furry and Robert Oppenheimer. If one quantizes a real scalar field, then one finds that there is only one kind of annihilation operator; therefore, real scalar fields describe neutral bosons. Since complex scalar fields admit two different kinds of annihilation operators, which are related by conjugation, such fields describe charged bosons.

Feynman–Stueckelberg interpretation

By considering the propagation of the negative energy modes of the electron field backward in time, Ernst Stueckelberg reached a pictorial understanding of the fact that the particle and antiparticle have equal mass **m** and spin **J** but opposite charges **q**. This allowed him to rewrite perturbation theory precisely in the form of diagrams. Richard Feynman later gave an independent systematic derivation of these diagrams from a particle formalism, and they are now called Feynman diagrams. Each line of a diagram represents a particle propagating either backward or forward in time. This technique is the most widespread method of computing amplitudes in quantum field theory today.

Since this picture was first developed by Ernst Stueckelberg, and acquired its modern form in Feynman's work, it is called the *Feynman-Stueckelberg interpretation* of antiparticles to honor both scientists.

As a consequence of this interpretation, Villata argued that the assumption of antimatter as CPT-transformed matter would imply that the gravitational interaction between matter and antimatter is repulsive.[5]

6.1.5 See also

- Gravitational interaction of antimatter
- Parity, charge conjugation and time reversal symmetry.
- CP violations and the baryon asymmetry of the universe.
- Quantum field theory and the list of particles
- Baryogenesis

6.1.6 References

[1] http://news.nationalgeographic.com/news/2010/11/101118-antimatter-trapped-engines-bombs-nature-science-cern/

[2] Weinberg, Steve. *The quantum theory of fields, Volume 1 : Foundations*. p. 14. ISBN 0-521-55001-7.

[3] Dirac, Paul (1930). "A Theory of Electrons and Protons". *Proceedings of the Royal Society A* **126** (801): 360–365. Bibcode:1930RSPSA.126..360D. doi:10.1098/rspa.1930.0013.

[4] Sodickson, L.; W. Bowman; J. Stephenson (1961). "Single-Quantum Annihilation of Positrons". *Physical Review* **124** (6): 1851–1861. Bibcode:1961PhRv..124.1851S. doi:10.1103/PhysRev.124.1851.

[5] M. Villata, CPT symmetry and antimatter gravity in general relativity, 2011, EPL (Europhysics Letters) 94, 20001

- Feynman, R. P. (1987). "The reason for antiparticles". In R. P. Feynman and S. Weinberg. *The 1986 Dirac memorial lectures*. Cambridge University Press. ISBN 0-521-34000-4.

- Weinberg, S. (1995). *The Quantum Theory of Fields, Volume 1: Foundations*. Cambridge University Press. ISBN 0-521-55001-7.

6.2 Beta decay

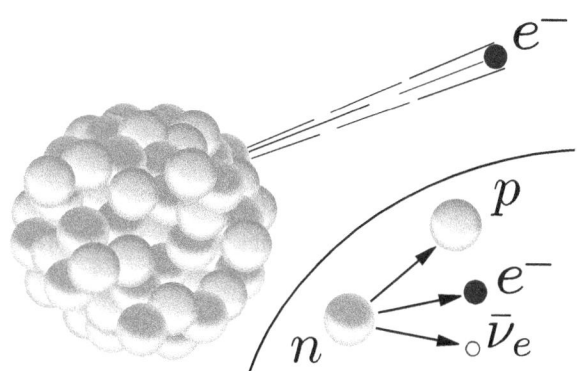

$\beta-$ decay in an atomic nucleus (the accompanying antineutrino is omitted). The inset shows beta decay of a free neutron. In both processes, the intermediate emission of a virtual $W-$ boson (which then decays to electron and antineutrino) is not shown.

In nuclear physics, **beta decay** (β-decay) is a type of radioactive decay in which a proton is transformed into a neutron, or vice versa, inside an atomic nucleus. This process allows the atom to move closer to the optimal ratio of protons and neutrons. As a result of this transformation, the nucleus emits a detectable beta particle, which is an electron or positron.[1]

Beta decay is mediated by the weak force. There are two types of beta decay, known as *beta minus* and *beta plus*. Beta minus (β^-) decay produces an electron and electron antineutrino, while beta plus (β^+) decay produces a positron and electron neutrino; β^+ decay is thus also known as positron emission.[2]

An example of electron emission (β^- decay) is the decay of carbon-14 into nitrogen-14:

$$^{14}_{6}C \rightarrow {}^{14}_{7}N + e^- + \bar{\nu}_e$$

In this form of decay, the original element becomes a new chemical element in a process known as nuclear transmutation. This new element has an unchanged mass number A, but an atomic number Z that is increased by one. As in all nuclear decays, the decaying element (in this case $^{14}_{6}C$) is known as the *parent nuclide* while the resulting element (in this case $^{14}_{7}N$) is known as the *daughter nuclide*. The emitted electron or positron is known as a beta particle.

An example of positron emission ($\beta^{*}+$ decay) is the decay of magnesium-23 into sodium-23:

$$^{23}_{12}Mg \rightarrow\ ^{23}_{11}Na + e+ + \nu_e$$

In contrast to $\beta^{*}-$ decay, $\beta^{*}+$ decay is accompanied by the emission of an electron neutrino and a positron. $\beta^{*}+$ decay also results in nuclear transmutation, with the resulting element having an atomic number that is decreased by one.

Electron capture is sometimes included as a type of beta decay, because the basic nuclear process, mediated by the weak force, is the same. In electron capture, an inner atomic electron is captured by a proton in the nucleus, transforming it into a neutron, and an electron neutrino is released. An example of electron capture is the decay of krypton-81 into bromine-81:

$$^{81}_{36}Kr + e- \rightarrow\ ^{81}_{35}Br + \nu_e$$

Electron capture is a competing (simultaneous) decay process for all nuclei that can undergo $\beta^{*}+$ decay. The converse, however, is not true: electron capture is the *only* type of decay that is allowed in proton-rich nuclides that do not have sufficient energy to emit a positron and neutrino.*[3]

6.2.1 $\beta^{*}-$ decay

In $\beta-$ decay, the weak interaction converts an atomic nucleus into a nucleus with atomic number increased by one, while emitting an electron (e−) and an electron antineutrino (ν_e).

The generic equation is:

$$^{A}_{Z}X \rightarrow\ ^{A}_{Z+1}X' + e- + \nu_e^{*}[1]$$

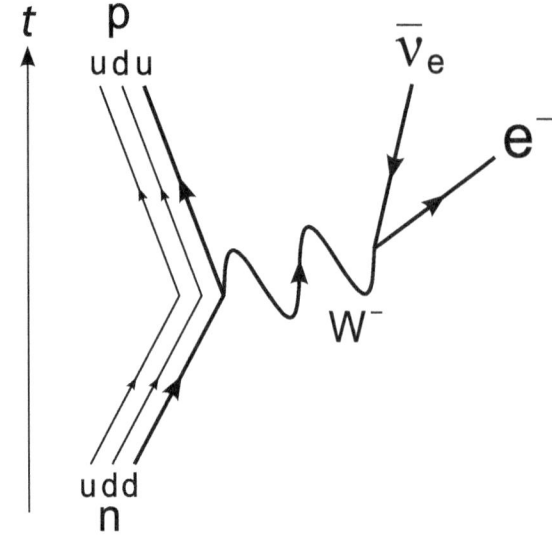

The Feynman diagram for $\beta-$ decay of a neutron into a proton, electron, and electron antineutrino via an intermediate $W-$ boson.

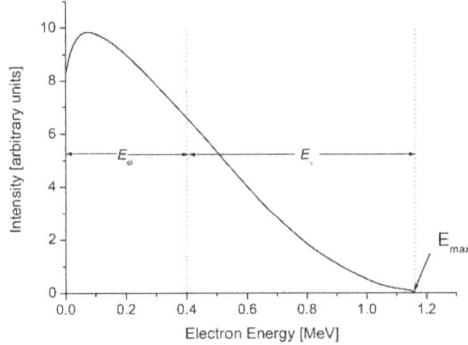

A beta spectrum, showing a typical division of energy between electron and antineutrino

where A and Z are the mass number and atomic number of the decaying nucleus, and X and X' are the initial and final elements, respectively.

Another example is when the free neutron ($^{1}_{0}n$) decays by $\beta-$ decay into a proton (p):

$$n \rightarrow p + e- + \nu_e.$$

At the fundamental level (as depicted in the Feynman diagram on the left), this is caused by the conversion of the negatively charged ($-\frac{1}{3}$ e) down quark to the positively charged ($+\frac{2}{3}$ e) up quark by emission of a $W-$ boson; the $W-$ boson subsequently decays into an electron and an electron antineutrino:

6.2. BETA DECAY

$$d \rightarrow u + e^- + \bar{\nu}_e.$$

The beta spectrum is a continuous spectrum: the total decay energy is divided between the electron and the antineutrino. In the figure to the right, this is shown, by way of example, for an electron of 0.4 MeV energy. In this example, the antineutrino then gets the remainder: 0.76 MeV, since the total decay energy is assumed to be 1.16 MeV.

β− decay generally occurs in neutron-rich nuclei.[4]

6.2.2 β*+ decay

Main article: Positron emission

In β+ decay, or "positron emission", the weak interaction converts an atomic nucleus into a nucleus with atomic number decreased by one, while emitting a positron (e+) and an electron neutrino (ν_e). The generic equation is:

$$^A_Z X \rightarrow\ ^A_{Z-1} X' + e^+ + \nu_e\ [1]$$

This may be considered as the decay of a proton inside the nucleus to a neutron

$$p \rightarrow n + e^+ + \nu_e\ [1]$$

However, β+ decay cannot occur in an isolated proton because it requires energy due to the mass of the neutron being greater than the mass of the proton. β+ decay can only happen inside nuclei when the daughter nucleus has a greater binding energy (and therefore a lower total energy) than the mother nucleus. The difference between these energies goes into the reaction of converting a proton into a neutron, a positron and a neutrino and into the kinetic energy of these particles. In an opposite process to negative beta decay, the weak interaction converts a proton into a neutron by converting an up quark into a down quark by having it emit a W+ or absorb a W−.

6.2.3 Electron capture (K-capture)

Main article: Electron capture

In all cases where β+ decay of a nucleus is allowed energetically, so is electron capture, the process in which the same nucleus captures an atomic electron with the emission of a neutrino:

$$^A_Z X + e^- \rightarrow\ ^A_{Z-1} X' + \nu_e$$

The emitted neutrino is mono-energetic. In proton-rich nuclei where the energy difference between initial and final states is less than $2m_e c^2$, β+ decay is not energetically possible, and electron capture is the sole decay mode.[3]

If the captured electron comes from the innermost shell of the atom, the K-shell, which has the highest probability to interact with the nucleus, the process is called K-capture.[5] If it comes from the L-shell, the process is called L-capture, etc.

6.2.4 Competition of beta decay types

Three types of beta decay in competition are illustrated by the single isotope copper-64 (29 protons, 35 neutrons), which has a half-life of about 12.7 hours. This isotope has one unpaired proton and one unpaired neutron, so either the proton or the neutron can decay. This particular nuclide (though not all nuclides in this situation) is almost equally likely to decay through proton decay by positron emission (18%) or electron capture (43%), as through neutron decay by electron emission (39%).

6.2.5 Helicity (polarization) of neutrinos, electrons and positrons emitted in beta decay

After the discovery of parity non-conservation (see history below), it was found that, in beta decay, electrons are emitted mostly with negative helicity, i.e., they move, naively speaking, like left-handed screws driven into a material (they have negative longitudinal polarization).[6] Conversely, positrons have mostly positive helicity, i.e., they move like right-handed screws. Neutrinos (emitted in positron decay) have positive helicity, while antineutrinos (emitted in electron decay) have negative helicity.[7]

The higher the energy of the particles, the higher their polarization.

6.2.6 Energy release

The Q value is defined as the total energy released in a given nuclear decay. In beta decay, Q is therefore also the sum of the kinetic energies of the emitted beta particle, neutrino, and recoiling nucleus. (Because of the large mass of the nucleus compared to that of the beta particle and neutrino, the kinetic energy of the recoiling nucleus can generally be neglected.) Beta particles can therefore be emitted with any kinetic energy ranging from 0 to Q.[1] A typical Q is around 1 MeV, but can range from a few keV to a few tens of MeV.

Since the rest mass of the electron is 511 keV, the most energetic beta particles are ultrarelativistic, with speeds very close to the speed of light.

β^- decay

Consider the generic equation for beta decay

$$^{A}_{Z}X \rightarrow {}^{A}_{Z+1}X' + e^- + \nu_e.$$

The Q value for this decay is

$$Q = \left[m_N \left({}^{A}_{Z}X \right) - m_N \left({}^{A}_{Z+1}X' \right) - m_e - m_{\bar{\nu}_e} \right] c^2$$

where $m_N \left({}^{A}_{Z}X \right)$ is the mass of the nucleus of the $^{A}_{Z}X$ atom, m_e is the mass of the electron, and $m_{\bar{\nu}_e}$ is the mass of the electron antineutrino. In other words, the total energy released is the mass energy of the initial nucleus, minus the mass energy of the final nucleus, electron, and antineutrino. The mass of the nucleus m_N is related to the standard atomic mass m by

$$m \left({}^{A}_{Z}X \right) c^2 = m_N \left({}^{A}_{Z}X \right) c^2 + Z m_e c^2 - \sum_{i=1}^{Z} B_i$$

That is, the total atomic mass is the mass of the nucleus, plus the mass of the electrons, minus the binding energy B_i of each electron. Substituting this into our original equation, while neglecting the nearly-zero antineutrino mass and difference in electron binding energy, which is very small for high-Z atoms, we have

$$Q = \left[m \left({}^{A}_{Z}X \right) - m \left({}^{A}_{Z+1}X' \right) \right] c^2$$

This energy is carried away as kinetic energy by the electron and neutrino.

Because the reaction will proceed only when the Q-value is positive, β^- decay can occur when the mass of atom $^{A}_{Z}X$ is greater than the mass of atom $^{A}_{Z+1}X'$.[8]

β^+ decay

The equations for β^+ decay are similar, with the generic equation

$$^{A}_{Z}X \rightarrow {}^{A}_{Z-1}X' + e^+ + \nu_e$$

giving

$$Q = \left[m_N \left({}^{A}_{Z}X \right) - m_N \left({}^{A}_{Z-1}X' \right) - m_e - m_{\nu_e} \right] c^2$$

However, in this equation, the electron masses do not cancel, and we are left with

$$Q = \left[m \left({}^{A}_{Z}X \right) - m \left({}^{A}_{Z-1}X' \right) - 2m_e \right] c^2$$

Because the reaction will proceed only when the Q-value is positive, β^+ decay can occur when the mass of atom $^{A}_{Z}X$ exceeds that of $^{A}_{Z-1}X'$ by at least twice the mass of the electron.[8]

Electron capture

The analogous calculation for electron capture must take into account the binding energy of the electrons. This is because the atom will be left in an excited state after capturing the electron, and the binding energy of the captured innermost electron is significant. Using the generic equation for electron capture

$$^{A}_{Z}X + e^- \rightarrow {}^{A}_{Z-1}X' + \nu_e$$

we have

$$Q = \left[m_N \left({}^{A}_{Z}X \right) + m_e - m_N \left({}^{A}_{Z-1}X' \right) - m_{\nu_e} \right] c^2$$

which simplifies to

$$Q = \left[m \left({}^{A}_{Z}X \right) - m \left({}^{A}_{Z-1}X' \right) \right] c^2 - B_n$$

where B_n is the binding energy of the captured electron.

Because the binding energy of the electron is much less than the mass of the electron, nuclei that can undergo β^+ decay can always also undergo electron capture, but the reverse is not true.[8]

6.2.7 Nuclear transmutation

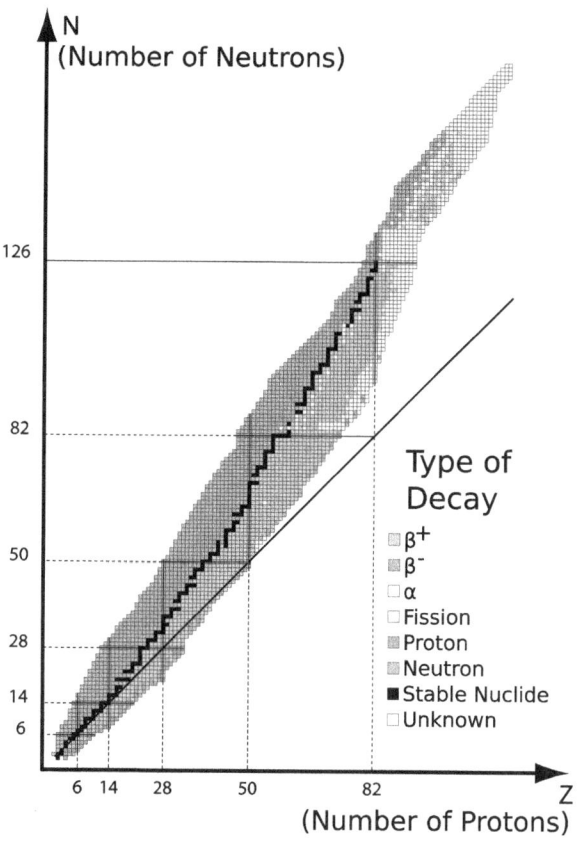

If the proton and neutron are part of an atomic nucleus, these decay processes transmute one chemical element into another. For example:

Beta decay does not change the number A of nucleons in the nucleus, but changes only its charge Z. Thus the set of all nuclides with the same A can be introduced; these *isobaric* nuclides may turn into each other via beta decay. Among them, several nuclides (at least one for any given mass number A) are beta stable, because they present local minima of the mass excess: if such a nucleus has (A, Z) numbers, the neighbour nuclei (A, Z−1) and (A, Z+1) have higher mass excess and can beta decay into (A, Z), but not vice versa. For all odd mass numbers A, there is only one known beta-stable isobar. For even A, there are up to three different beta-stable isobars experimentally known; for example, 96 40Zr, 96 42Mo, and 96 44Ru are all beta-stable. There are about 355 known beta-decay stable nuclides total.*[9]

Usually, unstable nuclides are clearly either "neutron rich" or "proton rich", with the former undergoing beta decay and the latter undergoing electron capture (or more rarely, due to the higher energy requirements, positron decay). However, in a few cases of odd-proton, odd-neutron radionuclides, it may be energetically favorable for the radionuclide to decay to an even-proton, even-neutron isobar either by undergoing beta-positive or beta-negative decay. An often-cited example is 64 29Cu, which decays by positron emission/electron capture 61% of the time to 64 28Ni, and 39% of the time by (negative) beta decay to 64 30Zn.*[10]

Most naturally occurring isotopes on Earth are beta stable. Those that are not have half-lives ranging from under a second to periods of time significantly greater than the age of the universe. One common example of a long-lived isotope is the odd-proton odd-neutron nuclide 40 19K, which undergoes all three types of beta decay ($\beta-$, $\beta+$ and electron capture) with a half-life of 1.277×10^9 years.*[11]

6.2.8 Double beta decay

Main article: Double beta decay

Some nuclei can undergo double beta decay ($\beta\beta$ decay) where the charge of the nucleus changes by two units. Double beta decay is difficult to study, as the process has an extremely long half-life. In nuclei for which both β decay and $\beta\beta$ decay are possible, the rarer $\beta\beta$ decay process is effectively impossible to observe. However, in nuclei where β decay is forbidden but $\beta\beta$ decay is allowed, the process can be seen and a half-life measured.*[12] Thus, $\beta\beta$ decay is usually studied only for beta stable nuclei. Like single beta decay, double beta decay does not change A; thus, at least one of the nuclides with some given A has to be stable with regard to both single and double beta decay.

"Ordinary" double beta decay results in the emission of two electrons and two antineutrinos. If neutrinos are Majorana particles (i.e., they are their own antiparticles), then a decay known as neutrinoless double beta decay will occur. Most neutrino physicists believe that neutrinoless double beta decay has never been observed.*[12]

6.2.9 Bound-state β^- decay

A very small minority of free neutron decays (about four per million) are so-called "two-body decays", in which the proton, electron and antineutrino are produced, but the electron fails to gain the 13.6 eV energy necessary to escape the proton, and therefore simply remains bound to it, as a neutral hydrogen atom.*[13] In this type of beta decay, in essence all of the neutron decay energy is carried off by the antineutrino.

For fully ionized atoms (bare nuclei), it is possible in likewise manner for electrons to fail to escape the atom, and to be emitted from the nucleus into low-lying atomic bound states (orbitals). This can not occur for neutral atoms whose low-lying bound states are already filled by electrons.

The phenomenon in fully ionized atoms was first observed for ^{163}Dy*66+ in 1992 by Jung et al. of the Darmstadt Heavy-Ion Research group. Although neutral ^{163}Dy is a stable isotope, the fully ionized ^{163}Dy*66+ undergoes β decay into the K and L shells with a half-life of 47 days.*[14]

Another possibility is that a fully ionized atom undergoes greatly accelerated β decay, as observed for ^{187}Re by Bosch et al., also at Darmstadt. Neutral ^{187}Re does undergo β decay with a half-life of 42×10^9 years, but for fully ionized ^{187}Re*75+ this is shortened by a factor of 10^9 to only 32.9 years.*[15] For comparison the variation of decay rates of other nuclear processes due to chemical environment is less than 1%.

6.2.10 Forbidden transitions

Beta decays can be classified according to the L-value of the emitted radiation. When $L > 0$, the decay is referred to as "forbidden". Nuclear selection rules require high L-values to be accompanied by changes in nuclear spin (J) and parity (π). The selection rules for the Lth forbidden transitions are:

$$\Delta J = L - 1, L, L+1; \Delta \pi = (-1)^L,$$

where $\Delta \pi = 1$ or -1 corresponds to no parity change or parity change, respectively. The special case of a transition between isobaric analogue states, where the structure of the final state is very similar to the structure of the initial state, is referred to as "superallowed" for beta decay, and proceeds very quickly. The following table lists the ΔJ and Δπ values for the first few values of L:

6.2.11 Fermi transitions

A **Fermi transition** is a beta decay in which the spins of the emitted electron (positron) and anti-neutrino (neutrino) couple to total spin $S = 0$, leading to an angular momentum change $\Delta J = 0$ between the initial and final states of the nucleus (assuming an allowed transition $\Delta L = 0$). In the non-relativistic limit, the nuclear part of the operator for a Fermi transition is given by

$$\mathcal{O}_F = G_V \sum_a \hat{\tau}_{a\pm}$$

with G_V the weak vector coupling constant, $\tau_\pm$ the isospin raising and lowering operators, and a running over all protons and neutrons in the nucleus.

6.2.12 Gamow-Teller transitions

A **Gamow-Teller transition** is a beta decay in which the spins of the emitted electron (positron) and anti-neutrino (neutrino) couple to total spin $S = 1$, leading to an angular momentum change $\Delta J = 0, \pm 1$ between the initial and final states of the nucleus (assuming an allowed transition). In this case, the nuclear part of the operator is given by

$$\mathcal{O}_{GT} = G_A \sum_a \hat{\sigma}_a \hat{\tau}_{a\pm}$$

with G_A the weak axial-vector coupling constant, and σ the spin Pauli matrices, which can produce a spin-flip in the decaying nucleon.

6.2.13 Beta emission spectrum

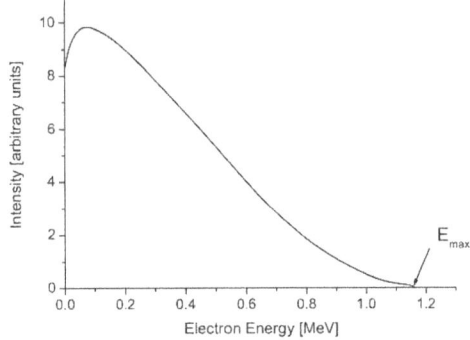

Beta spectrum of ^{210}Bi. E_{max} = Q = 1.16 MeV is the maximum energy

Beta decay can be considered as a perturbation as described in quantum mechanics, and thus Fermi's Golden Rule can be applied. This leads to an expression for the kinetic energy spectrum $N(T)$ of emitted betas as follows:*[16]

$$N(T) = C_L(T) F(Z, T) p E (Q - T)^2$$

where T is the kinetic energy, C_L is a shape function that depends on the forbiddenness of the decay (it is constant for allowed decays), $F(Z, T)$ is the Fermi Function (see below) with Z the charge of the final-state nucleus, $E = T + mc^2$ is the total energy, $p = \sqrt{(E/c)^2 - (mc)^2}$ is the momentum, and Q is the Q value of the decay. The kinetic energy of the emitted neutrino is given approximately by Q minus the kinetic energy of the beta.

As an example, the beta decay spectrum of ^{210}Bi (originally called RaE) is shown to the right.

Fermi function

The Fermi function that appears in the beta spectrum formula accounts for the Coulomb attraction / repulsion between the emitted beta and the final state nucleus. Approximating the associated wavefunctions to be spherically symmetric, the Fermi function can be analytically calculated to be:[17]

$$F(Z, T) = \frac{2(1 + S)}{\Gamma(1 + 2S)^2}(2p\rho)^{2S-2} e^{\pi \eta} |\Gamma(S + i\eta)|^2,$$

where $S = \sqrt{1 - \alpha^2 Z^2}$ (α is the fine-structure constant), $\eta = \pm \alpha Z E/pc$ (+ for electrons, − for positrons), $\varrho = r_N/\hbar$ (r_N is the radius of the final state nucleus), and Γ is the Gamma function.

For non-relativistic betas ($Q \ll m_e c^2$), this expression can be approximated by:[18]

$$F(Z, T) \approx \frac{2\pi\eta}{1 - e^{-2\pi\eta}}.$$

Other approximations can be found in the literature.[19][20]

Kurie plot

A **Kurie plot** (also known as a **Fermi–Kurie plot**) is a graph used in studying beta decay developed by Franz N. D. Kurie, in which the square root of the number of beta particles whose momenta (or energy) lie within a certain narrow range, divided by the Fermi function, is plotted against beta-particle energy.[21][22] It is a straight line for allowed transitions and some forbidden transitions, in accord with the Fermi beta-decay theory. The energy-axis (x-axis) intercept of a Kurie plot corresponds to the maximum energy imparted to the electron/positron (the decay's Q-value). With a Kurie plot one can find the limit on the effective mass of a neutrino.[23]

6.2.14 History

Discovery and characterization of β* − decay

Radioactivity was discovered in 1896 by Henri Becquerel in uranium, and subsequently observed by Marie and Pierre Curie in thorium and in the new elements polonium and radium. In 1899, Ernest Rutherford separated radioactive emissions into two types: alpha and beta (now beta minus), based on penetration of objects and ability to cause ionization. Alpha rays could be stopped by thin sheets of paper or aluminium, whereas beta rays could penetrate several millimetres of aluminium. (In 1900, Paul Villard identified a still more penetrating type of radiation, which Rutherford identified as a fundamentally new type in 1903, and termed gamma rays).

In 1900, Becquerel measured the mass-to-charge ratio (*m/e*) for beta particles by the method of J.J. Thomson used to study cathode rays and identify the electron. He found that *m/e* for a beta particle is the same as for Thomson's electron, and therefore suggested that the beta particle is in fact an electron.

In 1901, Rutherford and Frederick Soddy showed that alpha and beta radioactivity involves the transmutation of atoms into atoms of other chemical elements. In 1913, after the products of more radioactive decays were known, Soddy and Kazimierz Fajans independently proposed their radioactive displacement law, which states that beta (i.e., β−) emission from one element produces another element one place to the right in the periodic table, while alpha emission produces an element two places to the left.

Neutrinos in beta decay

Historically, the study of beta decay provided the first physical evidence of the neutrino. Measurements of the beta particle (electron) kinetic energy spectrum in 1911 by Lise Meitner and Otto Hahn and in 1913 by Jean Danysz showed multiple lines on a diffuse background, offering the first hint of a continuous spectrum.[24] In 1914, James Chadwick used a magnetic spectrometer with one of Hans Geiger's new counters to make a more accurate measurement and showed that the spectrum was continuous.[24][25] This was in apparent contradiction to the law of conservation of energy, since if beta decay were simply electron emission as assumed at the time, then the energy of the emitted electron should equal the energy difference between the initial and final nuclear states and lead to a narrow energy distribution, as observed for both alpha and gamma decay.[26] For beta decay, however, the observed broad continuous spectrum suggested that energy is lost in the beta decay process.

In 1920–1927, Charles Drummond Ellis (along with James Chadwick and colleagues) further established that the beta decay spectrum is continuous, ending all controversies. It also had an effective upper bound in energy, which was a severe blow to Bohr's suggestion that conservation of energy might be true only in a statistical sense, and might be violated in any given decay. Now the problem of how to account for the variability of energy in known beta decay products, as well as for conservation of momentum and angular momentum in the process, became acute.

A second problem related to the conservation of angular momentum. Molecular band spectra showed that the nuclear spin of nitrogen-14 is 1 (i.e. equal to the reduced Planck constant), and more generally that the spin is integral for nuclei of even mass number and half-integral for nuclei of odd mass number, as later explained by the proton-neutron model of the nucleus.[26] Beta decay leaves the mass number unchanged, so that the change of nuclear spin must be an integer. However the electron spin is 1/2, so that angular momentum would not be conserved if beta decay were simply electron emission.

In a famous letter written in 1930, Wolfgang Pauli suggested that, in addition to electrons and protons, atomic nuclei also contained an extremely light neutral particle, which he called the neutron. He suggested that this "neutron" was also emitted during beta decay (thus accounting for the known missing energy, momentum, and angular momentum) and had simply not yet been observed. In 1931, Enrico Fermi renamed Pauli's "neutron" to neutrino and, in 1934, he published a very successful model of beta decay in which neutrinos were produced. The neutrino interaction with matter was so weak that detecting it proved a severe experimental challenge, which was finally met in 1956 in the Cowan–Reines neutrino experiment.[27] However, the properties of neutrinos were (with a few minor modifications) as predicted by Pauli and Fermi.

Discovery of other types of beta decay

In 1934, Frédéric and Irène Joliot-Curie bombarded aluminium with alpha particles to effect the nuclear reaction 4 2He + 27 13Al → 30 15P + 1 0n, and observed that the product isotope 30 15P emits a positron identical to those found in cosmic rays by Carl David Anderson in 1932. This was the first example of β+ decay (positron emission), which they termed artificial radioactivity since 30 15P is a short-lived nuclide which does not exist in nature.

The theory of electron capture was first discussed by Gian-Carlo Wick in a 1934 paper, and then developed by Hideki Yukawa and others. K-electron capture was first observed in 1937 by Luis Alvarez, in the nuclide ^{48}V.[28][29][30] Alvarez went on to study electron capture in ^{67}Ga and other nuclides.[28][31][32]

Non-conservation of parity

In 1956, Chien-Shiung Wu and coworkers proved in the Wu experiment that parity is not conserved in beta decay.[33][34] This surprising fact had been postulated shortly before in an article by Tsung-Dao Lee and Chen Ning Yang.[35]

6.2.15 See also

- Double beta decay
- Electron capture
- Neutrino
- Alpha decay
- Betavoltaics
- Particle radiation
- Radionuclide
- Tritium illumination, a form of fluorescent lighting powered by beta decay
- Pandemonium effect
- Total absorption spectroscopy

6.2.16 References

- Tuli, J. K. (2011). *Nuclear Wallet Cards* (PDF) (8th ed.). Brookhaven National Laboratory.

[1] Konya, J.; Nagy, N. M. (2012). *Nuclear and Radiochemistry*. Elsevier. pp. 74–75. ISBN 978-0-12-391487-3.

[2] Basdevant, Jean-Louis; Rich, James; Spiro, Michael (2005). *Fundamentals in Nuclear Physics: From Nuclear Structure to Cosmology*. Springer. ISBN 978-0387016726.

[3] Zuber, Kai (2011). *Neutrino Physics* (2 ed.). CRC Press. p. 466. ISBN 9781420064711.

[4] Loveland, Walter D. (2005). *Modern Nuclear Chemistry*. Wiley. p. 232. ISBN 0471115320.

[5] Tatjana Jevremovic (21 April 2009). *Nuclear Principles in Engineering*. Springer Science & Business Media. p. 201. ISBN 978-0-387-85608-7.

[6] H. Frauenfelder, R. Bobone, E. Von Goeler, N. Levine, H. R. Lewis, R. N. Peacock, A. Rossi and G. DePasquali, Physical Review 106 (1957) 386

[7] E. J. Konopinski and M. E. Rose, *The Theory of nuclear Beta Decay*, in: *Alpha-, Beta- and Gamma-Ray Spectroscopy*, ed. by Kai Siegbahn, Vol. 2, North-Holland Publishing Company, Amsterdam, 1966

[8] Kenneth S. Krane (5 November 1987). *Introductory Nuclear Physics*. Wiley. ISBN 978-0-471-80553-3.

[9] "Interactive Chart of Nuclides". National Nuclear Data Center, Brookhaven National Laboratory. Retrieved 2014-09-18.

[10] "WWW Table of Radioactive Isotopes, Copper 64". *LBNL Isotopes Project*. Lawrence Berkeley National Laboratory. Retrieved 2014-09-18.

[11] "WWW Table of Radioactive Isotopes, Potassium 40". *LBNL Isotopes Project*. Lawrence Berkeley National Laboratory. Retrieved 2014-09-18.

[12] S.M. Bilenky (October 5, 2010). "Neutrinoless double beta-decay". *Physics of Particles and Nuclei* **41** (5). arXiv:1001.1946. Bibcode:2010PPN....41..690B. doi:10.1134/S1063779610050035.

[13] An Overview Of Neutron Decay J. Byrne in Quark-Mixing, CKM Unitarity (H.Abele and D.Mund, 2002), see p.XV

[14] Jung, M. et al. (1992). "First observation of bound-state β^-– decay". *Physical Review Letters* **69** (15): 2164–2167. Bibcode:1992PhRvL..69.2164J. doi:10.1103/PhysRevLett.69.2164. PMID 10046415.

[15] Bosch, F. et al. (1996). "Observation of bound-state beta minus decay of fully ionized ^{187}Re: ^{187}Re–^{187}Os Cosmochronometry". *Physical Review Letters* **77** (26): 5190–5193. Bibcode:1996PhRvL..77.5190B. doi:10.1103/PhysRevLett.77.5190. PMID 10062738.

[16] Nave, C. R. "Energy and Momentum Spectra for Beta Decay". *HyperPhysics*. Retrieved 2013-03-09.

[17] Fermi, E. (1934). "Versuch einer Theorie der β-Strahlen. I". *Zeitschrift für Physik* **88** (3–4): 161–177. Bibcode:1934ZPhy...88..161F. doi:10.1007/BF01351864.

[18] Mott, N. F.; Massey, H. S. W. (1933). *The Theory of Atomic Collisions*. Clarendon Press. LCCN 34001940.

[19] Venkataramaiah, P.; Gopala, K.; Basavaraju, A.; Suryanarayana, S. S.; Sanjeeviah, H. (1985). "A simple relation for the Fermi function". *Journal of Physics G* **11** (3): 359–364. Bibcode:1985JPhG...11..359V. doi:10.1088/0305-4616/11/3/014.

[20] Schenter, G. K.; Vogel, P. (1983). "A simple approximation of the fermi function in nuclear beta decay". *Nuclear Science and Engineering* **83** (3): 393–396. OSTI 5307377.

[21] Kurie, F. N. D.; Richardson, J. R.; Paxton, H. C. (1936). "The Radiations Emitted from Artificially Produced Radioactive Substances. I. The Upper Limits and Shapes of the β-Ray Spectra from Several Elements". *Physical Review* **49** (5): 368–381. Bibcode:1936PhRv...49..368K. doi:10.1103/PhysRev.49.368.

[22] Kurie, F. N. D. (1948). "On the Use of the Kurie Plot". *Physical Review* **73** (10): 1207. Bibcode:1948PhRv...73.1207K. doi:10.1103/PhysRev.73.1207.

[23] Rodejohann, Werner (2012). "Neutrinoless double beta decay and neutrino physics". arXiv:1206.2560v2.

[24] Jensen, Carsten (2000). *Controversy and Consensus: Nuclear Beta Decay 1911-1934*. Birkhäuser Verlag. ISBN 3-7643-5313-9.

[25] Chadwick, James (1914). "Intensitätsverteilung im magnetischen Spektren der β-Strahlen von Radium B + C". *Verhandlungen der Deutschen Physikalischen Gesellschaft* (in German) (Deutsche Physikalische Gesellschaft) **16**: 383–391.

[26] Brown, Laurie M. (1978). "The idea of the neutrino". *Physics Today* **31** (9): 23–8. Bibcode:1978PhT....31i..23B. doi:10.1063/1.2995181.

[27] C. L Cowan Jr., F. Reines, F. B. Harrison, H. W. Kruse, A. D McGuire (July 20, 1956). "Detection of the Free Neutrino: a Confirmation". *Science* **124** (3212): 103–4. Bibcode:1956Sci...124..103C. doi:10.1126/science.124.3212.103. PMID 17796274.

[28] Segré, E. (1987). "K-Electron Capture by Nuclei". In Trower, P. W. *Discovering Alvarez: Selected Works of Luis W. Alvarez*. University of Chicago Press. pp. 11–12. ISBN 978-0-226-81304-2.

[29] "The Nobel Prize in Physics 1968: Luis Alvarez". The Nobel Foundation. Retrieved 2009-10-07.

[30] Alvarez, L. W. (1937). "Nuclear K Electron Capture". *Physical Review* **52** (2): 134–135. Bibcode:1937PhRv...52..134A. doi:10.1103/PhysRev.52.134.

[31] Alvarez, L. W. (1938). "Electron Capture and Internal Conversion in Gallium 67". *Physical Review* **53** (7): 606. Bibcode:1938PhRv...53..606A. doi:10.1103/PhysRev.53.606.

[32] Alvarez, L. W. (1938). "The Capture of Orbital Electrons by Nuclei". *Physical Review* **54** (7): 486–497. Bibcode:1938PhRv...54..486A. doi:10.1103/PhysRev.54.486.

[33] C. S. Wu; E. Ambler; R. W. Hayward; D. D. Hoppes; R. P. Hudson (1957). "Experimental Test of Parity Conservation in Beta Decay". *Physical Review* **105**: 1413–1415. Bibcode:1957PhRv..105.1413W. doi:10.1103/PhysRev.105.1413.

[34] http://blogs.scientificamerican.
com/guest-blog/2013/10/15/
channeling-ada-lovelace-chien-shiung-wu-courageous-hero-of-physics/

[35] T. D. Lee, C. N. Yang (1956). "Question of Parity Conservation in Weak Interactions". *Physical Review* **104**: 254–258. Bibcode:1956PhRv..104..254L. doi:10.1103/PhysRev.104.254.

6.2.17 Bibliography

- Sin-Itiro Tomonaga (1997). *The Story of Spin*. University of Chicago Press.

6.2.18 External links

-  **The Live Chart of Nuclides - IAEA** with filter on decay type

- Definition of Beta Disintegration (Decay) at Science Dictionary

6.3 CKM matrix

In the Standard Model of particle physics, the **Cabibbo–Kobayashi–Maskawa matrix** (**CKM matrix**, **quark mixing matrix**, sometimes also called **KM matrix**) is a unitary matrix which contains information on the strength of flavour-changing weak decays. Technically, it specifies the mismatch of quantum states of quarks when they propagate freely and when they take part in the weak interactions. It is important in the understanding of CP violation. This matrix was introduced for three generations of quarks by Makoto Kobayashi and Toshihide Maskawa, adding one generation to the matrix previously introduced by Nicola Cabibbo. This matrix is also an extension of the GIM mechanism, which only includes two of the three current families of quarks.

6.3.1 The matrix

In 1963, Nicola Cabibbo introduced the Cabibbo angle (θ_c) to preserve the universality of the weak interaction.[*][1] Cabibbo was inspired by previous work by Murray Gell-Mann and Maurice Lévy,[*][2] on the effectively rotated non-strange and strange vector and axial weak currents, which he references.[*][3]

In light of current knowledge (quarks were not yet theorized), the Cabibbo angle is related to the relative probability that down and strange quarks decay into up quarks

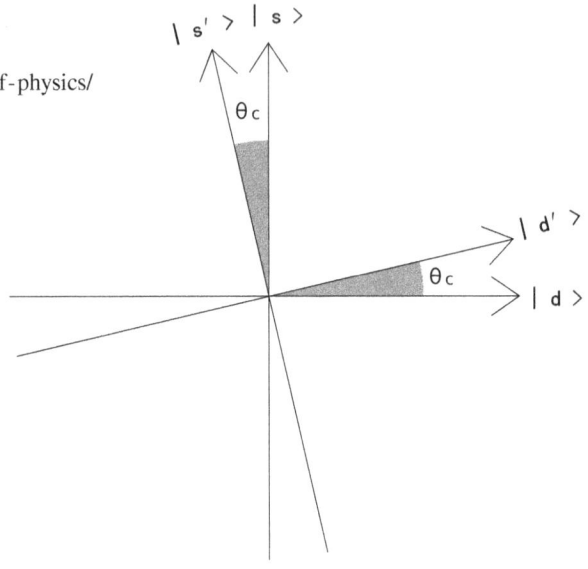

The Cabibbo angle represents the rotation of the mass eigenstate vector space formed by the mass eigenstates $|d\rangle, |s\rangle$ into the weak eigenstate vector space formed by the weak eigenstates $|d'\rangle, |s'\rangle$. $\theta_C = 13.02°$.

($|V_{ud}|^2$ and $|V_{us}|^2$ respectively). In particle physics parlance, the object that couples to the up quark via charged-current weak interaction is a superposition of down-type quarks, here denoted by d'.[*][4] Mathematically this is:

$$d' = V_{ud}d + V_{us}s,$$

or using the Cabibbo angle:

$$d' = \cos\theta_c d + \sin\theta_c s.$$

Using the currently accepted values for $|V_{ud}|$ and $|V_{us}|$ (see below), the Cabibbo angle can be calculated using

$$\tan\theta_c = \frac{|V_{us}|}{|V_{ud}|} = \frac{0.22534}{0.97427} \rightarrow \theta_c = 13.02°.$$

When the charm quark was discovered in 1974, it was noticed that the down and strange quark could decay into either the up or charm quark, leading to two sets of equations:

$$d' = V_{ud}d + V_{us}s;$$
$$s' = V_{cd}d + V_{cs}s,$$

or using the Cabibbo angle:

$$d' = \cos\theta_c d + \sin\theta_c s;$$

6.3. CKM MATRIX

$s' = -\sin\theta_c d + \cos\theta_c s.$

This can also be written in matrix notation as:

$$\begin{bmatrix} d' \\ s' \end{bmatrix} = \begin{bmatrix} V_{ud} & V_{us} \\ V_{cd} & V_{cs} \end{bmatrix} \begin{bmatrix} d \\ s \end{bmatrix},$$

or using the Cabibbo angle

$$\begin{bmatrix} d' \\ s' \end{bmatrix} = \begin{bmatrix} \cos\theta_c & \sin\theta_c \\ -\sin\theta_c & \cos\theta_c \end{bmatrix} \begin{bmatrix} d \\ s \end{bmatrix},$$

where the various $|V_{ij}|^2$ represent the probability that the quark of j flavor decays into a quark of i flavor. This 2×2 rotation matrix is called the Cabibbo matrix.

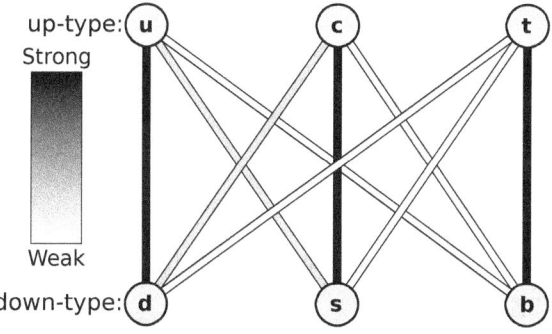

A pictorial representation of the six quarks' decay modes, with mass increasing from left to right.

Observing that CP-violation could not be explained in a four-quark model, Kobayashi and Maskawa generalized the Cabibbo matrix into the Cabibbo–Kobayashi–Maskawa matrix (or CKM matrix) to keep track of the weak decays of three generations of quarks:[*][5]

$$\begin{bmatrix} d' \\ s' \\ b' \end{bmatrix} = \begin{bmatrix} V_{ud} & V_{us} & V_{ub} \\ V_{cd} & V_{cs} & V_{cb} \\ V_{td} & V_{ts} & V_{tb} \end{bmatrix} \begin{bmatrix} d \\ s \\ b \end{bmatrix}.$$

On the left is the weak interaction doublet partners of up-type quarks, and on the right is the CKM matrix along with a vector of mass eigenstates of down-type quarks. The CKM matrix describes the probability of a transition from one quark i to another quark j. These transitions are proportional to $|V_{ij}|^2$.

Currently, the best determination of the magnitudes of the CKM matrix elements is:[*][6]

$$\begin{bmatrix} |V_{ud}| & |V_{us}| & |V_{ub}| \\ |V_{cd}| & |V_{cs}| & |V_{cb}| \\ |V_{td}| & |V_{ts}| & |V_{tb}| \end{bmatrix} = \begin{bmatrix} 0.97427 \pm 0.00015 & 0.22534 \pm 0.00065 & 0.00351^{+0.00015}_{-0.00014} \\ 0.22520 \pm 0.00065 & 0.97344 \pm 0.00016 & 0.0412^{+0.0011}_{-0.0005} \\ 0.00867^{+0.00029}_{-0.00031} & 0.0404^{+0.0011}_{-0.0005} & 0.999146^{+0.000021}_{-0.000046} \end{bmatrix}$$

Note that the choice of usage of down-type quarks in the definition is purely arbitrary and does not represent some sort of deep physical asymmetry between up-type and down-type quarks. We could just as easily define the matrix the other way around, describing weak interaction partners of mass eigenstates of up-type quarks, u', c' and t', in terms of u, c, and t. Since the CKM matrix is unitary (and therefore its inverse is the same as its conjugate transpose), we would obtain essentially the same matrix.

6.3.2 Counting

To proceed further, it is necessary to count the number of parameters in this matrix, V which appear in experiments, and therefore are physically important. If there are N generations of quarks ($2N$ flavours) then

- An $N \times N$ unitary matrix (that is, a matrix V such that $VV^{*\dagger} = I$, where $V^{*\dagger}$ is the conjugate transpose of V and I is the identity matrix) requires N^2 real parameters to be specified.

- $2N - 1$ of these parameters are not physically significant, because one phase can be absorbed into each quark field (both of the mass eigenstates, and of the weak eigenstates), but an overall common phase is unobservable. Hence, the total number of free variables independent of the choice of the phases of basis vectors is $N^2 - (2N - 1) = (N - 1)^2$.

  - Of these, $N(N - 1)/2$ are rotation angles called quark *mixing angles*.
  - The remaining $(N - 1)(N - 2)/2$ are complex phases, which cause CP violation.

For the case $N = 2$, there is only one parameter which is a mixing angle between two generations of quarks. Historically, this was the first version of CKM matrix when only two generations were known. It is called the **Cabibbo angle** after its inventor Nicola Cabibbo.

For the Standard Model case ($N = 3$), there are three mixing angles and one CP-violating complex phase.[*][7]

6.3.3 Observations and predictions

Cabibbo's idea originated from a need to explain two observed phenomena:

1. the transitions $u \leftrightarrow d$, $e \leftrightarrow \nu_e$, and $\mu \leftrightarrow \nu_\mu$ had similar amplitudes.
2. the transitions with change in strangeness $\Delta S = 1$ had amplitudes equal to 1/4 of those with $\Delta S = 0$.

Cabibbo's solution consisted of postulating weak universality to resolve the first issue, along with a mixing angle θ_c, now called the *Cabibbo angle*, between the *d* and *s* quarks to resolve the second.

For two generations of quarks, there are no CP violating phases, as shown by the counting of the previous section. Since CP violations were seen in neutral kaon decays already in 1964, the emergence of the Standard Model soon after was a clear signal of the existence of a third generation of quarks, as pointed out in 1973 by Kobayashi and Maskawa. The discovery of the bottom quark at Fermilab (by Leon Lederman's group) in 1976 therefore immediately started off the search for the missing third-generation quark, the top quark.

Note, however, that the specific values of the angles are *not* a prediction of the standard model: they are open, unfixed parameters. At this time, there is no generally accepted theory that explains why the measured values are what they are.

6.3.4 Weak universality

The constraints of unitarity of the CKM-matrix on the diagonal terms can be written as

$$\sum_k |V_{ik}|^2 = \sum_i |V_{ik}|^2 = 1$$

for all generations *i*. This implies that the sum of all couplings of any of the up-type quarks to all the down-type quarks is the same for all generations. This relation is called *weak universality* and was first pointed out by Nicola Cabibbo in 1967. Theoretically it is a consequence of the fact that all SU(2) doublets couple with the same strength to the vector bosons of weak interactions. It has been subjected to continuing experimental tests.

6.3.5 The unitarity triangles

The remaining constraints of unitarity of the CKM-matrix can be written in the form

$$\sum_k V_{ik} V_{jk}^* = 0.$$

For any fixed and different *i* and *j*, this is a constraint on three complex numbers, one for each *k*, which says that these numbers form the sides of a triangle in the complex plane. There are six choices of *i* and *j* (three independent), and hence six such triangles, each of which is called a *unitary triangle*. Their shapes can be very different, but they all have the same area, which can be related to the CP violating phase. The area vanishes for the specific parameters in the Standard Model for which there would be no CP violation. The orientation of the triangles depend on the phases of the quark fields.

Since the three sides of the triangles are open to direct experiment, as are the three angles, a class of tests of the Standard Model is to check that the triangle closes. This is the purpose of a modern series of experiments under way at the Japanese BELLE and the American BaBar experiments, as well as at LHCb in CERN, Switzerland.

6.3.6 Parameterizations

Four independent parameters are required to fully define the CKM matrix. Many parameterizations have been proposed, and three of the most common ones are shown below.

KM parameters

The original parameterization of Kobayashi and Maskawa used three angles (θ_1, θ_2, θ_3) and a CP-violating phase (δ).[*][5] Cosines and sines of the angles are denoted c_i and s_i, respectively. θ_1 is the Cabibbo angle.

$$\begin{bmatrix} c_1 & -s_1 c_3 & -s_1 s_3 \\ s_1 c_2 & c_1 c_2 c_3 - s_2 s_3 e^{i\delta} & c_1 c_2 s_3 + s_2 c_3 e^{i\delta} \\ s_1 s_2 & c_1 s_2 c_3 + c_2 s_3 e^{i\delta} & c_1 s_2 s_3 - c_2 c_3 e^{i\delta} \end{bmatrix}.$$

"Standard" parameters

A "standard" parameterization of the CKM matrix uses three Euler angles (θ_{12}, θ_{23}, θ_{13}) and one CP-violating phase (δ_{13}).[*][8] Couplings between quark generation *i* and *j* vanish if $\theta_{ij} = 0$. Cosines and sines of the angles are denoted c_{ij} and s_{ij}, respectively. θ_{12} is the Cabibbo angle.

$$\begin{bmatrix} 1 & 0 & 0 \\ 0 & c_{23} & s_{23} \\ 0 & -s_{23} & c_{23} \end{bmatrix} \begin{bmatrix} c_{13} & 0 & s_{13}e^{-i\delta_{13}} \\ 0 & 1 & 0 \\ -s_{13}e^{i\delta_{13}} & 0 & c_{13} \end{bmatrix} \begin{bmatrix} c_{12} & s_{12} & 0 \\ -s_{12} & c_{12} & 0 \\ 0 & 0 & 1 \end{bmatrix}$$

$$= \begin{bmatrix} c_{12}c_{13} & s_{12}c_{13} & s_{13}e^- \\ -s_{12}c_{23} - c_{12}s_{23}s_{13}e^{i\delta_{13}} & c_{12}c_{23} - s_{12}s_{23}s_{13}e^{i\delta_{13}} & s_{23}c \\ s_{12}s_{23} - c_{12}c_{23}s_{13}e^{i\delta_{13}} & -c_{12}s_{23} - s_{12}c_{23}s_{13}e^{i\delta_{13}} & c_{23}c \end{bmatrix}$$

The currently best known values for the standard parameters are:[*][9]

$\theta_{12} = 13.04 \pm 0.05°$, $\theta_{13} = 0.201 \pm 0.011°$, $\theta_{23} = 2.38 \pm 0.06°$, and $\delta_{13} = 1.20 \pm 0.08$ rad.

Wolfenstein parameters

A third parameterization of the CKM matrix was introduced by Lincoln Wolfenstein with the four parameters λ, A, ϱ, and η.*[10] The four Wolfenstein parameters have the property that all are of order 1 and are related to the "standard" parameterization:

$$\lambda = s_{12}$$
$$A\lambda^2 = s_{23}$$
$$A\lambda^3(\varrho - i\eta) = s_{13}e^{*} - i\delta$$

The Wolfenstein parameterization of the CKM matrix, is an approximation of the standard parameterization. To order λ^3, it is:

$$\begin{bmatrix} 1 - \lambda^2/2 & \lambda & A\lambda^3(\rho - i\eta) \\ -\lambda & 1 - \lambda^2/2 & A\lambda^2 \\ A\lambda^3(1 - \rho - i\eta) & -A\lambda^2 & 1 \end{bmatrix}.$$

The CP violation can be determined by measuring $\varrho - i\eta$.

Using the values of the previous section for the CKM matrix, the best determination of the Wolfenstein parameters is:*[11]

$\lambda = 0.2257 + 0.0009$
-0.0010, $A = 0.814 + 0.021$
-0.022, $\varrho = 0.135 + 0.031$
-0.016, and $\eta = 0.349 + 0.015$
-0.017.

6.3.7 Nobel Prize

In 2008, Kobayashi and Maskawa shared one half of the Nobel Prize in Physics "for the discovery of the origin of the broken symmetry which predicts the existence of at least three families of quarks in nature".*[12] Some physicists were reported to harbor bitter feelings about the fact that the Nobel Prize committee failed to reward the work of Cabibbo, whose prior work was closely related to that of Kobayashi and Maskawa.*[13] Asked for a reaction on the prize, Cabibbo preferred to give no comment.*[14]

6.3.8 See also

- Formulation of the Standard Model and CP violations.

- Quantum chromodynamics, flavour and strong CP problem.

- Weinberg angle, a similar angle for Z and photon mixing.

- Pontecorvo–Maki–Nakagawa–Sakata matrix, the equivalent mixing matrix for neutrinos.

- Koide formula

6.3.9 References

[1] N. Cabibbo (1963). "Unitary Symmetry and Leptonic Decays". *Physical Review Letters* **10** (12): 531–533. Bibcode:1963PhRvL..10..531C. doi:10.1103/PhysRevLett.10.531.

[2] M. Gell-Mann, M. Lévy (1960). "The Axial Vector Current in Beta Decay". *Il Nuovo Cimento* **16** (4): 705–726. doi:10.1007/BF02859738.

[3] L. Maiani (2009). "Sul Premio Nobel Per La Fisica 2008" (PDF). *Il Nuovo Saggiatore* **25** (1–2): 78.

[4] I.S. Hughes (1991). "Chapter 11.1 – Cabibbo Mixing". *Elementary Particles* (3rd ed.). Cambridge University Press. pp. 242–243. ISBN 0-521-40402-9.

[5] M. Kobayashi, T. Maskawa; Maskawa (1973). "CP-Violation in the Renormalizable Theory of Weak Interaction". *Progress of Theoretical Physics* **49** (2): 652–657. Bibcode:1973PThPh..49..652K. doi:10.1143/PTP.49.652.

[6] J. Beringer; Arguin, J. -F.; Barnett, R. M.; Copic, K.; Dahl, O.; Groom, D. E.; Lin, C. -J.; Lys, J.; Murayama, H.; Wohl, C. G.; Yao, W. -M.; Zyla, P. A.; Amsler, C.; Antonelli, M.; Asner, D. M.; Baer, H.; Band, H. R.; Basaglia, T.; Bauer, C. W.; Beatty, J. J.; Belousov, V. I.; Bergren, E.; Bernardi, G.; Bertl, W.; Bethke, S.; Bichsel, H.; Biebel, O.; Blucher, E.; Blusk, S. et al. (2012). "Review of Particles Physics: The CKM Quark-Mixing Matrix" (PDF). *Physical Review D* **80** (1): 1–1526 [162]. Bibcode:2012PhRvD..86a0001B. doi:10.1103/PhysRevD.86.010001.

[7] J.C. Baez (6 March 2005). "Neutrinos and the Mysterious Maki-Nakagawa-Sakata Matrix". Retrieved 2009-01-04. In fact, the Maki–Nakagawa–Sakata matrix actually affects the behavior of all leptons, not just neutrinos. Furthermore, a similar trick works for quarks – but then the matrix U is called the Cabibbo–Kobayashi–Maskawa matrix.

[8] L.L. Chau and W.-Y. Keung (1984). "Comments on the Parametrization of the Kobayashi-Maskawa Matrix". *Physical Review Letters* **53** (19): 1802. Bibcode:1984PhRvL..53.1802C. doi:10.1103/PhysRevLett.53.1802.

[9] Values obtained from values of Wolfenstein parameters in the 2008 *Review of Particle Physics*.

[10] L. Wolfenstein (1983). "Parametrization of the Kobayashi-Maskawa Matrix". *Physical Review Letters* **51** (21): 1945. Bibcode:1983PhRvL..51.1945W. doi:10.1103/PhysRevLett.51.1945.

[11] C. Amsler; Doser, M.; Antonelli, M.; Asner, D.M.; Babu, K.S.; Baer, H.; Band, H.R.; Barnett, R.M.; Bergren, E.; Beringer, J.; Bernardi, G.; Bertl, W.; Bichsel, H.; Biebel, O.; Bloch, P.; Blucher, E.; Blusk, S.; Cahn, R.N.; Carena, M.; Caso, C.; Ceccucci, A.; Chakraborty, D.; Chen, M.-C.; Chivukula, R.S.; Cowan, G.; Dahl, O.; d'Ambrosio, G.; Damour, T.; De Gouvêa, A. et al. (2008). "Review of Particles Physics: The CKM Quark-Mixing Matrix" (PDF). *Physics Letters B* **667**: 1–1340. Bibcode:2008PhLB..667....1P. doi:10.1016/j.physletb.2008.07.018.

[12] "The Nobel Prize in Physics 2008" (Press release). The Nobel Foundation. 7 October 2008. Retrieved 2009-11-24.

[13] V. Jamieson (7 October 2008). "Physics Nobel Snubs key Researcher". *New Scientist*. Retrieved 2009-11-24.

[14] "Nobel, l'amarezza dei fisici italiani". *Corriere della Sera* (in Italian). 7 October 2008. Retrieved 2009-11-24.

6.3.10 Further reading

- D.J Griffiths (2008). *Introduction to Elementary Particles* (2nd ed.). John Wiley & Sons. ISBN 978-3-527-40601-2.

- B. Povh et al. (1995). *Particles and Nuclei: An Introduction to the Physical Concepts*. Springer. ISBN 3-540-20168-8.

- I.I. Bigi, A.I. Sanda (2000). *CP violation*. Cambridge University Press. ISBN 0-521-44349-0.

- Particle Data Group: The CKM quark-mixing matrix

- Particle Data Group: CP violation in meson decays

- The Babar experiment at SLAC and the BELLE experiment at KEK Japan

6.4 Color–flavor locking

Color–flavor locking (CFL) is a phenomenon that is expected to occur in ultra-high-density strange matter, a form of quark matter. The quarks form Cooper pairs, whose color properties are correlated with their flavor properties in a one-to-one correspondence between three color pairs and three flavor pairs. According to the Standard Model of particle physics, the color-flavor-locked phase is the highest-density phase of three-flavor matter.[*][1]

6.4.1 Color-flavor-locked Cooper pairing

If each quark is represented as ψ_i^α, with color index α taking values 1, 2, 3 corresponding to red, green, and blue, and flavor index i taking values 1, 2, 3 corresponding to up, down, and strange, then the color-flavor-locked pattern of Cooper pairing is [*][2]

$$\langle \psi_i^\alpha C \gamma_5 \psi_j^\beta \rangle \propto \delta_i^\alpha \delta_j^\beta - \delta_j^\alpha \delta_i^\beta = \epsilon^{\alpha\beta A} \epsilon_{ijA}$$

This means that a Cooper pair of an up quark and a down quark must have colors red and green, and so on. This pairing pattern is special because it leaves a large unbroken symmetry group.

6.4.2 Physical properties

The CFL phase has several remarkable properties.

- It breaks chiral symmetry.

- It is a superfluid.

- It is an electromagnetic insulator, in which there is a "rotated" photon, containing a small admixture of one of the gluons.

- It has the same symmetries as sufficiently dense hyperonic matter.

There are several variants of the CFL phase, representing distortions of the pairing structure in response to external stresses such as a difference between the mass of the strange quark and the mass of the up and down quarks.

6.4.3 References

[1] M. Alford, K. Rajagopal, T. Schäfer, A. Schmitt (2008). "Color superconductivity in dense quark matter". *Reviews of Modern Physics* **80** (4): 1455. arXiv:0709.4635. Bibcode:2008RvMP...80.1455A. doi:10.1103/RevModPhys.80.1455.

[2] M. Alford, K. Rajagopal and F. Wilczek (1998). "QCD at Finite Baryon Density: Nucleon Droplets and Color Superconductivity". *Physics Letters B* **422**: 247. arXiv:hep-ph/9711395. Bibcode:1998PhLB..422..247A. doi:10.1016/S0370-2693(98)00051-3.

6.5 Constituent quark

A **constituent quark** is a current quark with a covering.

In the low energy limit of QCD, a description by means of perturbation theory is not possible. Here, no Asymptotic freedom exists, but the interactions between valence quarks and sea quarks gain strongly on significance. Part of the effects of virtual quarks and virtual gluons in the 'sea' can be assigned to a quark so well that the term 'constituent quark' seems appropriate.

According to the Feynman diagrams, constituent quarks seem to be 'dressed' current quarks, i.e. current quarks surrounded by a cloud of virtual quarks and gluons. This cloud in the end explains the large constituent-quark masses.

Definition: Constituent quarks are valence quarks for which the correlations for the description of hadrons by means of gluons and sea-quarks are put into effective quark masses of these valence quarks.

The effective quark mass is called **constituent quark mass**. Hadrons consist of 'glued' constituent quarks. The use of positions for the light constituent quarks is not exactly unproblematic in the description of hadrons.

6.5.1 Binding energy

The **quantum chromodynamic binding energy** of a valence quark in a hadron is the amount of energy required to make the hadron spontaneously emit a meson containing the valence quark. This is the same as the constituent quark mass.

Note that the following values are model dependent.

Describing hadrons using non-relativistic quantum mechanics becomes difficult.

6.5.2 References

[1] Griffiths, David (2008). *Introduction to Elementary Particles*. WILEY-VCH. p. 135.

6.6 Current quark

Current quarks (also called **naked quarks** or **bare quarks**) are defined as the constituent quark cores (constituent quarks with no covering) of a valence quark.*[1]*:150-151

If, in one constituent quark, the current quark is hit inside the covering with large force, it accelerates through the covering and leaves it behind. In addition, current quarks possess one asymptotic freedom within the perturbation theory described limits. In quantum chromodynamics, the mass of the current quarks carries the designation current quark mass.

The **local term** plays no more role for the description of the hadrons with the light current quarks:

A description is only possible with the help of relativistic quantum mechanics.

6.6.1 Current quark mass

The **current quark mass** is also called the mass of the 'naked' quarks. The mass of the current quark is reduced by the term of the constituent quark covering mass.

The current quark mass is a logical consequence of the mathematical formalism of the QFT, thus it is from a not descriptive origin. The current quark masses of the light current quarks are much smaller than the constituent quark masses. Reason for this is the missing of the mass of the constituent quark covering.

The current quark mass is a parameter to compute sufficiently small color charges.

Definition: The current quark mass means the mass of the constituent quark mass reduced by the mass of the respective constituent quark covering.

There is almost no difference between current quark mass and constituent quark mass for the heavy quarks (c,b,t). This is not so for the light quarks (u, d, s).

The comparison of the results of the computations with the experimental data supplies the values for the current quark masses.

6.6.2 References

[1] Timothy Paul Smith (2003). *Hidden Worlds: Hunting for Quarks in Ordinary Matter*. Princeton University Press. ISBN 0-691-05773-7.

6.7 CP violation

In particle physics, **CP violation** (CP standing for **charge parity**) is a violation of the postulated **CP-symmetry** (or **charge conjugation parity symmetry**): the combination of C-symmetry (charge conjugation symmetry) and P-symmetry (parity symmetry). CP-symmetry states that the laws of physics should be the same if a particle is interchanged with its antiparticle (C symmetry), and when its spatial coordinates are inverted ("mirror" or P symmetry). The discovery of CP violation in 1964 in the decays of neutral kaons resulted in the Nobel Prize in Physics in 1980 for its discoverers James Cronin and Val Fitch.

It plays an important role both in the attempts of cosmology to explain the dominance of matter over antimatter in the present Universe, and in the study of weak interactions in particle physics.

6.7.1 CP-symmetry

CP-symmetry, often called just *CP*, is the product of two symmetries: C for charge conjugation, which transforms a particle into its antiparticle, and P for parity, which creates the mirror image of a physical system. The strong interaction and electromagnetic interaction seem to be invariant under the combined CP transformation operation, but this symmetry is slightly violated during certain types of weak decay. Historically, CP-symmetry was proposed to restore order after the discovery of parity violation in the 1950s.

The idea behind parity symmetry is that the equations of particle physics are invariant under mirror inversion. This leads to the prediction that the mirror image of a reaction (such as a chemical reaction or radioactive decay) occurs at the same rate as the original reaction. Parity symmetry appears to be valid for all reactions involving electromagnetism and strong interactions. Until 1956, parity conservation was believed to be one of the fundamental geometric conservation laws (along with conservation of energy and conservation of momentum). However, in 1956 a careful critical review of the existing experimental data by theoretical physicists Tsung-Dao Lee and Chen Ning Yang revealed that while parity conservation had been verified in decays by the strong or electromagnetic interactions, it was untested in the weak interaction. They proposed several possible direct experimental tests. The first test based on beta decay of cobalt-60 nuclei was carried out in 1956 by a group led by Chien-Shiung Wu, and demonstrated conclusively that weak interactions violate the P symmetry or, as the analogy goes, some reactions did not occur as often as their mirror image.

Overall, the symmetry of a quantum mechanical system can be restored if another symmetry S can be found such that the combined symmetry PS remains unbroken. This rather subtle point about the structure of Hilbert space was realized shortly after the discovery of P violation, and it was proposed that charge conjugation was the desired symmetry to restore order.

Simply speaking, charge conjugation is a symmetry between particles and antiparticles, and so CP-symmetry was proposed in 1957 by Lev Landau as the true symmetry between matter and antimatter. In other words, a process in which all particles are exchanged with their antiparticles was assumed to be equivalent to the mirror image of the original process.

CP violation in the Standard Model

"Direct" CP violation is allowed in the Standard Model if a complex phase appears in the CKM matrix describing quark mixing, or the PMNS matrix describing neutrino mixing. A necessary condition for the appearance of the complex phase is the presence of at least three generations of quarks (if fewer generations are present, the complex phase parameter can be absorbed into redefinitions of the quark fields).

The reason why such a complex phase causes CP violation is not immediately obvious, but can be seen as follows. Consider any given particles (or sets of particles) a and b, and their antiparticles $\bar{a}$ and $\bar{b}$. Now consider the processes $a \to b$ and the corresponding antiparticle process $\bar{a} \to \bar{b}$, and denote their amplitudes M and $\bar{M}$ respectively. Before CP violation, these terms must be the *same* complex number. We can separate the magnitude and phase by writing $M = |M|e^{i\theta}$. If a phase term is introduced from (e.g.) the CKM matrix, denote it $e^{i\phi}$. Note that $\bar{M}$ contains the conjugate matrix to M, so it picks up a phase term $e^{-i\phi}$. Now we have:

$$M = |M|e^{i\theta}e^{i\phi}$$
$$\bar{M} = |M|e^{i\theta}e^{-i\phi}$$

Physically measurable reaction rates are proportional to $|M|^2$, thus so far nothing is different. However, consider that there are *two different routes* (e.g. intermediate states) for $a \to b$. Now we have:

$$M = |M_1|e^{i\theta_1}e^{i\phi_1} + |M_2|e^{i\theta_2}e^{i\phi_2}$$
$$\bar{M} = |M_1|e^{i\theta_1}e^{-i\phi_1} + |M_2|e^{i\theta_2}e^{-i\phi_2}$$

Some further calculation gives:

$$|M|^2 - |\bar{M}|^2 = 4|M_1||M_2|\sin(\theta_1 - \theta_2)\sin(\phi_1 - \phi_2)$$

Thus, we see that a complex phase gives rise to processes that proceed at different rates for particles and antiparticles, and CP is violated.

6.7.2 Experimental status

Indirect CP violation

In 1964, James Cronin, Val Fitch and coworkers provided clear evidence (which was first announced at the 12th ICHEP conference in Dubna) that CP-symmetry could be

6.7. CP VIOLATION

broken. This work[1] won them the 1980 Nobel Prize. This discovery showed that weak interactions violate not only the charge-conjugation symmetry C between particles and antiparticles and the P or parity, but also their combination. The discovery shocked particle physics and opened the door to questions still at the core of particle physics and of cosmology today. The lack of an exact CP-symmetry, but also the fact that it is so nearly a symmetry, created a great puzzle.

Only a weaker version of the symmetry could be preserved by physical phenomena, which was CPT symmetry. Besides C and P, there is a third operation, time reversal (T), which corresponds to reversal of motion. Invariance under time reversal implies that whenever a motion is allowed by the laws of physics, the reversed motion is also an allowed one. The combination of CPT is thought to constitute an exact symmetry of all types of fundamental interactions. Because of the CPT symmetry, a violation of the CP-symmetry is equivalent to a violation of the T symmetry. CP violation implied nonconservation of T, provided that the long-held CPT theorem was valid. In this theorem, regarded as one of the basic principles of quantum field theory, charge conjugation, parity, and time reversal are applied together.

Direct CP violation

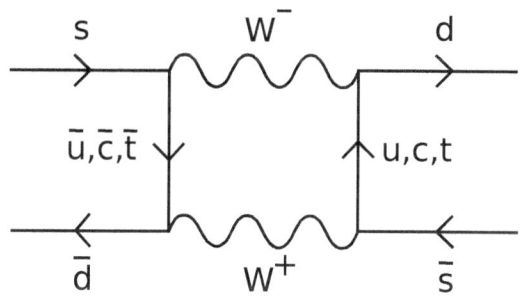

Kaon oscillation box diagram

The kind of CP violation discovered in 1964 was linked to the fact that neutral kaons can transform into their antiparticles (in which each quark is replaced with the other's antiquark) and vice versa, but such transformation does not occur with exactly the same probability in both directions; this is called *indirect* CP violation. Despite many searches, no other manifestation of CP violation was discovered until the 1990s, when the NA31 experiment at CERN suggested evidence for CP violation in the decay process of the very same neutral kaons (*direct* CP violation). The observation was somewhat controversial, and final proof for it came in 1999 from the KTeV experiment at Fermilab[2] and the NA48 experiment at CERN.[3]

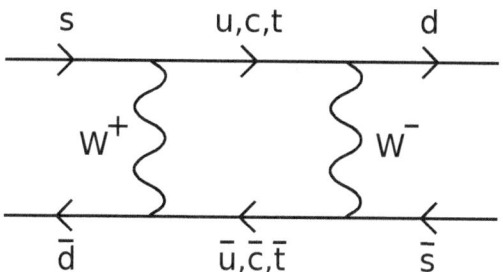

The two box diagrams above are the Feynman diagrams providing the leading contributions to the amplitude of K0-K0 oscillation

In 2001, a new generation of experiments, including the BaBar Experiment at the Stanford Linear Accelerator Center (SLAC)[4] and the Belle Experiment at the High Energy Accelerator Research Organisation (KEK)[5] in Japan, observed direct CP violation in a different system, namely in decays of the B mesons.[6] By now a large number of CP violation processes in B meson decays have been discovered. Before these "B-factory" experiments, there was a logical possibility that all CP violation was confined to kaon physics. However, this raised the question of why it's *not* extended to the strong force, and furthermore, why this is not predicted in the unextended Standard Model, despite the model being undeniably accurate with "normal" phenomena.

In 2011, a first indication of CP violation in decays of neutral D mesons was reported by the LHCb experiment at CERN.[7]

6.7.3 Strong CP problem

There is no experimentally known violation of the CP-symmetry in quantum chromodynamics. As there is no known reason for it to be conserved in QCD specifically, this is a "fine tuning" problem known as the strong CP problem.

QCD does not violate the CP-symmetry as easily as the electroweak theory; unlike the electroweak theory in which the gauge fields couple to chiral currents constructed from the fermionic fields, the gluons couple to vector currents. Experiments do not indicate any CP violation in the QCD sector. For example, a generic CP violation in the strongly interacting sector would create the electric dipole moment of the neutron which would be comparable to 10^{-18} e·m while the experimental upper bound is roughly one trillionth that size.

This is a problem because at the end, there are natural terms in the QCD Lagrangian that are able to break the CP-symmetry.

$$\mathcal{L} = -\frac{1}{4}F_{\mu\nu}F^{\mu\nu} - \frac{n_f g^2 \theta}{32\pi^2}F_{\mu\nu}\tilde{F}^{\mu\nu} + \bar{\psi}(i\gamma^\mu D_\mu - me^{i\theta'\gamma_5})\psi$$

For a nonzero choice of the θ angle and the chiral quark mass phase θ' one expects the CP-symmetry to be violated. One usually assumes that the chiral quark mass phase can be converted to a contribution to the total effective $\bar{\theta}$ angle, but it remains to be explained why this angle is extremely small instead of being of order one; the particular value of the θ angle that must be very close to zero (in this case) is an example of a fine-tuning problem in physics, and is typically solved by physics beyond the Standard Model.

There are several proposed solutions to solve the strong CP problem. The most well-known is Peccei–Quinn theory, involving new scalar particles called axions. A newer, more radical approach not requiring the axion is a theory involving two time dimensions first proposed in 1998 by Bars, Deliduman, and Andreev.[8]

Little CP problem

The little CP problem is a term coined by Lisa Randall. It refers to an issue related to the enhanced new physics contributions to the electric dipole moment (EDM) of the neutron in flavor anarchic models.[9]

6.7.4 CP violation and the matter–antimatter imbalance

Main article: Baryogenesis

The universe is made chiefly of matter, rather than consisting of equal parts of matter and antimatter as might be expected. It can be demonstrated that, to create an imbalance in matter and antimatter from an initial condition of balance, the Sakharov conditions must be satisfied, one of which is the existence of CP violation during the extreme conditions of the first seconds after the Big Bang. Explanations which do not involve CP violation are less plausible, since they rely on the assumption that the matter–antimatter imbalance was present at the beginning, or on other admittedly exotic assumptions.

The Big Bang should have produced equal amounts of matter and antimatter if CP-symmetry was preserved; as such, there should have been total cancellation of both—protons should have cancelled with antiprotons, electrons with positrons, neutrons with antineutrons, and so on. This would have resulted in a sea of radiation in the universe with no matter. Since this is not the case, after the Big Bang, physical laws must have acted differently for matter and antimatter, i.e. violating CP-symmetry.

The Standard Model contains at least three sources of CP violation. The first of these, involving the Cabibbo–Kobayashi–Maskawa matrix in the quark sector, has been observed experimentally and can only account for a small portion of the CP violation required to explain the matter-antimatter asymmetry. The strong interaction should also violate CP, in principle, but the failure to observe the electric dipole moment of the neutron in experiments suggests that any CP violation in the strong sector is also too small to account for the necessary CP violation in the early universe. The third source of CP violation is the Pontecorvo–Maki–Nakagawa–Sakata matrix in the lepton sector. Current neutrino experiments are not yet sensitive enough to allow experimental observation of CP violation in the lepton sector, but the NOvA experiment currently under construction could observe some small fraction of possible CP violating phases and proposed neutrino experiments Hyper-Kamiokande and LBNE will be sensitive to a relatively large fraction of CP violating phases. Further into the future, a neutrino factory could be sensitive to nearly all possible CP violating phases. If neutrinos are Majorana fermions, the PMNS matrix could have two independent CP violating phases leading to a fourth source of CP violation within the Standard Model. The experimental evidence for Majorana neutrinos would be the observation of neutrinoless double-beta decay. As of September 2013, the best limits come from the GERDA experiment. CP violation in the lepton sector generates a matter-antimatter asymmetry through a process called leptogenesis. This could become the preferred explanation in the Standard Model for the matter-antimatter asymmetry of the universe once CP violation is experimentally confirmed in the lepton sector.

If CP violation in the lepton sector is experimentally determined to be too small to account for matter-antimatter asymmetry, some new physics beyond the Standard Model would be required to explain additional sources of CP violation. Fortunately, it is generally the case that adding new particles and/or interactions to the Standard Model introduces new sources of CP violation since CP is not a symmetry of nature.

6.7.5 See also

- B-factory
- LHCb
- BTeV experiment
- Cabibbo–Kobayashi–Maskawa matrix
- Penguin diagram

- Neutral particle oscillation

6.7.6 References

[1] "Evidence for the 2π Decay of the K^0_2 Meson System". *Physical Review Letters* **13**: 138. 1964. Bibcode:1964PhRvL..13..138C. doi:10.1103/PhysRevLett.13.138.

[2] "Observation of Direct CP Violation in $K_{S,L} \to \pi\pi$ Decays". *Physical Review Letters* **83**: 22. 1999. arXiv:hep-ex/9905060. Bibcode:1999PhRvL..83...22A. doi:10.1103/PhysRevLett.83.22.

[3] NA48 Collaboration, V. Fanti, A. Lai, D. Marras, L. Musa et al. (1999). "A new measurement of direct CP violation in two pion decays of the neutral kaon". *Physics Letters B* **465** (1–4): 335–348. arXiv:hep-ex/9909022. Bibcode:1999PhLB..465..335F. doi:10.1016/S0370-2693(99)01030-8.

[4] "Measurement of CP-Violating Asymmetries in B^0 Decays to CP Eigenstates". *Physical Review Letters* **86**: 2515. 2001. arXiv:hep-ex/0102030. Bibcode:2001PhRvL..86.2515A. doi:10.1103/PhysRevLett.86.2515.

[5] "Observation of Large CP Violation in the Neutral B Meson System". *Physical Review Letters* **87**. 2001. arXiv:hep-ex/0107061. Bibcode:2001PhRvL..87i1802A. doi:10.1103/PhysRevLett.87.091802.

[6] Rodgers, Peter (August 2001). "Where did all the antimatter go?". *Physics World*. p. 11.

[7] Carbone, A. (2012). "A search for time-integrated CP violation in $D^0 \to h^{*-}h^{*+}$ decays". arXiv:1210.8257.

[8] I. Bars; C. Deliduman; O. Andreev (1998). "Gauged Duality, Conformal Symmetry, and Spacetime with Two Times". *Physical Review D* **58** (6): 066004. arXiv:hep-th/9803188. Bibcode:1998PhRvD..58f6004B. doi:10.1103/PhysRevD.58.066004.

[9] Kadosh, Avihay; Pallante, Elisabetta (2011). "CP violation and FCNC in a warped A_4 flavor model". *Journal of High Energy Physics* **2011** (6). arXiv:1101.5420. Bibcode:2011JHEP...06..121K. doi:10.1007/JHEP06(2011)121.

6.7.7 Further reading

- Sozzi, M.S. (2008). *Discrete symmetries and CP violation*. Oxford University Press. ISBN 978-0-19-929666-8.
- G. C. Branco, L. Lavoura and J. P. Silva (1999). *CP violation*. Clarendon Press. ISBN 0-19-850399-7.
- I. Bigi and A. Sanda (1999). *CP violation*. Cambridge University Press. ISBN 0-521-44349-0.
- Michael Beyer, ed. (2002). *CP Violation in Particle, Nuclear and Astrophysics*. Springer. ISBN 3-540-43705-3. (*A collection of essays introducing the subject, with an emphasis on experimental results.*)
- L. Wolfenstein (1989). *CP violation*. North–Holland Publishing. ISBN 0-444-88081-X. (*A compilation of reprints of numerous important papers on the topic, including papers by T.D. Lee, Cronin, Fitch, Kobayashi and Maskawa, and many others.*)
- David J. Griffiths (1987). *Introduction to Elementary Particles*. John Wiley & Sons. ISBN 0-471-60386-4.
- Bigi, I. (1997). "CP Violation —An Essential Mystery in Nature's Grand Design". *Surveys of High Energy Physics* **12**: 269–336. arXiv:hep-ph/9712475. Bibcode:1997hep.ph...12475B. doi:10.1080/01422419808228861.
- Mark Trodden (1998). "Electroweak Baryogenesis". *Reviews of Modern Physics* **71** (5): 1463. arXiv:hep-ph/9803479. Bibcode:1999RvMP...71.1463T. doi:10.1103/RevModPhys.71.1463.
- Davide Castelvecchi. "What is direct CP-violation?". SLAC. Retrieved 2009-07-01.

6.7.8 External links

- Cern Courier article

6.8 Fermion

In particle physics, a **fermion** (a name coined by Paul Dirac[1] from the surname of Enrico Fermi) is any particle characterized by Fermi–Dirac statistics. These particles obey the Pauli exclusion principle. Fermions include all quarks and leptons, as well as any composite particle made of an odd number of these, such as all baryons and many atoms and nuclei. Fermions differ from bosons, which obey Bose–Einstein statistics.

A fermion can be an elementary particle, such as the electron, or it can be a composite particle, such as the proton. According to the spin-statistics theorem in any reasonable relativistic quantum field theory, particles with integer spin are bosons, while particles with half-integer spin are neutrons fermions.

Besides this spin characteristic, fermions have another specific property: they possess conserved baryon or lepton quantum numbers. Therefore what is usually referred as the spin statistics relation is in fact a spin statistics-quantum number relation.[2]

Enrico Fermi

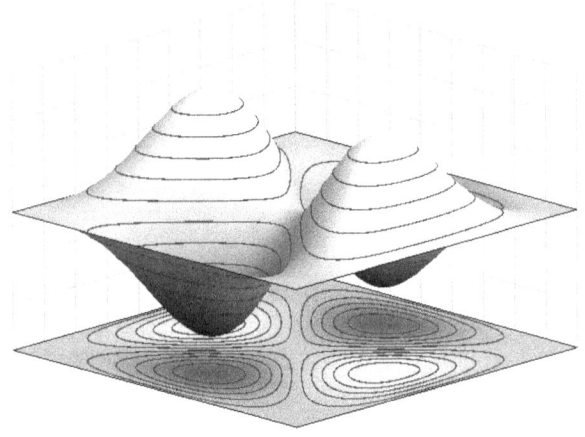

Antisymmetric wavefunction for a (fermionic) 2-particle state in an infinite square well potential.

As a consequence of the Pauli exclusion principle, only one fermion can occupy a particular quantum state at any given time. If multiple fermions have the same spatial probability distribution, then at least one property of each fermion, such as its spin, must be different. Fermions are usually associated with matter, whereas bosons are generally force carrier particles, although in the current state of particle physics the distinction between the two concepts is unclear. At low temperature fermions show superfluidity for uncharged particles and superconductivity for charged particles. Composite fermions, such as protons and neutrons, are the key building blocks of everyday matter. Weakly interacting fermions can also display bosonic behavior under extreme conditions, such as superconductivity.

6.8.1 Elementary fermions

The Standard Model recognizes two types of elementary fermions, quarks and leptons. In all, the model distinguishes 24 different fermions. There are six quarks (up, down, strange, charm, bottom and top quarks), and six leptons (electron, electron neutrino, muon, muon neutrino, tau particle and tau neutrino), along with the corresponding antiparticle of each of these.

Mathematically, fermions come in three types - Weyl fermions (massless), Dirac fermions (massive), and Majorana fermions (each its own antiparticle). Most Standard Model fermions are believed to be Dirac fermions, although it is unknown at this time whether the neutrinos are Dirac or Majorana fermions. Dirac fermions can be treated as a combination of two Weyl fermions.*[3]*:106 So far there is no known example of Weyl fermion in particle physics. In July 2015, Weyl fermions have been experimentally realized in Weyl semimetals.

6.8.2 Composite fermions

See also: List of particles § Composite particles

Composite particles (such as hadrons, nuclei, and atoms) can be bosons or fermions depending on their constituents. More precisely, because of the relation between spin and statistics, a particle containing an odd number of fermions is itself a fermion. It will have half-integer spin.

Examples include the following:

- A baryon, such as the proton or neutron, contains three fermionic quarks and thus it is a fermion.
- The nucleus of a carbon-13 atom contains six protons and seven neutrons and is therefore a fermion.
- The atom helium-3 (^{3}He) is made of two protons, one neutron, and two electrons, and therefore it is a fermion.

The number of bosons within a composite particle made up of simple particles bound with a potential has no effect on whether it is a boson or a fermion.

Fermionic or bosonic behavior of a composite particle (or system) is only seen at large (compared to size of the system) distances. At proximity, where spatial structure begins to be important, a composite particle (or system) behaves according to its constituent makeup.

Fermions can exhibit bosonic behavior when they become loosely bound in pairs. This is the origin of superconductivity and the superfluidity of helium-3: in superconducting materials, electrons interact through the exchange of phonons, forming Cooper pairs, while in helium-3, Cooper pairs are formed via spin fluctuations.

The quasiparticles of the fractional quantum Hall effect are also known as composite fermions, which are electrons with an even number of quantized vortices attached to them.

Skyrmions

Main article: Skyrmion

In a quantum field theory, there can be field configurations of bosons which are topologically twisted. These are coherent states (or solitons) which behave like a particle, and they can be fermionic even if all the constituent particles are bosons. This was discovered by Tony Skyrme in the early 1960s, so fermions made of bosons are named skyrmions after him.

Skyrme's original example involved fields which take values on a three-dimensional sphere, the original nonlinear sigma model which describes the large distance behavior of pions. In Skyrme's model, reproduced in the large N or string approximation to quantum chromodynamics (QCD), the proton and neutron are fermionic topological solitons of the pion field.

Whereas Skyrme's example involved pion physics, there is a much more familiar example in quantum electrodynamics with a magnetic monopole. A bosonic monopole with the smallest possible magnetic charge and a bosonic version of the electron will form a fermionic dyon.

The analogy between the Skyrme field and the Higgs field of the electroweak sector has been used[4] to postulate that all fermions are skyrmions. This could explain why all known fermions have baryon or lepton quantum numbers and provide a physical mechanism for the Pauli exclusion principle.

6.8.3 See also

6.8.4 Notes

[1] Notes on Dirac's lecture *Developments in Atomic Theory* at Le Palais de la Découverte, 6 December 1945, UK-NATARCHI Dirac Papers BW83/2/257889. See note 64 on page 331 in "The Strangest Man: The Hidden Life of Paul Dirac, Mystic of the Atom" by Graham Farmelo

[2] Physical Review D volume 87, page 0550003, year 2013, author Weiner, Richard M., title "Spin-statistics-quantum number connection and supersymmetry" arxiv:1302.0969

[3] T. Morii; C. S. Lim; S. N. Mukherjee (1 January 2004). *The Physics of the Standard Model and Beyond*. World Scientific. ISBN 978-981-279-560-1.

[4] Weiner, Richard M. (2010). "The Mysteries of Fermions". *International Journal of Theoretical Physics* **49** (5): 1174–1180. arXiv:0901.3816. Bibcode:2010IJTP...49.1174W. doi:10.1007/s10773-010-0292-7.

6.9 Gell-Mann–Nishijima formula

The **Gell-Mann–Nishijima formula** (sometimes known as the **NNG formula**) relates the baryon number B, the strangeness S, the isospin I_3 of hadrons to the charge Q. It was originally given by Kazuhiko Nishijima and Tadao Nakano in 1953,[1] and led to the proposal of strangeness as a concept, which Nishijima originally called "eta-charge" after the eta meson.[2] Murray Gell-Mann proposed the formula independently in 1956.[3] The modern version of the formula relates all flavour quantum numbers (isospin up and down, strangeness, charm, bottomness, and topness) with the baryon number and the electric charge.

6.9.1 Formula

The original form of the Gell-Mann–Nishijima formula is:

$$Q = I_3 + \frac{1}{2}(B + S).$$

This equation was originally based on empirical experiments. It is now understood as a result of the quark model. In particular, the electric charge Q of a particle is related to its isospin I_3 and its hypercharge Y via the relation:

$$Q = I_3 + \frac{1}{2}Y.$$

Since the discovery of charm, top, and bottom quark flavors, this formula has been generalized. It now takes the form:

$$Q = I_3 + \frac{1}{2}(B + S + C + B' + T)$$

where Q is the charge, I_3 the 3rd-component of the isospin, B the baryon number, and S, C, B', T are the strangeness, charm, bottomness and topness numbers.

Expressed in terms of quark content, these would become:

$$Q = \frac{2}{3}[(n_u - n_{\bar{u}}) + (n_c - n_{\bar{c}}) + (n_t - n_{\bar{t}})] - \frac{1}{3}[(n_d - n_{\bar{d}})$$
$$B = \frac{1}{3}[(n_u - n_{\bar{u}}) + (n_c - n_{\bar{c}}) + (n_t - n_{\bar{t}}) + (n_d - n_{\bar{d}}) + (n_s - n_{\bar{s}}) + (n_b - n_{\bar{b}})]$$
$$I_3 = \tfrac{1}{2}[(n_u - n_{\bar{u}}) - (n_d - n_{\bar{d}})]$$
$$S = -(n_s - n_{\bar{s}}); \; C = +(n_c - n_{\bar{c}}); \; B' = -(n_b - n_{\bar{b}}); \; T = +(n_t - n_{\bar{t}})$$

By convention, the flavor quantum numbers (strangeness, charm, bottomness, and topness) carry the same sign as the electric charge of the particle. So, since the strange and bottom quarks have a negative charge, they have flavor quantum numbers equal to −1. And since the charm and top quarks have positive electric charge, their flavor quantum numbers are +1.

6.9.2 References

[1] Nakano, T; Nishijima, N (1953). "Charge Independence for V-particles". *Progress of Theoretical Physics* **10** (5): 581. Bibcode:1953PThPh..10..581N. doi:10.1143/PTP.10.581.

[2] Nishijima, K (1955). "Charge Independence Theory of V Particles". *Progress of Theoretical Physics* **13** (3): 285. Bibcode:1955PThPh..13..285N. doi:10.1143/PTP.13.285.

[3] Gell-Mann, M (1956). "The Interpretation of the New Particles as Displaced Charged Multiplets". *Il Nuovo Cimento* **4** (S2): 848. doi:10.1007/BF02748000.

6.9.3 Further reading

- Griffiths, DJ (2008). *Introduction to Elementary Particles* (2nd ed.). Wiley-VCH. ISBN 978-3-527-40601-2.

6.10 Generation

In particle physics, a **generation** (or **family**) is a division of the elementary particles. Between generations, particles differ by their (flavour) quantum number and mass, but their interactions are identical.

There are three generations according to the Standard Model of particle physics. Each generation is divided into two types of leptons and two types of quarks. The two leptons may be classified into one with electric charge −1 (electron-like) and one neutral (neutrino); the two quarks may be classified into one with charge $-\tfrac{1}{3}$ (down-type) and one with charge $+\tfrac{2}{3}$ (up-type).

6.10.1 Overview

Each member of a higher generation has greater mass than the corresponding particle of the previous generation, with the possible exception of the neutrinos (whose small but non-zero masses have not been accurately determined). For example, the first-generation electron has a mass of only 0.511 MeV/c^2, the second-generation muon has a mass of 106 MeV/c^2, and the third-generation tau has a mass of 1777 MeV/c^2 (almost twice as heavy as a proton). This mass hierarchy causes particles of higher generations to decay to the first generation, which explains why everyday matter (atoms) is made of particles from the first generation. Electrons surround a nucleus made of protons and neutrons, which contain up and down quarks. The second and third generations of charged particles do not occur in normal matter and are only seen in extremely high-energy environments such as cosmic rays or particle accelerators. The term *generation* was first introduced by Haim Harari in Les Houches Summer School, 1976.[1] [2]

Neutrinos of all generations stream throughout the universe but rarely interact with normal matter.[3] It is hoped that a comprehensive understanding of the relationship between the generations of the leptons may eventually explain the ratio of masses of the fundamental particles, and shed further light on the nature of mass generally, from a quantum perspective.[4]

6.10.2 Fourth generation

Fourth and further generations are considered to be unlikely. Some of the arguments against the possibility of a fourth generation are based on the subtle modifications of precision electroweak observables that extra generations would induce; such modifications are strongly disfavored by measurements. Furthermore, a fourth generation with a "light" neutrino (one with a mass less than about 45 GeV/c^2) has been ruled out by measurements of the widths of the Z boson at CERN's Large Electron–Positron Collider (LEP).[5] Nonetheless, searches at high-energy colliders for particles from a fourth generation continue, but as yet no evidence has been observed.[6] In such searches, fourth-generation particles are denoted by the same symbols as third-generation ones with an added prime (e.g. b' and t').

According to the results of the statistical analysis by researchers from CERN, and Humboldt University of Berlin, the existence of further fermions can be excluded with a probability of 99.99999% (5.3 sigma). The researchers combined latest data collected by the particle accelerators LHC and Tevatron with many known measurements results relating to particles, such as the Z-boson or the top-quark.

The most important data used for this analysis come from the discovery of the Higgs particle. In the Standard Model, the Higgs particle gives all other particles their mass. As additional fermions were not detected directly in accelerator experiments, they have to be heavier than the fermions known so far. Hence, these fermions would also interact with the Higgs particle more strongly. This interaction would have modified the properties of the Higgs particle such that this particle would not have been detected.[*][7]

6.10.3 See also

- Metric expansion of space
- Spacetime
- Supersymmetry
- World line

6.10.4 References

[1] Harari, H. (1977). "Beyond charm". In Balian, R.; Llewellyn-Smith, C.H. *Weak and Electromagnetic Interactions at High Energy, Les Houches, France, Jul 5- Aug 14, 1976*. Les Houches Summer School Proceedings **29**. North-Holland. p. 613.

[2] Harari H. (1977). "Three generations of quarks and leptons" (PDF). In E. van Goeler, Weinstein R. (eds.). *Proceedings of the XII Rencontre de Moriond*. p. 170. SLAC-PUB-1974.

[3] "Experiment confirms famous physics model" (Press release). MIT News Office. 18 April 2007.

[4] M.H. Mac Gregor (2006). "A 'Muon Mass Tree' with α-quantized Lepton, Quark, and Hadron Masses". arXiv:hep-ph/0607233 [hep-ph].

[5] D. Decamp *et al.* (ALEPH collaboration) (1989). "Determination of the number of light neutrino species". *Physics Letters B* **231** (4): 519. Bibcode:1989PhLB..231..519D. doi:10.1016/0370-2693(89)90704-1.

[6] C. Amsler *et al.* (Particle Data Group) (2008). "Review of Particle Physics: b′ (4th Generation) Quarks, Searches for" (PDF). *Physics Letters B* **667** (1): 1–1340. Bibcode:2008PhLB..667....1P. doi:10.1016/j.physletb.2008.07.018.

[7] *12 matter particles suffice in nature* Dec 13, 2012 Phys.Org

6.11 Gluon

Gluons /ˈɡluːɒnz/ are elementary particles that act as the exchange particles (or gauge bosons) for the strong force between quarks, analogous to the exchange of photons in the electromagnetic force between two charged particles.[*][6]

In technical terms, gluons are vector gauge bosons that mediate strong interactions of quarks in quantum chromodynamics (QCD). Gluons themselves carry the color charge of the strong interaction. This is unlike the photon, which mediates the electromagnetic interaction but lacks an electric charge. Gluons therefore participate in the strong interaction in addition to mediating it, making QCD significantly harder to analyze than QED (quantum electrodynamics).

6.11.1 Properties

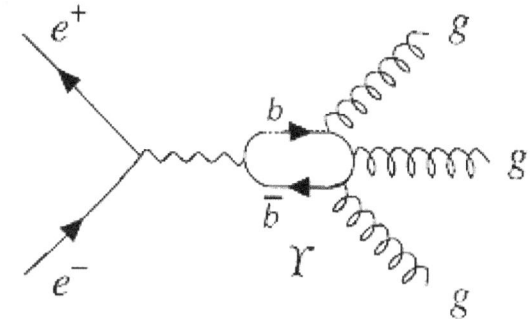

Diagram 2: $e^+ + e^- \to \Upsilon(9.46) \to 3g$

The gluon is a vector boson; like the photon, it has a spin of 1. While massive spin-1 particles have three polarization states, massless gauge bosons like the gluon have only two polarization states because gauge invariance requires the polarization to be transverse. In quantum field theory, unbroken gauge invariance requires that gauge bosons have zero mass (experiment limits the gluon's rest mass to less than a few meV/c^2). The gluon has negative intrinsic parity.

6.11.2 Numerology of gluons

Unlike the single photon of QED or the three W and Z bosons of the weak interaction, there are eight independent types of gluon in QCD.

This may be difficult to understand intuitively. Quarks carry three types of color charge; antiquarks carry three types of anticolor. Gluons may be thought of as carrying both color and anticolor, but to correctly understand how they are combined, it is necessary to consider the mathematics of color charge in more detail.

Color charge and superposition

In quantum mechanics, the states of particles may be added according to the principle of superposition; that is, they may be in a "combined state" with a *probability*, if some particular quantity is measured, of giving several different outcomes. A relevant illustration in the case at hand would be a gluon with a color state described by:

$$(r\bar{b} + b\bar{r})/\sqrt{2}.$$

This is read as "red–antiblue plus blue–antired". (The factor of the square root of two is required for normalization, a detail that is not crucial to understand in this discussion.) If one were somehow able to make a direct measurement of the color of a gluon in this state, there would be a 50% chance of it having red-antiblue color charge and a 50% chance of blue-antired color charge.

Color singlet states

It is often said that the stable strongly interacting particles (such as the proton and the neutron, i.e. hadrons) observed in nature are "colorless", but more precisely they are in a "color singlet" state, which is mathematically analogous to a *spin* singlet state.[*][7] Such states allow interaction with other color singlets, but not with other color states; because long-range gluon interactions do not exist, this illustrates that gluons in the singlet state do not exist either.[*][7]

The color singlet state is:[*][7]

$$(r\bar{r} + b\bar{b} + g\bar{g})/\sqrt{3}.$$

In words, if one could measure the color of the state, there would be equal probabilities of it being red-antired, blue-antiblue, or green-antigreen.

Eight gluon colors

There are eight remaining independent color states, which correspond to the "eight types" or "eight colors" of gluons. Because states can be mixed together as discussed above, there are many ways of presenting these states, which are known as the "color octet". One commonly used list is:[*][7]

These are equivalent to the Gell-Mann matrices; the translation between the two is that red-antired is the upper-left matrix entry, red-antiblue is the upper middle entry, blue-antigreen is the middle right entry, and so on. The critical feature of these particular eight states is that they are linearly independent, and also independent of the singlet state; there is no way to add any combination of states to produce any other. (It is also impossible to add them to make rr, gg, or bb[*][8] otherwise the forbidden singlet state could also be made.) There are many other possible choices, but all are mathematically equivalent, at least equally complex, and give the same physical results.

Group theory details

Technically, QCD is a gauge theory with SU(3) gauge symmetry. Quarks are introduced as spinor fields in N_f flavors, each in the fundamental representation (triplet, denoted **3**) of the color gauge group, SU(3). The gluons are vector fields in the adjoint representation (octets, denoted **8**) of color SU(3). For a general gauge group, the number of force-carriers (like photons or gluons) is always equal to the dimension of the adjoint representation. For the simple case of SU(N), the dimension of this representation is $N^2 - 1$.

In terms of group theory, the assertion that there are no color singlet gluons is simply the statement that quantum chromodynamics has an SU(3) rather than a U(3) symmetry. There is no known *a priori* reason for one group to be preferred over the other, but as discussed above, the experimental evidence supports SU(3).[*][7] The U(1) group for electromagnetic field combines with a slightly more complicated group known as SU(2),S stands for "special", which means the corresponding matrices have derterminant 1.

6.11.3 Confinement

Main article: Color confinement

Since gluons themselves carry color charge, they participate in strong interactions. These gluon-gluon interactions constrain color fields to string-like objects called "flux tubes", which exert constant force when stretched. Due to this force, quarks are confined within composite particles called hadrons. This effectively limits the range of the strong interaction to 1×10[*]−15 meters, roughly the size of an atomic nucleus. Beyond a certain distance, the energy of the flux tube binding two quarks increases linearly. At a large enough distance, it becomes energetically more favorable to pull a quark-antiquark pair out of the vacuum rather than increase the length of the flux tube.

Gluons also share this property of being confined within hadrons. One consequence is that gluons are not directly involved in the nuclear forces between hadrons. The force mediators for these are other hadrons called mesons.

Although in the normal phase of QCD single gluons may not travel freely, it is predicted that there exist hadrons that are formed entirely of gluons —called glueballs. There are also conjectures about other exotic hadrons in which real gluons (as opposed to virtual ones found in ordinary hadrons) would be primary constituents. Beyond the normal phase of QCD (at extreme temperatures and pressures), quark–gluon plasma forms. In such a plasma there are no hadrons; quarks and gluons become free particles.

6.11.4 Experimental observations

Quarks and gluons (colored) manifest themselves by fragmenting into more quarks and gluons, which in turn hadronize into normal (colorless) particles, correlated in jets. As shown in 1978 summer conferences[2] the PLUTO detector at the electron-positron collider DORIS (DESY) produced the first evidence that the hadronic decays of the very narrow resonance $\Upsilon(9.46)$ could be interpreted as three-jet event topologies produced by three gluons. Later published analyses by the same experiment confirmed this interpretation and also the spin 1 nature of the gluon[9][10] (see also the recollection[2] and PLUTO experiments).

In summer 1979 at higher energies at the electron-positron collider PETRA (DESY) again three-jet topologies were observed, now interpreted as qq gluon bremsstrahlung, now clearly visible, by TASSO,[11] MARK-J[12] and PLUTO experiments[13] (later in 1980 also by JADE[14]). The spin 1 of the gluon was confirmed in 1980 by TASSO[15] and PLUTO experiments[16] (see also the review[3]). In 1991 a subsequent experiment at the LEP storage ring at CERN again confirmed this result.[17]

The gluons play an important role in the elementary strong interactions between quarks and gluons, described by QCD and studied particularly at the electron-proton collider HERA at DESY. The number and momentum distribution of the gluons in the proton (gluon density) have been measured by two experiments, H1 and ZEUS,[18] in the years 1996 till today (2012). The gluon contribution to the proton spin has been studied by the HERMES experiment at HERA.[19] The gluon density in the proton (when behaving hadronically) also has been measured.[20]

Color confinement is verified by the failure of free quark searches (searches of fractional charges). Quarks are normally produced in pairs (quark + antiquark) to compensate the quantum color and flavor numbers; however at Fermilab single production of top quarks has been shown (technically this still involves a pair production, but quark and antiquark are of different flavor).[21] No glueball has been demonstrated.

Deconfinement was claimed in 2000 at CERN SPS[22] in heavy-ion collisions, and it implies a new state of matter: quark–gluon plasma, less interacting than in the nucleus, almost as in a liquid. It was found at the Relativistic Heavy Ion Collider (RHIC) at Brookhaven in the years 2004–2010 by four contemporaneous experiments.[23] A quark–gluon plasma state has been confirmed at the CERN Large Hadron Collider (LHC) by the three experiments ALICE, ATLAS and CMS in 2010.[24]

6.11.5 See also

- Quark
- Hadron
- Meson
- Gauge boson
- Quark model
- Quantum chromodynamics
- Quark–gluon plasma
- Color confinement
- Glueball
- Gluon field
- Gluon field strength tensor
- Exotic hadrons
- Standard Model
- Three-jet events
- Deep inelastic scattering

6.11.6 References

[1] M. Gell-Mann (1962). "Symmetries of Baryons and Mesons". *Physical Review* **125** (3): 1067–1084. Bibcode:1962PhRv..125.1067G. doi:10.1103/PhysRev.125.1067.

[2] B.R. Stella and H.-J. Meyer (2011). "Υ(9.46 GeV) and the gluon discovery (a critical recollection of PLUTO results)". *European Physical Journal H* **36** (2): 203–243. arXiv:1008.1869v3. Bibcode:2011EPJH...36..203S. doi:10.1140/epjh/e2011-10029-3.

[3] P. Söding (2010). "On the discovery of the gluon". *European Physical Journal H* **35** (1): 3–28. Bibcode:2010EPJH...35....3S. doi:10.1140/epjh/e2010-00002-5.

[4] W.-M. Yao et al. (2006). "Review of Particle Physics" (PDF). *Journal of Physics G* **33**: 1. arXiv:astro-ph/0601168. Bibcode:2006JPhG...33....1Y. doi:10.1088/0954-3899/33/1/001.

[5] F. Yndurain (1995). "Limits on the mass of the gluon". *Physics Letters B* **345** (4): 524. Bibcode:1995PhLB..345..524Y. doi:10.1016/0370-2693(94)01677-5.

[6] C.R. Nave. "The Color Force". *HyperPhysics*. Georgia State University, Department of Physics. Retrieved 2012-04-02.

[7] David Griffiths (1987). *Introduction to Elementary Particles*. John Wiley & Sons. pp. 280–281. ISBN 0-471-60386-4.

[8] J. Baez. "Why are there eight gluons and not nine?". Retrieved 2009-09-13.

[9] Ch. Berger *et al.* (PLUTO Collaboration) (1979). "Jet analysis of the $\Upsilon(9.46)$ decay into charged hadrons". *Physics Letters B* **82** (3–4): 449. Bibcode:1979PhLB...82..449B. doi:10.1016/0370-2693(79)90265-X.

[10] Ch. Berger *et al.* (PLUTO Collaboration) (1981). "Topology of the Υ decay". *Zeitschrift für Physik C* **8** (2): 101. Bibcode:1981ZPhyC...8..101B. doi:10.1007/BF01547873.

[11] R. Brandelik *et al.* (TASSO collaboration) (1979). "Evidence for Planar Events in e^+e^- Annihilation at High Energies". *Physics Letters B* **86** (2): 243–249. Bibcode:1979PhLB...86..243B. doi:10.1016/0370-2693(79)90830-X.

[12] D.P. Barber *et al.* (MARK-J collaboration) (1979). "Discovery of Three-Jet Events and a Test of Quantum Chromodynamics at PETRA". *Physical Review Letters* **43** (12): 830. Bibcode:1979PhRvL..43..830B. doi:10.1103/PhysRevLett.43.830.

[13] Ch. Berger *et al.* (PLUTO Collaboration) (1979). "Evidence for Gluon Bremsstrahlung in e^+e^- Annihilations at High Energies". *Physics Letters B* **86** (3–4): 418. Bibcode:1979PhLB...86..418B. doi:10.1016/0370-2693(79)90869-4.

[14] W. Bartel *et al.* (JADE Collaboration) (1980). "Observation of planar three-jet events in e^+e^- annihilation and evidence for gluon bremsstrahlung". *Physics Letters B* **91**: 142. Bibcode:1980PhLB...91..142B. doi:10.1016/0370-2693(80)90680-2.

[15] R. Brandelik *et al.* (TASSO Collaboration) (1980). "Evidence for a spin-1 gluon in three-jet events". *Physics Letters B* **97** (3–4): 453. Bibcode:1980PhLB...97..453B. doi:10.1016/0370-2693(80)90639-5.

[16] Ch. Berger *et al.* (PLUTO Collaboration) (1980). "A study of multi-jet events in e^+e^- annihilation". *Physics Letters B* **97** (3–4): 459. Bibcode:1980PhLB...97..459B. doi:10.1016/0370-2693(80)90640-1.

[17] G. Alexander *et al.* (OPAL Collaboration) (1991). "Measurement of Three-Jet Distributions Sensitive to the Gluon Spin in e^+e^- Annihilations at $\sqrt{s} = 91$ GeV". *Zeitschrift für Physik C* **52** (4): 543. Bibcode:1991ZPhyC..52..543A. doi:10.1007/BF01562326.

[18] L. Lindeman (H1 and ZEUS collaborations) (1997). "Proton structure functions and gluon density at HERA". *Nuclear Physics B Proceedings Supplements* **64**: 179–183. Bibcode:1998NuPhS..64..179L. doi:10.1016/S0920-5632(97)01057-8.

[19] http://www-hermes.desy.de

[20] C. Adloff *et al.* (H1 collaboration) (1999). "Charged particle cross sections in the photoproduction and extraction of the gluon density in the photon". *European Physical Journal C* **10**: 363–372. arXiv:hep-ex/9810020. Bibcode:1999EPJC...10..363H. doi:10.1007/s100520050761.

[21] M. Chalmers (6 March 2009). "Top result for Tevatron". *Physics World*. Retrieved 2012-04-02.

[22] M.C. Abreu et al. (2000). "Evidence for deconfinement of quark and antiquark from the J/Ψ suppression pattern measured in Pb-Pb collisions at the CERN SpS". *Physics Letters B* **477**: 28–36. Bibcode:2000PhLB..477...28A. doi:10.1016/S0370-2693(00)00237-9.

[23] D. Overbye (15 February 2010). "In Brookhaven Collider, Scientists Briefly Break a Law of Nature". *New York Times*. Retrieved 2012-04-02.

[24] "LHC experiments bring new insight into primordial universe" (Press release). CERN. 26 November 2010. Retrieved 2012-04-02.

6.11.7 Further reading

- A. Ali and G. Kramer (2011). "JETS and QCD: A historical review of the discovery of the quark and gluon jets and its impact on QCD". *European Physical Journal H* **36** (2): 245–326. arXiv:1012.2288. Bibcode:2011EPJH...36..245A. doi:10.1140/epjh/e2011-10047-1.

6.12 Hadronization

In particle physics, **hadronization** (or **hadronisation**) is the process of the formation of hadrons out of quarks and gluons. This occurs after high-energy collisions in a particle collider in which free quarks or gluons are created. Due to postulated colour confinement, these cannot exist individually. In the Standard Model they combine with quarks and antiquarks spontaneously created from the vacuum to form hadrons. The QCD (Quantum Chromodynamics) of the hadronization process are not yet fully understood, but are modeled and parameterized in a number of phenomenological studies, including the Lund string model and in various long-range QCD approximation schemes.*[1]*[2]*[3]

The tight cone of particles created by the hadronization of a single quark is called a jet. In particle detectors, jets are observed rather than quarks, whose existence must be inferred. The models and approximation schemes and their predicted Jet hadronization, or **fragmentation**, have been extensively compared with measurement in a number of high energy particle physics experiments; e.g. TASSO,*[4] OPAL,*[5] H1.*[6]

Hadronization also occurred shortly after the Big Bang when the quark–gluon plasma cooled to the temperature below which free quarks and gluons cannot exist (about 170 MeV). The quarks and gluons then combined into hadrons.

A top quark, however, has a mean lifetime of $5\times 10^*-25$ seconds, which is shorter than the time scale at which the strong force of QCD acts, so a top quark decays before it can hadronize, allowing physicists to observe a "bare quark."*[7] Thus, they have not been observed as components of any observed hadron, while all other quarks have been observed only as components of hadrons.

6.12.1 Hadronization simulation and models

Hadronization can be explored using Monte Carlo simulation. After the particle shower has terminated, partons with virtualities on the order of the cut off scale remain. From this point on, the parton is in the low momentum transfer, long-distance regime in which non-perturbative effects become important. The most dominant of these effects is hadronization, which converts partons into observable hadrons. No exact theory for hadronization is known but there are two successful models for parameterization.

The scale at which partons are given to the hadronization is fixed by the Shower Monte Carlo program. Hadronization models typically start at some predefined scale of their own. This can cause significant issue if not set up properly within the Shower Monte Carlo. Common choices of Shower Monte Carlo are PYTHIA and HERWIG. Each of these correspond to one of the two parameterization models.

6.12.2 References

[1] Yu. L. Dokshitzer, V. A. Khoze, A. H. Mueller and S. I. Troyan, *Basics of Perturbative QCD* Editions Frontieres (1991)

[2] A. Bassetto, M. Ciafaloni, G. Marchesini and A. H. Mueller, *Nucl. Phys.* 207B (1982) 189

[3] A. H. Mueller, *Phys. Lett.* 104B (1981) 161

[4] TASSO Collaboration, W. Braunschweig et al., *Zeit. Phys.* C47 (1990) 187

[5] OPAL Collaboration, M.Z. Akrawy et al., *Phys. Lett.* 247B (1990) 617.

[6] H1 Collaboration, S. Aid et al., "A Study of the fragmentation of quarks in e- p collisions at HERA." *Nucl.Phys.B* 445:3-24,1995.

[7] Abazov, et al., "Evidence for the Production of Single Top Quarks", Fermilab-Pub08/056-E (2008)

- Greco, V.; Ko, C. M.; Lévai, P. (2003). "Parton Coalescence and the Antiproton/Pion Anomaly at RHIC". *Physical Review Letters* **90** (20): 202302. arXiv:nucl-th/0301093. Bibcode:2003PhRvL..90t2302G. doi:10.1103/PhysRevLett.90.202302. PMID 12785885.

- Fries, R. J.; Müller, B.; Nonaka, C.; Bass, SA (2003). "Hadronization in Heavy-Ion Collisions: Recombination and Fragmentation of Partons Hadronization in Heavy-Ion Collisions". *Physical Review Letters* **90** (20): 202303. arXiv:nucl-th/0301087. Bibcode:2003PhRvL..90t2303F. doi:10.1103/PhysRevLett.90.202303. PMID 12785886.

6.13 Lepton

For other uses, see Lepton (disambiguation).

A **lepton** is an elementary, half-integer spin (spin $\frac{1}{2}$) particle that does not undergo strong interactions, but is subject to the Pauli exclusion principle.*[1] The best known of all leptons is the electron, which is directly tied to all chemical properties. Two main classes of leptons exist: charged leptons (also known as the *electron-like* leptons), and neutral leptons (better known as neutrinos). Charged leptons can combine with other particles to form various composite particles such as atoms and positronium, while neutrinos rarely interact with anything, and are consequently rarely observed.

There are six types of leptons, known as *flavours*, forming three *generations*.*[2] The first generation is the *electronic leptons*, comprising the electron (e−) and electron neutrino (ν e); the second is the *muonic leptons*, comprising the muon (μ−) and muon neutrino (ν μ); and the third is the *tauonic leptons*, comprising the tau (τ−) and the tau neutrino (ν τ). Electrons have the least mass of all the charged leptons. The heavier muons and taus will rapidly change into electrons through a process of particle decay: the transformation from a higher mass state to a lower mass state. Thus electrons are stable and the most common charged lepton in the universe, whereas muons and taus can only be produced in high energy collisions (such as those involving cosmic rays and those carried out in particle accelerators).

Leptons have various intrinsic properties, including electric charge, spin, and mass. Unlike quarks however, leptons are not subject to the strong interaction, but they are subject to the other three fundamental interactions: gravitation, electromagnetism (excluding neutrinos, which are electrically neutral), and the weak interaction. For every lepton flavor there is a corresponding type of antiparticle, known as antilepton, that differs from the lepton only in that some of its properties have equal magnitude but opposite sign. However, according to certain theories, neutrinos may be their own antiparticle, but it is not currently known whether this is the case or not.

The first charged lepton, the electron, was theorized in the mid-19th century by several scientists[*][3][*][4][*][5] and was discovered in 1897 by J. J. Thomson.[*][6] The next lepton to be observed was the muon, discovered by Carl D. Anderson in 1936, which was classified as a meson at the time.[*][7] After investigation, it was realized that the muon did not have the expected properties of a meson, but rather behaved like an electron, only with higher mass. It took until 1947 for the concept of "leptons" as a family of particle to be proposed.[*][8] The first neutrino, the electron neutrino, was proposed by Wolfgang Pauli in 1930 to explain certain characteristics of beta decay.[*][8] It was first observed in the Cowan–Reines neutrino experiment conducted by Clyde Cowan and Frederick Reines in 1956.[*][8][*][9] The muon neutrino was discovered in 1962 by Leon M. Lederman, Melvin Schwartz and Jack Steinberger,[*][10] and the tau discovered between 1974 and 1977 by Martin Lewis Perl and his colleagues from the Stanford Linear Accelerator Center and Lawrence Berkeley National Laboratory.[*][11] The tau neutrino remained elusive until July 2000, when the DONUT collaboration from Fermilab announced its discovery.[*][12][*][13]

Leptons are an important part of the Standard Model. Electrons are one of the components of atoms, alongside protons and neutrons. Exotic atoms with muons and taus instead of electrons can also be synthesized, as well as lepton–antilepton particles such as positronium.

6.13.1 Etymology

The name *lepton* comes from the Greek λεπτός *leptós*, "fine, small, thin" (neuter form: λεπτόν *leptón*);[*][14][*][15] the earliest attested form of the word is the Mycenaean Greek 𐀩𐀠𐀵, *re-po-to*, written in Linear B syllabic script.[*][16] *Lepton* was first used by physicist Léon Rosenfeld in 1948:[*][17]

> Following a suggestion of Prof. C. Møller, I adopt —as a pendant to "nucleon" —the denomination "lepton" (from λεπτός, small, thin, delicate) to denote a particle of small mass.

The etymology incorrectly implies that all the leptons are of small mass. When Rosenfeld named them, the only known leptons were electrons and muons, which are in fact of small mass —the mass of an electron (0.511 MeV/c^2)[*][18] and the mass of a muon (with a value of 105.7 MeV/c^2)[*][19] are fractions of the mass of the "heavy" proton (938.3 MeV/c^2).[*][20] However, the mass of the tau (discovered in the mid 1970s) (1777 MeV/c^2)[*][21] is nearly twice that of the proton, and about 3,500 times that of the electron.

6.13.2 History

See also: Electron § Discovery, Muon § History and Tau (particle) § History

The first lepton identified was the electron, discovered

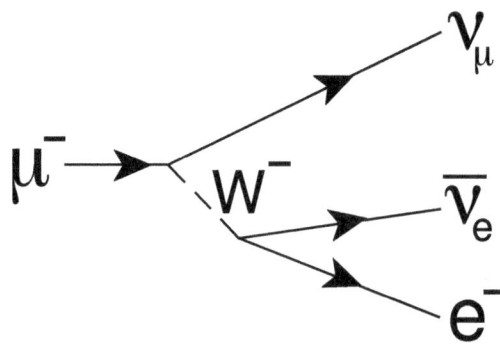

A muon transmutes into a muon neutrino by emitting a W− boson. The W− boson subsequently decays into an electron and an electron antineutrino.

by J.J. Thomson and his team of British physicists in 1897.[*][22][*][23] Then in 1930 Wolfgang Pauli postulated the electron neutrino to preserve conservation of energy, conservation of momentum, and conservation of angular momentum in beta decay.[*][24] Pauli theorized that an undetected particle was carrying away the difference between the energy, momentum, and angular momentum of the initial and observed final particles. The electron neutrino was simply called the neutrino, as it was not yet known that neutrinos came in different flavours (or different "generations").

Nearly 40 years after the discovery of the electron, the muon was discovered by Carl D. Anderson in 1936. Due to its mass, it was initially categorized as a meson rather than a lepton.*[25] It later became clear that the muon was much more similar to the electron than to mesons, as muons do not undergo the strong interaction, and thus the muon was reclassified: electrons, muons, and the (electron) neutrino were grouped into a new group of particles – the leptons. In 1962 Leon M. Lederman, Melvin Schwartz and Jack Steinberger showed that more than one type of neutrino exists by first detecting interactions of the muon neutrino, which earned them the 1988 Nobel Prize, although by then the different flavours of neutrino had already been theorized.*[26]

The tau was first detected in a series of experiments between 1974 and 1977 by Martin Lewis Perl with his colleagues at the SLAC LBL group.*[27] Like the electron and the muon, it too was expected to have an associated neutrino. The first evidence for tau neutrinos came from the observation of "missing" energy and momentum in tau decay, analogous to the "missing" energy and momentum in beta decay leading to the discovery of the electron neutrino. The first detection of tau neutrino interactions was announced in 2000 by the DONUT collaboration at Fermilab, making it the latest particle of the Standard Model to have been directly observed,*[28] apart from the Higgs boson, which probably has been discovered in 2012.

Although all present data is consistent with three generations of leptons, some particle physicists are searching for a fourth generation. The current lower limit on the mass of such a fourth charged lepton is 100.8 GeV/c^2,*[29] while its associated neutrino would have a mass of at least 45.0 GeV/c^2.*[30]

6.13.3 Properties

Spin and chirality

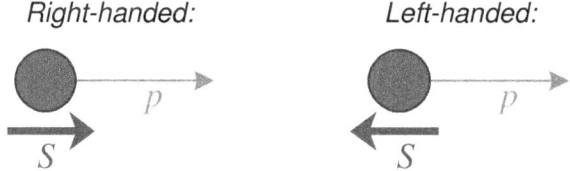

Left-handed and right-handed helicities

Leptons are spin-$\frac{1}{2}$ particles. The spin-statistics theorem thus implies that they are fermions and thus that they are subject to the Pauli exclusion principle; no two leptons of the same species can be in exactly the same state at the same time. Furthermore, it means that a lepton can have only two possible spin states, namely up or down.

A closely related property is chirality, which in turn is closely related to a more easily visualized property called helicity. The helicity of a particle is the direction of its spin relative to its momentum; particles with spin in the same direction as their momentum are called *right-handed* and otherwise they are called *left-handed*. When a particle is mass-less, the direction of its momentum relative to its spin is frame independent, while for massive particles it is possible to 'overtake' the particle by a Lorentz transformation flipping the helicity. Chirality is a technical property (defined through the transformation behaviour under the Poincaré group) that agrees with helicity for (approximately) massless particles and is still well defined for massive particles.

In many quantum field theories—such as quantum electrodynamics and quantum chromodynamics—left and right-handed fermions are identical. However in the Standard Model left-handed and right-handed fermions are treated asymmetrically. Only left-handed fermions participate in the weak interaction, while there are no right-handed neutrinos. This is an example of parity violation. In the literature left-handed fields are often denoted by a capital L subscript (e.g. e–$_L$) and right-handed fields are denoted by a capital R subscript.

Electromagnetic interaction

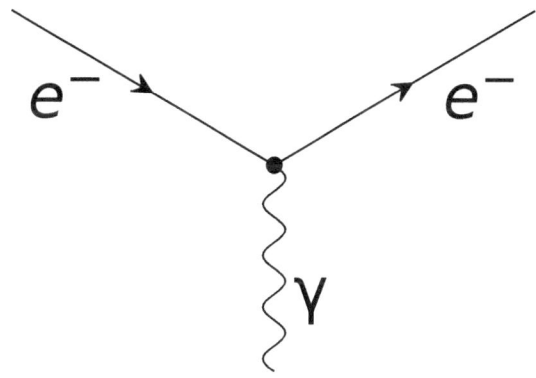

Lepton–photon interaction

One of the most prominent properties of leptons is their electric charge, Q. The electric charge determines the strength of their electromagnetic interactions. It determines the strength of the electric field generated by the particle (see Coulomb's law) and how strongly the particle reacts to an external electric or magnetic field (see Lorentz force). Each generation contains one lepton with $Q = -e$ (conventionally the charge of a particle is expressed in units of the elementary charge) and one lepton with zero electric charge. The lepton with electric charge is commonly simply referred to as a 'charged lepton' while the neutral lepton is called a neutrino. For example the first generation consists of the electron e– with a negative electric charge and the electrically neutral electron neutrino ν e.

In the language of quantum field theory the electromagnetic interaction of the charged leptons is expressed by the fact that the particles interact with the quantum of the electromagnetic field, the photon. The Feynman diagram of the electron-photon interaction is shown on the right.

Because leptons possess an intrinsic rotation in the form of their spin, charged leptons generate a magnetic field. The size of their magnetic dipole moment μ is given by,

$$\mu = g \frac{Q\hbar}{4m},$$

where m is the mass of the lepton and g is the so-called g-factor for the lepton. First order approximation quantum mechanics predicts that the g-factor is 2 for all leptons. However, higher order quantum effects caused by loops in Feynman diagrams introduce corrections to this value. These corrections, referred to as the anomalous magnetic dipole moment, are very sensitive to the details of a quantum field theory model and thus provide the opportunity for precision tests of the standard model. The theoretical and measured values for the electron anomalous magnetic dipole moment are within agreement within eight significant figures.[*][31]

Weak Interaction

In the Standard Model the left-handed charged lepton and the left-handed neutrino are arranged in doublet (ν e_L, e–$_L$) that transforms in the spinor representation ($T = \frac{1}{2}$) of the weak isospin SU(2) gauge symmetry. This means that these particles are eigenstates of the isospin projection T_3 with eigenvalues $\frac{1}{2}$ and $-\frac{1}{2}$ respectively. In the meantime, the right-handed charged lepton transforms as a weak isospin scalar ($T = 0$) and thus does not participate in the weak interaction, while there is no right-handed neutrino at all.

The Higgs mechanism recombines the gauge fields of the weak isospin SU(2) and the weak hypercharge U(1) symmetries to three massive vector bosons (W+, W–, Z0) mediating the weak interaction, and one massless vector boson, the photon, responsible for the electromagnetic interaction. The electric charge Q can be calculated from the isospin projection T_3 and weak hypercharge Y_W through the Gell-Mann–Nishijima formula,

$$Q = T_3 + Y_W/2$$

To recover the observed electric charges for all particles the left-handed weak isospin doublet (ν e_L, e–$_L$) must thus have $Y_W = -1$, while the right-handed isospin scalar e– R must have $Y_W = -2$. The interaction of the leptons with the massive weak interaction vector bosons is shown in the figure on the left.

Mass

In the Standard Model each lepton starts out with no intrinsic mass. The charged leptons (i.e. the electron, muon, and tau) obtain an effective mass through interaction with the Higgs field, but the neutrinos remain massless. For technical reasons the masslessness of the neutrinos implies that there is no mixing of the different generations of charged leptons as there is for quarks. This is in close agreement with current experimental observations.[*][32]

However, it is known from experiments – most prominently from observed neutrino oscillations[*][33] – that neutrinos do in fact have some very small mass, probably less than 2 eV/c^2.[*][34] This implies the existence of physics beyond the Standard Model. The currently most favoured extension is the so-called seesaw mechanism, which would explain both why the left-handed neutrinos are so light compared to the corresponding charged leptons, and why we have not yet seen any right-handed neutrinos.

Leptonic numbers

Main article: Lepton number

The members of each generation's weak isospin doublet are assigned leptonic numbers that are conserved under the Standard Model.[*][35] Electrons and electron neutrinos have an *electronic number* of $L_e = 1$, while muons and muon neutrinos have a *muonic number* of $L_\mu = 1$, while tau particles and tau neutrinos have a *tauonic number* of $L_\tau = 1$. The antileptons have their respective generation's leptonic numbers of –1.

Conservation of the leptonic numbers means that the number of leptons of the same type remains the same, when particles interact. This implies that leptons and antileptons must be created in pairs of a single generation. For example, the following processes are allowed under conservation of leptonic numbers:

$$\begin{pmatrix} \nu_e \\ e^- \end{pmatrix}, \begin{pmatrix} \nu_\mu \\ \mu^- \end{pmatrix}, \begin{pmatrix} \nu_\tau \\ \tau^- \end{pmatrix}$$

Each generation forms a weak isospin doublet.

e- + e+ → γ + γ,

τ- + τ+ → Z0 + Z0,

but not these:

γ → e- + μ+,

W- → e- + ν τ,

Z0 → μ- + τ+.

However, neutrino oscillations are known to violate the conservation of the individual leptonic numbers. Such a violation is considered to be smoking gun evidence for physics beyond the Standard Model. A much stronger conservation law is the conservation of the total number of leptons (L), conserved even in the case of neutrino oscillations, but even it is still violated by a tiny amount by the chiral anomaly.

6.13.4 Universality

The coupling of the leptons to gauge bosons are flavour-independent (i.e., the interactions between leptons and gauge bosons are the same for all leptons).[35] This property is called *lepton universality* and has been tested in measurements of the tau and muon lifetimes and of Z boson partial decay widths, particularly at the Stanford Linear Collider (SLC) and Large Electron-Positron Collider (LEP) experiments.[36]:241–243[37]:138

The decay rate (Γ) of muons through the process μ- → e- + ν

e + ν

μ is approximately given by an expression of the form (see muon decay for more details)[35]

$$\Gamma\left(\mu^- \to e^- + \bar{\nu}_e + \nu_\mu\right) = K_1 G_F^2 m_\mu^5,$$

where K_1 is some constant, and G_F is the Fermi coupling constant. The decay rate of tau particles through the process τ- → e- + ν

e + ν

τ is given by an expression of the same form[35]

$$\Gamma\left(\tau^- \to e^- + \bar{\nu}_e + \nu_\tau\right) = K_2 G_F^2 m_\tau^5,$$

where K_2 is some constant. Muon–Tauon universality implies that $K_1 = K_2$. On the other hand, electron–muon universality implies[35]

$$\Gamma\left(\tau^- \to e^- + \bar{\nu}_e + \nu_\tau\right) = \Gamma\left(\tau^- \to \mu^- + \bar{\nu}_\mu + \nu_\tau\right).$$

This explains why the branching ratios for the electronic mode (17.85%) and muonic (17.36%) mode of tau decay are equal (within error).[21]

Universality also accounts for the ratio of muon and tau lifetimes. The lifetime of a lepton (τ_l) is related to the decay rate by[35]

$$\tau_l = \frac{B\left(l^- \to e^- + \bar{\nu}_e + \nu_l\right)}{\Gamma\left(l^- \to e^- + \bar{\nu}_e + \nu_l\right)},$$

where $B(x \to y)$ and $\Gamma(x \to y)$ denotes the branching ratios and the resonance width of the process x → y.

The ratio of tau and muon lifetime is thus given by[35]

$$\frac{\tau_\tau}{\tau_\mu} = \frac{B\left(\tau^- \to e^- + \bar{\nu}_e + \nu_\tau\right)}{B\left(\mu^- \to e^- + \bar{\nu}_e + \nu_\mu\right)} \left(\frac{m_\mu}{m_\tau}\right)^5.$$

Using the values of the 2008 *Review of Particle Physics* for the branching ratios of muons[19] and tau[21] yields a lifetime ratio of ~1.29×10*–7, comparable to the measured lifetime ratio of ~1.32×10*–7. The difference is due to K_1 and K_2 not actually being constants; they depend on the mass of leptons.

6.13.5 Table of leptons

6.13.6 See also

- Koide formula
- List of particles
- Preons – hypothetical particles which were once postulated to be subcomponents of quarks and leptons

6.13.7 Notes

[1] "Lepton (physics)". *Encyclopædia Britannica*. Retrieved 2010-09-29.

[2] R. Nave. "Leptons". *HyperPhysics*. Georgia State University, Department of Physics and Astronomy. Retrieved 2010-09-29.

[3] W.V. Farrar (1969). "Richard Laming and the Coal-Gas Industry, with His Views on the Structure of Matter". *Annals of Science* **25** (3): 243–254. doi:10.1080/00033796900200141.

[4] T. Arabatzis (2006). *Representing Electrons: A Biographical Approach to Theoretical Entities*. University of Chicago Press. pp. 70–74. ISBN 0-226-02421-0.

[5] J.Z. Buchwald, A. Warwick (2001). *Histories of the Electron: The Birth of Microphysics*. MIT Press. pp. 195–203. ISBN 0-262-52424-4.

[6] J.J. Thomson (1897). "Cathode Rays". *Philosophical Magazine* **44** (269): 293. doi:10.1080/14786449708621070.

[7] S.H. Neddermeyer, C.D. Anderson; Anderson (1937). "Note on the Nature of Cosmic-Ray Particles". *Physical Review* **51** (10): 884–886. Bibcode:1937PhRv...51..884N. doi:10.1103/PhysRev.51.884.

[8] "The Reines-Cowan Experiments: Detecting the Poltergeist" (PDF). *Los Alamos Science* **25**: 3. 1997. Retrieved 2010-02-10.

[9] F. Reines, C.L. Cowan, Jr.; Cowan (1956). "The Neutrino". *Nature* **178** (4531): 446. Bibcode:1956Natur.178..446R. doi:10.1038/178446a0.

[10] G. Danby; Gaillard, J-M.; Goulianos, K.; Lederman, L.; Mistry, N.; Schwartz, M.; Steinberger, J. et al. (1962). "Observation of high-energy neutrino reactions and the existence of two kinds of neutrinos". *Physical Review Letters* **9**: 36. Bibcode:1962PhRvL...9...36D. doi:10.1103/PhysRevLett.9.36.

[11] M.L. Perl; Abrams, G.; Boyarski, A.; Breidenbach, M.; Briggs, D.; Bulos, F.; Chinowsky, W.; Dakin, J.; Feldman, G.; Friedberg, C.; Fryberger, D.; Goldhaber, G.; Hanson, G.; Heile, F.; Jean-Marie, B.; Kadyk, J.; Larsen, R.; Litke, A.; Lüke, D.; Lulu, B.; Lüth, V.; Lyon, D.; Morehouse, C.; Paterson, J.; Pierre, F.; Pun, T.; Rapidis, P.; Richter, B.; Sadoulet, B. et al. (1975). "Evidence for Anomalous Lepton Production in e+e− Annihilation". *Physical Review Letters* **35** (22): 1489. Bibcode:1975PhRvL..35.1489P. doi:10.1103/PhysRevLett.35.1489.

[12] "Physicists Find First Direct Evidence for Tau Neutrino at Fermilab" (Press release). Fermilab. 20 July 2000.

[13] K. Kodama *et al.* (DONUT Collaboration); Kodama; Ushida; Andreopoulos; Saoulidou; Tzanakos; Yager; Baller; Boehnlein; Freeman; Lundberg; Morfin; Rameika; Yun; Song; Yoon; Chung; Berghaus; Kubantsev; Reay; Sidwell; Stanton; Yoshida; Aoki; Hara; Rhee; Ciampa; Erickson; Graham et al. (2001). "Observation of tau neutrino interactions". *Physics Letters B* **504** (3): 218. arXiv:hep-ex/0012035. Bibcode:2001PhLB..504..218D. doi:10.1016/S0370-2693(01)00307-0.

[14] "lepton". *Online Etymology Dictionary*.

[15] λεπτός. Liddell, Henry George; Scott, Robert; *A Greek–English Lexicon* at the Perseus Project.

[16] Found on the KN L 693 and PY Un 1322 tablets. "The Linear B word re-po-to". Palaeolexicon. Word study tool of ancient languages. Raymoure, K.A. "re-po-to". *Minoan Linear A & Mycenaean Linear B*. Deaditerranean. "KN 693 L (103)". "PY 1322 Un + fr. (Cii)". *DĀMOS: Database of Mycenaean at Oslo*. University of Oslo.

[17] L. Rosenfeld (1948)

[18] C. Amsler *et al.* (2008): Particle listings – e−

[19] C. Amsler *et al.* (2008): Particle listings – μ−

[20] C. Amsler *et al.* (2008): Particle listings – p+

[21] C. Amsler *et al.* (2008): Particle listings – τ−

[22] S. Weinberg (2003)

[23] R. Wilson (1997)

[24] K. Riesselmann (2007)

[25] S.H. Neddermeyer, C.D. Anderson (1937)

[26] I.V. Anicin (2005)

[27] M.L. Perl et al. (1975)

[28] K. Kodama (2001)

[29] C. Amsler *et al.* (2008) Heavy Charged Leptons Searches

[30] C. Amsler *et al.* (2008) Searches for Heavy Neutral Leptons

[31] M.E. Peskin, D.V. Schroeder (1995), p. 197

[32] M.E. Peskin, D.V. Schroeder (1995), p. 27

[33] Y. Fukuda *et al.* (1998)

[34] C.Amsler et al. (2008): Particle listings – Neutrino properties

[35] B.R. Martin, G. Shaw (1992)

[36] J. P. Cumalat (1993). *Physics in Collision 12*. Atlantica Séguier Frontières. ISBN 978-2-86332-129-4.

[37] G Fraser (1 January 1998). *The Particle Century*. CRC Press. ISBN 978-1-4200-5033-2.

[38] J. Peltoniemi, J. Sarkamo (2005)

6.13.8 References

- C. Amsler *et al.* (Particle Data Group); Amsler; Doser; Antonelli; Asner; Babu; Baer; Band; Barnett; Bergren; Beringer; Bernardi; Bertl; Bichsel; Biebel; Bloch; Blucher; Blusk; Cahn; Carena; Caso; Ceccucci; Chakraborty; Chen; Chivukula; Cowan; Dahl; d'Ambrosio; Damour et al. (2008). "Review of Particle Physics". *Physics Letters B* **667**: 1. Bibcode:2008PhLB..667....1P. doi:10.1016/j.physletb.2008.07.018.

- I.V. Anicin (2005). "The Neutrino – Its Past, Present and Future". *SFIN (Institute of Physics, Belgrade) year XV, Series A: Conferences, No. A2 (2002) 3–59*: 3172. arXiv:physics/0503172. Bibcode:2005physics...3172A.

- Y. Fukuda; Hayakawa, T.; Ichihara, E.; Inoue, K.; Ishihara, K.; Ishino, H.; Itow, Y.; Kajita, T. et al. (1998). "Evidence for Oscillation of Atmospheric Neutrinos". *Physical Review Letters* **81** (8): 1562–1567. arXiv:hep-ex/9807003. Bibcode:1998PhRvL..81.1562F. doi:10.1103/PhysRevLett.81.1562.

- K. Kodama; Ushida, N.; Andreopoulos, C.; Saoulidou, N.; Tzanakos, G.; Yager, P.; Baller, B.; Boehnlein, D.; Freeman, W.; Lundberg, B.; Morfin, J.; Rameika, R.; Yun, J.C.; Song, J.S.; Yoon, C.S.; Chung, S.H.; Berghaus, P.; Kubantsev, M.; Reay, N.W.; Sidwell, R.; Stanton, N.; Yoshida, S.; Aoki, S.; Hara, T.; Rhee, J.T.; Ciampa, D.; Erickson, C.; Graham, M.; Heller, K. et al. (2001). "Observation of tau neutrino interactions". *Physics Letters B* **504** (3): 218. arXiv:hep-ex/0012035. Bibcode:2001PhLB..504..218D. doi:10.1016/S0370-2693(01)00307-0.

- B.R. Martin, G. Shaw (1992). "Chapter 2 – Leptons, quarks and hadrons". *Particle Physics*. John Wiley & Sons. pp. 23–47. ISBN 0-471-92358-3.

- S.H. Neddermeyer, C.D. Anderson; Anderson (1937). "Note on the Nature of Cosmic-Ray Particles". *Physical Review* **51** (10): 884–886. Bibcode:1937PhRv...51..884N. doi:10.1103/PhysRev.51.884.

- J. Peltoniemi, J. Sarkamo (2005). "Laboratory measurements and limits for neutrino properties". *The Ultimate Neutrino Page*. Retrieved 2008-11-07.

- M.L. Perl; Abrams, G.; Boyarski, A.; Breidenbach, M.; Briggs, D.; Bulos, F.; Chinowsky, W.; Dakin, J. et al. (1975). "Evidence for Anomalous Lepton Production in e^+–e^- Annihilation". *Physical Review Letters* **35** (22): 1489–1492. Bibcode:1975PhRvL..35.1489P. doi:10.1103/PhysRevLett.35.1489.

- M.E. Peskin, D.V. Schroeder (1995). *Introduction to Quantum Field Theory*. Westview Press. ISBN 0-201-50397-2.

- K. Riesselmann (2007). "Logbook: Neutrino Invention". *Symmetry Magazine* **4** (2).

- L. Rosenfeld (1948). *Nuclear Forces*. Interscience Publishers. p. xvii.

- R. Shankar (1994). "Chapter 2 – Rotational Invariance and Angular Momentum". *Principles of Quantum Mechanics* (2nd ed.). Springer. pp. 305–352. ISBN 978-0-306-44790-7.

- S. Weinberg (2003). *The Discovery of Subatomic Particles*. Cambridge University Press. ISBN 0-521-82351-X.

- R. Wilson (1997). *Astronomy Through the Ages: The Story of the Human Attempt to Understand the Universe*. CRC Press. p. 138. ISBN 0-7484-0748-0.

6.13.9 External links

- Particle Data Group homepage. The PDG compiles authoritative information on particle properties.

- Leptons, a summary of leptons from *Hyperphysics*.

6.14 Leptoquark

Leptoquarks are hypothetical particles that carry information between quarks and leptons of a given generation that allow quarks and leptons to interact. They are color-triplet bosons that carry both lepton and baryon numbers. They are encountered in various extensions of the Standard Model, such as technicolor theories or GUTs based on Pati–Salam model, SU(5) or E_6, etc. Their quantum numbers like spin, (fractional) electric charge and weak isospin vary among theories.

Leptoquarks, predicted to be nearly as heavy as an atom of lead, could only be created at high energies, and would decay rapidly. A third generation leptoquark, for example, might decay into a bottom quark and a tau lepton. Some theorists propose that the 'leptoquark' observed by HERA and DESY could be a new force that bonds positrons and quarks or be examples of preons found at high energies.[1] Leptoquarks could explain the reason for the three generations of matter. Furthermore, leptoquarks could explain why the same number of quarks and leptons exist and many other similarities between the quark and the lepton sectors. At high energies, when leptons that do not feel the strong force and quarks that cannot be separately observed because of the strong force become one, it could form a more fundamental particle and describe a higher symmetry. There would be three kinds of leptoquarks made of the leptons and quarks of each generation.

The LHeC project to add an electron ring to collide bunches with the existing LHC proton ring is proposed as a project to look for higher-generation leptoquarks.[2]

6.14.1 Existence

In 1997, an excess of events at the HERA accelerator created a stir in the particle physics community, because one possible explanation of the excess was the involvement of leptoquarks. However, more recent studies performed both at HERA and at the Tevatron with larger samples of data ruled out this possibility for masses of the leptoquark up to 275-325 GeV.[3] Second generation leptoquarks were also looked for and not found.[4] For leptoquarks to be proven to exist, the missing energy in particle collisions attributed to neutrinos would have to be excessively energetic. It is likely that the creation of leptoquarks would mimic the creation of massive quarks.[5]

6.14.2 See also

- Quark–lepton complementarity
- X and Y bosons

6.14.3 References

[1] Scientific American

[2] Birmingham LHeC project page

[3] H1 Collaboration; Andreev, V.; Anthonis, T.; Aplin, S.; Asmone, A.; Astvatsatourov, A.; Babaev, A.; Backovic, S.; Bähr, J.; Baghdasaryan, A.; Baranov, P.; Barrelet, E.; Bartel, W.; Baudrand, S.; Baumgartner, S.; Becker, J.; Beckingham, M.; Behnke, O.; Behrendt, O.; Belousov, A.; Berger, Ch.; Berger, N.; Bizot, J.C.; Boenig, M.-O.; Boudry, V.; Bracinik, J.; Brandt, G.; Brisson, V.; Brown, D.P. et al. (2005). "Search for Leptoquark Bosons in ep Collisions at HERA" . *Physics Letters B* **629**: 9–19. arXiv:hep-ex/0506044. Bibcode:2005PhLB..629....9H. doi:10.1016/j.physletb.2005.09.048.

[4] The Search for Leptoquarks.

[5] Search for Third Generation Leptoquarks

6.15 November Revolution

The **J/ψ** (**J/Psi**) **meson** or **psion** [1] is a subatomic particle, a flavor-neutral meson consisting of a charm quark and a charm antiquark. Mesons formed by a bound state of a charm quark and a charm anti-quark are generally known as "charmonium". The J/ψ is the first excited state of charmonium (i.e. the form of the charmonium with the second-smallest rest mass). The J/ψ has a rest mass of 3.0969 GeV/c^2, and a mean lifetime of 7.2×10^{-21} s. This lifetime was about a thousand[2] times longer than expected.

Its discovery was made independently by two research groups, one at the Stanford Linear Accelerator Center, headed by Burton Richter, and one at the Brookhaven National Laboratory, headed by Samuel Ting of MIT. They discovered they had actually found the same particle, and both announced their discoveries on 11 November 1974. The importance of this discovery is highlighted by the fact that the subsequent, rapid changes in high-energy physics at the time have become collectively known as the "**November Revolution**". Richter and Ting were rewarded for their shared discovery with the 1976 Nobel Prize in Physics.

6.15.1 Background to discovery

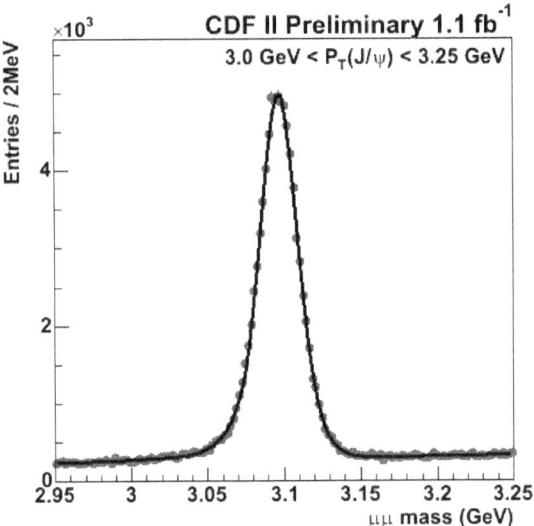

J/Ψ Production at Fermilab

The background to the discovery of the J/ψ was both theoretical and experimental. In the 1960s, the first quark models of elementary particle physics were proposed, which said that protons, neutrons and all other baryons, and also all mesons, are made from three kinds of fractionally-charged particles, the "quarks", that come in three different types or "flavors", called *up*, *down*, and *strange*. Despite the impressive ability of quark models to bring order to the "elementary particle zoo", their status was considered something like mathematical fiction at the time, a simple artifact of deeper physical reasons.

Starting in 1969, deep inelastic scattering experiments at SLAC revealed surprising experimental evidence for particles inside of protons. Whether these were quarks or something else was not known at first. Many experiments were needed to fully identify the properties of the subprotonic components. To a first approximation, they were indeed the already-described quarks.

On the theoretical front, gauge theories with broken symmetry became the first fully viable contenders for explaining the weak interaction after Gerardus 't Hooft discovered in 1971 how to calculate with them beyond tree level. The first experimental evidence for these electroweak unification theories was the discovery of the weak neutral current in 1973. Gauge theories with quarks became a viable contender for the strong interaction in 1973 when the concept of asymptotic freedom was identified.

However, a naive mixture of electroweak theory and the quark model led to calculations about known decay modes that contradicted observation: in particular, it predicted Z boson-mediated *flavor-changing* decays of a strange quark into a down quark, which were not observed. A 1970 idea of Sheldon Glashow, John Iliopoulos, and Luciano Maiani, known as the GIM mechanism, showed that the flavor-changing decays would be eliminated if there were a fourth quark, *charm*, that paired with the strange quark. This work led, by the summer of 1974, to theoretical predictions of what a charm/anticharm meson would be like. These predictions were ignored. The work of Richter and Ting was done for other reasons, mostly to explore new energy regimes.

6.15.2 The name

Because of the nearly simultaneous discovery, the J/ψ is the only particle to have a two-letter name. Richter named it "SP", after the SPEAR accelerator used at SLAC; however, none of his coworkers liked that name. After consulting with Greek-born Leo Resvanis to see which Greek letters were still available, and rejecting "iota" because its name implies insignificance, Richter chose "psi" – a name which, as Gerson Goldhaber pointed out, contains the original name "SP", but in reverse order.[3] Coincidentally, later spark chamber pictures often resembled the psi shape. Ting assigned the name "J" to it, which is one letter removed from "K", the name of the already-known strange meson; possibly by coincidence, "J" strongly resembles the Chinese character for Ting's name (丁). (Cf. the naming of Gallium.) J is also the first letter of Ting's oldest daughter's name, Jeanne.

Since the scientific community considered it unjust to give one of the two discoverers priority, most subsequent publications have referred to the particle as the "J/ψ".

The first excited state of the J/ψ was called the ψ'; it is now called the ψ(2S), indicating its quantum state. The next excited state was called the ψ"; it is now called ψ(3770), indicating mass in MeV. Other vector charm-ant07charm states are denoted similarly with ψ and the quantum state (if known) or the mass.[4] The "J" is not used, since Richter's group alone first found excited states.

The name *charmonium* is used for the J/ψ and other charm-antcharm bound states. This is by analogy with positronium, which also consists of a particle and its antiparticle (an electron and positron in the case of positronium).

6.15.3 J/ψ melting

In a hot QCD medium, when the temperature is raised well beyond the Hagedorn temperature, the J/ψ and its excitations are expected to melt.*[5] This is one of the predicted signals of the formation of the quark–gluon plasma. Heavy-ion experiments at CERN's Super Proton Synchrotron and at BNL's Relativistic Heavy Ion Collider have studied this phenomenon without a conclusive outcome as of 2009. This is due to the requirement that the disappearance of J/ψ mesons is evaluated with respect to the baseline provided by the total production of all charm quark-containing subatomic particles, and because it is widely expected that some J/ψ are produced and/or destroyed at time of QGP hadronization. Thus, there is uncertainty in the prevailing conditions at the initial collisions.

In fact, instead of suppression, enhanced production of J/ψ is expected*[6] in heavy ion experiments at LHC where the quark-combinant production mechanism should be dominant given the large abundance of charm quarks in the QGP. Aside of J/ψ, charmed B mesons (B
c), offer a signature that indicates that quarks move freely and bind at-will when combining to form hadrons.*[7]*[8]

6.15.4 Decay modes

Hadronic decay modes of J/ψ are strongly suppressed because of the OZI Rule. This effect strongly increases the lifetime of the particle and thereby gives it its very narrow decay width of just 93.2±2.1 keV. Because of this strong suppression, electromagnetic decays begin to compete with hadronic decays. This is why the J/ψ has a significant branching fraction to leptons.

6.15.5 See also

- OZI Rule
- List of multiple discoveries

6.15.6 Notes

[1] http://books.google.com.au/books?id= 8AD3GDoVaMkC&pg=PA462&dq=psion+ meson+-wikipedia&hl=en&sa=X&ei= y40kVKjHOI-A8QXpioC4Bw&ved=0CC8Q6AEwAA# v=onepage&q=psion%20meson%20-wikipedia&f=false retrieved 25 September 2014

[2] "Shared Physics prize for elementary particle" (Press release). The Royal Swedish Academy of Sciences. 18 October 1976. Retrieved 2012-04-23.

[3] Zielinski, L (8 August 2006). "Physics Folklore". QuarkNet. Retrieved 2009-04-13.

[4] Roos, M; Wohl, CG; (Particle Data Group) (2004). "Naming schemes for hadrons" (PDF). Retrieved 2009-04-13.

[5] Matsui, T; Satz, H (1986). "J/ψ suppression by quark-gluon plasma formation". *Physics Letters B* **178** (4): 416–422. Bibcode:1986PhLB..178..416M. doi:10.1016/0370-2693(86)91404-8.

[6] Thews, R. L.; Schroedter, M.; Rafelski, J. (2001). "Enhanced J/ψ production in deconfined quark matter". *Physical Review C* **63** (5): 054905. arXiv:hep-ph/0007323. Bibcode:2001PhRvC..63e4905T. doi:10.1103/PhysRevC.63.054905.

[7] Schroedter, M.; Thews, R. L.; Rafelski, J. (2000). "B_c-meson production in ultrarelativistic nuclear collisions". *Physical Review C* **62** (2): 024905. arXiv:hep-ph/0004041. Bibcode:2000PhRvC..62b4905S. doi:10.1103/PhysRevC.62.024905.

[8] Fulcher, L. P.; Rafelski, J.; Thews, R. L. (1999). "B_c mesons as a signal of deconfinement". arXiv:hep-ph/9905201 [hep-ph].

6.15.7 References

- Glashow, S. L.; Iliopoulos, J.; Maiani, L. (1970). "Weak Interactions with Lepton-Hadron Symmetry". *Physical Review D* **2** (7): 1285–1292. Bibcode:1970PhRvD...2.1285G. doi:10.1103/PhysRevD.2.1285.

- Aubert, J. et al. (1974). "Experimental Observation of a Heavy Particle J". *Physical Review Letters* **33** (23): 1404–1406. Bibcode:1974PhRvL..33.1404A. doi:10.1103/PhysRevLett.33.1404.

- Augustin, J. et al. (1974). "Discovery of a Narrow Resonance in $e^* + e^* -$ Annihilation". *Physical Review Letters* **33** (23): 1406–1408. Bibcode:1974PhRvL..33.1406A. doi:10.1103/PhysRevLett.33.1406.

- Bobra, M. (2005). "Logbook: J/ψ particle". *Symmetry Magazine* **2** (7): 34.

- Yao, W.-M. (Particle Data Group) et al. (2006). "Review of Particle Physics: Naming Scheme for Hadrons" (PDF). *Journal of Physics G* **33**: 108. doi:10.1088/0954-3899/33/1/001.

6.16 Parton

In particle physics, the **parton model** was proposed by Richard Feynman in 1969 as a way to analyze high-energy hadron collisions.*[1] It was later recognized that partons describe the same objects now more commonly referred to as quarks and gluons. Therefore a more detailed presentation of the properties and physical theories pertaining indirectly to partons can be found under quarks.

6.16.1 Model

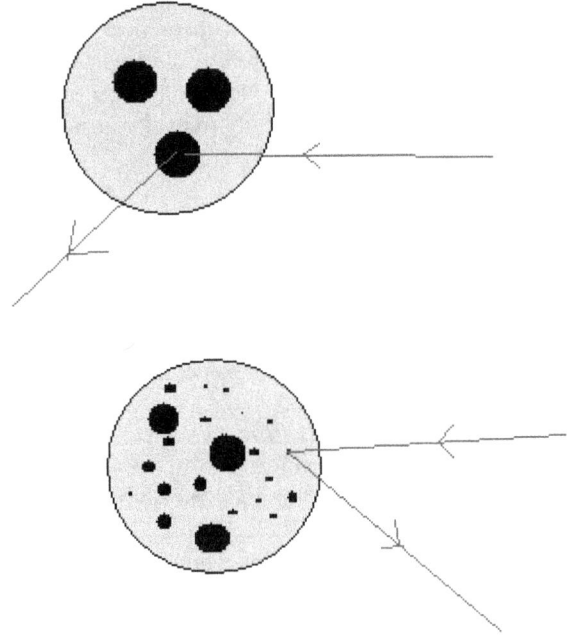

The scattering particle only sees the valence partons. At higher energies, the scattering particles also detects the sea partons.

In this model, a hadron (for example, a proton) is composed of a number of point-like constituents, termed "partons". Additionally, the hadron is in a reference frame where it has infinite momentum —a valid approximation at high energies. Thus, parton motion is slowed by time dilation, and the hadron charge distribution is Lorentz-contracted, so incoming particles will be scattered "instantaneously and incoherently". The parton model was immediately applied to electron-proton deep inelastic scattering by Bjorken and Paschos.*[2] Later, with the experimental observation of Bjorken scaling, the validation of the quark model, and the confirmation of asymptotic freedom in quantum chromodynamics, partons were matched to quarks and gluons. The parton model remains a justifiable approximation at high energies, and others have extended the theory over the years.

In the parton model, partons are defined with respect to a physical scale (as probed by the inverse of the momentum transfer). For instance, a quark parton at one length scale can turn out to be a superposition of a quark parton state with a quark parton and a gluon parton state together with other states with more partons at a smaller length scale. Similarly, a gluon parton at one scale can resolve into a superposition of a gluon parton state, a gluon parton and quark-antiquark partons state and other multiparton states. Because of this, the number of partons in a hadron actually goes up with momentum transfer. At low energies (i.e. large length scales), a baryon contains three valence partons (quarks) and a meson contains two valence partons (a quark and an antiquark parton). At higher energies, however, observations show *sea partons* (nonvalence partons) in addition to valence partons.

6.16.2 Parton distribution functions

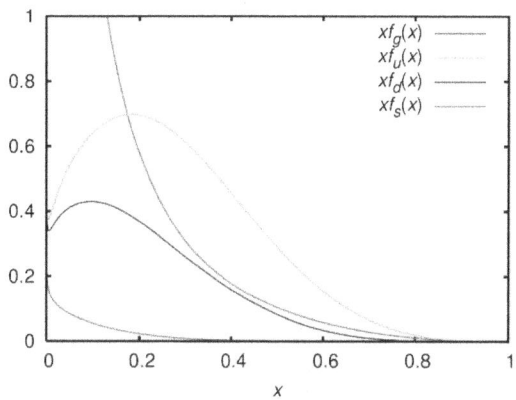

The CTEQ6 parton distribution functions in the MS renormalization scheme and Q = 2 GeV for gluons (red), up (green), down (blue), and strange (violet) quarks. Plotted is the product of longitudinal momentum fraction x *and the distribution functions* f *versus* x.

A parton distribution function within so called *collinear factorization* is defined as the probability density for finding a particle with a certain longitudinal momentum fraction x at resolution scale Q^2. Because of the inherent nonperturbative nature of partons which can not be observed as free particles, parton densities cannot be fully obtained by perturbative QCD. Within QCD one can, however, study variation of parton density with resolution scale provided by external probe. Such scale is for instance provided by a virtual photon with virtuality Q^2 or by a jet. Due to the limitations in present lattice QCD calculations, the known parton distribution functions are instead obtained by fitting observables to experimental data.

Experimentally determined parton distribution functions are available from various groups worldwide. The major unpolarized data sets are:

- *ABM* by S. Alekhin, J. Bluemlein, S. Moch

- *CTEQ*, from the CTEQ Collaboration

- *GRV/GJR*, from M. Glück, P. Jimenez-Delgado, E. Reya, and A. Vogt

- *HERA* PDFs, by H1 and ZEUS collaborations from the Deutsches Elektronen-Synchrotron center (DESY) in Germany

- *MRST/MSTW*, from A. D. Martin, R. G. Roberts, W. J. Stirling, R. S. Thorne, and G. Watt

- *NNPDF*, from the NNPDF Collaboration

The *LHAPDF* *[3] library provides a unified and easy-to-use Fortran/C++ interface to all major PDF sets.

Generalized parton distributions (GPDs) are a more recent approach to better understand hadron structure by representing the parton distributions as functions of more variables, such as the transverse momentum and spin of the parton. Early names included "non-forward", "non-diagonal" or "skewed" parton distributions. They are accessed through exclusive processes for which all particles are detected in the final state. Ordinary parton distribution functions are recovered by setting to zero (forward limit) the extra variables in the generalized parton distributions. Other rules show that the electric form factor, the magnetic form factor, or even the form factors associated to the energy-momentum tensor are also included in the GPDs. A full 3-dimensional image of partons inside hadrons can also be obtained from GPDs.*[4]

6.16.3 References

[1] Feynman, R. P. (1969). "The Behavior of Hadron Collisions at Extreme Energies". *High Energy Collisions: Third International Conference at Stony Brook, N.Y.* Gordon & Breach. pp. 237–249. ISBN 978-0-677-13950-0.

[2] Bjorken, J.; Paschos, E. (1969). "Inelastic Electron-Proton and γ-Proton Scattering and the Structure of the Nucleon". *Physical Review* **185** (5): 1975–1982. Bibcode:1969PhRv..185.1975B. doi:10.1103/PhysRev.185.1975.

[3] Whalley, M. R.; Bourilkov, D; Group, R. C. (2005). "The Les Houches accord PDFs (LHAPDF) and LHAGLUE". p. 8110. arXiv:hep-ph/0508110. Bibcode:2005hep.ph....8110W.

[4] Belitsky, A. V.; Radyushkin, A. V. (2005). "Unraveling hadron structure with generalized parton distributions". *Physics Reports* **418**: 1–387. arXiv:hep-ph/0504030. Bibcode:2005PhR...418....1B. doi:10.1016/j.physrep.2005.06.002.

6.16.4 Further reading

- Glück, M.; Reya, E.; Vogt, A. (1998). "Dynamical Parton Distributions Revisited". *European Physical Journal C* **5** (3): 461. arXiv:hep-ph/9806404. Bibcode:1998EPJC....5..461G. doi:10.1007/s100529800978.

- Hoodbhoy, P. A. (2006). "Generalized Parton Distributions" (PDF). National Center for Physics and Quaid-e-Azam University. Retrieved 2011-04-06.

- Ji, X. (2004). "Generalized Parton Distributions" (PDF). *Annual Review of Nuclear and Particle Science* **54**: 413–450. Bibcode:2004ARNPS..54..413J. doi:10.1146/annurev.nucl.54.070103.181302.

- Kretzer, S.; Lai, H.; Olness, F.; Tung, W. (2004). "CTEQ6 Parton Distributions with Heavy Quark Mass Effects". *Physical Review D* **69** (11): 114005. arXiv:hep-ph/0307022. Bibcode:2004PhRvD..69k4005K. doi:10.1103/PhysRevD.69.114005.

- Martin, A. D.; Roberts, R. G.; Stirling, W. J.; Thorne, R. S. (2005). "Parton distributions incorporating QED contributions". *European Physical Journal C* **39** (2): 155–161. arXiv:hep-ph/0411040. Bibcode:2005EPJC...39..155M. doi:10.1140/epjc/s2004-02088-7.

6.16.5 External links

- Parton distribution functions – from HEPDATA: The Durham HEP Databases
- CTEQ6 parton distribution functions

6.17 Preon

For the protein diseases, see Prion. For the Freon trade name, see Chlorofluorocarbon.

In particle physics, **preons** are "point-like" particles, conceived to be subcomponents of quarks and leptons.[*][1] The word was coined by Jogesh Pati and Abdus Salam in 1974. Interest in preon models peaked in the 1980s but has slowed as the Standard Model of particle physics continues to describe the physics mostly successfully, and no direct experimental evidence for lepton and quark compositeness has been found.

Note that in the hadronic sector there are some intriguing open questions and some effects considered anomalies within the Standard Model. For example, four very important open questions are the proton spin puzzle, the EMC effect, the distributions of electric charges inside the nucleons as found by Hofstadter in 1956, and the ad hoc CKM matrix elements.

6.17.1 Background

Before the Standard Model (SM) was developed in the 1970s (the key elements of the Standard Model known as quarks were proposed by Murray Gell-Mann and George Zweig in 1964), physicists observed hundreds of different kinds of particles in particle accelerators. These were organized into relationships on their physical properties in a largely ad-hoc system of hierarchies, not entirely unlike the way taxonomy grouped animals based on their physical features. Not surprisingly, the huge number of particles was referred to as the "particle zoo".

The Standard Model, which is now the prevailing model of particle physics, dramatically simplified this picture by showing that most of the observed particles were mesons, which are combinations of two quarks, or baryons which are combinations of three quarks, plus a handful of other particles. The particles being seen in the ever-more-powerful accelerators were, according to the theory, typically nothing more than combinations of these quarks.

Within the Standard Model, there are several different types of particles. One of these, the quarks, has six different types, of which there are three varieties in each (dubbed "colors", red, green, and blue, giving rise to quantum chromodynamics). Additionally, there are six different types of what are known as leptons. Of these six leptons, there are three charged particles: the electron, muon, and tau. The neutrinos comprise the other three leptons, and for each neutrino there is a corresponding member from the other set of three leptons. In the Standard Model, there are also bosons, including the photons; W^*+, W^*-, and Z bosons; gluons and the Higgs boson; and an open space left for the graviton. Almost all of these particles come in "left-handed" and "right-handed" versions (see *chirality*). The quarks, leptons and W boson all have antiparticles with opposite electric charge.

The Standard Model also has a number of problems which have not been entirely solved. In particular, no successful theory of gravitation based on a particle theory has yet been proposed. Although the Model assumes the existence of a graviton, all attempts to produce a consistent theory based on them have failed. Additionally, mass remains a mystery in the Standard Model. Additionally Kalman [*][2] notes that according to the concept of atomism, the fundamental building blocks of nature are invisible and indivisible bits of matter that are ungenerated and indestructible. Quarks are not indestructible, some can decay into other quarks. Thus on fundamental grounds- quarks must be composed of fundamental quantities-preons. Although the mass of each successive particle follows certain patterns, predictions of the rest mass of most particles cannot be made precisely, except for the masses of almost all baryons which have been recently described very well by the model of de Souza.[*][3] The Higgs boson explains why particles show inertial mass (but does not explain rest mass).

The Standard Model also has problems predicting the large scale structure of the universe. For instance, the SM generally predicts equal amounts of matter and antimatter in the universe, something that is observably not the case. A number of attempts have been made to "fix" this through a variety of mechanisms, but to date none have won widespread support. Likewise, basic adaptations of the Model suggest the presence of proton decay, which has not yet been observed.

Preon theory is motivated by a desire to replicate the achievements of the periodic table, and the later Standard Model which tamed the "particle zoo", by finding more fundamental answers to the huge number of arbitrary constants present in the Standard Model. It is one of several models to have been put forward in an attempt to provide a more fundamental explanation of the results in experimental and theoretical particle physics. The preon model has attracted comparatively little interest to date among the particle physics community.

6.17.2 Motivations

Preon research is motivated by the desire to explain already known facts (retrodiction), which include

- To reduce the large number of particles, many that differ only in charge, to a smaller number of more fundamental particles. For example, the electron and positron are identical except for charge, and preon research is motivated by explaining that electrons and positrons are composed of similar preons with the relevant difference accounting for charge. The hope is to reproduce the reductionist strategy that has worked for the periodic table of elements.

- To explain the three generations of fermions.

- To calculate parameters that are currently unexplained by the Standard Model, such as particle masses, electric charges, and color charges, and reduce the number of experimental input parameters required by the Standard Model.

- To provide reasons for the very large differences in energy-masses observed in supposedly fundamental particles, from the electron neutrino to the top quark.

- To provide alternative explanations for the electroweak symmetry breaking without invoking a Higgs field, which in turn possibly needs a supersymmetry to correct the theoretical problems involved with the Higgs field. Supersymmetry itself has theoretical problems.

- To account for neutrino oscillation and mass.

- The desire to make new nontrivial predictions, for example, to provide possible cold dark matter candidates.

- To explain why there exists only the observed variety of particle species and not something else and to reproduce only these observed particles (since the prediction of non-observed particles is one of the major theoretical problems, as, for example, with supersymmetry).

6.17.3 History

A number of physicists have attempted to develop a theory of "pre-quarks" (from which the name *preon* derives) in an effort to justify theoretically the many parts of the Standard Model that are known only through experimental data.

Other names which have been used for these proposed fundamental particles (or particles intermediate between the most fundamental particles and those observed in the Standard Model) include *prequarks*, *subquarks*, *maons*,[*][4] *alphons*, *quinks*, *rishons*, *tweedles*, *helons*, *haplons*, *Y-particles*,[*][5] and *primons*.[*][6] *Preon* is the leading name in the physics community.

Efforts to develop a substructure date at least as far back as 1974 with a paper by Pati and Salam in *Physical Review*.[*][7] Other attempts include a 1977 paper by Terazawa, Chikashige and Akama,[*][8] similar, but independent, 1979 papers by Ne'eman,[*][9] Harari,[*][10] and Shupe,[*][11] a 1981 paper by Fritzsch and Mandelbaum,[*][12] and a 1992 book by D'Souza and Kalman.[*][1] None of these has gained wide acceptance in the physics world. However, in a recent work[*][13] de Souza has shown that his model describes well all weak decays of hadrons according to selection rules dictated by a quantum number derived from his compositeness model. In his model leptons are elementary particles and each quark is composed of two *primons*, and thus, all quarks are described by four *primons*. Therefore, there is no need for the Standard Model Higgs boson and each quark mass is derived from the interaction between each pair of *primons* by means of three Higgs-like bosons. In his 1989 Nobel Prize acceptance lecture, Hans Dehmelt described a most fundamental elementary particle, with definable properties, which he called the *cosmon*, as the likely end result of a long but finite chain of increasingly more elementary particles.*[14]

Each of the preon models postulates a set of fewer fundamental particles than those of the Standard Model, together with the rules governing how those fundamental particles operate. Based on these rules, the preon models try to explain the Standard Model, often predicting small discrepancies with this model and generating new particles and certain phenomena, which do not belong to the Standard Model. The Rishon model illustrates some of the typical efforts in the field.

Many of the preon models theorize that the apparent imbalance of matter and antimatter in the universe is in fact illusory, with large quantities of preon level antimatter confined within more complex structures.

Many preon models either do not account for the Higgs boson or rule it out, and propose that electro-weak symmetry is broken not by a scalar Higgs field but by composite preons.*[15] For example, Fredriksson preon theory does not need the Higgs boson, and explains the electro-weak breaking as the rearrangement of preons, rather than a Higgs-mediated field. In fact, Fredriksson preon model and de Souza model predict that the Standard Model Higgs boson does not exist.

When the term "preon" was coined, it was primarily to explain the two families of spin-1/2 fermions: leptons and quarks. More-recent preon models also account for spin-1 bosons, and are still called "preons".

6.17.4 Rishon model

Main article: Rishon model

The *rishon model* (RM) is the earliest effort to develop a preon model to explain the phenomenon appearing in the Standard Model (SM) of particle physics. It was first developed by Haim Harari and Michael A. Shupe (independently of each other), and later expanded by Harari and his then-student Nathan Seiberg.

The model has two kinds of fundamental particles called **rishons** (which means "primary" in Hebrew). They are **T** ("Third" since it has an electric charge of $1/3\ e$, or Tohu which means "unformed" in Hebrew Genesis) and **V** ("Vanishes", since it is electrically neutral, or Vohu which means "void" in Hebrew Genesis). All leptons and all flavours of quarks are three-rishon ordered triplets. These groups of three rishons have spin-½.

6.17.5 Criticisms

The mass paradox

One preon model started as an internal paper at the Collider Detector at Fermilab (CDF) around 1994. The paper was written after an unexpected and inexplicable excess of jets with energies above 200 GeV were detected in the 1992–1993 running period. However, scattering experiments have shown that quarks and leptons are "pointlike" down to distance scales of less than 10^*-18 m (or 1/1000 of a proton diameter). The momentum uncertainty of a preon (of whatever mass) confined to a box of this size is about 200 GeV/c, 50,000 times larger than the rest mass of an up-quark and 400,000 times larger than the rest mass of an electron.

Heisenberg's uncertainty principle states that $\Delta x \Delta p \geq \hbar/2$ and thus anything confined to a box smaller than Δx would have a momentum uncertainty proportionally greater. Thus, the preon model proposed particles smaller than the elementary particles they make up, since the momentum uncertainty Δp should be greater than the particles themselves. And so the preon model represents a mass paradox: How could quarks or electrons be made of smaller particles that would have many orders of magnitude greater mass-energies arising from their enormous momenta? This paradox is resolved by postulating a large binding force between preons cancelling their mass-energies.

Constraints

Any candidate preon theory must address particle chirality and the 't Hooft Chiral anomaly constraints, and would ideally have simpler theoretical structure than the Standard Model itself.

6.17.6 Conflicts with observed physics

Preon models propose additional unobserved forces or dynamics to account for the observed properties of elementary particles, which may have implications in conflict with observation.

For example, now that the LHC's observation of a Higgs boson is confirmed, the observation contradicts the predictions of many preon models that did not include it.

Preon theories require that quarks and electrons should have a finite size. It is possible that the Large Hadron Collider will observe this when raised to higher energies.

6.17.7 Popular culture

- In the 1948 reprint/edit of his 1930 novel *Skylark Three*, E. E. Smith postulated a series of 'subelectrons of the first and second type' with the latter being fundamental particles that were associated with the gravitation force. While this may not have been an element of the original novel (the scientific basis of some of the other novels in the series was revised extensively due to the additional eighteen years of scientific development), even the edited publication may be the first, or one of the first, mentions of the possibility that electrons are not fundamental particles.

- In the novelized version of the 1982 motion picture *Star Trek II: The Wrath of Khan*, written by Vonda McIntyre, two of Dr. Carol Marcus' Genesis project team, Vance Madison and Delwyn March, have studied sub-elementary particles they've named "boojums" and "snarks", in a field they jokingly call "kindergarten physics" because it is lower than "elementary" (analogy to school levels).

- James P. Hogan's novel *Voyage from Yesteryear* discussed preons (called *tweedles*), the physics of which became central to the plot. Hogan's "tweedle" physics was patently derived from the Rishon model.

6.17.8 See also

- Technicolor (physics)
- Preon star
- Preon-degenerate matter

6.17.9 Notes

[1] D'Souza, I.A.; Kalman, C.S. (1992). *Preons: Models of Leptons, Quarks and Gauge Bosons as Composite Objects*. World Scientific. ISBN 978-981-02-1019-9.

[2] Kalman, C. S. (2005). *Nuclear Physics B (Proc. Suppl.)* **142**: 235–237. Missing or empty |title= (help)

[3] de Souza, M.E. (2010). "Calculation of almost all energy levels of baryons". *Papers in Physics* **3**: 030003–1. doi:10.4279/PIP.030003.

[4] Overbye, D. (5 December 2006). "China Pursues Major Role in Particle Physics". *The New York Times*. Retrieved 2011-09-12.

[5] Yershov, V.N. (2005). "Equilibrium Configurations of Tripolar Charges". *Few-Body Systems* **37** (1–2): 79–106. arXiv:physics/0609185. Bibcode:2005FBS....37...79Y. doi:10.1007/s00601-004-0070-2.

[6] de Souza, M.E. (2005). "The Ultimate Division of Matter". *Scientia Plena* **1** (4): 83.

[7] Pati, J.C.; Salam, A. (1974). "Lepton number as the fourth "color"". *Physical Review D* **10**: 275–289. Bibcode:1974PhRvD..10..275P. doi:10.1103/PhysRevD.10.275.

with erratum published as *Physical Review D* **11** (3): 703. 1975. Bibcode:1975PhRvD..11..703P. doi:10.1103/PhysRevD.11.703.2. Missing or empty |title= (help)

[8] Terazawa, H.; Chikashige, Y.; Akama, K. (1977). "Unified model of the Nambu-Jona-Lasinio type for all elementary particles". *Physical Review D* **15** (2): 480–487. Bibcode:1977PhRvD..15..480T. doi:10.1103/PhysRevD.15.480.

[9] Ne'eman, Y. (1979). "Irreducible gauge theory of a consolidated Weinberg-Salam model". *Physics Letters B* **81** (2): 190–194. Bibcode:1979PhLB...81..190N. doi:10.1016/0370-2693(79)90521-5.

[10] Harari, H. (1979). "A schematic model of quarks and leptons" (PDF). *Physics Letters B* **86**: 83–6. Bibcode:1979PhLB...86...83H. doi:10.1016/0370-2693(79)90626-9.

[11] Shupe, M.A. (1979). "A composite model of leptons and quarks". *Physics Letters B* **86**: 87–92. Bibcode:1979PhLB...86...87S. doi:10.1016/0370-2693(79)90627-0.

[12] Fritzsch, H.; Mandelbaum, G. (1981). "Weak interactions as manifestations of the substructure of leptons and quarks". *Physics Letters B* **102** (5): 319. Bibcode:1981PhLB..102..319F. doi:10.1016/0370-2693(81)90626-2.

[13] de Souza, M.E. (2008). "Weak decays of hadrons reveal compositeness of quarks". *Scientia Plena* **4** (6): 064801–1.

[14] Dehmelt, H.G. (1989). "Experiments with an Isolated Subatomic Particle at Rest". *Nobel Lecture*. The Nobel Foundation. See also references therein.

[15] Dugne, J.-J.; Fredriksson, S.; Hansson, J.; Predazzi, E. (1997). "Higgs pain? Take a preon!". arXiv:hep-ph/9709227 [hep-ph].

6.17.10 Further reading

- Ball, P. (2007). "Splitting the quark". *Nature*. doi:10.1038/news.2007.292.

- Have We Hit Bottom Yet?- an article about preons and minuteness

6.18 QCD matter

Quark matter or **QCD matter** refers to any of a number of theorized phases of matter whose degrees of freedom include quarks and gluons. These theoretical phases would occur at extremely high temperatures and densities, billions of times higher than can be produced in equilibrium in laboratories. Under such extreme conditions, the familiar structure of matter, where the basic constituents are nuclei (consisting of nucleons which are bound states of quarks) and electrons, is disrupted. In quark matter it is more appropriate to treat the quarks themselves as the basic degrees of freedom.

In the standard model of particle physics, the strong force is described by the theory of quantum chromodynamics (QCD). At ordinary temperatures or densities this force just confines the quarks into composite particles (hadrons) of size around 10^*-15 m = 1 femtometer = 1 fm (corresponding to the QCD energy scale $\Lambda_{QCD} \approx 200$ MeV) and its effects are not noticeable at longer distances. However, when the temperature reaches the QCD energy scale (T of order 10^{12} kelvins) or the density rises to the point where the average inter-quark separation is less than 1 fm (quark chemical potential μ around 400 MeV), the hadrons are melted into their constituent quarks, and the strong interaction becomes the dominant feature of the physics. Such phases are called quark matter or QCD matter.

The strength of the color force makes the properties of quark matter unlike gas or plasma, instead leading to a state of matter more reminiscent of a liquid. At high densities, quark matter is a Fermi liquid, but is predicted to exhibit color superconductivity at high densities and temperatures below 10^{12} K.

6.18.1 Occurrence

Natural occurrence

- In the early universe, at high temperatures according to the Big Bang theory, when the universe was only a few tens of microseconds old, the phase of matter took the form of a hot phase of quark matter called the quark–gluon plasma (QGP).
- Compact stars (neutron stars). A neutron star is much cooler than 10^{12} K, but it is compressed by its own mass to such high densities, that it is reasonable to surmise that quark matter may exist in the core.[*][1] Compact stars composed mostly or entirely of quark matter are called quark stars or strange stars, yet at this time no star with properties expected of these objects has been observed.

- Strangelets. These are theoretically postulated (but as yet unobserved) lumps of strange matter comprising nearly equal amounts of up, down and strange quarks.

- Cosmic ray impacts. Cosmic rays comprise also high energy atomic nuclei, particularly that of iron. Laboratory experiments suggest that interaction with heavy noble gas in the upper atmosphere would lead to quark–gluon plasma formation.

Laboratory experiments

- Heavy-ion collisions at very high energies can produce small short-lived regions of space whose energy density is comparable to that of the 20-micro-second-old universe. This has been achieved by colliding heavy nuclei at high speeds, and a first time claim of formation of quark–gluon plasma came from the SPS accelerator at CERN in February 2000.[*][2] This work has been continued at more powerful accelerators, such as RHIC at Brookhaven National Laboratory in the USA, and as of 2010 at the LHC at CERN located in the border area of Switzerland & France. There is good evidence that the quark–gluon plasma has also been produced at RHIC.[*][3]

6.18.2 Thermodynamics

The context for understanding the thermodynamics of quark matter is the standard model of particle physics, which contains six different flavors of quarks, as well as leptons like electrons and neutrinos. These interact via the strong interaction, electromagnetism, and also the weak interaction which allows one flavor of quark to turn into another. Electromagnetic interactions occur between particles that carry electrical charge; strong interactions occur between particles that carry color charge.

The correct thermodynamic treatment of quark matter depends on the physical context. For large quantities that exist for long periods of time (the "thermodynamic limit"), we must take into account the fact that the only conserved charges in the standard model are quark number (equivalent to baryon number), electric charge, the eight color charges, and lepton number. Each of these can have an associated chemical potential. However, large volumes of matter must be electrically and color-neutral, which determines the electric and color charge chemical potentials. This leaves a three-dimensional phase space, parameterized by quark chemical potential, lepton chemical potential, and temperature.

In compact stars quark matter would occupy cubic kilometers and exist for millions of years, so the thermodynamic limit is appropriate. However, the neutrinos escape, violating lepton number, so the phase space for quark matter in compact stars only has two dimensions, temperature (T) and quark number chemical potential μ. A strangelet is not in the thermodynamic limit of large volume, so it is like an exotic nucleus: it may carry electric charge.

A heavy-ion collision is in neither the thermodynamic limit of large volumes nor long times. Putting aside questions of whether it is sufficiently equilibrated for thermodynamics to be applicable, there is certainly not enough time for weak interactions to occur, so flavor is conserved, and there are independent chemical potentials for all six quark flavors. The initial conditions (the impact parameter of the collision, the number of up and down quarks in the colliding nuclei, and the fact that they contain no quarks of other flavors) determine the chemical potentials. (Reference for this section:,[*4][*5]).

6.18.3 Phase diagram

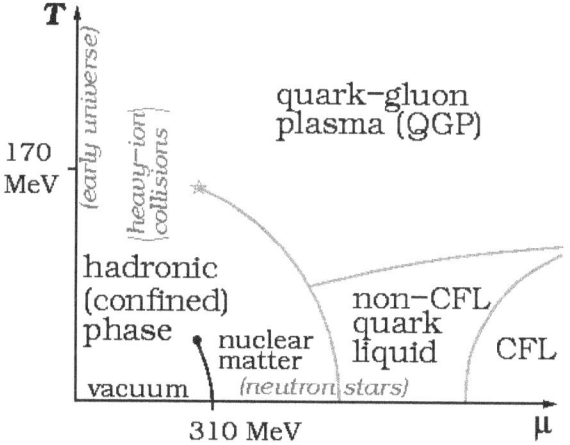

Conjectured form of the phase diagram of QCD matter, with temperature as ordinate (in mega-electron volts) and quark chemical potential as abscissa.[*4]

The phase diagram of quark matter is not well known, either experimentally or theoretically. A commonly conjectured form of the phase diagram is shown in the figure.[*4] It is applicable to matter in a compact star, where the only relevant thermodynamic potentials are quark chemical potential μ and temperature T. For guidance it also shows the typical values of μ and T in heavy-ion collisions and in the early universe. For readers who are not familiar with the concept of a chemical potential, it is helpful to think of μ as a measure of the imbalance between quarks and antiquarks in the system. Higher μ means a stronger bias favoring quarks over antiquarks. At low temperatures there are no antiquarks, and then higher μ generally means a higher density of quarks.

Ordinary atomic matter as we know it is really a mixed phase, droplets of nuclear matter (nuclei) surrounded by vacuum, which exists at the low-temperature phase boundary between vacuum and nuclear matter, at μ = 310 MeV and T close to zero. If we increase the quark density (i.e. increase μ) keeping the temperature low, we move into a phase of more and more compressed nuclear matter. Following this path corresponds to burrowing more and more deeply into a neutron star. Eventually, at an unknown critical value of μ, there is a transition to quark matter. At ultra-high densities we expect to find the color-flavor-locked (CFL) phase of color-superconducting quark matter. At intermediate densities we expect some other phases (labelled "non-CFL quark liquid" in the figure) whose nature is presently unknown,.[*4][*5] They might be other forms of color-superconducting quark matter, or something different.

Now, imagine starting at the bottom left corner of the phase diagram, in the vacuum where $\mu = T = 0$. If we heat up the system without introducing any preference for quarks over antiquarks, this corresponds to moving vertically upwards along the T axis. At first, quarks are still confined and we create a gas of hadrons (pions, mostly). Then around T = 150 MeV there is a crossover to the quark gluon plasma: thermal fluctuations break up the pions, and we find a gas of quarks, antiquarks, and gluons, as well as lighter particles such as photons, electrons, positrons, etc. Following this path corresponds to travelling far back in time (so to say), to the state of the universe shortly after the big bang (where there was a very tiny preference for quarks over antiquarks).

The line that rises up from the nuclear/quark matter transition and then bends back towards the T axis, with its end marked by a star, is the conjectured boundary between confined and unconfined phases. Until recently it was also believed to be a boundary between phases where chiral symmetry is broken (low temperature and density) and phases where it is unbroken (high temperature and density). It is now known that the CFL phase exhibits chiral symmetry breaking, and other quark matter phases may also break chiral symmetry, so it is not clear whether this is really a chiral transition line. The line ends at the "chiral critical point", marked by a star in this figure, which is a special temperature and density at which striking physical phenomena, analogous to critical opalescence, are expected. (Reference for this section:,[*4][*5][*6]).

For a complete description of phase diagram it is required that one must have complete understanding of dense, strongly interacting hadronic matter and strongly interacting quark matter from some underlying theory e.g. quantum chromodynamics (QCD). However because such a description requires the proper understanding of QCD in its non-perturbative regime, which is still far from being completely understood, any theoretical advance remains very challenging.

6.18.4 Theoretical challenges: calculation techniques

The phase structure of quark matter remains mostly conjectural because it is difficult to perform calculations predicting the properties of quark matter. The reason is that QCD, the theory describing the dominant interaction between quarks, is strongly coupled at the densities and temperatures of greatest physical interest, and hence it is very hard to obtain any predictions from it. Here are brief descriptions of some of the standard approaches.

Lattice gauge theory

The only first-principles calculational tool currently available is lattice QCD, i.e. brute-force computer calculations. Because of a technical obstacle known as the fermion sign problem, this method can only be used at low density and high temperature ($\mu < T$), and it predicts that the crossover to the quark–gluon plasma will occur around $T = 150$ MeV [*][7] However, it cannot be used to investigate the interesting color-superconducting phase structure at high density and low temperature.[*][8]

Weak coupling theory

Because QCD is asymptotically free it becomes weakly coupled at unrealistically high densities, and diagrammatic methods can be used.[*][5] Such methods show that the CFL phase occurs at very high density. At high temperatures, however, diagrammatic methods are still not under full control.

Models

To obtain a rough idea of what phases might occur, one can use a model that has some of the same properties as QCD, but is easier to manipulate. Many physicists use Nambu-Jona-Lasinio models, which contain no gluons, and replace the strong interaction with a four-fermion interaction. Mean-field methods are commonly used to analyse the phases. Another approach is the bag model, in which the effects of confinement are simulated by an additive energy density that penalizes unconfined quark matter.

Effective theories

Many physicists simply give up on a microscopic approach, and make informed guesses of the expected phases (perhaps based on NJL model results). For each phase, they then write down an effective theory for the low-energy excitations, in terms of a small number of parameters, and use it to make predictions that could allow those parameters to be fixed by experimental observations.[*][6]

Other approaches

Main article: AdS/QCD

There are other methods that are sometimes used to shed light on QCD, but for various reasons have not yet yielded useful results in studying quark matter.

1/N expansion Treat the number of colors N, which is actually 3, as a large number, and expand in powers of $1/N$. It turns out that at high density the higher-order corrections are large, and the expansion gives misleading results.[*][4]

Supersymmetry Adding scalar quarks (squarks) and fermionic gluons (gluinos) to the theory makes it more tractable, but the thermodynamics of quark matter depends crucially on the fact that only fermions can carry quark number, and on the number of degrees of freedom in general.

6.18.5 Experimental challenges

Experimentally, it is hard to map the phase diagram of quark matter because it has been rather difficult to learn how to tune to high enough temperatures and density in the laboratory experiment using collisions of relativistic heavy ions as experimental tools. However, these collisions ultimately will provide information about the crossover from hadronic matter to QGP. It has been suggested that the observations of compact stars may also constrain the information about the high-density low-temperature region. Models of the cooling, spin-down, and precession of these stars offer information about the relevant properties of their interior. As observations become more precise, physicists hope to learn more.[*][4]

One of the natural subjects for future research is the search for the exact location of the chiral critical point. Some ambitious lattice QCD calculations may have found evidence for it, and future calculations will clarify the situation. Heavy-ion collisions might be able to measure its position experimentally, but this will require scanning across a range of values of μ and T.*[9]

6.18.6 See also

- Color–flavor locking
- Lattice QCD
- Quantum chromodynamics
- Quark–gluon plasma
- Quark star
- Strange matter
- Strangeness production
- 1/N expansion

6.18.7 References

[1] Shapiro and Teukolsky: *Black Holes, White Dwarfs and Neutron Stars: The Physics of Compact Objects*, Wiley 2008

[2] Ulrich Heinz; Maurice Jacob (2000). "Evidence for a New State of Matter: an Assessment of the Results from the CERN Lead Beam Programme". arXiv:nucl-th/0002042 [nucl-th].

[3] Berndt Müller (2005). "Quark Matter 2005 -- Theoretical Summary". arXiv:nucl-th/0508062 [nucl-th].

[4] Alford, Mark G.; Schmitt, Andreas; Rajagopal, Krishna; Schäfer, Thomas (2008). "Color superconductivity in dense quark matter". *Review of Modern Physics* **80** (4): 1455–1515. arXiv:0709.4635. Bibcode:2008RvMP...80.1455A. doi:10.1103/RevModPhys.80.1455.

[5] Rischke, D (2004). "The quark–gluon plasma in equilibrium". *Progress in Particle and Nuclear Physics* **52**: 197. arXiv:nucl-th/0305030. Bibcode:2004PrPNP..52..197R. doi:10.1016/j.ppnp.2003.09.002.

[6] T. Schäfer (2004). "Quark matter". In A. B. Santra. *Mesons and Quarks*. 14th National Nuclear Physics Summer School. Alpha Science International. arXiv:hep-ph/0304281. ISBN 978-81-7319-589-1.

[7] P. Petreczky (2012). "Lattice QCD at non-zero temperature". *J.Phys.* G **39** (9): 093002. arXiv:1203.5320. Bibcode:2012JPhG...39i3002P. doi:10.1088/0954-3899/39/9/093002.

[8] Christian Schmidt (2006). "Lattice QCD at Finite Density". *PoS LAT2006*: 021. arXiv:hep-lat/0610116. Bibcode:2006slft.confE..21S.

[9] Rajagopal, K (1999). "Mapping the QCD phase diagram". *Nuclear Physics A* **661**: 150–161. arXiv:hep-ph/9908360. Bibcode:1999NuPhA.661..150R. doi:10.1016/S0375-9474(99)85017-9.

6.18.8 Further reading

- S. Hands (2001). "The phase diagram of QCD". *Contemporary Physics* **42** (4): 209. arXiv:physics/0105022. Bibcode:2001ConPh..42..209H. doi:10.1080/00107510110063843.

- K. Rajagopal (2001). "Free the quarks" (PDF). *Beam Line* **32** (2): 9–15.

6.18.9 External links

- Virtual Journal on QCD Matter
- RHIC finds Exotic Antimatter

6.19 Quark epoch

In physical cosmology the **quark epoch** was the period in the evolution of the early universe when the fundamental interactions of gravitation, electromagnetism, the strong interaction and the weak interaction had taken their present forms, but the temperature of the universe was still too high to allow quarks to bind together to form hadrons. The quark epoch began approximately 10^{-12} seconds after the Big Bang, when the preceding electroweak epoch ended as the electroweak interaction separated into the weak interaction and electromagnetism. During the quark epoch the universe was filled with a dense, hot quark–gluon plasma, containing quarks, leptons and their antiparticles. Collisions between particles were too energetic to allow quarks to combine into mesons or baryons. The quark epoch ended when the universe was about 10^{-6} seconds old, when the average energy of particle interactions had fallen below the binding energy of hadrons. The following period, when quarks became confined within hadrons, is known as the hadron epoch.

6.19.1 See also

- Chronology of the universe
- Cosmogony

6.19.2 References

- Allday, Jonathan (2002). *Quarks, Leptons and the Big Bang* (Second ed.). ISBN 978-0-7503-0806-9.

- *Physics 175: Stars and Galazies - The Big Bang, Matter and Energy*; Ithaca College, New York

6.20 Quark star

A **quark star** is a hypothetical type of compact exotic star composed of quark matter. These are ultra-dense phases of degenerate matter theorized to form inside particularly massive neutron stars.

The existence of quark stars has not been confirmed neither theoretically nor astronomically. The equation of state of quark matter is uncertain, as is the transition point between neutron-degenerate matter and quark matter. Theoretical uncertainties have precluded making predictions from first principles. Experimentally, the behaviour of quark matter is currently being actively studied with particle colliders, although this can only produce hot quark-gluon plasma blobs the size of an atomic nucleus, and they decay immediately after formation. There are no known artificial methods to produce or store cold quark matter as it would be found in quark stars.

6.20.1 Creation

It is theorized that when the neutron-degenerate matter, which makes up neutron stars, is put under sufficient pressure from the star's own gravity or the initial supernova creating it, the individual neutrons break down into their constituent quarks (up quarks and down quarks), forming what is known as quark matter. This conversion might be confined to the neutron star's center or it might transform the entire star, depending on the physical circumstances.[1] Such a star is known as a quark star.

Stability and strange quark matter

Ordinary quark matter consisting of up and down quarks (also referred to as u and d quarks) has a very high Fermi energy compared to ordinary atomic matter and is only stable under extreme temperatures and/or pressures. This suggests that only quark stars comprising neutron stars with a quark matter core will be stable, while quark stars consisting entirely of ordinary quark matter will be highly unstable and dissolve spontaneously.[2][3]

It has been shown that the high Fermi energy making ordinary quark matter unstable at low temperatures and pressures can be lowered substantially by the transformation of a sufficient number of u and d quarks into strange quarks, as strange quarks are, relatively speaking, a very heavy type of quark particle.[2] This kind of quark matter is known specifically as strange quark matter and it is speculated and subject to current scientific investigation whether it might in fact be stable under the conditions of interstellar space (i.e. near zero external pressure and temperature). If this is the case (known as the Bodmer–Witten assumption), quark stars made entirely of quark matter would be stable if they quickly transform into strange quark matter.[4]

Strange stars

Quark stars made of strange quark matter are known as strange stars, and they form a subgroup under the quark star category.[4]

Strange stars might exist without regard of the Bodmer-Witten assumption of stability at near-zero temperatures and pressures, as strange quark matter might form and remain stable at the core of neutron stars, in the same way as ordinary quark matter could.[1] Such strange stars will naturally have a crust layer of neutron star material. The depth of the crust layer will depend on the physical conditions and circumstances of the entire star and on the properties of strange quark matter in general. Stars partially made up of quark matter (including strange quark matter) are also referred to as hybrid stars.[5][6]

Theoretical investigations have revealed that quark stars might not only be produced from neutron stars and powerful supernovas, but they could also be created in the early cosmic phase separations following the Big Bang.[2] If these primordial quark stars transform into strange quark matter before the external temperature and pressure conditions of the early Universe makes them unstable, they might turn out stable, if the Bodmer–Witten assumption holds true. Such primordial strange stars could survive to this day.[2]

6.20.2 Characteristics

If the conversion of neutron-degenerate matter to (strange) quark matter is total, a quark star can to some extent be imagined as a single gigantic hadron. But this "hadron" will be bound by gravity, rather than the strong force that binds ordinary hadrons.

Strange stars

Recent theoretical research has found mechanisms by which quark stars with "strange quark nuggets" may decrease the objects' electric fields and densities from previous theoretical expectations, causing such stars to appear very much like—nearly indistinguishable from—neutron stars. However, the investigating team of Prashanth Jaikumar, Sanjay Reddy, and Andrew W. Steiner made some fundamental assumptions, that led to uncertainties in their results large enough that the case is not finally settled. More research, both observational and theoretical, remains to be done on strange stars in the future.*[7]

Other theoretical work*[8] contends that, "A sharp interface between quark matter and the vacuum would have very different properties from the surface of a neutron star"; and, addressing key parameters like surface tension and electrical forces that were neglected in the original study, the results show that as long as the surface tension is below a low critical value, the large strangelets are indeed unstable to fragmentation and strange stars naturally come with complex strangelet crusts, analogous to those of neutron stars.

6.20.3 Other theorized quark formations

- Jaffe 1977, suggested a four-quark state with strangeness (qsqs).

- Jaffe 1977 suggested the H dibaryon, a six-quark state with equal numbers of up-, down-, and strange quarks (represented as uuddss or udsuds).

- Bound multi-quark systems with heavy quarks (QQqq).

- In 1987, a pentaquark state was first proposed with a charm anti-quark (qqqsc).

- Pentaquark state with an antistrange quark and four light quarks consisting of up- and down-quarks only (qqqqs).

- Light pentaquarks are grouped within an antidecuplet, the lightest candidate, Θ*+.

   - This can also be described by the diquark model of Jaffe and Wilczek (QCD).

- Θ*++ and antiparticle Θ*--.

- Doubly strange pentaquark (ssddu), member of the light pentaquark antidecuplet.

- Charmed pentaquark $Θ_c(3100)$ (uuddc) state was detected by the H1 collaboration.*[9]

- Tetra quark particles might form inside neutron stars and under other extreme conditions. In 2008, 2013 and 2014 the tetra quark particle of Z(4430), was discovered and investigated in laboratories on Earth.*[10]

6.20.4 Observed overdense neutron stars

Statistically, the probability of a neutron star being a quark star is low, so in the Milky Way Galaxy there would only be a small population of quark stars. If it is correct however, that overdense neutron stars can turn into quark stars, that makes the possible number of quark stars higher than was originally thought, as we would be looking for the wrong type of star.

Quark stars and strange stars are entirely hypothetical as of 2011, but observations released by the Chandra X-ray Observatory on April 10, 2002 detected two candidates, designated RX J1856.5-3754 and 3C58, which had previously been thought to be neutron stars. Based on the known laws of physics, the former appeared much smaller and the latter much colder than it should be, suggesting that they are composed of material denser than neutron-degenerate matter. However, these observations are met with skepticism by researchers who say the results were not conclusive;*[11] and since the late 2000s, the possibility that RX J1856 is a quark star has been excluded (see RX J1856.5-3754).

Another star, XTE J1739-285,*[12] has been observed by a team led by Philip Kaaret of the University of Iowa and reported as a possible candidate.

In 2006, Y. L. Yue et al., from Peking University, suggested that PSR B0943+10 may in fact be a low-mass quark star.*[13]

It was reported in 2008 that observations of supernovae SN2006gy, SN2005gj and SN2005ap also suggest the existence of quark stars.*[14] It has been suggested that the collapsed core of supernova SN1987A may be a quark star.*[15]*[16]

6.20.5 See also

- Quark-nova

- Quantum chromodynamics

- Neutron stars – neutron matter – neutron-degenerate matter – neutron

- Deconfinement

- Tolman–Oppenheimer–Volkoff limit on the mass of a neutron star.

- Compact star

   - Exotic star
   - Neutron star
   - Pulsar
   - Magnetar

- White dwarf
- Stellar black hole
- Degenerate matter
  - QCD matter
  - Quark–gluon plasma
  - Strangelet
  - Quark matter
  - Neutronium
  - Preon matter

6.20.6 References

[1] Shapiro and Teukolsky: *Black Holes, White Dwarfs and Neutron Stars: The Physics of Compact Objects*, Wiley 2008

[2] Witten, Edward (1984). "Cosmic separation of phases". *Physical Review D* **30** (2): 272–285. Bibcode:1984PhRvD..30..272W. doi:10.1103/PhysRevD.30.272.

[3] E. Farhi; R. L. Jaffe (1984). "Strange matter". *Physical Review D* **30**: 2379. Bibcode:1984PhRvD..30.2379F. doi:10.1103/PhysRevD.30.2379.

[4] Fridolin Weber et al. (1994). "Strange-matter Stars". *Proceedings: Strangeness and Quark Matter*. World Scientific.

[5] Mark G. Alford, Sophia Han and Madappa Prakash (2013). "Generic conditions for stable hybrid stars". *Phys. Rev. D* **88**, 083013. arXiv:1302.4732. Bibcode:2013PhRvD..88h3013A. doi:10.1103/PhysRevD.88.083013.

[6] Ashok Goyal (March 2014). "Hybrid stars" (PDF). *Pramana - journal of physics, Vol. 62, No. 3* (Indian Academy of Sciences): 753–56. Retrieved 22 March 2014.

[7] Jaikumar, Prashanth; Reddy, Sanjay; Steiner, Andrew (2006). "Strange Star Surface: A Crust with Nuggets". *Physical Review Letters* **96** (4). arXiv:nucl-th/0507055. Bibcode:2006PhRvL..96d1101J. doi:10.1103/PhysRevLett.96.041101.

[8] Alford, Mark; Rajagopal, Krishna; Reddy, Sanjay; Steiner, Andrew (2006). "Stability of strange star crusts and strangelets". *Physical Review D* **73** (11). arXiv:hep-ph/0604134. Bibcode:2006PhRvD..73k4016A. doi:10.1103/PhysRevD.73.114016.

[9] H1 Collaboration; Aktas, A.; Andreev, V.; Anthonis, T.; Asmone, A.; Babaev, A.; Backovic, S.; Bähr, J. et al. (2004). "Evidence for a Narrow Anti-Charmed Baryon State of mass". *Physics Letters B* **588**: 17–28. arXiv:hep-ex/0403017. Bibcode:2004PhLB..588...17A. doi:10.1016/j.physletb.2004.03.012.

[10] Brian Koberlein (April 10, 2014). "How CERN's Discovery of Exotic Particles May Affect Astrophysics". Universe Today. Retrieved 14 April 2014./

[11] Truemper, J. E.; Burwitz, V.; Haberl, F.; Zavlin, V. E. (June 2004). "The puzzles of RX J1856.5-3754: neutron star or quark star?". *Nuclear Physics B Proceedings Supplements* **132**: 560–565. arXiv:astro-ph/0312600. Bibcode:2004NuPhS.132..560T. doi:10.1016/j.nuclphysbps.2004.04.094.

[12] Fastest spinning star may have exotic heart

[13] http://cds.cern.ch/record/935794/files/0603468.pdf?version=2

[14] Astronomy Now Online – Second Supernovae Point to Quark Stars

[15] Chan; Cheng; Harko; Lau; Lin; Suen; Tian (2009). "Could the compact remnant of SN 1987A be a quark star?". *Astrophysical Journal* **695**: 732–746. arXiv:0902.0653. Bibcode:2009ApJ...695..732C. doi:10.1088/0004-637X/695/1/732.

[16] Quark star may hold secret to early universe *New Scientist*

- Quark star on arxiv.org

6.20.7 External links

- Jaffe, R. (1977). "Perhaps a Stable Dihyperon". *Physical Review Letters* **38** (5): 195–198. Bibcode:1977PhRvL..38..195J. doi:10.1103/PhysRevLett.38.195.
- Neutron Star/Quark Star Interior (image to print)
- Quark star glimmers, *Nature*, April 11, 2002.
- Debate sparked on quark stars, *CERN Courier* **42**, #5.
- Wish Upon a Quark Star, Paul Beck, *Popular Science*, June 2002.
- Drake; Marshall; Dreizler; Freeman; Fruscione; Juda; Kashyap; Nicastro et al. (2002). "Is RX J185635-375 a Quark Star?". *Astrophysical Journal* **572** (2): 996–1001. arXiv:astro-ph/0204159. Bibcode:2002ApJ...572..996D. doi:10.1086/340368.
- Perhaps a 1,700 years old quark star in SNR MSH 15-52
- Curious About Astronomy: What process would bring about a quark star?
- RX J185635-375: Candidate Quark Star, Astronomy Picture of the Day, April 14, 2002.

- Quarks or Quirky Neutron Stars?, Mark K. Anderson, *Wired News*, April 19, 2002.
- Strange Quark Stars, Ask an Astrophysicist, question submitted April 12, 2002.
- Seeing "Strange" Stars, physorg.com, February 8, 2006.
- Quark Stars Could Produce Biggest Bang, spacedaily.com, June 7, 2006.
- Meissner Effect in Strange Quark Stars, Brian Niebergal, web page, University of Calgary.
- Irina Sagert; Mirjam Wietoska; Jurgen Schaffner-Bielich (2006). "Strange Exotic States and Compact Stars". *Journal of Physics G* **32** (12): S241–S249. arXiv:astro-ph/0608317. Bibcode:2006JPhG...32S.241S. doi:10.1088/0954-3899/32/12/S30.
- Quark Stars Involved in New Theory of Brightest Supernovae – The first-ever evidence of a neutron star collapsing into a quark star is announced, *Space.com*, 3 June 2008
- Quark Stars, Alternate View Column AV-114, John G. Cramer, Published in the November-2002 issue of Analog Science Fiction & Fact Magazine

6.21 Quantum chromodynamics

In theoretical physics, **quantum chromodynamics (QCD)** is the theory of strong interactions, a fundamental force describing the interactions between quarks and gluons which make up hadrons such as the proton, neutron and pion. QCD is a type of quantum field theory called a non-abelian gauge theory with symmetry group SU(3). The QCD analog of electric charge is a property called *color*. Gluons are the force carrier of the theory, like photons are for the electromagnetic force in quantum electrodynamics. The theory is an important part of the Standard Model of particle physics. A huge body of experimental evidence for QCD has been gathered over the years.

QCD enjoys two peculiar properties:

- **Confinement**, which means that the force between quarks does not diminish as they are separated. Because of this, when you do separate a quark from other quarks, the energy in the gluon field is enough to create another quark pair; they are thus forever bound into hadrons such as the proton and the neutron or the pion and kaon. Although analytically unproven, confinement is widely believed to be true because it explains the consistent failure of free quark searches, and it is easy to demonstrate in lattice QCD.
- **Asymptotic freedom**, which means that in very high-energy reactions, quarks and gluons interact very weakly creating a quark–gluon plasma. This prediction of QCD was first discovered in the early 1970s by David Politzer and by Frank Wilczek and David Gross. For this work they were awarded the 2004 Nobel Prize in Physics.

The phase transition temperature between these two properties has been measured by the ALICE experiment to be well above 160 MeV.*[1] Below this temperature, confinement is dominant, while above it, asymptotic freedom becomes dominant.

6.21.1 Terminology

The word *quark* was coined by American physicist Murray Gell-Mann (b. 1929) in its present sense. It originally comes from the phrase "Three quarks for Muster Mark" in *Finnegans Wake* by James Joyce. On June 27, 1978, Gell-Mann wrote a private letter to the editor of the *Oxford English Dictionary*, in which he related that he had been influenced by Joyce's words: "The allusion to three quarks seemed perfect." (Originally, only three quarks had been discovered.) Gell-Mann, however, wanted to pronounce the word to rhyme with "fork" rather than with "park", as Joyce seemed to indicate by rhyming words in the vicinity such as *Mark*. Gell-Mann got around that "by supposing that one ingredient of the line 'Three quarks for Muster Mark' was a cry of 'Three quarts for Mister ...' heard in H.C. Earwicker's pub", a plausible suggestion given the complex punning in Joyce's novel.*[2]

The three kinds of charge in QCD (as opposed to one in quantum electrodynamics or QED) are usually referred to as "color charge" by loose analogy to the three kinds of color (red, green and blue) perceived by humans. Other than this nomenclature, the quantum parameter "color" is completely unrelated to the everyday, familiar phenomenon of color.

Since the theory of electric charge is dubbed "electrodynamics", the Greek word "chroma" Χρώμα (meaning color) is applied to the theory of color charge, "chromodynamics".

6.21.2 History

With the invention of bubble chambers and spark chambers in the 1950s, experimental particle physics discovered a large and ever-growing number of particles called hadrons. It seemed that such a large number of particles could not all be fundamental. First, the particles were classified by charge and isospin by Eugene Wigner and Werner Heisenberg; then, in 1953, according to strangeness by Murray Gell-Mann and Kazuhiko Nishijima. To gain greater insight, the hadrons were sorted into groups having similar properties and masses using the *eightfold way*, invented in 1961 by Gell-Mann and Yuval Ne'eman. Gell-Mann and George Zweig, correcting an earlier approach of Shoichi Sakata, went on to propose in 1963 that the structure of the groups could be explained by the existence of three flavors of smaller particles inside the hadrons: the quarks.

Perhaps the first remark that quarks should possess an additional quantum number was made[*][3] as a short footnote in the preprint of Boris Struminsky[*][4] in connection with Ω^*– hyperon composed of three strange quarks with parallel spins (this situation was peculiar, because since quarks are fermions, such combination is forbidden by the Pauli exclusion principle):

> Three identical quarks cannot form an antisymmetric S-state. In order to realize an antisymmetric orbital S-state, it is necessary for the quark to have an additional quantum number.
> —B. V. Struminsky, *Magnetic moments of barions in the quark model*, JINR-Preprint P-1939, Dubna, Submitted on January 7, 1965

Boris Struminsky was a PhD student of Nikolay Bogolyubov. The problem considered in this preprint was suggested by Nikolay Bogolyubov, who advised Boris Struminsky in this research.[*][4] In the beginning of 1965, Nikolay Bogolyubov, Boris Struminsky and Albert Tavkhelidze wrote a preprint with a more detailed discussion of the additional quark quantum degree of freedom.[*][5] This work was also presented by Albert Tavchelidze without obtaining consent of his collaborators for doing so at an international conference in Trieste (Italy), in May 1965.[*][6][*][7]

A similar mysterious situation was with the Δ^*++ baryon; in the quark model, it is composed of three up quarks with parallel spins. In 1965, Moo-Young Han with Yoichiro Nambu and Oscar W. Greenberg independently resolved the problem by proposing that quarks possess an additional SU(3) gauge degree of freedom, later called color charge. Han and Nambu noted that quarks might interact via an octet of vector gauge bosons: the gluons.

Since free quark searches consistently failed to turn up any evidence for the new particles, and because an elementary particle back then was *defined* as a particle which could be separated and isolated, Gell-Mann often said that quarks were merely convenient mathematical constructs, not real particles. The meaning of this statement was usually clear in context: He meant quarks are confined, but he also was implying that the strong interactions could probably not be fully described by quantum field theory.

Richard Feynman argued that high energy experiments showed quarks are real particles: he called them *partons* (since they were parts of hadrons). By particles, Feynman meant objects which travel along paths, elementary particles in a field theory.

The difference between Feynman's and Gell-Mann's approaches reflected a deep split in the theoretical physics community. Feynman thought the quarks have a distribution of position or momentum, like any other particle, and he (correctly) believed that the diffusion of parton momentum explained diffractive scattering. Although Gell-Mann believed that certain quark charges could be localized, he was open to the possibility that the quarks themselves could not be localized because space and time break down. This was the more radical approach of S-matrix theory.

James Bjorken proposed that pointlike partons would imply certain relations should hold in deep inelastic scattering of electrons and protons, which were spectacularly verified in experiments at SLAC in 1969. This led physicists to abandon the S-matrix approach for the strong interactions.

The discovery of asymptotic freedom in the strong interactions by David Gross, David Politzer and Frank Wilczek allowed physicists to make precise predictions of the results of many high energy experiments using the quantum field theory technique of perturbation theory. Evidence of gluons was discovered in three-jet events at PETRA in 1979. These experiments became more and more precise, culminating in the verification of perturbative QCD at the level of a few percent at the LEP in CERN.

The other side of asymptotic freedom is confinement. Since the force between color charges does not decrease with distance, it is believed that quarks and gluons can never be liberated from hadrons. This aspect of the theory is verified within lattice QCD computations, but is not mathematically proven. One of the Millennium Prize Problems announced by the Clay Mathematics Institute requires a claimant to produce such a proof. Other aspects of non-perturbative QCD are the exploration of phases of quark matter, including the quark–gluon plasma.

The relation between the short-distance particle limit and the confining long-distance limit is one of the topics recently explored using string theory, the modern form of S-matrix theory.[*][8][*][9]

6.21.3 Theory

Some definitions

Every field theory of particle physics is based on certain symmetries of nature whose existence is deduced from observations. These can be

- local symmetries, that is the symmetry acts independently at each point in spacetime. Each such symmetry is the basis of a gauge theory and requires the introduction of its own gauge bosons.
- global symmetries, which are symmetries whose operations must be simultaneously applied to all points of spacetime.

QCD is a gauge theory of the SU(3) gauge group obtained by taking the color charge to define a local symmetry.

Since the strong interaction does not discriminate between different flavors of quark, QCD has approximate **flavor symmetry**, which is broken by the differing masses of the quarks.

There are additional global symmetries whose definitions require the notion of chirality, discrimination between left and right-handed. If the spin of a particle has a positive projection on its direction of motion then it is called left-handed; otherwise, it is right-handed. Chirality and handedness are not the same, but become approximately equivalent at high energies.

- **Chiral** symmetries involve independent transformations of these two types of particle.
- **Vector** symmetries (also called diagonal symmetries) mean the same transformation is applied on the two chiralities.
- **Axial** symmetries are those in which one transformation is applied on left-handed particles and the inverse on the right-handed particles.

Additional remarks: duality

As mentioned, *asymptotic freedom* means that at large energy – this corresponds also to *short distances* – there is practically no interaction between the particles. This is in contrast – more precisely one would say *dual* – to what one is used to, since usually one connects the absence of interactions with *large* distances. However, as already mentioned in the original paper of Franz Wegner,*[10] a solid state theorist who introduced 1971 simple gauge invariant lattice models, the high-temperature behaviour of the *original model*, e.g. the strong decay of correlations at large distances, corresponds to the low-temperature behaviour of the (usually ordered!) *dual model*, namely the asymptotic decay of non-trivial correlations, e.g. short-range deviations from almost perfect arrangements, for short distances. Here, in contrast to Wegner, we have only the dual model, which is that one described in this article.*[11]

Symmetry groups

The color group SU(3) corresponds to the local symmetry whose gauging gives rise to QCD. The electric charge labels a representation of the local symmetry group U(1) which is gauged to give QED: this is an abelian group. If one considers a version of QCD with N_f flavors of massless quarks, then there is a global (chiral) flavor symmetry group $SU_L(N_f) \times SU_R(N_f) \times U_B(1) \times U_A(1)$. The chiral symmetry is spontaneously broken by the QCD vacuum to the vector (L+R) $SU_V(N_f)$ with the formation of a chiral condensate. The vector symmetry, $U_B(1)$ corresponds to the baryon number of quarks and is an exact symmetry. The axial symmetry $U_A(1)$ is exact in the classical theory, but broken in the quantum theory, an occurrence called an anomaly. Gluon field configurations called instantons are closely related to this anomaly.

There are two different types of SU(3) symmetry: there is the symmetry that acts on the different colors of quarks, and this is an exact gauge symmetry mediated by the gluons, and there is also a flavor symmetry which rotates different flavors of quarks to each other, or *flavor SU(3)*. Flavor SU(3) is an approximate symmetry of the vacuum of QCD, and is not a fundamental symmetry at all. It is an accidental consequence of the small mass of the three lightest quarks.

In the QCD vacuum there are vacuum condensates of all the quarks whose mass is less than the QCD scale. This includes the up and down quarks, and to a lesser extent the strange quark, but not any of the others. The vacuum is symmetric under SU(2) isospin rotations of up and down, and to a lesser extent under rotations of up, down and strange, or full flavor group SU(3), and the observed particles make isospin and SU(3) multiplets.

The approximate flavor symmetries do have associated gauge bosons, observed particles like the rho and the omega, but these particles are nothing like the gluons and they are not massless. They are emergent gauge bosons in an approximate string description of QCD.

Lagrangian

The dynamics of the quarks and gluons are controlled by the quantum chromodynamics Lagrangian. The gauge invariant QCD Lagrangian is where $\psi_i(x)$ is the quark field, a dynamical function of spacetime, in the fundamental representation of the SU(3) gauge group, indexed by $i, j, \ldots$; $\mathcal{A}_\mu^a(x)$ are the gluon fields, also dynamical functions of spacetime, in the adjoint representation of the SU(3) gauge group, indexed by $a, b, \ldots$ The $\gamma^*\mu$ are Dirac matrices connecting the spinor representation to the vector representation of the Lorentz group.

The symbol $G_{\mu\nu}^a$ represents the gauge invariant gluon field strength tensor, analogous to the electromagnetic field strength tensor, $F^*\mu\nu$, in quantum electrodynamics. It is given by:*[12]

$$G_{\mu\nu}^a = \partial_\mu \mathcal{A}_\nu^a - \partial_\nu \mathcal{A}_\mu^a + g f^{abc} \mathcal{A}_\mu^b \mathcal{A}_\nu^c,$$

where f_{abc} are the structure constants of SU(3). Note that the rules to move-up or pull-down the a, b, or c indexes are trivial, (+, ..., +), so that $f^*abc = f_{abc} = f^*a_{bc}$ whereas for the μ or ν indexes one has the non-trivial *relativistic* rules, corresponding e.g. to the metric signature (+ − − −).

The constants m and g control the quark mass and coupling constants of the theory, subject to renormalization in the full quantum theory.

An important theoretical notion concerning the final term of the above Lagrangian is the *Wilson loop* variable. This loop variable plays a most important role in discretized forms of the QCD (see lattice QCD), and more generally, it distinguishes confined and deconfined states of a gauge theory. It was introduced by the Nobel prize winner Kenneth G. Wilson and is treated in a separate article.

Fields

Quarks are massive spin-1/2 fermions which carry a color charge whose gauging is the content of QCD. Quarks are represented by Dirac fields in the fundamental representation **3** of the gauge group SU(3). They also carry electric charge (either −1/3 or 2/3) and participate in weak interactions as part of weak isospin doublets. They carry global quantum numbers including the baryon number, which is 1/3 for each quark, hypercharge and one of the flavor quantum numbers.

Gluons are spin-1 bosons which also carry color charges, since they lie in the adjoint representation **8** of SU(3). They have no electric charge, do not participate in the weak interactions, and have no flavor. They lie in the singlet representation **1** of all these symmetry groups.

Every quark has its own antiquark. The charge of each antiquark is exactly the opposite of the corresponding quark.

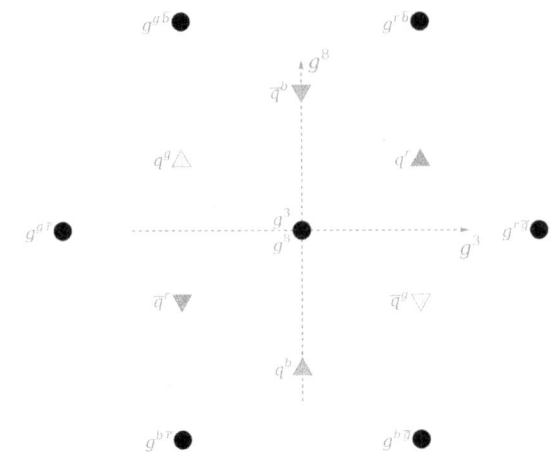

The pattern of strong charges for the three colors of quark, three antiquarks, and eight gluons (with two of zero charge overlapping).

Dynamics

According to the rules of quantum field theory, and the associated Feynman diagrams, the above theory gives rise to three basic interactions: a quark may emit (or absorb) a gluon, a gluon may emit (or absorb) a gluon, and two gluons may directly interact. This contrasts with QED, in which only the first kind of interaction occurs, since photons have no charge. Diagrams involving Faddeev–Popov ghosts must be considered too (except in the unitarity gauge).

Area law and confinement

Detailed computations with the above-mentioned Lagrangian*[13] show that the effective potential between a quark and its anti-quark in a meson contains a term $\propto r$, which represents some kind of "stiffness" of the interaction between the particle and its anti-particle at large distances, similar to the entropic elasticity of a rubber band (see below). This leads to *confinement* *[14] of the quarks to the interior of hadrons, i.e. mesons and nucleons, with typical radii R_c, corresponding to former "Bag models" of the hadrons*[15]. The order of magnitude of the "bag radius" is 1 fm (= 10^*-15 m). Moreover, the above-mentioned stiffness is quantitatively related to the so-called "area law" behaviour of the expectation value of the Wilson loop product P_W of the ordered coupling constants around a closed loop W; i.e. $\langle P_W \rangle$ is proportional to the *area* enclosed by the loop. For this behaviour the non-abelian behaviour of the gauge group is essential.

6.21.4 Methods

Further analysis of the content of the theory is complicated. Various techniques have been developed to work with QCD. Some of them are discussed briefly below.

Perturbative QCD

Main article: Perturbative QCD

This approach is based on asymptotic freedom, which allows perturbation theory to be used accurately in experiments performed at very high energies. Although limited in scope, this approach has resulted in the most precise tests of QCD to date.

Lattice QCD

Main article: Lattice QCD

Among non-perturbative approaches to QCD, the most well established one is lattice QCD. This approach uses a discrete set of spacetime points (called the lattice) to reduce the analytically intractable path integrals of the continuum theory to a very difficult numerical computation which is then carried out on supercomputers like the QCDOC which was constructed for precisely this purpose. While it is a slow and resource-intensive approach, it has wide applicability, giving insight into parts of the theory inaccessible by other means, in particular into the explicit forces acting between quarks and antiquarks in a meson. However, the numerical sign problem makes it difficult to use lattice methods to study QCD at high density and low temperature (e.g. nuclear matter or the interior of neutron stars).

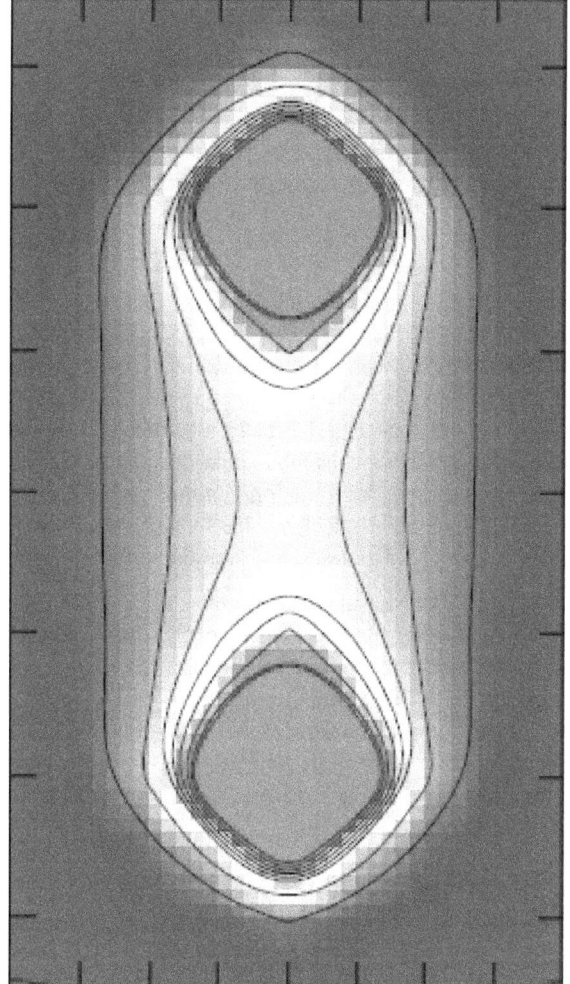

A quark and an antiquark (red color) are glued together (green color) to form a meson (result of a lattice QCD simulation by M. Cardoso et al.[16])*

1/N expansion

Main article: 1/N expansion

A well-known approximation scheme, the 1/N expansion, starts from the premise that the number of colors is infinite, and makes a series of corrections to account for the fact that it is not. Until now, it has been the source of qualitative insight rather than a method for quantitative predictions. Modern variants include the AdS/CFT approach.

Effective theories

For specific problems effective theories may be written down which give qualitatively correct results in certain limits. In the best of cases, these may then be obtained as systematic expansions in some parameter of the QCD Lagrangian. One such effective field theory is chiral perturbation theory or ChiPT, which is the QCD effective theory at low energies. More precisely, it is a low energy expansion based on the spontaneous chiral symmetry breaking of QCD, which is an exact symmetry when quark masses are equal to zero, but for the u,d and s quark, which have small mass, it is still a good approximate symmetry. Depending on the number of quarks which are treated as light, one uses either SU(2) ChiPT or SU(3) ChiPT . Other effective theories are heavy quark effective theory (which expands around heavy quark mass near infinity), and soft-collinear effective theory (which expands around large ratios of energy scales). In addition to effective theories, models like the Nambu–Jona-Lasinio model and the chiral model are often used when discussing general features.

QCD sum rules

Main article: QCD sum rules

Based on an Operator product expansion one can derive sets of relations that connect different observables with each other.

Nambu–Jona-Lasinio model

In one of his recent works, Kei-Ichi Kondo derived as a low-energy limit of QCD, a theory linked to the Nambu–Jona-Lasinio model since it is basically a particular non-local version of the Polyakov–Nambu–Jona-Lasinio model.*[17] The later being in its local version, nothing but the Nambu–Jona-Lasinio model in which one has included the Polyakov loop effect, in order to describe a 'certain confinement'.

The Nambu–Jona-Lasinio model in itself is, among many other things, used because it is a 'relatively simple' model of chiral symmetry breaking, phenomenon present up to certain conditions (Chiral limit i.e. massless fermions) in QCD itself. In this model, however, there is no confinement. In particular, the energy of an isolated quark in the physical vacuum turns out well defined and finite.

6.21.5 Experimental tests

The notion of quark flavors was prompted by the necessity of explaining the properties of hadrons during the development of the quark model. The notion of color was necessitated by the puzzle of the Δ++. This has been dealt with in the section on the history of QCD.

The first evidence for quarks as real constituent elements of hadrons was obtained in deep inelastic scattering experiments at SLAC. The first evidence for gluons came in three jet events at PETRA.

Several good quantitative tests of perturbative QCD exist:

- The running of the QCD coupling as deduced from many observations
- Scaling violation in polarized and unpolarized deep inelastic scattering
- Vector boson production at colliders (this includes the Drell-Yan process)
- Jet cross sections in colliders
- Event shape observables at the LEP
- Heavy-quark production in colliders

Quantitative tests of non-perturbative QCD are fewer, because the predictions are harder to make. The best is probably the running of the QCD coupling as probed through lattice computations of heavy-quarkonium spectra. There is a recent claim about the mass of the heavy meson B_c. Other non-perturbative tests are currently at the level of 5% at best. Continuing work on masses and form factors of hadrons and their weak matrix elements are promising candidates for future quantitative tests. The whole subject of quark matter and the quark–gluon plasma is a non-perturbative test bed for QCD which still remains to be properly exploited.

One qualitative prediction of QCD is that there exist composite particles made solely of gluons called glueballs that have not yet been definitively observed experimentally. A definitive observation of a glueball with the properties predicted by QCD would strongly confirm the theory. In principle, if glueballs could be definitively ruled out, this would be a serious experimental blow to QCD. But, as of 2013, scientists are unable to confirm or deny the existence of glueballs definitively, despite the fact that particle accelerators have sufficient energy to generate them.

6.21.6 Cross-relations to solid state physics

There are unexpected cross-relations to solid state physics. For example, the notion of gauge invariance forms the basis of the well-known Mattis spin glasses,*[18] which are systems with the usual spin degrees of freedom $s_i = \pm 1$ for $i = 1,...,N$, with the special fixed "random" couplings $J_{i,k} = \epsilon_i J_0 \epsilon_k$. Here the ϵ_i and ϵ_k quantities can independently and "randomly" take the values ± 1, which corresponds to a most-simple gauge transformation ($s_i \to s_i \cdot \epsilon_i \quad J_{i,k} \to \epsilon_i J_{i,k} \epsilon_k \quad s_k \to s_k \cdot \epsilon_k$). This means that thermodynamic expectation values of measurable quantities, e.g. of the energy $\mathcal{H} := -\sum s_i J_{i,k} s_k$, are invariant.

However, here the *coupling degrees of freedom* $J_{i,k}$, which in the QCD correspond to the *gluons*, are "frozen" to fixed values (quenching). In contrast, in the QCD they "fluctuate" (annealing), and through the large number of gauge degrees of freedom the entropy plays an important role (see below).

For positive J_0 the thermodynamics of the Mattis spin glass corresponds in fact simply to a "ferromagnet in disguise", just because these systems have no "frustration" at all. This term is a basic measure in spin glass theory.*[19] Quantitatively it is identical with the loop product $P_W := J_{i,k} J_{k,l} ... J_{n,m} J_{m,i}$ along a closed loop W. However, for a Mattis spin glass – in contrast to "genuine" spin glasses – the quantity P_W never becomes negative.

The basic notion "frustration" of the spin-glass is actually similar to the Wilson loop quantity of the QCD. The only difference is again that in the QCD one is dealing with SU(3) matrices, and that one is dealing with a "fluctuating" quantity. Energetically, perfect absence of frustration should be non-favorable and atypical for a spin glass, which means that one should add the loop product to the Hamiltonian, by some kind of term representing a "punishment". In the QCD the Wilson loop is essential for the Lagrangian rightaway.

The relation between the QCD and "disordered magnetic systems" (the spin glasses belong to them) were additionally stressed in a paper by Fradkin, Huberman und Shenker,*[20] which also stresses the notion of duality.

A further analogy consists in the already mentioned similarity to polymer physics, where, analogously to Wilson Loops, so-called "entangled nets" appear, which are important for the formation of the entropy-elasticity (force proportional to the length) of a rubber band. The non-abelian character of the SU(3) corresponds thereby to the non-trivial "chemical links", which glue different loop segments together, and "asymptotic freedom" means in the polymer analogy simply the fact that in the short-wave limit, i.e. for $0 \leftarrow \lambda_w \ll R_c$ (where R_c is a characteristic correlation length for the glued loops, corresponding to the above-mentioned "bag radius", while λ_w is the wavelength of an excitation) any non-trivial correlation vanishes totally, as if the system had crystallized.*[21]

There is also a correspondence between confinement in QCD – the fact that the color field is only different from zero in the interior of hadrons – and the behaviour of the usual magnetic field in the theory of type-II superconductors: there the magnetism is confined to the interiour of the Abrikosov flux-line lattice,*[22] i.e., the London penetration depth λ of that theory is analogous to the confinement radius R_c of quantum chromodynamics. Mathematically, this correspondendence is supported by the second term, $\propto g G_\mu^a \bar{\psi}_i \gamma^\mu T_{ij}^a \psi_j$, on the r.h.s. of the Lagrangian.

6.21.7 See also

- For overviews, see Standard Model, its field theoretical formulation, strong interactions, quarks and gluons, hadrons, confinement, QCD matter, or quark–gluon plasma.

- For details, see gauge theory, quantization procedure including BRST quantization and Faddeev–Popov ghosts. A more general category is quantum field theory.

- For techniques, see Lattice QCD, 1/N expansion, perturbative QCD, Soft-collinear effective theory, heavy quark effective theory, chiral models, and the Nambu and Jona-Lasinio model.

- For experiments, see quark search experiments, deep inelastic scattering, jet physics, quark–gluon plasma.

- Symmetry in quantum mechanics

6.21.8 References

[1] "Alice Physics". 26 August 2015.

[2] Gell-Mann, Murray (1995). *The Quark and the Jaguar*. Owl Books. ISBN 978-0-8050-7253-2.

[3] Fyodor Tkachov (2009). "A contribution to the history of quarks: Boris Struminsky's 1965 JINR publication". arXiv:0904.0343 [physics.hist-ph].

[4] B. V. Struminsky, Magnetic moments of barions in the quark model. JINR-Preprint P-1939, Dubna, Russia. Submitted on January 7, 1965.

[5] N. Bogolubov, B. Struminsky, A. Tavkhelidze. On composite models in the theory of elementary particles. JINR Preprint D-1968, Dubna 1965.

[6] A. Tavkhelidze. Proc. Seminar on High Energy Physics and Elementary Particles, Trieste, 1965, Vienna IAEA, 1965, p. 763.

[7] V. A. Matveev and A. N. Tavkhelidze (INR, RAS, Moscow) The quantum number color, colored quarks and QCD (Dedicated to the 40th Anniversary of the Discovery of the Quantum Number Color). Report presented at the 99th Session of the JINR Scientific Council, Dubna, 19–20 January 2006.

[8] J. Polchinski, M. Strassler (2002). "Hard Scattering and Gauge/String duality". *Physical Review Letters* **88** (3): 31601. arXiv:hep-th/0109174. Bibcode:2002PhRvL..88c1601P. doi:10.1103/PhysRevLett.88.031601. PMID 11801052.

[9] Brower, Richard C.; Mathur, Samir D.; Chung-I Tan (2000). "Glueball Spectrum for QCD from AdS Supergravity Duality". *Nuclear Physics B* **587**: 249–276. arXiv:hep-th/0003115. Bibcode:2000NuPhB.587..249B. doi:10.1016/S0550-3213(00)00435-1.

[10] F. Wegner, *Duality in Generalized Ising Models and Phase Transitions without Local Order Parameter*, J. Math. Phys. **12** (1971) 2259–2272.

> Reprinted in Claudio Rebbi (ed.), *Lattice Gauge Theories and Monte Carlo Simulations*, World Scientific, Singapore (1983), p. 60–73. Abstract:

[11] Perhaps one can guess that in the "original" model mainly the quarks would fluctuate, whereas in the present one, the "dual" model, mainly the gluons do.

[12] M. Eidemüller, H.G. Dosch, M. Jamin (1999). "The field strength correlator from QCD sum rules". *Nucl.Phys.Proc.Suppl.86:421–425,2000* (Heidelberg, Germany). arXiv:hep-ph/9908318.

[13] See all standard textbooks on the QCD, e.g., those noted above

[14] Only at extremely large pressures and or temperatures, e.g. for $T \cong 5 \cdot 10^{12}$ K or larger, *confinement* gives way to a quark–gluon plasma.

[15] Kenneth A. Johnson, "The bag model of quark confinement", Scientific American, July 1979

[16] M. Cardoso et al., "Lattice QCD computation of the colour fields for the static hybrid quark–gluon–antiquark system, and microscopic study of the Casimir scaling", Phys. Rev. D 81, 034504 (2010)).

[17] Kei-Ichi Kondo (2010). "Toward a first-principle derivation of confinement and chiral-symmetry-breaking crossover transitions in QCD". *Physical Review D* **82** (6): 065024. arXiv:1005.0314v2. Bibcode:2010PhRvD..82f5024K. doi:10.1103/PhysRevD.82.065024.

[18] D.C. Mattis, Phys. Lett. 56a (1976) 421

[19] J. Vanninemus and G. Toulouse, J. Phys. C 10 (1977) 537

[20] E. Fradkin, B.A. Huberman, S. Shenker, *Gauge Symmetries in random magnetic systems*, Phys. Rev. B 18 (1978) 4783–4794,

[21] A. Bergmann, A. Owen, "Dielectric relaxation spectroscopy of poly[(R)−3-Hydroxybutyrate] (PHD) during crystallization", Polymer International 53 (7) (2004) 863–868,

[22] Mathematically, the flux-line lattices are described by Emil Artin's braid group, which is nonabelian, since one braid can wind around another one.

6.21.9 Further reading

- Greiner, Walter;Schäfer, Andreas (1994). *Quantum Chromodynamics*. Springer. ISBN 0-387-57103-5.

- Halzen, Francis; Martin, Alan (1984). *Quarks & Leptons: An Introductory Course in Modern Particle Physics*. John Wiley & Sons. ISBN 0-471-88741-2.

- Creutz, Michael (1985). *Quarks, Gluons and Lattices*. Cambridge University Press. ISBN 978-0-521-31535-7.

6.21.10 External links

- Frank Wilczek (2000). "QCD made simple" (PDF). *Physics Today* **53** (8): 22–28. doi:10.1063/1.1310117.

- Particle data group

- The millennium prize for proving confinement

- Ab Initio Determination of Light Hadron Masses

- Andreas S Kronfeld *The Weight of the World Is Quantum Chromodynamics*

- Andreas S Kronfeld *Quantum chromodynamics with advanced computing*

- Standard model gets right answer

- Quantum Chromodynamics

6.22 Quarkonium

In particle physics, **quarkonium** (from quark + onium, pl. **quarkonia**) designates a flavorless meson whose constituents are a quark and its own antiquark. Examples of quarkonia are the J/ψ meson (an example of **charmonium**, cc) and the ϒ meson (**bottomonium**, bb). Because of the high mass of the top quark, **toponium** does not exist, since the top quark decays through the electroweak interaction before a bound state can form. Usually quarkonium refers only to charmonium and bottomonium, and not to any of the lighter quark–antiquark states. This usage is because the lighter quarks (up, down, and strange) are much less massive than the heavier quarks, and so the physical states actually seen in experiments (η, η′, and π⁰ mesons) are quantum mechanical mixtures of the light quark states. The much larger mass differences between the charm and bottom quarks and the lighter quarks results in states that are well defined in terms of a quark–antiquark pair of a given flavor.

6.22.1 Charmonium states

See also: J/ψ meson

In the following table, the same particle can be named with the spectroscopic notation or with its mass. In some cases excitation series are used: Ψ' is the first excitation of Ψ (for historical reasons, this one is called J/ψ particle); Ψ" is a second excitation, and so on. That is, names in the same cell are synonymous.

Some of the states are predicted, but have not been identified; others are unconfirmed. The quantum numbers of the X(3872) particle have been measured recently by the LHCb experiment at CERN*[1] . This measurement shed some light on its identity, excluding the third option among the three envised, which are :

- a candidate for the 1^1D_2 state;
- a charmonium hybrid state;
- a $D^0\bar{D}^{*0}$ molecule.

In 2005, the BaBar experiment announced the discovery of a new state: Y(4260).*[2]*[3] CLEO and Belle have since corroborated these observations. At first, Y(4260) was thought to be a charmonium state, but the evidence suggests more exotic explanations, such as a D "molecule", a 4-quark construct, or a hybrid meson.

Notes:

** Needs confirmation.

*† Predicted, but not yet identified.

*† Interpretation as a 1^{*--} charmonium state not favored.

6.22.2 Bottomonium states

See also: Upsilon meson

In the following table, the same particle can be named with the spectroscopic notation or with its mass.

Some of the states are predicted, but have not been identified; others are unconfirmed.

Notes:

** Preliminary results. Confirmation needed.

The χ_b (3P) state was the first particle discovered in the Large Hadron Collider. The article about this discovery was first submitted to arXiv on 21 December 2011.*[4]*[5] On April 2012, Tevatron's DØ experiment confirms the result in a paper published in *Phys. Rev. D.**[6]*[7]

6.22.3 QCD and quarkonia

The computation of the properties of mesons in Quantum chromodynamics (QCD) is a fully non-perturbative one. As a result, the only general method available is a direct computation using lattice QCD (LQCD) techniques. However, other techniques are effective for heavy quarkonia as well.

The light quarks in a meson move at relativistic speeds, since the mass of the bound state is much larger than the mass of the quark. However, the speed of the charm and the bottom quarks in their respective quarkonia is sufficiently smaller, so that relativistic effects affect these states much less. It is estimated that the speed, **v**, is roughly 0.3 times the speed of light for charmonia and roughly 0.1 times the speed of light for bottomonia. The computation can then be approximated by an expansion in powers of **v**/**c** and **v**^{2}/**c**^{2}. This technique is called non-relativistic QCD (NRQCD).

NRQCD has also been quantized as a lattice gauge theory, which provides another technique for LQCD calculations to use. Good agreement with the bottomonium masses has been found, and this provides one of the best non-perturbative tests of LQCD. For charmonium masses the agreement is not as good, but the LQCD community is actively working on improving their techniques. Work is also being done on calculations of such properties as widths of quarkonia states and transition rates between the states.

An early, but still effective, technique uses models of the *effective* potential to calculate masses of quarkonia states. In this technique, one uses the fact that the motion of the quarks that comprise the quarkonium state is non-relativistic to assume that they move in a static potential, much like non-relativistic models of the hydrogen atom. One of the most popular potential models is the so-called *Cornell potential*

$$V(r) = -\tfrac{a}{r} + br \text{ *[8]}$$

where r is the effective radius of the quarkonium state, a and b are parameters. This potential has two parts. The first part, a/r corresponds to the potential induced by one-gluon exchange between the quark and its anti-quark, and is known as the *Coulombic* part of the potential, since its $1/r$ form is identical to the well-known Coulombic potential induced by the electromagnetic force. The second part, br, is known as the *confinement* part of the potential, and parameterizes the poorly understood non-perturbative effects of QCD. Generally, when using this approach, a convenient form for the wave function of the quarks is taken, and then a and b are determined by fitting the results of the calculations to the masses of well-measured quarkonium states. Relativistic and other effects can be incorporated into this approach by adding extra terms to the potential, much in the same way that they are for the hydrogen atom in non-relativistic quantum mechanics. This form has been derived from QCD up to $\mathcal{O}(\Lambda^3_{\text{QCD}}r^2)$ by Y. Sumino in 2003.*[9] It is popular because it allows for accurate predictions of quarkonia parameters without a lengthy lattice computation, and provides a separation between the short-distance *Coulombic* effects and the long-distance *confinement* effects that can be useful in understanding the quark/anti-quark force generated by QCD.

Quarkonia have been suggested as a diagnostic tool of the formation of the quark–gluon plasma: both disappearance and enhancement of their formation depending on the yield of heavy quarks in plasma can occur.

6.22.4 See also

- Onium
- OZI Rule
- J/ψ meson
- Phi meson
- Upsilon meson
- Theta meson
- Non-relativistic QCD
- Lattice QCD
- Quantum chromodynamics

6.22.5 References

[1] LHCb collaboration; Aaij, R.; Abellan Beteta, C.; Adeva, B.; Adinolfi, M.; Adrover, C.; Affolder, A.; Ajaltouni, Z. et al. (February 2013). "Determination of the X(3872) meson quantum numbers". *Physical Review Letters* **1302** (22): 6269. arXiv:1302.6269. Bibcode:2013PhRvL.110v2001A. doi:10.1103/PhysRevLett.110.222001.

[2] "A new particle discovered by BaBar experiment". Istituto Nazionale di Fisica Nucleare. 6 July 2005. Retrieved 2010-03-06.

[3] B. Aubert *et al.* (BaBar Collaboration) (2005). "Observation of a broad structure in the $\pi^*+\pi^*-J/\psi$ mass spectrum around 4.26 GeV/c^2". *Physical Review Letters* **95** (14): 142001. arXiv:hep-ex/0506081. Bibcode:2005PhRvL..95n2001A. doi:10.1103/PhysRevLett.95.142001.

[4] ATLAS Collaboration (2012). "Observation of a new χ b state in radiative transitions to Υ(1S) and Υ(2S) at ATLAS". arXiv:1112.5154v4 [hep-ex].

[5] Jonathan Amos (2011-12-22). "LHC reports discovery of its first new particle". BBC.

[6] *Tevatron experiment confirms LHC discovery of Chi-b (P3) particle*

[7] *Observation of a narrow mass state decaying into Υ(1S) + γ in pp collisions at 1.96 TeV*

[8] Hee Sok Chung; Jungil Lee; Daekyoung Kang (2008). "Cornell Potential Parameters for S-wave Heavy Quarkonia". *Journal of the Korean Physical Society* **52** (4): 1151. arXiv:0803.3116. Bibcode:2008JKPS...52.1151C. doi:10.3938/jkps.52.1151.

[9] Y. Sumino (2003). "QCD potential as a "Coulomb-plus-linear" potential". *Phys. Lett. B* **571**: 173–183. arXiv:hep-ph/0303120. doi:10.1016/j.physletb.2003.05.010.

6.23 Quark–lepton complementarity

The **quark–lepton complementarity** (**QLC**) is a possible fundamental symmetry between quarks and leptons. First proposed in 1990 by Foot and Lew,*[1] it assumes that leptons as well as quarks come in three "colors". Such theory may reproduce the Standard Model at low energies, and hence quark–lepton symmetry may be realized in nature.

6.23.1 Possible evidence for QLC

Recent neutrino experiments confirm that the Pontecorvo–Maki–Nakagawa–Sakata matrix U_{PMNS} contains large mixing angles. For example, atmospheric measurements of particle decay yield θPMNS 23 ≈ 45°, while solar experiments yield θPMNS 12 ≈ 34°. These results should be compared with θPMNS 13 which is small,*[2] and with the quark mixing angles in the Cabibbo–Kobayashi–Maskawa matrix U_{CKM}. The disparity that nature indicates between quark and lepton mixing angles has been viewed in terms of a "quark–lepton complementarity" which can be expressed in the relations

$$\theta_{12}^{PMNS} + \theta_{12}^{CKM} \simeq 45°,$$

$$\theta_{23}^{PMNS} + \theta_{23}^{CKM} \simeq 45°.$$

Possible consequences of QLC have been investigated in the literature and in particular a simple correspondence between the PMNS and CKM matrices have been proposed and analyzed in terms of a correlation matrix. The correlation matrix V_M is simply defined as the product of the CKM and PMNS matrices:

$$V_M = U_{CKM} \cdot U_{PMNS},$$

Unitarity implies:

$$U_{PMNS} = U_{CKM}^\dagger V_M.$$

6.23.2 Open questions

One may ask where do the large lepton mixings come from? Is this information implicit in the form of the V_M matrix? This question has been widely investigated in the literature, but its answer is still open. Furthermore in some Grand Unification Theories (GUTs) the direct QLC correlation between the CKM and the PMNS mixing matrix can be obtained. In this class of models, the V_M matrix is determined by the heavy Majorana neutrino mass matrix.

Despite the naive relations between the PMNS and CKM angles, a detailed analysis shows that the correlation matrix is phenomenologically compatible with a tribimaximal pattern, and only marginally with a bimaximal pattern. It is possible to include bimaximal forms of the correlation matrix V_M in models with renormalization effects that are relevant, however, only in particular cases with $\tan\beta > 40$ and with quasi-degenerate neutrino masses.

6.23.3 See also

- Leptoquark

6.23.4 References

[1] R. Foot, H. Lew (1990). "Quark-lepton-symmetric model". *Physical Review D* **41** (11): 3502–3505. Bibcode:1990PhRvD..41.3502F. doi:10.1103/PhysRevD.41.3502.

[2] F. P. An et al. [DAYA-BAY Collaboration], Phys. Rev. Lett. 108, 171803 (2012) [arXiv:1203.1669 [hep-ex]] http://arxiv.org/abs/arXiv:1203.1669

- B.C. Chauhan, M. Picariello, J. Pulido, E. Torrente-Lujan (2007). "Quark-lepton complementarity, neutrino and standard model data predict θPMNS 13 = (9+1 −2)°". *European Physical Journal C* **50** (3): 573–578. arXiv:hep-ph/0605032. Bibcode:2007EPJC...50..573C. doi:10.1140/epjc/s10052-007-0212-z.

- K.M. Patel (2010). "An $SO(10) \times S_4$ Model of Quark-Lepton Complementarity;". *Physics Letters B* **695**: 225. arXiv:1008.5061. Bibcode:2011PhLB..695..225P. doi:10.1016/j.physletb.2010.11.024.

6.24 Standard Model

This article is about the Standard Model of particle physics. For other uses, see Standard model (disambiguation).
This article is a non-mathematical general overview of the Standard Model. For a mathematical description, see the article Standard Model (mathematical formulation).
For the Standard Model of Big Bang cosmology, Lambda-CDM model.

The **Standard Model** of particle physics is a theory con-

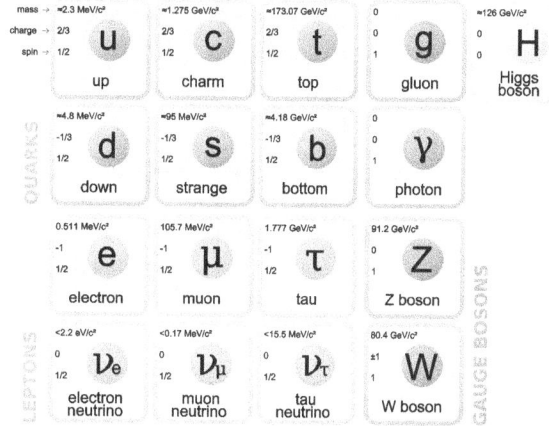

The Standard Model of elementary particles (more schematic depiction), with the three generations of matter, gauge bosons in the fourth column, and the Higgs boson in the fifth.

cerning the electromagnetic, weak, and strong nuclear interactions, as well as classifying all the subatomic particles known. It was developed throughout the latter half of the 20th century, as a collaborative effort of scientists around the world.*[1] The current formulation was finalized in the mid-1970s upon experimental confirmation of the existence of quarks. Since then, discoveries of the top quark (1995), the tau neutrino (2000), and more recently the Higgs boson (2013), have given further credence to the Standard Model. Because of its success in explaining a wide variety of experimental results, the Standard Model is sometimes regarded as a "theory of almost everything".

Although the Standard Model is believed to be theoretically self-consistent*[2] and has demonstrated huge and continued successes in providing experimental predictions, it does leave some phenomena unexplained and it falls short of being a complete theory of fundamental interactions. It does not incorporate the full theory of gravitation*[3] as described by general relativity, or account for the accelerating expansion of the universe (as possibly described by dark energy). The model does not contain any viable dark matter particle that possesses all of the required properties deduced from observational cosmology. It also does not incorporate neutrino oscillations (and their non-zero masses).

The development of the Standard Model was driven by theoretical and experimental particle physicists alike. For theorists, the Standard Model is a paradigm of a quantum field theory, which exhibits a wide range of physics including spontaneous symmetry breaking, anomalies, non-perturbative behavior, etc. It is used as a basis for building more exotic models that incorporate hypothetical particles, extra dimensions, and elaborate symmetries (such as supersymmetry) in an attempt to explain experimental results at variance with the Standard Model, such as the existence of dark matter and neutrino oscillations.

6.24.1 Historical background

The first step towards the Standard Model was Sheldon Glashow's discovery in 1961 of a way to combine the electromagnetic and weak interactions.[*][4] In 1967 Steven Weinberg[*][5] and Abdus Salam[*][6] incorporated the Higgs mechanism[*][7][*][8][*][9] into Glashow's electroweak theory, giving it its modern form.

The Higgs mechanism is believed to give rise to the masses of all the elementary particles in the Standard Model. This includes the masses of the W and Z bosons, and the masses of the fermions, i.e. the quarks and leptons.

After the neutral weak currents caused by Z boson exchange were discovered at CERN in 1973,[*][10][*][11][*][12][*][13] the electroweak theory became widely accepted and Glashow, Salam, and Weinberg shared the 1979 Nobel Prize in Physics for discovering it. The W and Z bosons were discovered experimentally in 1981, and their masses were found to be as the Standard Model predicted.

The theory of the strong interaction, to which many contributed, acquired its modern form around 1973–74, when experiments confirmed that the hadrons were composed of fractionally charged quarks.

6.24.2 Overview

At present, matter and energy are best understood in terms of the kinematics and interactions of elementary particles. To date, physics has reduced the laws governing the behavior and interaction of all known forms of matter and energy to a small set of fundamental laws and theories. A major goal of physics is to find the "common ground" that would unite all of these theories into one integrated theory of everything, of which all the other known laws would be special cases, and from which the behavior of all matter and energy could be derived (at least in principle).[*][14]

6.24.3 Particle content

The Standard Model includes members of several classes of elementary particles (fermions, gauge bosons, and the Higgs boson), which in turn can be distinguished by other characteristics, such as color charge.

Fermions

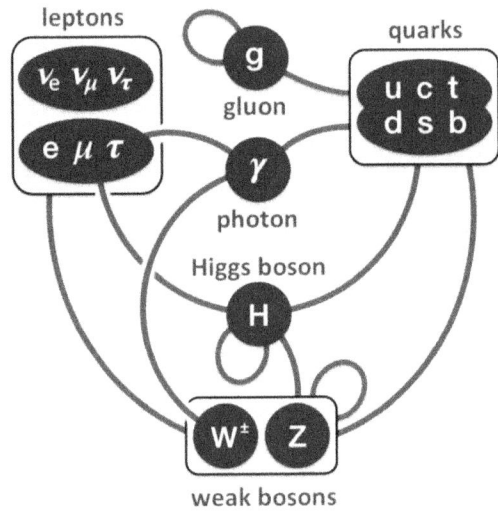

Summary of interactions between particles described by the Standard Model.

The Standard Model includes 12 elementary particles of spin-½ known as fermions. According to the spin-statistics theorem, fermions respect the Pauli exclusion principle. Each fermion has a corresponding antiparticle.

The fermions of the Standard Model are classified according to how they interact (or equivalently, by what charges they carry). There are six quarks (up, down, charm, strange, top, bottom), and six leptons (electron, electron neutrino, muon, muon neutrino, tau, tau neutrino). Pairs from each classification are grouped together to form a generation, with corresponding particles exhibiting similar physical behavior (see table).

The defining property of the quarks is that they carry color charge, and hence, interact via the strong interaction. A phenomenon called color confinement results in quarks being very strongly bound to one another, forming color-neutral composite particles (hadrons) containing either a quark and an antiquark (mesons) or three quarks (baryons). The familiar proton and the neutron are the two baryons having the smallest mass. Quarks also carry electric charge and weak isospin. Hence they interact with other fermions both electromagnetically and via the weak interaction.

6.24. STANDARD MODEL

The remaining six fermions do not carry colour charge and are called leptons. The three neutrinos do not carry electric charge either, so their motion is directly influenced only by the weak nuclear force, which makes them notoriously difficult to detect. However, by virtue of carrying an electric charge, the electron, muon, and tau all interact electromagnetically.

Each member of a generation has greater mass than the corresponding particles of lower generations. The first generation charged particles do not decay; hence all ordinary (baryonic) matter is made of such particles. Specifically, all atoms consist of electrons orbiting around atomic nuclei, ultimately constituted of up and down quarks. Second and third generations charged particles, on the other hand, decay with very short half lives, and are observed only in very high-energy environments. Neutrinos of all generations also do not decay, and pervade the universe, but rarely interact with baryonic matter.

Gauge bosons

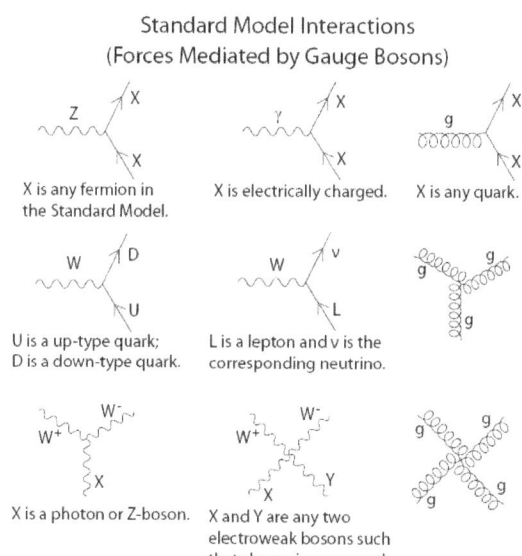

The above interactions form the basis of the standard model. Feynman diagrams in the standard model are built from these vertices. Modifications involving Higgs boson interactions and neutrino oscillations are omitted. The charge of the W bosons is dictated by the fermions they interact with; the conjugate of each listed vertex (i.e. reversing the direction of arrows) is also allowed.

In the Standard Model, gauge bosons are defined as force carriers that mediate the strong, weak, and electromagnetic fundamental interactions.

Interactions in physics are the ways that particles influence other particles. At a macroscopic level, electromagnetism allows particles to interact with one another via electric and magnetic fields, and gravitation allows particles with mass to attract one another in accordance with Einstein's theory of general relativity. The Standard Model explains such forces as resulting from matter particles exchanging other particles, generally referred to as *force mediating particles*. When a force-mediating particle is exchanged, at a macroscopic level the effect is equivalent to a force influencing both of them, and the particle is therefore said to have *mediated* (i.e., been the agent of) that force. The Feynman diagram calculations, which are a graphical representation of the perturbation theory approximation, invoke "force mediating particles", and when applied to analyze high-energy scattering experiments are in reasonable agreement with the data. However, perturbation theory (and with it the concept of a "force-mediating particle") fails in other situations. These include low-energy quantum chromodynamics, bound states, and solitons.

The gauge bosons of the Standard Model all have spin (as do matter particles). The value of the spin is 1, making them bosons. As a result, they do not follow the Pauli exclusion principle that constrains fermions: thus bosons (e.g. photons) do not have a theoretical limit on their spatial density (number per volume). The different types of gauge bosons are described below.

- Photons mediate the electromagnetic force between electrically charged particles. The photon is massless and is well-described by the theory of quantum electrodynamics.

- The W+, W−, and Z gauge bosons mediate the weak interactions between particles of different flavors (all quarks and leptons). They are massive, with the Z being more massive than the W±. The weak interactions involving the W± exclusively act on *left-handed* particles and *right-handed* antiparticles. Furthermore, the W± carries an electric charge of +1 and −1 and couples to the electromagnetic interaction. The electrically neutral Z boson interacts with both left-handed particles and antiparticles. These three gauge bosons along with the photons are grouped together, as collectively mediating the electroweak interaction.

- The eight gluons mediate the strong interactions between color charged particles (the quarks). Gluons are massless. The eightfold multiplicity of gluons is labeled by a combination of color and anticolor charge (e.g. red–antigreen).*[nb 1] Because the gluons have an effective color charge, they can also interact among themselves. The gluons and their interactions are described by the theory of quantum chromodynamics.

The interactions between all the particles described by the

Standard Model are summarized by the diagrams on the right of this section.

Higgs boson

Main article: Higgs boson

The Higgs particle is a massive scalar elementary particle theorized by Robert Brout, François Englert, Peter Higgs, Gerald Guralnik, C. R. Hagen, and Tom Kibble in 1964 (see 1964 PRL symmetry breaking papers) and is a key building block in the Standard Model.[7][8][9][15] It has no intrinsic spin, and for that reason is classified as a boson (like the gauge bosons, which have integer spin).

The Higgs boson plays a unique role in the Standard Model, by explaining why the other elementary particles, except the photon and gluon, are massive. In particular, the Higgs boson explains why the photon has no mass, while the W and Z bosons are very heavy. Elementary particle masses, and the differences between electromagnetism (mediated by the photon) and the weak force (mediated by the W and Z bosons), are critical to many aspects of the structure of microscopic (and hence macroscopic) matter. In electroweak theory, the Higgs boson generates the masses of the leptons (electron, muon, and tau) and quarks. As the Higgs boson is massive, it must interact with itself.

Because the Higgs boson is a very massive particle and also decays almost immediately when created, only a very high-energy particle accelerator can observe and record it. Experiments to confirm and determine the nature of the Higgs boson using the Large Hadron Collider (LHC) at CERN began in early 2010, and were performed at Fermilab's Tevatron until its closure in late 2011. Mathematical consistency of the Standard Model requires that any mechanism capable of generating the masses of elementary particles become visible at energies above 1.4 TeV;[16] therefore, the LHC (designed to collide two 7 to 8 TeV proton beams) was built to answer the question of whether the Higgs boson actually exists.[17]

On 4 July 2012, the two main experiments at the LHC (ATLAS and CMS) both reported independently that they found a new particle with a mass of about 125 GeV/c^2 (about 133 proton masses, on the order of 10^{-25} kg), which is "consistent with the Higgs boson." Although it has several properties similar to the predicted "simplest" Higgs,[18] they acknowledged that further work would be needed to conclude that it is indeed the Higgs boson, and exactly which version of the Standard Model Higgs is best supported if confirmed.[19][20][21][22][23]

On 14 March 2013 the Higgs Boson was tentatively confirmed to exist.[24]

Total particle count

Counting particles by a rule that distinguishes between particles and their corresponding antiparticles, and among the many color states of quarks and gluons, gives a total of 61 elementary particles.[25]

6.24.4 Theoretical aspects

Main article: Standard Model (mathematical formulation)

Construction of the Standard Model Lagrangian

Technically, quantum field theory provides the mathematical framework for the Standard Model, in which a Lagrangian controls the dynamics and kinematics of the theory. Each kind of particle is described in terms of a dynamical field that pervades space-time. The construction of the Standard Model proceeds following the modern method of constructing most field theories: by first postulating a set of symmetries of the system, and then by writing down the most general renormalizable Lagrangian from its particle (field) content that observes these symmetries.

The global Poincaré symmetry is postulated for all relativistic quantum field theories. It consists of the familiar translational symmetry, rotational symmetry and the inertial reference frame invariance central to the theory of special relativity. The local SU(3)×SU(2)×U(1) gauge symmetry is an internal symmetry that essentially defines the Standard Model. Roughly, the three factors of the gauge symmetry give rise to the three fundamental interactions. The fields fall into different representations of the various symmetry groups of the Standard Model (see table). Upon writing the most general Lagrangian, one finds that the dynamics depend on 19 parameters, whose numerical values are established by experiment. The parameters are summarized in the table above (note: with the Higgs mass is at 125 GeV, the Higgs self-coupling strength $\lambda \sim 1/8$).

Quantum chromodynamics sector Main article: Quantum chromodynamics

The quantum chromodynamics (QCD) sector defines the interactions between quarks and gluons, with SU(3) symmetry, generated by T^a. Since leptons do not interact with gluons, they are not affected by this sector. The Dirac Lagrangian of the quarks coupled to the gluon fields is given by

6.24. STANDARD MODEL

$$\mathcal{L}_{QCD} = i\overline{U}(\partial_\mu - ig_s G_\mu^a T^a)\gamma^\mu U + i\overline{D}(\partial_\mu - ig_s G_\mu^a T^a)\gamma^\mu D.$$

G_μ^a is the SU(3) gauge field containing the gluons, γ^μ are the Dirac matrices, D and U are the Dirac spinors associated with up- and down-type quarks, and g_s is the strong coupling constant.

Electroweak sector Main article: Electroweak interaction

The electroweak sector is a Yang–Mills gauge theory with the simple symmetry group U(1)×SU(2)$_L$,

$$\mathcal{L}_{EW} = \sum_\psi \bar{\psi}\gamma^\mu \left(i\partial_\mu - g'\frac{1}{2}Y_W B_\mu - g\frac{1}{2}\vec{\tau}_L \vec{W}_\mu \right)\psi$$

where B_μ is the U(1) gauge field; Y_W is the weak hypercharge—the generator of the U(1) group; $\vec{W}_\mu$ is the three-component SU(2) gauge field; $\vec{\tau}_L$ are the Pauli matrices—infinitesimal generators of the SU(2) group. The subscript L indicates that they only act on left fermions; g' and g are coupling constants.

Higgs sector Main article: Higgs mechanism

In the Standard Model, the Higgs field is a complex scalar of the group SU(2)$_L$:

$$\varphi = \frac{1}{\sqrt{2}}\begin{pmatrix} \varphi^+ \\ \varphi^0 \end{pmatrix},$$

where the indices + and 0 indicate the electric charge (Q) of the components. The weak isospin (Y_W) of both components is 1.

Before symmetry breaking, the Higgs Lagrangian is:

$$\mathcal{L}_H = \varphi^\dagger \left(\partial^\mu - \frac{i}{2}\left(g'Y_W B^\mu + g\vec{\tau}\vec{W}^\mu\right)\right)\left(\partial_\mu + \frac{i}{2}\left(g'Y_W B_\mu + g\vec{\tau}\vec{W}_\mu\right)\right)\varphi - \frac{\lambda^2}{4}(\varphi^\dagger\varphi - v^2)^2,$$

which can also be written as:

$$\mathcal{L}_H = \left|\left(\partial_\mu + \frac{i}{2}\left(g'Y_W B_\mu + g\vec{\tau}\vec{W}_\mu\right)\right)\varphi\right|^2 - \frac{\lambda^2}{4}(\varphi^\dagger\varphi - v^2)^2$$

6.24.5 Fundamental forces

Main article: Fundamental interaction

The Standard Model classified all four fundamental forces in nature. In the Standard Model, a force is described as an exchange of bosons between the objects affected, such as a photon for the electromagnetic force and a gluon for the strong interaction. Those particles are called force carriers.*[26]

6.24.6 Tests and predictions

The Standard Model (SM) predicted the existence of the W and Z bosons, gluon, and the top and charm quarks before these particles were observed. Their predicted properties were experimentally confirmed with good precision. To give an idea of the success of the SM, the following table compares the measured masses of the W and Z bosons with the masses predicted by the SM:

The SM also makes several predictions about the decay of Z bosons, which have been experimentally confirmed by the Large Electron-Positron Collider at CERN.

In May 2012 BaBar Collaboration reported that their recently analyzed data may suggest possible flaws in the Standard Model of particle physics.*[28]*[29] These data show that a particular type of particle decay called "B to D-star-tau-nu" happens more often than the Standard Model says it should. In this type of decay, a particle called the B-bar meson decays into a D meson, an antineutrino and a tau-lepton. While the level of certainty of the excess (3.4 sigma) is not enough to claim a break from the Standard Model, the results are a potential sign of something amiss and are likely to impact existing theories, including those attempting to deduce the properties of Higgs bosons.*[30]

On December 13, 2012, physicists reported the constancy, over space and time, of a basic physical constant of nature that supports the *standard model of physics*. The scientists, studying methanol molecules in a distant galaxy, found the change (Δμ/μ) in the proton-to-electron mass ratio μ to be equal to "(0.0 ± 1.0) × 10*−7 at redshift z = 0.89" and consistent with "a null result".*[31]*[32]

6.24.7 Challenges

See also: Physics beyond the Standard Model

Self-consistency of the Standard Model (currently formulated as a non-abelian gauge theory quantized through path-integrals) has not been mathematically proven. While regularized versions useful for approximate computations (for example lattice gauge theory) exist, it is not known whether they converge (in the sense of S-matrix elements) in the limit that the regulator is removed. A key question related to the consistency is the Yang–Mills existence and mass gap problem.

Experiments indicate that neutrinos have mass, which the classic Standard Model did not allow.*[33] To accommodate this finding, the classic Standard Model can be modified to include neutrino mass.

If one insists on using only Standard Model particles, this can be achieved by adding a non-renormalizable interaction of leptons with the Higgs boson.*[34] On a fundamental level, such an interaction emerges in the seesaw mechanism where heavy right-handed neutrinos are added to the theory. This is natural in the left-right symmetric extension of the Standard Model*[35]*[36] and in certain grand unified theories.*[37] As long as new physics appears below or around 10^{14} GeV, the neutrino masses can be of the right order of magnitude.

Theoretical and experimental research has attempted to extend the Standard Model into a Unified field theory or a Theory of everything, a complete theory explaining all physical phenomena including constants. Inadequacies of the Standard Model that motivate such research include:

- It does not attempt to explain gravitation, although a theoretical particle known as a graviton would help explain it, and unlike for the strong and electroweak interactions of the Standard Model, there is no known way of describing general relativity, the canonical theory of gravitation, consistently in terms of quantum field theory. The reason for this is, among other things, that quantum field theories of gravity generally break down before reaching the Planck scale. As a consequence, we have no reliable theory for the very early universe;

- Some consider it to be *ad hoc* and inelegant, requiring 19 numerical constants whose values are unrelated and arbitrary. Although the Standard Model, as it now stands, can explain why neutrinos have masses, the specifics of neutrino mass are still unclear. It is believed that explaining neutrino mass will require an additional 7 or 8 constants, which are also arbitrary parameters;

- The Higgs mechanism gives rise to the hierarchy problem if some new physics (coupled to the Higgs) is present at high energy scales. In these cases in order for the weak scale to be much smaller than the Planck scale, severe fine tuning of the parameters is required; there are, however, other scenarios that include quantum gravity in which such fine tuning can be avoided.*[38]There are also issues of Quantum triviality, which suggests that it may not be possible to create a consistent quantum field theory involving elementary scalar particles.

- It should be modified so as to be consistent with the emerging "Standard Model of cosmology." In particular, the Standard Model cannot explain the observed amount of cold dark matter (CDM) and gives contributions to dark energy which are many orders of magnitude too large. It is also difficult to accommodate the observed predominance of matter over antimatter (matter/antimatter asymmetry). The isotropy and homogeneity of the visible universe over large distances seems to require a mechanism like cosmic inflation, which would also constitute an extension of the Standard Model.

- The existence of ultra-high-energy cosmic rays are difficult to explain under the Standard Model.

Currently, no proposed Theory of Everything has been widely accepted or verified.

6.24.8 See also

- Fundamental interaction:
  - Quantum electrodynamics
  - Strong interaction: Color charge, Quantum chromodynamics, Quark model
  - Weak interaction: Electroweak theory, Fermi theory of beta decay, Weak hypercharge, Weak isospin

- Gauge theory: Nontechnical introduction to gauge theory

- Generation

- Higgs mechanism: Higgs boson, Higgsless model

- J. C. Ward

- J. J. Sakurai Prize for Theoretical Particle Physics

- Lagrangian

- Open questions: BTeV experiment, CP violation, Neutrino masses, Quark matter, Quantum triviality

- Penguin diagram

- Quantum field theory

- Standard Model: Mathematical formulation of, Physics beyond the Standard Model

6.24.9 Notes and references

[1] Technically, there are nine such color–anticolor combinations. However, there is one color-symmetric combination that can be constructed out of a linear superposition of the nine combinations, reducing the count to eight.

6.24.10 References

[1] R. Oerter (2006). *The Theory of Almost Everything: The Standard Model, the Unsung Triumph of Modern Physics* (Kindle ed.). Penguin Group. p. 2. ISBN 0-13-236678-9.

[2] In fact, there are mathematical issues regarding quantum field theories still under debate (see e.g. Landau pole), but the predictions extracted from the Standard Model by current methods applicable to current experiments are all self-consistent. For a further discussion see e.g. Chapter 25 of R. Mann (2010). *An Introduction to Particle Physics and the Standard Model.* CRC Press. ISBN 978-1-4200-8298-2.

[3] Sean Carroll, Ph.D., Cal Tech, 2007, The Teaching Company, *Dark Matter, Dark Energy: The Dark Side of the Universe*, Guidebook Part 2 page 59, Accessed Oct. 7, 2013, "...Standard Model of Particle Physics: The modern theory of elementary particles and their interactions ... It does not, strictly speaking, include gravity, although it's often convenient to include gravitons among the known particles of nature..."

[4] S.L. Glashow (1961). "Partial-symmetries of weak interactions". *Nuclear Physics* **22** (4): 579–588. Bibcode:1961NucPh..22..579G. doi:10.1016/0029-5582(61)90469-2.

[5] S. Weinberg (1967). "A Model of Leptons". *Physical Review Letters* **19** (21): 1264–1266. Bibcode:1967PhRvL..19.1264W. doi:10.1103/PhysRevLett.19.1264.

[6] A. Salam (1968). N. Svartholm, ed. *Elementary Particle Physics: Relativistic Groups and Analyticity.* Eighth Nobel Symposium. Stockholm: Almquvist and Wiksell. p. 367.

[7] F. Englert, R. Brout (1964). "Broken Symmetry and the Mass of Gauge Vector Mesons". *Physical Review Letters* **13** (9): 321–323. Bibcode:1964PhRvL..13..321E. doi:10.1103/PhysRevLett.13.321.

[8] P.W. Higgs (1964). "Broken Symmetries and the Masses of Gauge Bosons". *Physical Review Letters* **13** (16): 508–509. Bibcode:1964PhRvL..13..508H. doi:10.1103/PhysRevLett.13.508.

[9] G.S. Guralnik, C.R. Hagen, T.W.B. Kibble (1964). "Global Conservation Laws and Massless Particles". *Physical Review Letters* **13** (20): 585–587. Bibcode:1964PhRvL..13..585G. doi:10.1103/PhysRevLett.13.585.

[10] F.J. Hasert et al. (1973). "Search for elastic muon-neutrino electron scattering". *Physics Letters B* **46** (1): 121. Bibcode:1973PhLB...46..121H. doi:10.1016/0370-2693(73)90494-2.

[11] F.J. Hasert et al. (1973). "Observation of neutrino-like interactions without muon or electron in the Gargamelle neutrino experiment". *Physics Letters B* **46** (1): 138. Bibcode:1973PhLB...46..138H. doi:10.1016/0370-2693(73)90499-1.

[12] F.J. Hasert et al. (1974). "Observation of neutrino-like interactions without muon or electron in the Gargamelle neutrino experiment". *Nuclear Physics B* **73** (1): 1. Bibcode:1974NuPhB..73....1H. doi:10.1016/0550-3213(74)90038-8.

[13] D. Haidt (4 October 2004). "The discovery of the weak neutral currents". *CERN Courier.* Retrieved 8 May 2008.

[14] "Details can be worked out if the situation is simple enough for us to make an approximation, which is almost never, but often we can understand more or less what is happening." from *The Feynman Lectures on Physics*, Vol 1. pp. 2–7

[15] G.S. Guralnik (2009). "The History of the Guralnik, Hagen and Kibble development of the Theory of Spontaneous Symmetry Breaking and Gauge Particles". *International Journal of Modern Physics A* **24** (14): 2601–2627. arXiv:0907.3466. Bibcode:2009IJMPA..24.2601G. doi:10.1142/S0217751X09045431.

[16] B.W. Lee, C. Quigg, H.B. Thacker (1977). "Weak interactions at very high energies: The role of the Higgs-boson mass". *Physical Review D* **16** (5): 1519–1531. Bibcode:1977PhRvD..16.1519L. doi:10.1103/PhysRevD.16.1519.

[17] "Huge $10 billion collider resumes hunt for 'God particle'". CNN. 11 November 2009. Retrieved 2010-05-04.

[18] M. Strassler (10 July 2012). "Higgs Discovery: Is it a Higgs?". Retrieved 2013-08-06.

[19] "CERN experiments observe particle consistent with long-sought Higgs boson". CERN. 4 July 2012. Retrieved 2012-07-04.

[20] "Observation of a New Particle with a Mass of 125 GeV". CERN. 4 July 2012. Retrieved 2012-07-05.

[21] "ATLAS Experiment". ATLAS. 1 January 2006. Retrieved 2012-07-05.

[22] "Confirmed: CERN discovers new particle likely to be the Higgs boson". *YouTube.* Russia Today. 4 July 2012. Retrieved 2013-08-06.

[23] D. Overbye (4 July 2012). "A New Particle Could Be Physics' Holy Grail". *New York Times.* Retrieved 2012-07-04.

[24] "New results indicate that new particle is a Higgs boson". CERN. 14 March 2013. Retrieved 2013-08-06.

[25] S. Braibant, G. Giacomelli, M. Spurio (2009). *Particles and Fundamental Interactions: An Introduction to Particle Physics*. Springer. pp. 313–314. ISBN 978-94-007-2463-1.

[26] http://home.web.cern.ch/about/physics/standard-model Official CERN website

[27] http://www.pha.jhu.edu/~{}dfehling/particle.gif

[28] "BABAR Data in Tension with the Standard Model". SLAC. 31 May 2012. Retrieved 2013-08-06.

[29] BaBar Collaboration (2012). "Evidence for an excess of B → D*(*) τ*− ν$_\tau$ decays". *Physical Review Letters* **109** (10): 101802. arXiv:1205.5442. Bibcode:2012PhRvL.109j1802L. doi:10.1103/PhysRevLett.109.101802.

[30] "BaBar data hint at cracks in the Standard Model". e! Science News. 18 June 2012. Retrieved 2013-08-06.

[31] J. Bagdonaite et al. (2012). "A Stringent Limit on a Drifting Proton-to-Electron Mass Ratio from Alcohol in the Early Universe". *Science* **339** (6115): 46. Bibcode:2013Sci...339...46B. doi:10.1126/science.1224898.

[32] C. Moskowitz (13 December 2012). "Phew! Universe's Constant Has Stayed Constant". Space.com. Retrieved 2012-12-14.

[33] "Particle chameleon caught in the act of changing". CERN. 31 May 2010. Retrieved 2012-07-05.

[34] S. Weinberg (1979). "Baryon and Lepton Nonconserving Processes". *Physical Review Letters* **43** (21): 1566. Bibcode:1979PhRvL..43.1566W. doi:10.1103/PhysRevLett.43.1566.

[35] P. Minkowski (1977). "μ → e γ at a Rate of One Out of 10^9 Muon Decays?". *Physics Letters B* **67** (4): 421. Bibcode:1977PhLB...67..421M. doi:10.1016/0370-2693(77)90435-X.

[36] R. N. Mohapatra, G. Senjanovic (1980). "Neutrino Mass and Spontaneous Parity Nonconservation". *Physical Review Letters* **44** (14): 912–915. Bibcode:1980PhRvL..44..912M. doi:10.1103/PhysRevLett.44.912.

[37] M. Gell-Mann, P. Ramond and R. Slansky (1979). F. van Nieuwenhuizen and D. Z. Freedman, ed. *Supergravity*. North Holland. pp. 315–321. ISBN 0-444-85438-X.

[38] Salvio, Strumia (2014-03-17). "Agravity". *JHEP* **1406** (2014) 080. arXiv:1403.4226. Bibcode:2014JHEP...06..080S. doi:10.1007/JHEP06(2014)080.

6.24.11 Further reading

- R. Oerter (2006). *The Theory of Almost Everything: The Standard Model, the Unsung Triumph of Modern Physics*. Plume.

- B.A. Schumm (2004). *Deep Down Things: The Breathtaking Beauty of Particle Physics*. Johns Hopkins University Press. ISBN 0-8018-7971-X.

- "The Standard Model of Particle Physics Interactive Graphic".

Introductory textbooks

- I. Aitchison, A. Hey (2003). *Gauge Theories in Particle Physics: A Practical Introduction*. Institute of Physics. ISBN 978-0-585-44550-2.

- W. Greiner, B. Müller (2000). *Gauge Theory of Weak Interactions*. Springer. ISBN 3-540-67672-4.

- G.D. Coughlan, J.E. Dodd, B.M. Gripaios (2006). *The Ideas of Particle Physics: An Introduction for Scientists*. Cambridge University Press.

- D.J. Griffiths (1987). *Introduction to Elementary Particles*. John Wiley & Sons. ISBN 0-471-60386-4.

- G.L. Kane (1987). *Modern Elementary Particle Physics*. Perseus Books. ISBN 0-201-11749-5.

Advanced textbooks

- T.P. Cheng, L.F. Li (2006). *Gauge theory of elementary particle physics*. Oxford University Press. ISBN 0-19-851961-3. Highlights the gauge theory aspects of the Standard Model.

- J.F. Donoghue, E. Golowich, B.R. Holstein (1994). *Dynamics of the Standard Model*. Cambridge University Press. ISBN 978-0-521-47652-2. Highlights dynamical and phenomenological aspects of the Standard Model.

- L. O'Raifeartaigh (1988). *Group structure of gauge theories*. Cambridge University Press. ISBN 0-521-34785-8.

- Nagashima Y. Elementary Particle Physics: Foundations of the Standard Model, Volume 2. (Wiley 2013) 920 раnуы

- Schwartz, M.D. Quantum Field Theory and the Standard Model (Cambridge University Press 2013) 952 pages

- Langacker P. The standard model and beyond. (CRC Press, 2010) 670 pages Highlights group-theoretical aspects of the Standard Model.

Journal articles

- E.S. Abers, B.W. Lee (1973). "Gauge theories". *Physics Reports* **9**: 1–141. Bibcode:1973PhR.....9....1A. doi:10.1016/0370-1573(73)90027-6.

- M. Baak et al. (2012). "The Electroweak Fit of the Standard Model after the Discovery of a New Boson at the LHC". *The European Physical Journal C* **72** (11). arXiv:1209.2716. Bibcode:2012EPJC...72.2205B. doi:10.1140/epjc/s10052-012-2205-9.

- Y. Hayato et al. (1999). "Search for Proton Decay through $p \to \nu K^{*+}$ in a Large Water Cherenkov Detector". *Physical Review Letters* **83** (8): 1529. arXiv:hep-ex/9904020. Bibcode:1999PhRvL..83.1529H. doi:10.1103/PhysRevLett.83.1529.

- S.F. Novaes (2000). "Standard Model: An Introduction". arXiv:hep-ph/0001283 [hep-ph].

- D.P. Roy (1999). "Basic Constituents of Matter and their Interactions — A Progress Report". arXiv:hep-ph/9912523 [hep-ph].

- F. Wilczek (2004). "The Universe Is A Strange Place". *Nuclear Physics B - Proceedings Supplements* **134**: 3. arXiv:astro-ph/0401347. Bibcode:2004NuPhS.134....3W. doi:10.1016/j.nuclphysbps.2004.08.001.

6.24.12 External links

- "The Standard Model explained in Detail by CERN's John Ellis" omega tau podcast.
- "LHC sees hint of lightweight Higgs boson" "New Scientist".
- "Standard Model may be found incomplete," *New Scientist*.
- "Observation of the Top Quark" at Fermilab.
- "The Standard Model Lagrangian." After electroweak symmetry breaking, with no explicit Higgs boson.
- "Standard Model Lagrangian" with explicit Higgs terms. PDF, PostScript, and LaTeX versions.
- "The particle adventure." Web tutorial.

- Nobes, Matthew (2002) "Introduction to the Standard Model of Particle Physics" on Kuro5hin: Part 1, Part 2, Part 3a, Part 3b.

- "The Standard Model" The Standard Model on the CERN web site explains how the basic building blocks of matter interact, governed by four fundamental forces.

6.25 Strong interaction

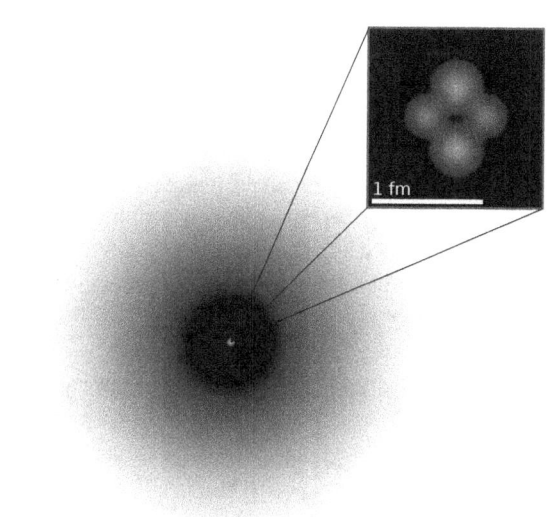

The nucleus of a helium atom. The two protons have the same charge, but still stay together due to the residual nuclear force

In particle physics, the **strong interaction** is the mechanism responsible for the strong nuclear force (also called the **strong force**, **nuclear strong force** or **colour force**), one of the four fundamental interactions of nature, the others being electromagnetism, the weak interaction and gravitation. Effective only at a distance of a femtometer, it is approximately 100 times stronger than electromagnetism, a million times stronger than the weak force interaction and 10^{38} times stronger than gravitation at that range.[*][1] It ensures the stability of ordinary matter, as it confines the quark elementary particles into hadron particles, such as the proton and neutron, the largest components of the mass of ordinary matter. Furthermore, most of the mass-energy of a common proton or neutron is in the form of the strong force field energy; the individual quarks provide only about 1% of the mass-energy of a proton.

The strong interaction is observable in two areas: on a larger scale (about 1 to 3 femtometers (fm)), it is the force that binds protons and neutrons (nucleons) together to form the nucleus of an atom. On the smaller scale (less than about 0.8 fm, the radius of a nucleon), it is the force (carried by gluons) that holds quarks together to form protons, neutrons, and other hadron particles. The strong force inherently has so high a strength that the energy of an object bound by the strong force (a hadron) is high enough to produce new massive particles. Thus, if hadrons are struck by high-energy particles, they give rise to new hadrons instead of emitting freely moving radiation (gluons). This property of the strong force is called colour confinement, and it prevents the free "emission" of the strong force: instead, in practice, jets of massive particles are observed.

In the context of binding protons and neutrons together to form atomic nuclei, the strong interaction is called the nuclear force (or *residual strong force*). In this case, it is the residuum of the strong interaction between the quarks that make up the protons and neutrons. As such, the residual strong interaction obeys a quite different distance-dependent behavior between nucleons, from when it is acting to bind quarks within nucleons. The binding energy that is partly released on the breakup of a nucleus is related to the residual strong force and is harnessed in nuclear power and fission-type nuclear weapons.*[2]*[3]

The strong interaction is thought to be mediated by massless particles called gluons, that are exchanged between quarks, antiquarks, and other gluons. Gluons, in turn, are thought to interact with quarks and gluons as all carry a type of charge called colour charge. Colour charge is analogous to electromagnetic charge, but it comes in three types rather than one (+/- red, +/- green, +/- blue) that results in a different type of force, with different rules of behavior. These rules are detailed in the theory of quantum chromodynamics (QCD), which is the theory of quark-gluon interactions.

Just after the Big Bang, and during the electroweak epoch, the electroweak force separated from the strong force. Although it is expected that a Grand Unified Theory exists to describe this, no such theory has been successfully formulated, and the unification remains an unsolved problem in physics.

6.25.1 History

Before the 1970s, physicists were uncertain about the binding mechanism of the atomic nucleus. It was known that the nucleus was composed of protons and neutrons and that protons possessed positive electric charge, while neutrons were electrically neutral. However, these facts seemed to contradict one another. By physical understanding at that time, positive charges would repel one another and the nucleus should therefore fly apart. However, this was never observed. New physics was needed to explain this phenomenon.

A stronger attractive force was postulated to explain how the atomic nucleus was bound together despite the protons' mutual electromagnetic repulsion. This hypothesized force was called the *strong force*, which was believed to be a fundamental force that acted on the protons and neutrons that make up the nucleus.

It was later discovered that protons and neutrons were not fundamental particles, but were made up of constituent particles called quarks. The strong attraction between nucleons was the side-effect of a more fundamental force that bound the quarks together in the protons and neutrons. The theory of quantum chromodynamics explains that quarks carry what is called a colour charge, although it has no relation to visible colour.*[4] Quarks with unlike colour charge attract one another as a result of the **strong interaction**, which is mediated by particles called gluons.

6.25.2 Details

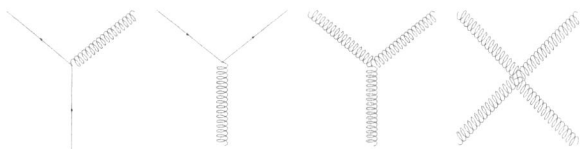

The fundamental couplings of the strong interaction, from left to right: gluon radiation, gluon splitting and gluon self-coupling.

The word *strong* is used since the strong interaction is the "strongest" of the four fundamental forces; its strength is around 10^2 times that of the electromagnetic force, some 10^6 times as great as that of the weak force, and about 10^{39} times that of gravitation, at a distance of a femtometer or less.

Behaviour of the strong force

The contemporary understanding of strong force is described by quantum chromodynamics (QCD), a part of the standard model of particle physics. Mathematically, QCD is a non-Abelian gauge theory based on a local (gauge) symmetry group called SU(3).

Quarks and gluons are the only fundamental particles that carry non-vanishing colour charge, and hence participate in strong interactions. The strong force itself acts directly only on elementary quark and gluon particles.

All quarks and gluons in QCD interact with each other through the strong force. The strength of interaction is parametrized by the strong coupling constant. This strength is modified by the gauge colour charge of the particle, a group theoretical property.

6.25. STRONG INTERACTION

The strong force acts between quarks. Unlike all other forces (electromagnetic, weak, and gravitational), the strong force does not diminish in strength with increasing distance. After a limiting distance (about the size of a hadron) has been reached, it remains at a strength of about 10,000 newtons, no matter how much farther the distance between the quarks.*[5] In QCD, this phenomenon is called colour confinement; it implies that only hadrons, not individual free quarks, can be observed. The explanation is that the amount of work done against a force of 10,000 newtons (about the weight of a one-metric ton mass on the surface of the Earth) is enough to create particle-antiparticle pairs within a very short distance of an interaction. In simple terms, the very energy applied to pull two quarks apart will create a pair of new quarks that will pair up with the original ones. The failure of all experiments that have searched for free quarks is considered to be evidence for this phenomenon.

The elementary quark and gluon particles affected are unobservable directly, but they instead emerge as jets of newly created hadrons, whenever energy is deposited into a quark-quark bond, as when a quark in a proton is struck by a very fast quark (in an impacting proton) during a particle accelerator experiment. However, quark–gluon plasmas have been observed.

Every quark in the universe does not attract every other quark in the above distance independent manner, since colour-confinement implies that the strong force acts without distance-diminishment only between pairs of single quarks, and that in collections of bound quarks (i.e., hadrons), the net colour-charge of the quarks cancels out, as seen from far away. Collections of quarks (hadrons) therefore appear (nearly) without colour-charge, and the strong force is therefore nearly absent between these hadrons (i.e., between baryons or mesons). However, the cancellation is not quite perfect. A small residual force remains (described below) known as the **residual strong force**. This residual force *does* diminish rapidly with distance, and is thus very short-range (effectively a few femtometers). It manifests as a force between the "colourless" hadrons, and is therefore sometimes known as the **strong nuclear force** or simply nuclear force.

Residual strong force

The residual effect of the strong force is called the nuclear force. The nuclear force acts between hadrons, such as mesons or the nucleons in atomic nuclei. This "residual strong force", acting indirectly, transmits gluons that form part of the virtual pi and rho mesons, which, in turn, transmit the nuclear force between nucleons.

The residual strong force is thus a minor residuum of the

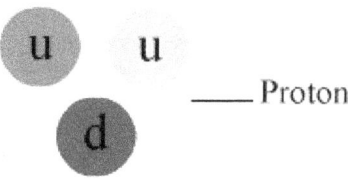

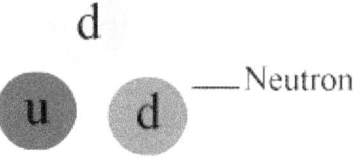

An animation of the nuclear force (or residual strong force) interaction between a proton and a neutron. The small coloured double circles are gluons, which can be seen binding the proton and neutron together. These gluons also hold the quark-antiquark combination called the pion together, and thus help transmit a residual part of the strong force even between colourless hadrons. Anticolours are shown as per this diagram. For a larger version, click here

strong force that binds quarks together into protons and neutrons. This same force is much weaker *between* neutrons and protons, because it is mostly neutralized *within* them, in the same way that electromagnetic forces between neutral atoms (van der Waals forces) are much weaker than the electromagnetic forces that hold the atoms internally together.*[6]

Unlike the strong force itself, the nuclear force, or residual strong force, *does* diminish in strength, and in fact diminishes rapidly with distance. The decrease is approximately as a negative exponential power of distance, though there is no simple expression known for this; see Yukawa potential. This fact, together with the less-rapid decrease of the disruptive electromagnetic force between protons with distance, causes the instability of larger atomic nuclei, such as all those with atomic numbers larger than 82 (the element lead).

6.25.3 See also

- Nuclear binding energy
- Colour charge

- Coupling constant
- Nuclear physics
- QCD matter
- Quantum field theory and Gauge theory
- Standard model of particle physics and Standard Model (mathematical formulation)
- Weak interaction, electromagnetism and gravity
- Intermolecular force
- Vortex
- Yukawa interaction

6.25.4 References

[1] Relative strength of interaction varies with distance. See for instance Matt Strassler's essay, "The strength of the known forces".

[2] on Binding energy: see Binding Energy, Mass Defect, Furry Elephant physics educational site, retr 2012 7 1

[3] on Binding energy: see Chapter 4 NUCLEAR PROCESSES, THE STRONG FORCE, M. Ragheb 1/27/2012, University of Illinois

[4] Feynman, R. P. (1985). *QED: The Strange Theory of Light and Matter*. Princeton University Press. p. 136. ISBN 0-691-08388-6. The idiot physicists, unable to come up with any wonderful Greek words anymore, call this type of polarization by the unfortunate name of 'colour,' which has nothing to do with colour in the normal sense.

[5] Fritzsch, op. cite, p. 164. The author states that the force between differently coloured quarks remains constant at any distance after they travel only a tiny distance from each other, and is equal to that need to raise one ton, which is 1000 kg x 9.8 m/s^2 = ~10,000 N.

[6] Fritzsch, H. (1983). *Quarks: The Stuff of Matter*. Basic Books. pp. 167–168. ISBN 978-0-465-06781-7.

6.25.5 Further reading

- Christman, J. R. (2001). "MISN-0-280: *The Strong Interaction*" (PDF). *Project PHYSNET*.
- Griffiths, David (1987). *Introduction to Elementary Particles*. John Wiley & Sons. ISBN 0-471-60386-4.
- Halzen, F.; Martin, A. D. (1984). *Quarks and Leptons: An Introductory Course in Modern Particle Physics*. John Wiley & Sons. ISBN 0-471-88741-2.
- Kane, G. L. (1987). *Modern Elementary Particle Physics*. Perseus Books. ISBN 0-201-11749-5.
- Morris, R. (2003). *The Last Sorcerers: The Path from Alchemy to the Periodic Table*. Joseph Henry Press. ISBN 0-309-50593-3.

6.25.6 External links

- Strong force at *Encyclopædia Britannica*

6.26 Weak interaction

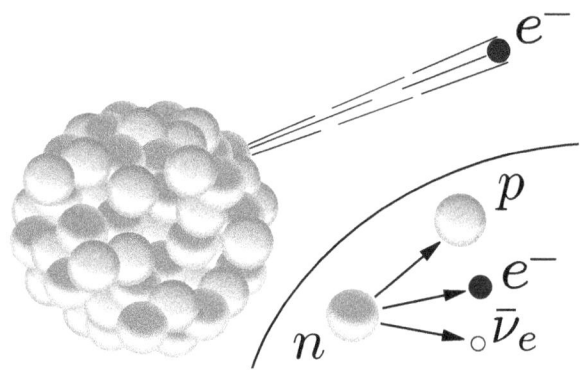

The radioactive beta decay is possible due to the weak interaction, which transforms a neutron into: a proton, an electron, and an electron antineutrino.

In particle physics, the **weak interaction** is the mechanism responsible for the **weak force** or **weak nuclear force**, one of the four known fundamental interactions of nature, alongside the strong interaction, electromagnetism, and gravitation. The weak interaction is responsible for the radioactive decay of subatomic particles, and it plays an essential role in nuclear fission. The theory of the weak interaction is sometimes called **quantum flavordynamics** (**QFD**), in analogy with the terms QCD and QED, but the term is rarely used because the weak force is best understood in terms of electro-weak theory (EWT).*[1]

In the Standard Model of particle physics, the weak interaction is caused by the emission or absorption of W and Z bosons. All known fermions interact through the weak interaction. Fermions are particles that have half-integer spin (one of the fundamental properties of particles). A fermion can be an elementary particle, such as the electron, or it can be a composite particle, such as the proton. The masses of W^*+, W^*-, and Z bosons are each far greater than that of protons or neutrons, consistent with the short range of the weak force. The force is termed *weak* because

6.26. WEAK INTERACTION

its field strength over a given distance is typically several orders of magnitude less than that of the strong nuclear force and electromagnetic force.

During the quark epoch, the electroweak force split into the electromagnetic and weak forces. Important examples of weak interaction include beta decay, and the production, from hydrogen, of deuterium needed to power the sun's thermonuclear process. Most fermions will decay by a weak interaction over time. Such decay also makes radiocarbon dating possible, as carbon-14 decays through the weak interaction to nitrogen-14. It can also create radioluminescence, commonly used in tritium illumination, and in the related field of betavoltaics.*[2]

Quarks, which make up composite particles like neutrons and protons, come in six "flavours" – up, down, strange, charm, top and bottom – which give those composite particles their properties. The weak interaction is unique in that it allows for quarks to swap their flavour for another. For example, during beta minus decay, a down quark decays into an up quark, converting a neutron to a proton. Also the weak interaction is the only fundamental interaction that breaks parity-symmetry, and similarly, the only one to break CP-symmetry.

6.26.1 History

In 1933, Enrico Fermi proposed the first theory of the weak interaction, known as Fermi's interaction. He suggested that beta decay could be explained by a four-fermion interaction, involving a contact force with no range.*[3]*[4]

However, it is better described as a non-contact force field having a finite range, albeit very short. In 1968, Sheldon Glashow, Abdus Salam and Steven Weinberg unified the electromagnetic force and the weak interaction by showing them to be two aspects of a single force, now termed the electro-weak force.

The existence of the W and Z bosons was not directly confirmed until 1983.

6.26.2 Properties

The weak interaction is unique in a number of respects:

1. It is the only interaction capable of changing the flavor of quarks (i.e., of changing one type of quark into another).

2. It is the only interaction that violates **P** or parity-symmetry. It is also the only one that violates **CP** symmetry.

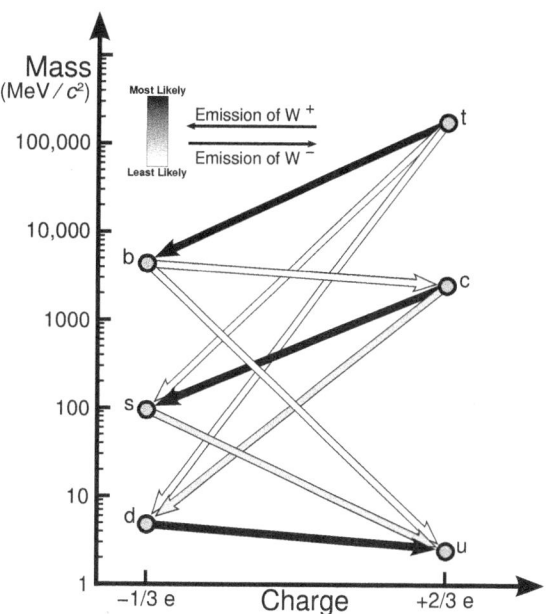

A diagram depicting the various decay routes due to the weak interaction and some indication of their likelihood. The intensity of the lines are given by the CKM parameters.

3. It is propagated by carrier particles (known as gauge bosons) that have significant masses, an unusual feature which is explained in the Standard Model by the Higgs mechanism.

Due to their large mass (approximately 90 GeV/c^2*[5]) these carrier particles, termed the W and Z bosons, are short-lived: they have a lifetime of under 1×10*−24 seconds.*[6] The weak interaction has a coupling constant (an indicator of interaction strength) of between 10*−7 and 10*−6, compared to the strong interaction's coupling constant of about 1 and the electromagnetic coupling constant of about 10*−2;*[7] consequently the weak interaction is weak in terms of strength.*[8] The weak interaction has a very short range (around 10*−17–10*−16 m*[8]).*[7] At distances around 10*−18 meters, the weak interaction has a strength of a similar magnitude to the electromagnetic force, but this starts to decrease exponentially with increasing distance. At distances of around 3×10*−17 m, the weak interaction is 10,000 times weaker than the electromagnetic.*[9]

The weak interaction affects all the fermions of the Standard Model, as well as the Higgs boson; neutrinos interact through gravity and the weak interaction only, and neutrinos were the original reason for the name *weak force*.*[8] The weak interaction does not produce bound states (nor does it involve binding energy) – something that gravity does on an astronomical scale, that the electromagnetic force does at the atomic level, and that the strong nuclear force does inside nuclei.*[10]

Its most noticeable effect is due to its first unique feature: flavor changing. A neutron, for example, is heavier than a proton (its sister nucleon), but it cannot decay into a proton without changing the flavor (type) of one of its two *down* quarks to *up*. Neither the strong interaction nor electromagnetism permit flavour changing, so this must proceed by **weak decay**; without weak decay, quark properties such as strangeness and charm (associated with the quarks of the same name) would also be conserved across all interactions. All mesons are unstable because of weak decay.*[11] In the process known as beta decay, a *down* quark in the neutron can change into an *up* quark by emitting a virtual W− boson which is then converted into an electron and an electron antineutrino.*[12] Another example is the electron capture, a common variant of radioactive decay, where a proton (up quark) and an electron within an atom interact, and are changed to a neutron (down quark) and an electron neutrino.

Due to the large mass of a boson, weak decay is much more unlikely than strong or electromagnetic decay, and hence occurs less rapidly. For example, a neutral pion (which decays electromagnetically) has a life of about 10^*-16 seconds, while a charged pion (which decays through the weak interaction) lives about 10^*-8 seconds, a hundred million times longer.*[13] In contrast, a free neutron (which also decays through the weak interaction) lives about 15 minutes.*[12]

Weak isospin and weak hypercharge

Main article: Weak isospin

All particles have a property called weak isospin (T_3), which serves as a quantum number and governs how that particle interacts in the weak interaction. Weak isospin therefore plays the same role in the weak interaction as electric charge does in electromagnetism, and color charge in the strong interaction. All fermions have a weak isospin value of either $+\frac{1}{2}$ or $-\frac{1}{2}$. For example, the up quark has a T_3 of $+\frac{1}{2}$ and the down quark $-\frac{1}{2}$. A quark never decays through the weak interaction into a quark of the same T_3: quarks with a T_3 of $+\frac{1}{2}$ decay into quarks with a T_3 of $-\frac{1}{2}$ and vice versa.

In any given interaction, weak isospin is conserved: the sum of the weak isospin numbers of the particles entering the interaction equals the sum of the weak isospin numbers of the particles exiting that interaction. For example, a (left-handed) π+, with a weak isospin of 1 normally decays into a ν
μ (+1/2) and a μ+ (as a right-handed antiparticle, +1/2).*[13]

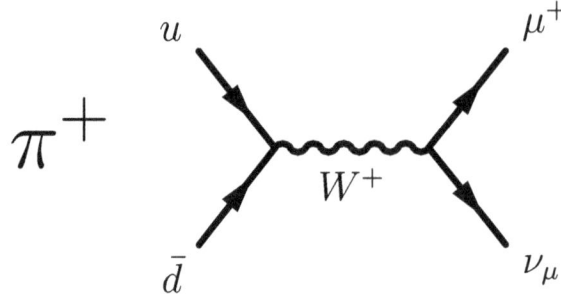

π+ decay through the weak interaction

Following the development of the electroweak theory, another property, weak hypercharge, was developed. It is dependent on a particle's electrical charge and weak isospin, and is defined as:

$$Y_W = 2(Q - T_3)$$

where Y_W is the weak hypercharge of a given type of particle, Q is its electrical charge (in elementary charge units) and T_3 is its weak isospin. Whereas some particles have a weak isospin of zero, all particles, except gluons, have non-zero weak hypercharge. Weak hypercharge is the generator of the U(1) component of the electroweak gauge group.

6.26.3 Interaction types

There are two types of weak interaction (called *vertices*). The first type is called the "charged-current interaction" because it is mediated by particles that carry an electric charge (the W+ or W− bosons), and is responsible for the beta decay phenomenon. The second type is called the "neutral-current interaction" because it is mediated by a neutral particle, the Z boson.

Charged-current interaction

In one type of charged current interaction, a charged lepton (such as an electron or a muon, having a charge of −1) can absorb a W+ boson (a particle with a charge of +1) and be thereby converted into a corresponding neutrino (with a charge of 0), where the type ("family") of neutrino (electron, muon or tau) is the same as the type of lepton in the interaction, for example:

$$\mu^- + W^+ \to \nu_\mu$$

Similarly, a down-type quark (*d* with a charge of $-\frac{1}{3}$) can be converted into an up-type quark (*u*, with a charge of

6.26. WEAK INTERACTION

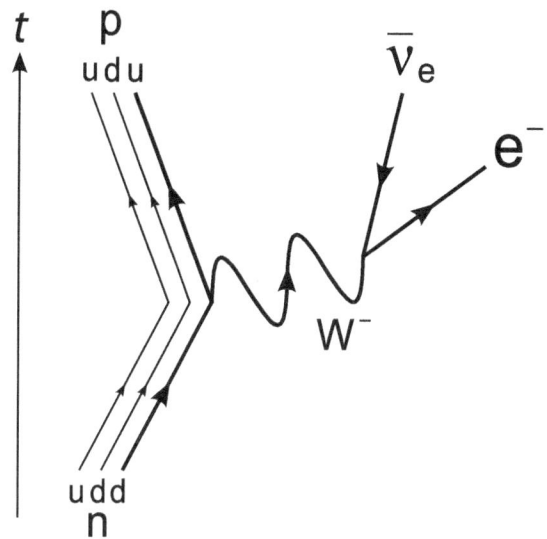

The Feynman diagram for beta-minus decay of a neutron into a proton, electron and electron anti-neutrino, via an intermediate heavy W− boson

$+^2/_3$), by emitting a W− boson or by absorbing a W+ boson. More precisely, the down-type quark becomes a quantum superposition of up-type quarks: that is to say, it has a possibility of becoming any one of the three up-type quarks, with the probabilities given in the CKM matrix tables. Conversely, an up-type quark can emit a W+ boson – or absorb a W− boson – and thereby be converted into a down-type quark, for example:

$$d \to u + W^-$$
$$d + W^+ \to u$$
$$c \to s + W^+$$
$$c + W^- \to s$$

The W boson is unstable so will rapidly decay, with a very short lifetime. For example:

$$W^- \to e^- + \bar{\nu}_e$$
$$W^+ \to e^+ + \nu_e$$

Decay of the W boson to other products can happen, with varying probabilities.[15]

In the so-called beta decay of a neutron (see picture, above), a down quark within the neutron emits a virtual W− boson and is thereby converted into an up quark, converting the neutron into a proton. Because of the energy involved in the process (i.e., the mass difference between the down quark and the up quark), the W− boson can only be converted into an electron and an electron-antineutrino.[16] At the quark level, the process can be represented as:

$$d \to u + e^- + \bar{\nu}_e$$

Neutral-current interaction

In neutral current interactions, a quark or a lepton (e.g., an electron or a muon) emits or absorbs a neutral Z boson. For example:

$$e^- \to e^- + Z^0$$

Like the W boson, the Z boson also decays rapidly,[15] for example:

$$Z^0 \to b + \bar{b}$$

6.26.4 Electroweak theory

Main article: Electroweak interaction

The Standard Model of particle physics describes the electromagnetic interaction and the weak interaction as two different aspects of a single electroweak interaction, the theory of which was developed around 1968 by Sheldon Glashow, Abdus Salam and Steven Weinberg. They were awarded the 1979 Nobel Prize in Physics for their work.[17] The Higgs mechanism provides an explanation for the presence of three massive gauge bosons (the three carriers of the weak interaction) and the massless photon of the electromagnetic interaction.[18]

According to the electroweak theory, at very high energies, the universe has four massless gauge boson fields similar to the photon and a complex scalar Higgs field doublet. However, at low energies, gauge symmetry is spontaneously broken down to the U(1) symmetry of electromagnetism (one of the Higgs fields acquires a vacuum expectation value). This symmetry breaking would produce three massless bosons, but they become integrated by three photon-like fields (through the Higgs mechanism) giving them mass. These three fields become the W+, W− and Z bosons of the weak interaction, while the fourth gauge field, which remains massless, is the photon of electromagnetism.[18]

This theory has made a number of predictions, including a prediction of the masses of the Z and W bosons before their discovery. On 4 July 2012, the CMS and the ATLAS experimental teams at the Large Hadron Collider independently announced that they had confirmed the formal discovery of a previously unknown boson of mass between 125–127 GeV/c^2, whose behaviour so far was "consistent with" a Higgs boson, while adding a cautious note that further data and analysis were needed before positively identifying the new boson as being a Higgs boson of some type. By 14 March 2013, the Higgs boson was tentatively confirmed to exist.*[19]

6.26.5 Violation of symmetry

Left- and right-handed particles: p is the particle's momentum and S is its spin. Note the lack of reflective symmetry between the states.

The laws of nature were long thought to remain the same under mirror reflection, the reversal of one spatial axis. The results of an experiment viewed via a mirror were expected to be identical to the results of a mirror-reflected copy of the experimental apparatus. This so-called law of parity conservation was known to be respected by classical gravitation, electromagnetism and the strong interaction; it was assumed to be a universal law.*[20] However, in the mid-1950s Chen Ning Yang and Tsung-Dao Lee suggested that the weak interaction might violate this law. Chien Shiung Wu and collaborators in 1957 discovered that the weak interaction violates parity, earning Yang and Lee the 1957 Nobel Prize in Physics.*[21]

Although the weak interaction used to be described by Fermi's theory, the discovery of parity violation and renormalization theory suggested that a new approach was needed. In 1957, Robert Marshak and George Sudarshan and, somewhat later, Richard Feynman and Murray Gell-Mann proposed a **V−A** (vector minus axial vector or left-handed) Lagrangian for weak interactions. In this theory, the weak interaction acts only on left-handed particles (and right-handed antiparticles). Since the mirror reflection of a left-handed particle is right-handed, this explains the maximal violation of parity. Interestingly, the **V−A** theory was developed before the discovery of the Z boson, so it did not include the right-handed fields that enter in the neutral current interaction.

However, this theory allowed a compound symmetry **CP** to be conserved. **CP** combines parity **P** (switching left to right) with charge conjugation **C** (switching particles with antiparticles). Physicists were again surprised when in 1964, James Cronin and Val Fitch provided clear evidence in kaon decays that CP symmetry could be broken too, winning them the 1980 Nobel Prize in Physics.*[22] In 1973, Makoto Kobayashi and Toshihide Maskawa showed that CP violation in the weak interaction required more than two generations of particles,*[23] effectively predicting the existence of a then unknown third generation. This discovery earned them half of the 2008 Nobel Prize in Physics.*[24] Unlike parity violation, CP violation occurs in only a small number of instances, but remains widely held as an answer to the difference between the amount of matter and antimatter in the universe; it thus forms one of Andrei Sakharov's three conditions for baryogenesis.*[25]

6.26.6 See also

- Weakless Universe – the postulate that weak interactions are not anthropically necessary
- Gravity
- Nuclear force
- Electromagnetism

6.26.7 References

Citations

[1] Griffiths, David (2009). *Introduction to Elementary Particles.* pp. 59–60. ISBN 978-3-527-40601-2.

[2] "The Nobel Prize in Physics 1979: Press Release". *NobelPrize.org*. Nobel Media. Retrieved 22 March 2011.

[3] Fermi, Enrico (1934). "Versuch einer Theorie der β-Strahlen. I". *Zeitschrift für Physik A* **88** (3–4): 161–177. Bibcode:1934ZPhy...88..161F. doi:10.1007/BF01351864.

[4] Wilson, Fred L. (December 1968). "Fermi's Theory of Beta Decay". *American Journal of Physics* **36** (12): 1150–1160. Bibcode:1968AmJPh..36.1150W. doi:10.1119/1.1974382.

[5] W.-M. Yao *et al.* (Particle Data Group) (2006). "Review of Particle Physics: Quarks" (PDF). *Journal of Physics G* **33**: 1–1232. arXiv:astro-ph/0601168. Bibcode:2006JPhG...33....1Y. doi:10.1088/0954-3899/33/1/001.

[6] Peter Watkins (1986). *Story of the W and Z.* Cambridge: Cambridge University Press. p. 70. ISBN 978-0-521-31875-4.

[7] "Coupling Constants for the Fundamental Forces". *HyperPhysics*. Georgia State University. Retrieved 2 March 2011.

[8] J. Christman (2001). "The Weak Interaction" (PDF). *Physnet*. Michigan State University.

[9] "Electroweak". *The Particle Adventure*. Particle Data Group. Retrieved 3 March 2011.

[10] Walter Greiner; Berndt Müller (2009). *Gauge Theory of Weak Interactions*. Springer. p. 2. ISBN 978-3-540-87842-1.

[11] Cottingham & Greenwood (1986, 2001), p.29

[12] Cottingham & Greenwood (1986, 2001), p.28

[13] Cottingham & Greenwood (1986, 2001), p.30

[14] Baez, John C.; Huerta, John (2009). "The Algebra of Grand Unified Theories". *Bull.Am.Math.Soc.* **0904**: 483–552. arXiv:0904.1556. Bibcode:2009arXiv0904.1556B. doi:10.1090/s0273-0979-10-01294-2. Retrieved 15 October 2013.

[15] K. Nakamura *et al.* (Particle Data Group) (2010). "Gauge and Higgs Bosons" (PDF). *Journal of Physics G* **37**. doi:10.1088/0954-3899/37/7a/075021.

[16] K. Nakamura *et al.* (Particle Data Group) (2010). "n" (PDF). *Journal of Physics G* **37**: 7. doi:10.1088/0954-3899/37/7a/075021.

[17] "The Nobel Prize in Physics 1979". *NobelPrize.org*. Nobel Media. Retrieved 26 February 2011.

[18] C. Amsler *et al.* (Particle Data Group) (2008). "Review of Particle Physics – Higgs Bosons: Theory and Searches" (PDF). *Physics Letters B* **667**: 1–6. Bibcode:2008PhLB..667....1P. doi:10.1016/j.physletb.2008.07.018.

[19] "New results indicate that new particle is a Higgs boson | CERN". Home.web.cern.ch. Retrieved 20 September 2013.

[20] Charles W. Carey (2006). "Lee, Tsung-Dao". *American scientists*. Facts on File Inc. p. 225. ISBN 9781438108070.

[21] "The Nobel Prize in Physics 1957". *NobelPrize.org*. Nobel Media. Retrieved 26 February 2011.

[22] "The Nobel Prize in Physics 1980". *NobelPrize.org*. Nobel Media. Retrieved 26 February 2011.

[23] M. Kobayashi, T. Maskawa (1973). "CP-Violation in the Renormalizable Theory of Weak Interaction". *Progress of Theoretical Physics* **49** (2): 652–657. Bibcode:1973PThPh..49..652K. doi:10.1143/PTP.49.652.

[24] "The Nobel Prize in Physics 1980". *NobelPrize.org*. Nobel Media. Retrieved 17 March 2011.

[25] Paul Langacker (2001) [1989]. "Cp Violation and Cosmology". In Cecilia Jarlskog. *CP violation*. London, River Edge: World Scientific Publishing Co. p. 552. ISBN 9789971505615.

General readers

- R. Oerter (2006). *The Theory of Almost Everything: The Standard Model, the Unsung Triumph of Modern Physics*. Plume. ISBN 978-0-13-236678-6.

- B.A. Schumm (2004). *Deep Down Things: The Breathtaking Beauty of Particle Physics*. Johns Hopkins University Press. ISBN 0-8018-7971-X.

Texts

- D.A. Bromley (2000). *Gauge Theory of Weak Interactions*. Springer. ISBN 3-540-67672-4.

- G.D. Coughlan, J.E. Dodd, B.M. Gripaios (2006). *The Ideas of Particle Physics: An Introduction for Scientists* (3rd ed.). Cambridge University Press. ISBN 978-0-521-67775-2.

- W. N. Cottingham; D. A. Greenwood (2001) [1986]. *An introduction to nuclear physics* (2nd ed.). Cambridge University Press. p. 30. ISBN 978-0-521-65733-4.

- D.J. Griffiths (1987). *Introduction to Elementary Particles*. John Wiley & Sons. ISBN 0-471-60386-4.

- G.L. Kane (1987). *Modern Elementary Particle Physics*. Perseus Books. ISBN 0-201-11749-5.

- D.H. Perkins (2000). *Introduction to High Energy Physics*. Cambridge University Press. ISBN 0-521-62196-8.

6.27 List of particles

This is a list of the different types of particles found or believed to exist in the whole of the universe. For individual lists of the different particles, see the list below.

6.27.1 Elementary particles

Main article: Elementary particle

Elementary particles are particles with no measurable internal structure; that is, they are not composed of other particles. They are the fundamental objects of quantum field theory. Many families and sub-families of elementary particles exist. Elementary particles are classified according to their spin. Fermions have half-integer spin while bosons have integer spin. All the particles of the Standard Model have been experimentally observed, recently including the Higgs boson.*[1]*[2]

Fermions

Main article: Fermion

Fermions are one of the two fundamental classes of particles, the other being bosons. Fermion particles are described by Fermi–Dirac statistics and have quantum numbers described by the Pauli exclusion principle. They include the quarks and leptons, as well as any composite particles consisting of an odd number of these, such as all baryons and many atoms and nuclei.

Fermions have half-integer spin; for all known elementary fermions this is $1/2$. All known fermions, except neutrinos, are also Dirac fermions; that is, each known fermion has its own distinct antiparticle. It is not known whether the neutrino is a Dirac fermion or a Majorana fermion.*[3] Fermions are the basic building blocks of all matter. They are classified according to whether they interact via the color force or not. In the Standard Model, there are 12 types of elementary fermions: six quarks and six leptons.

Quarks Main article: Quark

Quarks are the fundamental constituents of hadrons and interact via the strong interaction. Quarks are the only known carriers of fractional charge, but because they combine in groups of three (baryons) or in groups of two with antiquarks (mesons), only integer charge is observed in nature. Their respective antiparticles are the antiquarks, which are identical except for the fact that they carry the opposite electric charge (for example the up quark carries charge $+2/3$, while the up antiquark carries charge $-2/3$), color charge, and baryon number. There are six flavors of quarks; the three positively charged quarks are called "up-type quarks" and the three negatively charged quarks are called "down-type quarks".

Leptons Main article: Leptons

Leptons do not interact via the strong interaction. Their respective antiparticles are the antileptons which are identical, except for the fact that they carry the opposite electric charge and lepton number. The antiparticle of an electron is an antielectron, which is nearly always called a "positron" for historical reasons. There are six leptons in total; the three charged leptons are called "electron-like leptons", while the neutral leptons are called "neutrinos". Neutrinos are known to oscillate, so that neutrinos of definite flavor do not have definite mass, rather they exist in a superposition of mass eigenstates. The hypothetical heavy right-handed neutrino, called a "sterile neutrino", has been left off the list.

Bosons

Main article: Boson

Bosons are one of the two fundamental classes of particles, the other being fermions. Bosons are characterized by Bose–Einstein statistics and all have integer spins. Bosons may be either elementary, like photons and gluons, or composite, like mesons.

The fundamental forces of nature are mediated by gauge bosons, and mass is believed to be created by the Higgs field. According to the Standard Model the elementary bosons are:

The graviton is added to the list although it is not predicted by the Standard Model, but by other theories in the framework of quantum field theory. Furthermore, gravity is non-renormalizable. There are a total of eight independent gluons. The Higgs boson is postulated by the electroweak theory primarily to explain the origin of particle masses. In a process known as the "Higgs mechanism", the Higgs boson and the other gauge bosons in the Standard Model acquire mass via spontaneous symmetry breaking of the SU(2) gauge symmetry. The Minimal Supersymmetric Standard Model (MSSM) predicts several Higgs bosons. A new particle expected to be the Higgs boson was observed at the CERN/LHC on March 14, 2013, around the energy of 126.5GeV with an accuracy of close to five sigma (99.9999%, which is accepted as definitive). The Higgs mechanism giving mass to other particles has not been observed yet.

Hypothetical particles

Supersymmetric theories predict the existence of more particles, none of which have been confirmed experimentally as of 2014:

Note: just as the photon, Z boson and $W^*\pm$ bosons are superpositions of the B^0, W^0, W^1, and W^2 fields – the photino, zino, and wino*± are superpositions of the $bino^0$, wino⁰, wino¹, and wino² by definition.

No matter if one uses the original gauginos or this superpositions as a basis, the only predicted physical particles are neutralinos and charginos as a superposition of them together with the Higgsinos.

Other theories predict the existence of additional bosons:

Mirror particles are predicted by theories that restore parity symmetry.

"Magnetic monopole" is a generic name for particles with non-zero magnetic charge. They are predicted by some GUTs.

"Tachyon" is a generic name for hypothetical particles that travel faster than the speed of light and have an imaginary rest mass.

Preons were suggested as subparticles of quarks and leptons, but modern collider experiments have all but ruled out their existence.

Kaluza–Klein towers of particles are predicted by some models of extra dimensions. The extra-dimensional momentum is manifested as extra mass in four-dimensional spacetime.

6.27.2 Composite particles

Hadrons

Main article: Hadron

Hadrons are defined as strongly interacting composite particles. Hadrons are either:

- Composite fermions, in which case they are called baryons.
- Composite bosons, in which case they are called mesons.

Quark models, first proposed in 1964 independently by Murray Gell-Mann and George Zweig (who called quarks "aces"), describe the known hadrons as composed of valence quarks and/or antiquarks, tightly bound by the color force, which is mediated by gluons. A "sea" of virtual quark-antiquark pairs is also present in each hadron.

Baryons   See also: List of baryons

Ordinary baryons (composite fermions) contain three valence quarks or three valence antiquarks each.

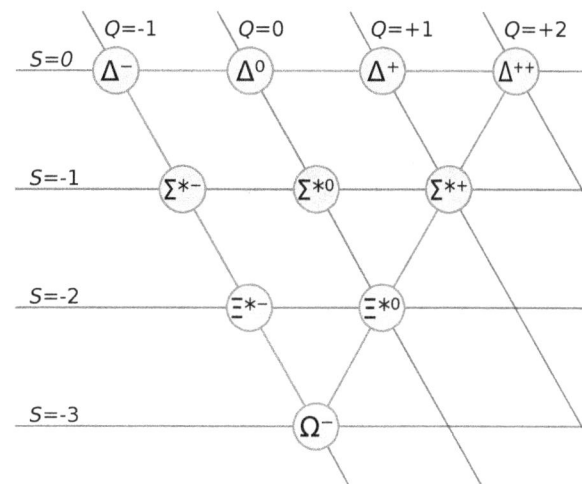

A combination of three u, d or s-quarks with a total spin of $3/2$ form the so-called "baryon decuplet".

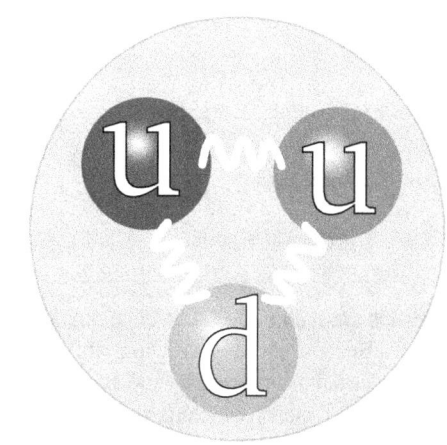

Proton quark structure: 2 up quarks and 1 down quark. The gluon tubes or flux tubes are now known to be Y shaped.

- Nucleons are the fermionic constituents of normal atomic nuclei:
  - Protons, composed of two up and one down quark (uud)
  - Neutrons, composed of two down and one up quark (ddu)
- Hyperons, such as the Λ, Σ, Ξ, and Ω particles, which contain one or more strange quarks, are short-lived and heavier than nucleons. Although not normally present in atomic nuclei, they can appear in short-lived hypernuclei.

- A number of charmed and bottom baryons have also been observed.

Some hints at the existence of exotic baryons have been found recently; however, negative results have also been reported. Their existence is uncertain.

- Pentaquarks consist of four valence quarks and one valence antiquark.

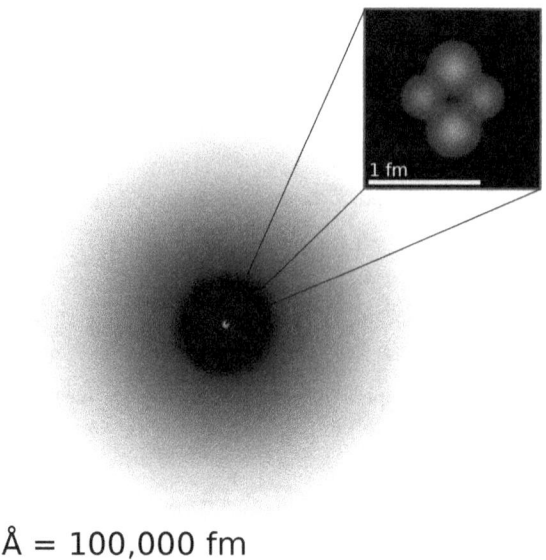

1 Å = 100,000 fm

A semi-accurate depiction of the helium atom. In the nucleus, the protons are in red and neutrons are in purple. In reality, the nucleus is also spherically symmetrical.

Nuclear reactions can change one nuclide into another. See table of nuclides for a complete list of isotopes.

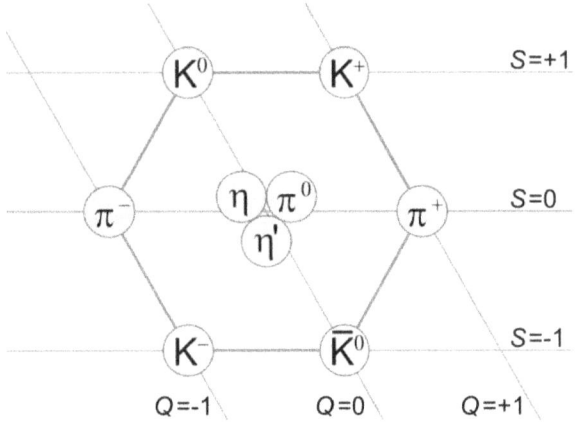

Mesons of spin 0 form a nonet

Mesons   See also: List of mesons

Ordinary mesons are made up of a valence quark and a valence antiquark. Because mesons have spin of 0 or 1 and are not themselves elementary particles, they are "composite" bosons. Examples of mesons include the pion, kaon, and the J/ψ. In quantum hydrodynamic models, mesons mediate the residual strong force between nucleons.

At one time or another, positive signatures have been reported for all of the following exotic mesons but their existences have yet to be confirmed.

- A tetraquark consists of two valence quarks and two valence antiquarks;
- A glueball is a bound state of gluons with no valence quarks;
- Hybrid mesons consist of one or more valence quark-antiquark pairs and one or more real gluons.

Atomic nuclei

Atomic nuclei consist of protons and neutrons. Each type of nucleus contains a specific number of protons and a specific number of neutrons, and is called a "nuclide" or "isotope".

Atoms

Atoms are the smallest neutral particles into which matter can be divided by chemical reactions. An atom consists of a small, heavy nucleus surrounded by a relatively large, light cloud of electrons. Each type of atom corresponds to a specific chemical element. To date, 118 elements have been discovered, while only the elements 1-112, 114, and 116 have received official names.

The atomic nucleus consists of protons and neutrons. Protons and neutrons are, in turn, made of quarks.

Molecules

Molecules are the smallest particles into which a non-elemental substance can be divided while maintaining the physical properties of the substance. Each type of molecule corresponds to a specific chemical compound. Molecules are a composite of two or more atoms. See list of compounds for a list of molecules.

6.27.3 Condensed matter

The field equations of condensed matter physics are remarkably similar to those of high energy particle physics. As a result, much of the theory of particle physics applies to condensed matter physics as well; in particular, there are a selection of field excitations, called quasi-particles, that can be created and explored. These include:

- Phonons are vibrational modes in a crystal lattice.
- Excitons are bound states of an electron and a hole.
- Plasmons are coherent excitations of a plasma.
- Polaritons are mixtures of photons with other quasi-particles.
- Polarons are moving, charged (quasi-) particles that are surrounded by ions in a material.
- Magnons are coherent excitations of electron spins in a material.

6.27.4 Other

- An anyon is a generalization of fermion and boson in two-dimensional systems like sheets of graphene that obeys braid statistics.
- A plekton is a theoretical kind of particle discussed as a generalization of the braid statistics of the anyon to dimension > 2.
- A WIMP (weakly interacting massive particle) is any one of a number of particles that might explain dark matter (such as the neutralino or the axion).
- The pomeron, used to explain the elastic scattering of hadrons and the location of Regge poles in Regge theory.
- The skyrmion, a topological solution of the pion field, used to model the low-energy properties of the nucleon, such as the axial vector current coupling and the mass.
- A genon is a particle existing in a closed timelike world line where spacetime is curled as in a Frank Tipler or Ronald Mallett time machine.
- A goldstone boson is a massless excitation of a field that has been spontaneously broken. The pions are quasi-goldstone bosons (quasi- because they are not exactly massless) of the broken chiral isospin symmetry of quantum chromodynamics.
- A goldstino is a goldstone fermion produced by the spontaneous breaking of supersymmetry.
- An instanton is a field configuration which is a local minimum of the Euclidean action. Instantons are used in nonperturbative calculations of tunneling rates.

- A dyon is a hypothetical particle with both electric and magnetic charges.
- A geon is an electromagnetic or gravitational wave which is held together in a confined region by the gravitational attraction of its own field energy.
- An inflaton is the generic name for an unidentified scalar particle responsible for the cosmic inflation.
- A spurion is the name given to a "particle" inserted mathematically into an isospin-violating decay in order to analyze it as though it conserved isospin.
- What is called "true muonium", a bound state of a muon and an antimuon, is a theoretical exotic atom which has never been observed.

6.27.5 Classification by speed

- A tardyon or bradyon travels slower than light and has a non-zero rest mass.
- A luxon travels at the speed of light and has no rest mass.
- A tachyon (mentioned above) is a hypothetical particle that travels faster than the speed of light and has an imaginary rest mass.

6.27.6 See also

- Acceleron
- List of baryons
- List of compounds for a list of molecules.
- List of fictional elements, materials, isotopes and atomic particles
- List of mesons
- Periodic table for an overview of atoms.
- Standard Model for the current theory of these particles.
- Table of nuclides
- Timeline of particle discoveries

6.27.7 References

[1] Observation of a new boson at a mass of 125 GeV with the CMS experiment at the LHC (2013). *arXiv:1207.7235*.

[2] Observation of a new particle in the search for the Standard Model Higgs boson with the ATLAS detector at the LHC (2012). *arXiv:1207.7214*.

[3] B. Kayser, *Two Questions About Neutrinos*, arXiv:1012.4469v1 [hep-ph] (2010).

[4] R. Maartens (2004). *Brane-World Gravity* (PDF). *Living Reviews in Relativity* **7**. p. 7. Also available in web format at http://www.livingreviews.org/lrr-2004-7.

- C. Amsler *et al.* (Particle Data Group) (2008). "Review of Particle Physics". *Physics Letters B* **667** (1–5): 1. Bibcode:2008PhLB..667....1P. doi:10.1016/j.physletb.2008.07.018. *(All information on this list, and more, can be found in the extensive, biannually-updated review by the Particle Data Group)*

6.28 Timeline of particle discoveries

This is a **timeline of subatomic particle discoveries**, including all particles thus far discovered which appear to be elementary (that is, indivisible) given the best available evidence. It also includes the discovery of composite particles and antiparticles that were of particular historical importance.

More specifically, the inclusion criteria are:

- Elementary particles from the Standard Model of particle physics that have so far been observed. The Standard Model is the most comprehensive existing model of particle behavior. All Standard Model particles including the Higgs boson have been verified, and all other observed particles are combinations of two or more Standard Model particles.

- Antiparticles which were historically important to the development of particle physics, specifically the positron and antiproton. The discovery of these particles required very different experimental methods from that of their ordinary matter counterparts, and provided evidence that *all* particles had antiparticles —an idea that is fundamental to quantum field theory, the modern mathematical framework for particle physics. In the case of most subsequent particle discoveries, the particle and its anti-particle were discovered essentially simultaneously.

- Composite particles which were the first particle discovered containing a particular elementary constituent, or whose discovery was critical to the understanding of particle physics.

6.28.1 See also

- List of mesons
- List of baryons
- List of particles
- physics

6.28.2 References

[1] Hockberger, P. E. (2002). "A history of ultraviolet photobiology for humans, animals and microorganisms". *Photochem. Photobiol.* **76** (6): 561–579. doi:10.1562/0031-8655(2002)0760561AHOUPF2.0.CO2. ISSN 0031-8655. PMID 12511035.

[2] The ozone layer protects humans from this. Lyman, T. (1914). "Victor Schumann". *Astrophysical Journal* **38**: 1–4. Bibcode:1914ApJ....39....1L. doi:10.1086/142050.

[3] W.C. Röntgen (1895). "Über ein neue Art von Strahlen. Vorläufige Mitteilung". *Sitzber. Physik. Med. Ges.* **137**: 1. as translated in A. Stanton (1896). "On a New Kind of Rays". *Nature* **53** (1369): 274–276. Bibcode:1896Natur..53R.274.. doi:10.1038/053274b0.

[4] J.J. Thomson (1897). "Cathode Rays". *Philosophical Magazine* **44** (269): 293–316. doi:10.1080/14786449708621070.

[5] E. Rutherford (1899). "Uranium Radiation and the Electrical Conduction Produced by it". *Philosophical Magazine* **47** (284): 109–163. doi:10.1080/14786449908621245.

[6] P. Villard (1900). "Sur la Réflexion et la Réfraction des Rayons Cathodiques et des Rayons Déviables du Radium". *Comptes Rendus de l'Académie des Sciences* **130**: 1010.

[7] E. Rutherford (1911). "The Scattering of α- and β- Particles by Matter and the Structure of the Atom". *Philosophical Magazine* **21** (125): 669–688. doi:10.1080/14786440508637080.

[8] E. Rutherford (1919). "Collision of α Particles with Light Atoms IV. An Anomalous Effect in Nitrogen". *Philosophical Magazine* **37**: 581.

[9] J. Chadwick (1932). "Possible Existence of a Neutron". *Nature* **129** (3252): 312. Bibcode:1932Natur.129Q.312C. doi:10.1038/129312a0.

[10] E. Rutherford (1920). "Nuclear Constitution of Atoms". *Proceedings of the Royal Society A* **97** (686): 374–400. Bibcode:1920RSPSA..97..374R. doi:10.1098/rspa.1920.0040.

[11] C.D. Anderson (1932). "The Apparent Existence of Easily Deflectable Positives". *Science* **76** (1967): 238–9. Bibcode:1932Sci....76..238A. doi:10.1126/science.76.1967.238. PMID 17731542.

[12] S.H. Neddermeyer, C.D. Anderson (1937). "Note on the nature of Cosmic-Ray Particles". *Physical Review* **51** (10): 884–886. Bibcode:1937PhRv...51..884N. doi:10.1103/PhysRev.51.884.

[13] M. Conversi, E. Pancini, O. Piccioni (1947). "On the Disintegration of Negative Muons". *Physical Review* **71** (3): 209–210. Bibcode:1947PhRv...71..209C. doi:10.1103/PhysRev.71.209.

[14] C.D. Anderson (1935). "On the Interaction of Elementary Particles". *Proceedings of the Physico-Mathematical Society of Japan* **17**: 48.

[15] G.D. Rochester, C.C. Butler (1947). "Evidence for the Existence of New Unstable Elementary Particles". *Nature* **160** (4077): 855–857. Bibcode:1947Natur.160..855R. doi:10.1038/160855a0.

[16] The Strange Quark

[17] O. Chamberlain, E. Segrè, C. Wiegand, T. Ypsilantis (1955). "Observation of Antiprotons". *Physical Review* **100** (3): 947–950. Bibcode:1955PhRv..100..947C. doi:10.1103/PhysRev.100.947.

[18] F. Reines, C.L. Cowan (1956). "The Neutrino". *Nature* **178** (4531): 446–449. Bibcode:1956Natur.178..446R. doi:10.1038/178446a0.

[19] G. Danby et al. (1962). "Observation of High-Energy Neutrino Reactions and the Existence of Two Kinds of Neutrinos". *Physical Review Letters* **9** (1): 36–44. Bibcode:1962PhRvL...9...36D. doi:10.1103/PhysRevLett.9.36.

[20] R. Nave. "The Xi Baryon". Hyperphysics. Retrieved 20 June 2009.

[21] E.D. Bloom et al. (1969). "High-Energy Inelastic e–p Scattering at 6° and 10°". *Physical Review Letters* **23** (16): 930–934. Bibcode:1969PhRvL..23..930B. doi:10.1103/PhysRevLett.23.930.

[22] M. Breidenbach et al. (1969). "Observed Behavior of Highly Inelastic Electron-Proton Scattering". *Physical Review Letters* **23** (16): 935–939. Bibcode:1969PhRvL..23..935B. doi:10.1103/PhysRevLett.23.935.

[23] J.J. Aubert et al. (1974). "Experimental Observation of a Heavy Particle J". *Physical Review Letters* **33** (23): 1404–1406. Bibcode:1974PhRvL..33.1404A. doi:10.1103/PhysRevLett.33.1404.

[24] J.-E. Augustin et al. (1974). "Discovery of a Narrow Resonance in e^*+e^*– Annihilation". *Physical Review Letters* **33** (23): 1406–1408. Bibcode:1974PhRvL..33.1406A. doi:10.1103/PhysRevLett.33.1406.

[25] B.J. Bjørken, S.L. Glashow (1964). "Elementary Particles and SU(4)". *Physics Letters* **11** (3): 255–257. Bibcode:1964PhL....11..255B. doi:10.1016/0031-9163(64)90433-0.

[26] M.L. Perl et al. (1975). "Evidence for Anomalous Lepton Production in e^*+e^*– Annihilation". *Physical Review Letters* **35** (22): 1489–1492. Bibcode:1975PhRvL..35.1489P. doi:10.1103/PhysRevLett.35.1489.

[27] S.W. Herb et al. (1977). "Observation of a Dimuon Resonance at 9.5 GeV in 400-GeV Proton-Nucleus Collisions". *Physical Review Letters* **39** (5): 252–255. Bibcode:1977PhRvL..39..252H. doi:10.1103/PhysRevLett.39.252.

[28] D.P. Barber et al. (1979). "Discovery of Three-Jet Events and a Test of Quantum Chromodynamics at PETRA". *Physical Review Letters* **43** (12): 830–833. Bibcode:1979PhRvL..43..830B. doi:10.1103/PhysRevLett.43.830.

[29] J.J. Aubert et al. (European Muon Collaboration) (1983). "The ratio of the nucleon structure functions F_2^*N for iron and deuterium". *Physics Letters B* **123** (3–4): 275–278. Bibcode:1983PhLB..123..275A. doi:10.1016/0370-2693(83)90437-9.

[30] G. Arnison et al. (UA1 collaboration) (1983). "Experimental observation of lepton pairs of invariant mass around 95 GeV/c^2 at the CERN SPS collider". *Physics Letters B* **126** (5): 398–410. Bibcode:1983PhLB..126..398A. doi:10.1016/0370-2693(83)90188-0.

[31] F. Abe et al. (CDF collaboration) (1995). "Observation of Top quark production in p–p Collisions with the Collider Detector at Fermilab". *Physical Review Letters* **74** (14): 2626–2631. arXiv:hep-ex/9503002. Bibcode:1995PhRvL..74.2626A. doi:10.1103/PhysRevLett.74.2626. PMID 10057978.

[32] S. Arabuchi et al. (D0 collaboration) (1995). "Observation of the Top Quark". *Physical Review Letters* **74** (14): 2632–2637. arXiv:hep-ex/9503003. Bibcode:1995PhRvL..74.2632A. doi:10.1103/PhysRevLett.74.2632. PMID 10057979.

[33] G. Baur et al. (1996). "Production of Antihydrogen". *Physics Letters B* **368** (3): 251–258. Bibcode:1996PhLB..368..251B. doi:10.1016/0370-2693(96)00005-6.

[34] "Physicists Find First Direct Evidence for Tau Neutrino at Fermilab" (Press release). Fermilab. 20 July 2000. Retrieved 20 March 2010.

[35] Boyle, Alan (4 July 2012). "Milestone in Higgs quest: Scientists find new particle". *MSNBC* (MSNBC). Retrieved 5 July 2012.

- V.V. Ezhela et al. (1996). *Particle Physics: One Hundred Years of Discoveries: An Annotated Chronological Bibliography*. Springer–Verlag. ISBN 1-56396-642-5.

Chapter 7

Appendix B – Selected biographies

7.1 James Bjorken

James Daniel "BJ"Bjorken (born 1934) is an American theoretical physicist. He was a Putnam Fellow in 1954, received a BS in physics from MIT in 1956, and obtained his PhD from Stanford University in 1959. He was a visiting scholar at the Institute for Advanced Study in the fall of 1962.*[1] Bjorken is Emeritus Professor at the Stanford Linear Accelerator Center, and was a member of the Theory Department of the Fermi National Accelerator Laboratory (1979–1989). He was awarded the Dirac Medal of the ICTP in 2004 and in 2015 the Wolf Prize in Physics.

Bjorken discovered what is known as *light-cone scaling*, (or "Bjorken scaling") a phenomenon in the deep inelastic scattering of light on strongly interacting particles, known as *hadrons* (such as protons and neutrons). This was critical to the recognition of quarks as actual elementary particles (rather than just convenient theoretical constructs), and led to the theory of strong interactions known as quantum chromodynamics. In Bjorken's picture, the quarks become point-like, observable objects at very short distances (high energies). Bjorken was also among the first to point out to the phenomena of jet quenching in heavy ion collisions in 1982.

Richard Feynman subsequently reformulated this concept into the parton model, used to understand the quark composition of hadrons at high energies.*[2] The predictions of Bjorken scaling were confirmed in the early late 1960s electroproduction experiments at SLAC, in which quarks were seen for the first time. The general idea, with small logarithmic modifications, is explained in quantum chromodynamics by "asymptotic freedom".

Bjorken co-authored, with Sidney Drell, a classic companion volume textbook on relativistic quantum mechanics and quantum fields.

7.1.1 Publications

Books

- J.D. Bjorken, S. Drell (1964). *Relativistic Quantum Mechanics*. McGraw-Hill. ISBN 0-07-005493-2.
- J.D. Bjorken, S. Drell (1965). *Relativistic Quantum Fields*. McGraw-Hill. ISBN 0-07-005494-0.

Papers

- J.D. Bjorken (1969). "Asymptotic Sum Rules at Infinite Momentum". *Physical Review* **179** (5): 1547–1553. Bibcode:1969PhRv..179.1547B. doi:10.1103/PhysRev.179.1547.
- J.D. Bjorken (1982). "Energy Loss of Energetic Partons in Quark-Gluon Plasma: Possible Excitation of High pT Jets in Hadron-Hadron Collisions." . *FERMILAB-Pub-82/59-THY*.

7.1.2 References

[1] Institute for Advanced Study: A Community of Scholars

[2] The Parton Model by P. Hansson, KTH, November 18, 2004 PDF file

- SLAC Bio
- APS bio
- Wolf prize

7.2 Nicola Cabibbo

Nicola Cabibbo (10 April 1935 – 16 August 2010*[1]) was an Italian physicist, best known for his work on the weak interaction. He was also the president of the Italian National Institute of Nuclear Physics from 1983 to 1992, and from 1993 until his death he was the president of the Pontifical Academy of Sciences.*[2] He was born in Rome.

7.2.1 Work

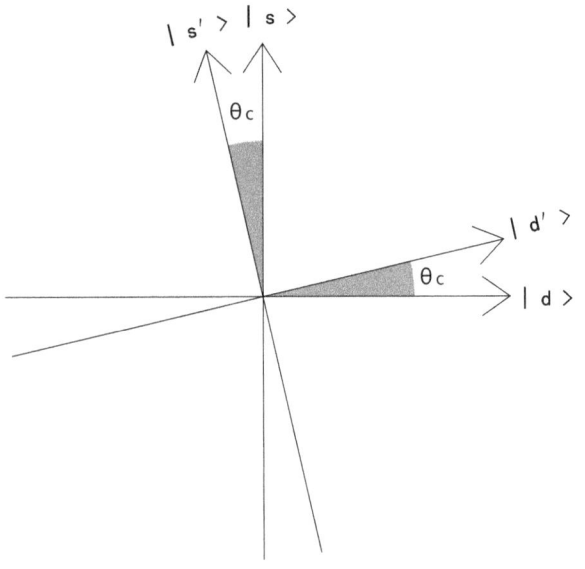

The Cabibbo angle represents the rotation of the mass eigenstate vector space formed by the mass eigenstates $|d\rangle$, $|s\rangle$ into the weak eigenstate vector space formed by the weak eigenstates $|d'\rangle$, $|s'\rangle$. The rotation angle is $\theta_C = 13.04°$.

Cabibbo's major work on the weak interaction originated from a need to explain two observed phenomena:

- The transitions between up and down quarks, between electrons and electron neutrinos, and between muons and muon neutrinos had similar likelihood of occurring (similar amplitudes); and

- The transitions with change in strangeness had amplitudes equal to one fourth of those with no change in strangeness.

Cabibbo addressed these issues, following Gell-Mann and Levy, by postulating weak universality, which involves a similarity in the weak interaction coupling strength between different generations of particles. He addressed the second issue with a mixing angle θ_C (now[*][3] called the Cabibbo angle), between the down and strange quarks. Modern measurements show that $\theta_C = 13.04°$.

Before the discovery of the third generation of quarks, this work was extended by Makoto Kobayashi and Toshihide Maskawa to the Cabibbo–Kobayashi–Maskawa matrix. In 2008, Kobayashi and Maskawa shared one half of the Nobel Prize in Physics for their work. Some physicists had bitter feelings that the Nobel Prize committee failed to reward Gell-Mann, Levy, and possibly Cabibbo for their part.[*][4][*][5] Asked for a reaction on the prize, Cabibbo preferred to give no comment. According to sources close to him, however, he was embittered.[*][6]

Later, Cabibbo researched applications of supercomputers to address problems in modern physics with the experiments APE 100 and APE 1000.

Cabibbo supported attempts to rehabilitate executed Italian philosopher Giordano Bruno, citing the apologies on Galileo Galilei as a possible model to correct the historical wrongs done by the Church.[*][7]

After his death in 2011, the Franklin Institute awarded him with the Benjamin Franklin Medal in Physics.[*][8]

7.2.2 Death

He died from respiratory problems in a Rome hospital on August 16, 2010 at the age of 75.

7.2.3 References

[1] "Morto il fisico Cabibbo Gli fu negato il Nobel". *Corriere della Sera* (in Italian). 16 August 2010. Retrieved 2010-08-16.

[2] Maiani, L. (2010). "Obituary: Nicola Cabibbo (1935–2010)". *Nature* **467** (7313): 284. doi:10.1038/467284a. PMID 20844530.

[3] Introduced by Murray Gell-Mann and Maurice Lévy, in M. Gell-Mann, M. Lévy (1960). "The Axial Vector Current in Beta Decay". *Il Nuovo Cimento* **16** (4): 705–726. doi:10.1007/BF02859738.and referenced by Cabbibo in his paper

[4] ⊠同民 (2013). " 与 2008 年⊠⊠⊠物理⊠失之交臂的物理学家". 物理双月刊 **35**. pp. 354–357.

[5] Valerie Jamieson (7 October 2008). "Physics Nobel snubs key researcher". *New Scientist*. Retrieved 2009-11-06.

[6] "Nobel, l'amarezza dei fisici italiani" (in Italian). Corriere della Sera. 7 October 2008. Retrieved 2009-11-06.

[7] "Un scientifique évoque la réhabilitation d'un théologien brûlé pour hérésie" (in French).

[8] "Benjamin Franklin Medal in Physics". Franklin Institute. 2011. Retrieved December 23, 2011.

7.2.4 External links

- Cabibbo biography from the Istituto e Museo di Storia della Scienza (Italian)

- Parisi, Giorgio (2011). "Nicola Cabibbo". *Physics Today* **64** (2): 59. Bibcode:2011PhT....64b..59P. doi:10.1063/1.3554322.

7.3 Richard Feynman

"Feynman" redirects here. For other uses, see Feynman (disambiguation).

Richard Phillips Feynman, (/ˈfaɪnmən/; May 11, 1918 – February 15, 1988) was an American theoretical physicist known for his work in the path integral formulation of quantum mechanics, the theory of quantum electrodynamics, and the physics of the superfluidity of supercooled liquid helium, as well as in particle physics for which he proposed the parton model. For his contributions to the development of quantum electrodynamics, Feynman, jointly with Julian Schwinger and Sin-Itiro Tomonaga, received the Nobel Prize in Physics in 1965. He developed a widely used pictorial representation scheme for the mathematical expressions governing the behavior of subatomic particles, which later became known as Feynman diagrams. During his lifetime, Feynman became one of the best-known scientists in the world. In a 1999 poll of 130 leading physicists worldwide by the British journal *Physics World* he was ranked as one of the ten greatest physicists of all time.*[4]

He assisted in the development of the atomic bomb during World War II and became known to a wide public in the 1980s as a member of the Rogers Commission, the panel that investigated the Space Shuttle Challenger disaster. In addition to his work in theoretical physics, Feynman has been credited with pioneering the field of quantum computing,*[5]*[6] and introducing the concept of nanotechnology. He held the Richard Chace Tolman professorship in theoretical physics at the California Institute of Technology.

Feynman was a keen popularizer of physics through both books and lectures, including a 1959 talk on top-down nanotechnology called *There's Plenty of Room at the Bottom*, and the three-volume publication of his undergraduate lectures, *The Feynman Lectures on Physics*. Feynman also became known through his semi-autobiographical books *Surely You're Joking, Mr. Feynman!* and *What Do You Care What Other People Think?* and books written about him, such as *Tuva or Bust!* and *Genius: The Life and Science of Richard Feynman* by James Gleick.

7.3.1 Early life

Richard Phillips Feynman was born on May 11, 1918, in Queens, New York City,*[7]*[8] the son of Lucille (née Phillips), a homemaker, and Melville Arthur Feynman, a sales manager.*[9] His family originated from Russia and Poland; both of his parents were Ashkenazi Jews.*[10] They were not religious, and by his youth Feynman described himself as an "avowed atheist".*[11] He also stated

"To select, for approbation the peculiar elements that come from some supposedly Jewish heredity is to open the door to all kinds of nonsense on racial theory," and adding "... at thirteen I was not only converted to other religious views, but I also stopped believing that the Jewish people are in any way 'the chosen people'." *[12] Later in his life, during a visit to the Jewish Theological Seminary, he remarked that he found the Talmud a "wonderful book" and "valuable" .*[13]

Feynman was a late talker, and by his third birthday had yet to utter a single word. He retained a Bronx accent as an adult.*[14]*[15] That accent was thick enough to be perceived as an affectation or exaggeration*[16]*[17] – so much so that his good friends Wolfgang Pauli and Hans Bethe once commented that Feynman spoke like a "bum" .*[16]

The young Feynman was heavily influenced by his father, who encouraged him to ask questions to challenge orthodox thinking, and who was always ready to teach Feynman something new. From his mother he gained the sense of humor that he had throughout his life. As a child, he had a talent for engineering, maintained an experimental laboratory in his home, and delighted in repairing radios. When he was in grade school, he created a home burglar alarm system while his parents were out for the day running errands.*[18]

When Richard was five years old, his mother gave birth to a younger brother, but this brother died at four weeks of age. Four years later, Richard gained a sister, Joan, and the family moved to Far Rockaway, Queens.*[9] Though separated by nine years, Joan and Richard were close, as they both shared a natural curiosity about the world. Their mother thought that women did not have the cranial capacity to comprehend such things. Despite their mother's disapproval of Joan's desire to study astronomy, Richard encouraged his sister to explore the universe. Joan eventually became an astrophysicist specializing in interactions between the Earth and the solar wind.*[19]

7.3.2 Education

Upon starting high school, Feynman was quickly promoted into a higher math class and an unspecified school-administered IQ test estimated his IQ at 123*[20] — high, but "merely respectable" according to biographer James Gleick;*[21] When he turned 15, he taught himself trigonometry, advanced algebra, infinite series, analytic geometry, and both differential and integral calculus.*[22] Before entering college, he was experimenting with and deriving mathematical topics such as the half-derivative using his own notation. In high school he was developing the mathematical intuition behind his Taylor series of mathematical operators.

His habit of direct characterization sometimes rattled more conventional thinkers; for example, one of his questions, when learning feline anatomy, was "Do you have a map of the cat?" (referring to an anatomical chart).*[23]

Feynman attended Far Rockaway High School, a school in Far Rockaway, Queens also attended by fellow Nobel laureates Burton Richter and Baruch Samuel Blumberg.*[24] A member of the Arista Honor Society, in his last year in high school Feynman won the New York University Math Championship; the large difference between his score and those of his closest competitors shocked the judges.

He applied to Columbia University but was not accepted because of their quota for the number of Jews admitted.*[9]*[25] Instead, he attended the Massachusetts Institute of Technology, where he received a bachelor's degree in 1939 and in the same year was named a Putnam Fellow.*[26]

He attained a perfect score on the graduate school entrance exams to Princeton University in mathematics and physics —an unprecedented feat—but did rather poorly on the history and English portions.*[27] Attendees at Feynman's first seminar included Albert Einstein, Wolfgang Pauli, and John von Neumann. He received a PhD from Princeton in 1942; his thesis advisor was John Archibald Wheeler. Feynman's thesis applied the principle of stationary action to problems of quantum mechanics, inspired by a desire to quantize the Wheeler–Feynman absorber theory of electrodynamics, laying the groundwork for the "path integral" approach and Feynman diagrams, and was titled "The Principle of Least Action in Quantum Mechanics".

> This was Richard Feynman nearing the crest of his powers. At twenty-three ... there was no physicist on earth who could match his exuberant command over the native materials of theoretical science. It was not just a facility at mathematics (though it had become clear ... that the mathematical machinery emerging from the Wheeler–Feynman collaboration was beyond Wheeler's own ability). Feynman seemed to possess a frightening ease with the substance behind the equations, like Albert Einstein at the same age, like the Soviet physicist Lev Landau—but few others.
>
> —James Gleick, *Genius: The Life and Science of Richard Feynman*

Feynman (center) with Robert Oppenheimer (right) relaxing at a Los Alamos social function during the Manhattan Project

7.3.3 Manhattan Project

At Princeton, the physicist Robert R. Wilson encouraged Feynman to participate in the Manhattan Project—the wartime U.S. Army project at Los Alamos developing the atomic bomb. Feynman said he was persuaded to join this effort to build it before Nazi Germany developed their own bomb. He was assigned to Hans Bethe's theoretical division and impressed Bethe enough to be made a group leader. He and Bethe developed the Bethe–Feynman formula for calculating the yield of a fission bomb, which built upon previous work by Robert Serber.

He immersed himself in work on the project, and was present at the Trinity bomb test. Feynman claimed to be the only person to see the explosion without the very dark glasses or welder's lenses provided, reasoning that it was safe to look through a truck windshield, as it would screen out the harmful ultraviolet radiation. On witnessing the blast, Feynman ducked towards the floor of his truck because of the immense brightness of the explosion, where he saw a temporary "purple splotch" afterimage of the event.*[28]

As a junior physicist, he was not central to the project. The greater part of his work was administering the computation group of human computers in the theoretical division (one of his students there, John G. Kemeny, later went on to co-design and co-specify the programming language BASIC). Later, with Nicholas Metropolis, he assisted in establishing the system for using IBM punched cards for computation.

Feynman's other work at Los Alamos included calculating neutron equations for the Los Alamos "Water Boiler", a small nuclear reactor, to measure how close an assembly of fissile material was to criticality. On completing this work he was transferred to the Oak Ridge facility, where he aided engineers in devising safety procedures for material storage so that criticality accidents (for example, due to sub-critical amounts of fissile material inadvertently stored in proximity on opposite sides of a wall) could be avoided. He also did theoretical work and calculations on the proposed uranium hydride bomb, which later proved not to be feasible.

Feynman was sought out by physicist Niels Bohr for one-on-one discussions. He later discovered the reason: most of the other physicists were too much in awe of Bohr to argue with him. Feynman had no such inhibitions, vigorously pointing out anything he considered to be flawed in Bohr's thinking. Feynman said he felt as much respect for Bohr as anyone else, but once anyone got him talking about physics, he would become so focused he forgot about social niceties.

Due to the top secret nature of the work, Los Alamos was isolated. In Feynman's own words, "There wasn't anything to *do* there." Bored, he indulged his curiosity by learning to pick the combination locks on cabinets and desks used to secure papers. Feynman played many jokes on colleagues. In one case he found the combination to a locked filing cabinet by trying the numbers he thought a physicist would use (it proved to be 27–18–28 after the base of natural logarithms, e = 2.71828...), and found that the three filing cabinets where a colleague kept a set of atomic bomb research notes all had the same combination.[*][29] He left a series of notes in the cabinets as a prank, which initially spooked his colleague, Frederic de Hoffmann, into thinking a spy or saboteur had gained access to atomic bomb secrets. On several occasions, Feynman drove to Albuquerque to see his ailing wife in a car borrowed from Klaus Fuchs, who was later discovered to be a real spy for the Soviets, transporting nuclear secrets in his car to Santa Fe.

On occasion, Feynman would find an isolated section of the mesa where he could drum in the style of American natives; "and maybe I would dance and chant, a little" . This did not go unnoticed, and rumors spread about a mysterious Indian drummer called "Injun Joe" . He also became a friend of the laboratory head, J. Robert Oppenheimer, who unsuccessfully tried to court him away from his other commitments after the war to work at the University of California, Berkeley.

Feynman alludes to his thoughts on the justification for getting involved in the Manhattan project in *The Pleasure of Finding Things Out*. He felt the possibility of Nazi Germany developing the bomb before the Allies was a compelling reason to help with its development for the U.S. He goes on to say that it was an error on his part not to reconsider the situation once Germany was defeated. In the same publication, Feynman also talks about his worries in the atomic bomb age, feeling for some considerable time that there was a high risk that the bomb would be used again soon, so that it was pointless to build for the future. Later he describes this period as a "depression" .

7.3.4 Early academic career

Following the completion of his PhD in 1942, Feynman held an appointment at the University of Wisconsin–Madison as an assistant professor of physics. The appointment was spent on leave for his involvement in the Manhattan project. In 1945, he received a letter from Dean Mark Ingraham of the College of Letters and Science requesting his return to UW to teach in the coming academic year. His appointment was not extended when he did not commit to return. In a talk given several years later at UW, Feynman quipped, "It's great to be back at the only university that ever had the good sense to fire me." [*][30]

After the war, Feynman declined an offer from the Institute for Advanced Study in Princeton, New Jersey, despite the presence there of such distinguished faculty members as Albert Einstein, Kurt Gödel and John von Neumann. Feynman followed Hans Bethe, instead, to Cornell University, where Feynman taught theoretical physics from 1945 to 1950. During a temporary depression following the destruction of Hiroshima by the bomb produced by the Manhattan Project, he focused on complex physics problems, not for utility, but for self-satisfaction. One of these was analyzing the physics of a twirling, nutating dish as it is moving through the air. His work during this period, which used equations of rotation to express various spinning speeds, proved important to his Nobel Prize–winning work, yet because he felt burned out and had turned his attention to less immediately practical problems, he was surprised by the offers of professorships from other renowned universities.

Despite yet another offer from the Institute for Advanced Study, Feynman rejected the Institute on the grounds that there were no teaching duties: Feynman felt that students were a source of inspiration and teaching was a diversion during uncreative spells. Because of this, the Institute for Advanced Study and Princeton University jointly offered him a package whereby he could teach at the university and also be at the institute. Feynman instead accepted an offer from the California Institute of Technology (Caltech)—and as he says in his book *Surely You're Joking Mr. Feynman!* —because a desire to live in a mild climate had firmly fixed itself in his mind while he was installing tire chains on his car in the middle of a snowstorm in Ithaca.

Feynman has been called the "Great Explainer" .[*][31] He gained a reputation for taking great care when giving explanations to his students and for making it a moral duty to make the topic accessible. His guiding principle was that, if a topic could not be explained in a freshman lecture, it was not yet fully understood. Feynman gained great pleasure[*][32] from coming up with such a "freshman-level" explanation, for example, of the connection between spin and statistics. What he said was that groups of particles with spin ½ "repel", whereas groups with integer spin "clump". This was a brilliantly simplified way of demonstrating how Fermi–Dirac statistics and Bose–Einstein statistics evolved as a consequence of studying how fermions and bosons behave under a rotation of 360°. This was also a question he pondered in his more advanced lectures, and to which he demonstrated the solution in the 1986 Dirac memorial lecture.*[33] In the same lecture, he further explained that antiparticles must exist, for if particles had only positive energies, they would not be restricted to a so-called "light cone".

He opposed rote learning or unthinking memorization and other teaching methods that emphasized form over function. *Clear thinking* and *clear presentation* were fundamental prerequisites for his attention. It could be perilous even to approach him when unprepared, and he did not forget the fools or pretenders.*[34]

7.3.5 Caltech years

The Feynman section at the Caltech bookstore

Feynman did significant work while at Caltech, including research in:

- Quantum electrodynamics. The theory for which Feynman won his Nobel Prize is known for its accurate predictions.*[35] This theory was begun in the earlier years during Feynman's work at Princeton as a graduate student and continued while he was at Cornell. This work consisted of two distinct formulations. The first is his path integral formulation, and the second is the formulation of Feynman diagrams. Both formulations contained his sum over histories method in which every possible path from one state to the next is considered, the final path being a sum over the possibilities (also referred to as sum-over-paths).*[36] For several years he lectured to students at Caltech on his path integral formulation of quantum theory. The second formulation of quantum electrodynamics (using Feynman diagrams) was specifically mentioned by the Nobel committee. The logical connection with the path integral formulation is interesting. Feynman did not prove that the rules for his diagrams followed mathematically from the path integral formulation. Some special cases were later proved by other people, but only in the real case, so the proofs do not work when spin is involved. The second formulation should be thought of as starting anew, but guided by the intuitive insight provided by the first formulation. Freeman Dyson published a paper in 1949, which added new rules to Feynman's that told how to implement renormalization. Students everywhere learned and used the powerful new tool that Feynman had created. Eventually computer programs were written to compute Feynman diagrams, providing a tool of unprecedented power. It is possible to write such programs because the Feynman diagrams constitute a formal language with a grammar. Marc Kac provided the formal proofs of the summation under history, showing that the parabolic partial differential equation can be reexpressed as a sum under different histories (that is, an expectation operator), what is now known as the Feynman–Kac formula, the use of which extends beyond physics to many applications of stochastic processes.*[37]

- Physics of the superfluidity of supercooled liquid helium, where helium seems to display a complete lack of viscosity when flowing. Feynman provided a quantum-mechanical explanation for the Soviet physicist Lev D. Landau's theory of superfluidity.*[38] Applying the Schrödinger equation to the question showed that the superfluid was displaying quantum mechanical behavior observable on a macroscopic scale. This helped with the problem of superconductivity but the solution eluded Feynman.*[39] It was solved with the BCS theory of superconductivity, proposed by John Bardeen, Leon Neil Cooper, and John Robert Schrieffer.

- A model of weak decay, which showed that the current coupling in the process is a combination of vector and axial currents (an example of weak decay is the decay of a neutron into an electron, a proton, and an anti-neutrino). Although E. C. George Sudarshan and Robert Marshak developed the theory nearly simultaneously, Feynman's collaboration with Murray Gell-Mann was seen as seminal because the weak interaction was neatly described by the vector and axial currents. It thus combined the 1933 beta decay theory of Enrico Fermi with an explanation of parity violation.

He also developed Feynman diagrams, a bookkeeping device that helps in conceptualizing and calculating interactions between particles in spacetime, including the interactions between electrons and their antimatter counterparts, positrons. This device allowed him, and later others, to approach time reversibility and other fundamental processes. Feynman's mental picture for these diagrams started with the *hard sphere* approximation, and the interactions could be thought of as *collisions* at first. It was not until decades later that physicists thought of analyzing the nodes of the Feynman diagrams more closely. Feynman painted Feynman diagrams on the exterior of his van.*[40]*[41]

From his diagrams of a small number of particles interacting in spacetime, Feynman could then model all of physics in terms of the spins of those particles and the range of coupling of the fundamental forces.*[42] Feynman attempted an explanation of the strong interactions governing nucleons scattering called the parton model. The parton model emerged as a complement to the quark model developed by his Caltech colleague Murray Gell-Mann. The relationship between the two models was murky; Gell-Mann referred to Feynman's partons derisively as "put-ons". In the mid-1960s, physicists believed that quarks were just a bookkeeping device for symmetry numbers, not real particles, as the statistics of the Omega-minus particle, if it were interpreted as three identical strange quarks bound together, seemed impossible if quarks were real. The Stanford linear accelerator deep inelastic scattering experiments of the late 1960s showed, analogously to Ernest Rutherford's experiment of scattering alpha particles on gold nuclei in 1911, that nucleons (protons and neutrons) contained point-like particles that scattered electrons. It was natural to identify these with quarks, but Feynman's parton model attempted to interpret the experimental data in a way that did not introduce additional hypotheses. For example, the data showed that some 45% of the energy momentum was carried by electrically-neutral particles in the nucleon. These electrically-neutral particles are now seen to be the gluons that carry the forces between the quarks and carry also the three-valued color quantum number that solves the Omega-minus problem. Feynman did not dispute the quark model; for example, when the fifth quark was discovered in 1977, Feynman immediately pointed out to his students that the discovery implied the existence of a sixth quark, which was discovered in the decade after his death.

After the success of quantum electrodynamics, Feynman turned to quantum gravity. By analogy with the photon, which has spin 1, he investigated the consequences of a free massless spin 2 field, and derived the Einstein field equation of general relativity, but little more.*[43] The computational device that Feynman discovered then for gravity, "ghosts", which are "particles" in the interior of his diagrams that have the "wrong" connection between spin and statistics, have proved invaluable in explaining the quantum particle behavior of the Yang–Mills theories, for example, QCD and the electro-weak theory.

Mention of Feynman's prize on the monument at the American Museum of Natural History in New York City. Because the monument is dedicated to American Laureates, Tomonaga is not mentioned.

Feynman was elected a Foreign Member of the Royal Society (ForMemRS) in 1965.*[2]*[7] At this time in the early 1960s, Feynman exhausted himself by working on multiple major projects at the same time, including a request, while at Caltech, to "spruce up" the teaching of undergraduates. After three years devoted to the task, he produced a series of lectures that eventually became *The Feynman Lectures on Physics*. He wanted a picture of a drumhead sprinkled with powder to show the modes of vibration at the beginning of the book. Concerned over the connections to drugs and rock and roll that could be made from the image, the publishers changed the cover to plain red, though they included a picture of him playing drums in the foreword. *The Feynman Lectures on Physics*[44] occupied two physicists, Robert B. Leighton and Matthew Sands, as part-time co-authors for several years. Even though the books were not adopted by most universities as textbooks, they continue to sell well because they provide a deep understanding of physics. Many of his lectures and miscellaneous talks were turned into other books, including *The Character of Physical Law*, *QED: The Strange Theory of Light and Matter*, *Statistical Mechanics*, *Lectures on Gravitation*, and the *Feynman Lectures on Computation*.

Feynman's students competed keenly for his attention; he was once awakened when a student solved a problem and dropped it in his mailbox; glimpsing the student sneaking across his lawn, he could not go back to sleep, and he read the student's solution. The next morning his breakfast was interrupted by another triumphant student, but Feynman informed him that he was too late.

Partly as a way to bring publicity to progress in physics, Feynman offered $1,000 prizes for two of his challenges in nanotechnology; one was claimed by William McLellan and the other by Tom Newman.*[45] He was also one of the first scientists to conceive the possibility of quantum computers.

In 1974, Feynman delivered the Caltech commencement address on the topic of *cargo cult science*, which has the semblance of science, but is only pseudoscience due to a lack of "a kind of scientific integrity, a principle of scientific thought that corresponds to a kind of utter honesty" on the part of the scientist. He instructed the graduating class that "The first principle is that you must not fool yourself —and you are the easiest person to fool. So you have to be very careful about that. After you've not fooled yourself, it's easy not to fool other scientists. You just have to be honest in a conventional way after that." *[46]

Richard Feynman at the Robert Treat Paine Estate in Waltham, MA, in 1984.

In 1984–86, he developed a variational method for the approximate calculation of path integrals, which has led to a powerful method of converting divergent perturbation expansions into convergent strong-coupling expansions (variational perturbation theory) and, as a consequence, to the most accurate determination*[47] of critical exponents measured in satellite experiments.*[48]

In the late 1980s, according to "Richard Feynman and the Connection Machine", Feynman played a crucial role in developing the first massively parallel computer, and in finding innovative uses for it in numerical computations, in building neural networks, as well as physical simulations using cellular automata (such as turbulent fluid flow), working with Stephen Wolfram at Caltech.*[49] His son Carl also played a role in the development of the original Connection Machine engineering; Feynman influencing the interconnects while his son worked on the software.

Feynman diagrams are now fundamental for string theory and M-theory, and have even been extended topologically.*[50] The *world-lines* of the diagrams have developed to become *tubes* to allow better modeling of more complicated objects such as *strings* and *membranes*. Shortly before his death, Feynman criticized string theory in an interview: "I don't like that they're not calculating anything," he said. "I don't like that they don't check their ideas. I don't like that for anything that disagrees with an experiment, they cook up an explanation—a fix-up to say, 'Well, it still might be true.'" These words have since been much-quoted by opponents of the string-theoretic direction for particle physics.*[38]

7.3.6 Challenger disaster

Main article: Space Shuttle Challenger disaster

Feynman played an important role on the Presidential

The 1986 Space Shuttle Challenger *disaster*

Rogers Commission, which investigated the *Challenger* disaster. During a televised hearing, Feynman demonstrated that the material used in the shuttle's O-rings became less resilient in cold weather by compressing a sample of the material in a clamp and immersing it in ice-cold water.*[51] The commission ultimately determined that the disaster was caused by the primary O-ring not properly sealing in unusually cold weather at Cape Canaveral.*[52]

Feynman devoted the latter half of his book *What Do You Care What Other People Think?* to his experience on the Rogers Commission, straying from his usual convention of brief, light-hearted anecdotes to deliver an extended and sober narrative. Feynman's account reveals a disconnect between NASA's engineers and executives that was far more striking than he expected. His interviews of NASA's high-ranking managers revealed startling misunderstandings of elementary concepts. For instance, NASA managers claimed that there was a 1 in 100,000 chance of a catastrophic failure aboard the shuttle, but Feynman discovered that NASA's own engineers estimated the chance of a catastrophe at closer to 1 in 200. He concluded that the space shuttle reliability estimate by NASA management was fantastically unrealistic, and he was particularly angered that NASA used these figures to recruit Christa McAuliffe into the Teacher-in-Space program. He warned in his appendix to the commission's report (which was included only after he threatened not to sign the report), "For a successful technology, reality must take precedence over public relations, for nature cannot be fooled." *[53]

A television documentary drama named *The Challenger* (US title: *The Challenger Disaster*), detailing Feynman's part in the investigation, was aired in 2013.*[54]

7.3.7 Cultural identification

Although born to and raised by parents who were Ashkenazi, Feynman was not only an atheist,*[55] but declined to be labelled Jewish. He routinely refused to be included in lists or books that classified people by race. He asked to not be included in Tina Levitan's *The Laureates: Jewish Winners of the Nobel Prize*, writing, "To select, for approbation the peculiar elements that come from some supposedly Jewish heredity is to open the door to all kinds of nonsense on racial theory," and adding "... at thirteen I was not only converted to other religious views, but I also stopped believing that the Jewish people are in any way 'the chosen people'." *[12]

7.3.8 Personal life

While researching for his PhD, Feynman married his first wife, Arline Greenbaum (often misspelled *Arlene*). They married knowing that Arline was seriously ill from tuberculosis, of which she died in 1945. In 1946, Feynman wrote a letter to her, but kept it sealed for the rest of his life.*[56] This portion of Feynman's life was portrayed in the 1996 film *Infinity*, which featured Feynman's daughter, Michelle, in a cameo role.

He married a second time in June 1952, to Mary Louise Bell of Neodesha, Kansas; this marriage was unsuccessful:

> He begins working calculus problems in his head as soon as he awakens. He did calculus while driving in his car, while sitting in the living room, and while lying in bed at night.
> —Mary Louise Bell divorce complaint*[3]

He later married Gweneth Howarth (1934–1989), who was from Ripponden, Yorkshire, and shared his enthusiasm for life and spirited adventure.*[40] Besides their home in Altadena, California, they had a beach house in Baja California, purchased with the prize money from Feynman's Nobel Prize, his one third share of $55,000. They remained married until Feynman's death. They had a son, Carl, in 1962, and adopted a daughter, Michelle, in 1968.*[40]

Feynman had a great deal of success teaching Carl, using, for example, discussions about ants and Martians as a device for gaining perspective on problems and issues. He was surprised to learn that the same teaching devices were not useful with Michelle.*[41] Mathematics was a common interest for father and son; they both entered the computer field as consultants and were involved in advancing a new method of using multiple computers to solve complex problems—later known as parallel computing. The Jet Propulsion Laboratory retained Feynman as a computational consultant during critical missions. One co-worker characterized Feynman as akin to Don Quixote at his desk, rather than at a computer workstation, ready to do battle with the windmills.

Feynman traveled to Brazil, where he gave courses at the CBPF (Brazilian Center for Physics Research) and near the end of his life schemed to visit the Russian land of Tuva, a dream that, because of Cold War bureaucratic problems, never became reality.*[57] The day after he died, a letter arrived for him from the Soviet government, giving him authorization to travel to Tuva. Out of his enthusiastic interest in reaching Tuva came the phrase "Tuva or Bust" (also the title of a book about his efforts to get there), which was tossed about frequently amongst his circle of friends in hope that they, one day, could see it firsthand. The documentary movie, *Genghis Blues*, mentions some of his attempts to communicate with Tuva and chronicles the successful journey there by his friends.

Responding to Hubert Humphrey's congratulation for his Nobel Prize, Feynman admitted to a long admiration for the then vice president.*[58]

Feynman took up drawing at one time and enjoyed some success under the pseudonym "Ofey", culminating in an exhibition of his work. He learned to play a metal percussion instrument (*frigideira*) in a samba style in Brazil, and participated in a samba school.

In addition, he had some degree of synesthesia for equations, explaining that the letters in certain mathematical functions appeared in color for him, even though invariably printed in standard black-and-white.*[59]

According to *Genius*, the James Gleick-authored biography, Feynman tried LSD during his professorship at Caltech.*[38] Somewhat embarrassed by his actions, he largely sidestepped the issue when dictating his anecdotes; he mentions it in passing in the "O Americano, Outra Vez" section, while the "Altered States" chapter in *Surely You're Joking, Mr. Feynman!* describes only marijuana and ketamine experiences at John Lilly's famed sensory deprivation tanks, as a way of studying consciousness.*[29] Feynman gave up alcohol when he began to show vague, early signs of alcoholism, as he did not want to do anything that could damage his brain—the same reason given in "O Americano, Outra Vez" for his reluctance to experiment with LSD.*[29]

In *Surely You're Joking, Mr. Feynman!*, he gives advice on the best way to pick up a girl in a hostess bar. At Caltech, he used a nude or topless bar as an office away from his usual office, making sketches or writing physics equations on paper placemats. When the county officials tried to close the place, all visitors except Feynman refused to testify in favor of the bar, fearing that their families or patrons would learn about their visits. Only Feynman accepted, and in court, he affirmed that the bar was a public need, stating that craftsmen, technicians, engineers, common workers, "and a physics professor" frequented the establishment. While the bar lost the court case, it was allowed to remain open as a similar case was pending appeal.*[29]

His friend Ralph Leighton described Feynman's quest to travel to Tannu Tuva in the book Tuva or Bust!; later Feynman's daughter Michelle would realize this journey.

Feynman has a minor acting role in the film *Anti-Clock* credited as "The Professor" .*[60]

7.3.9 Death

Feynman had two rare forms of cancer, liposarcoma and Waldenström's macroglobulinemia, dying shortly after a final attempt at surgery for the former on February 15, 1988, aged 69.*[38] His last words are noted as, "I'd hate to die twice. It's so boring." *[38]*[61]

7.3.10 Popular legacy

Actor Alan Alda commissioned playwright Peter Parnell to write a two-character play about a fictional day in the life of Feynman set two years before Feynman's death. The play, *QED*, which was based on writings about Richard Feynman's life during the 1990s, premiered at the Mark Taper Forum in Los Angeles in 2001. The play was then presented at the Vivian Beaumont Theater on Broadway, with both presentations starring Alda as Richard Feynman.*[62]

On May 4, 2005, the United States Postal Service issued the *American Scientists* commemorative set of four 37-cent self-adhesive stamps in several configurations. The scientists depicted were Richard Feynman, John von Neumann, Barbara McClintock, and Josiah Willard Gibbs. Feynman's stamp, sepia-toned, features a photograph of a 30-something Feynman and eight small Feynman diagrams.*[63] The stamps were designed by Victor Stabin under the artistic direction of Carl T. Herrman.*[64]

The main building for the Computing Division at Fermilab is named the "Feynman Computing Center" in his honor.*[65]

Real Time Opera premiered its opera *Feynman* at the Norfolk (CT) Chamber Music Festival in June 2005.*[66]

In a 1992 *New York Times* article on Feynman and his legacy, James Gleick recounts the story of how Murray Gell-Mann described what has become known as "The Feynman Algorithm" or "The Feynman Problem-Solving Algorithm" to a student: "The student asks Gell-Mann about Feynman's notes. Gell-Mann says no, Dick's methods are not the same as the methods used here. The student asks, well, what are Feynman's methods? Gell-Mann leans coyly against the blackboard and says: Dick's method is this. You write down the problem. You think very hard. (He shuts his eyes and presses his knuckles parodically to his forehead.) Then you write down the answer." *[67]

In 1998, a photograph of Richard Feynman giving a lecture was part of the poster series commissioned by Apple Inc. for their "Think Different" advertising campaign.*[68]

In 2011, Feynman was the subject of a biographical graphic novel entitled simply *Feynman*, written by Jim Ottaviani and illustrated by Leland Myrick.*[69]

In 2013, the BBC drama *The Challenger* depicted Feynman's role on the Rogers Commission in exposing the O-ring flaw in NASA's solid-rocket boosters (SRBs), itself based in part on Feynman's book *What Do You Care What Other People Think?*[70]*[71]

7.3.11 Bibliography

Selected scientific works

- Feynman, Richard P. (2000). Laurie M. Brown, ed. *Selected Papers of Richard Feynman: With Commentary*. 20th Century Physics. World Scientific. ISBN 978-981-02-4131-5.

- Feynman, Richard P. (1942). Laurie M. Brown, ed. *The Principle of Least Action in Quantum Mechanics*. PhD Dissertation, Princeton University. World Scientific (with title *Feynman's Thesis: a New Approach to Quantum Theory*) (published 2005). ISBN 978-981-256-380-4.

- Wheeler, John A.; Feynman, Richard P. (1945). "Interaction with the Absorber as the Mechanism of Radiation" . *Reviews of Modern Physics* 17

(2–3): 157–181. Bibcode:1945RvMP...17..157W. doi:10.1103/RevModPhys.17.157.

- Feynman, Richard P. (1946). *A Theorem and its Application to Finite Tampers*. Los Alamos Scientific Laboratory, Atomic Energy Commission. OSTI 4341197.

- Feynman, Richard P.; Welton, T. A. (1946). *Neutron Diffusion in a Space Lattice of Fissionable and Absorbing Materials*. Los Alamos Scientific Laboratory, Atomic Energy Commission. OSTI 4381097.

- Feynman, Richard P.; Metropolis, N.; Teller, E. (1947). *Equations of State of Elements Based on the Generalized Fermi-Thomas Theory*. Los Alamos Scientific Laboratory, Atomic Energy Commission. OSTI 4417654.

- Feynman, Richard P. (1948a). "Space-time approach to non-relativistic quantum mechanics". *Reviews of Modern Physics* **20** (2): 367–387. Bibcode:1948RvMP...20..367F. doi:10.1103/RevModPhys.20.367.

- Feynman, Richard P. (1948b). "Relativistic Cut-Off for Quantum Electrodynamics". *Physical Review* **74** (10): 1430–1438. Bibcode:1948PhRv...74.1430F. doi:10.1103/PhysRev.74.1430.

- Wheeler, John A.; Feynman, Richard P. (1949). "Classical Electrodynamics in Terms of Direct Interparticle Action". *Reviews of Modern Physics* **21** (3): 425–433. Bibcode:1949RvMP...21..425W. doi:10.1103/RevModPhys.21.425.

- Feynman, Richard P. (1949). "The theory of positrons". *Physical Review* **76** (6): 749–759. Bibcode:1949PhRv...76..749F. doi:10.1103/PhysRev.76.749.

- Feynman, Richard P. (1949b). "Space-Time Approach to Quantum Electrodynamic". *Physical Review* **76** (6): 769–789. Bibcode:1949PhRv...76..769F. doi:10.1103/PhysRev.76.769.

- Feynman, Richard P. (1950). "Mathematical formulation of the quantum theory of electromagnetic interaction". *Physical Review* **80** (3): 440–457. Bibcode:1950PhRv...80..440F. doi:10.1103/PhysRev.80.440.

- Feynman, Richard P. (1951). "An Operator Calculus Having Applications in Quantum Electrodynamics". *Physical Review* **84**: 108–128. Bibcode:1951PhRv...84..108F. doi:10.1103/PhysRev.84.108.

- Feynman, Richard P. (1953). "The λ-Transition in Liquid Helium". *Physical Review* **90** (6): 1116–1117. Bibcode:1953PhRv...90.1116F. doi:10.1103/PhysRev.90.1116.2.

- Feynman, Richard P.; de Hoffmann, F.; Serber, R. (1955). *Dispersion of the Neutron Emission in U235 Fission*. Los Alamos Scientific Laboratory, Atomic Energy Commission. OSTI 4354998.

- Feynman, Richard P. (1956). "Science and the Open Channel". *Science* (February 24, 1956) **123** (3191): 307. Bibcode:1956Sci...123..307F. doi:10.1126/science.123.3191.307. PMID 17774518.

- Cohen, M.; Feynman, Richard P. (1957). "Theory of Inelastic Scattering of Cold Neutrons from Liquid Helium". *Physical Review* **107**: 13–24. Bibcode:1957PhRv..107...13C. doi:10.1103/PhysRev.107.13.

- Feynman, Richard P.; Vernon, F. L.; Hellwarth, R. W. (1957). "Geometric representation of the Schrödinger equation for solving maser equations". *J. Appl. Phys* **28**: 49. Bibcode:1957JAP....28...49F. doi:10.1063/1.1722572.

- Feynman, Richard P. (1959). "Plenty of Room at the Bottom". Presentation to American Physical Society.

- Edgar, R. S.; Feynman, Richard P.; Klein, S.; Lielausis, I.; Steinberg, C. M. (1962). "Mapping experiments with r mutants of bacteriophage T4D". *Genetics* (February 1962) **47** (2): 179–86. PMC 1210321. PMID 13889186.

- Feynman, Richard P. (1966). "The Development of the Space-Time View of Quantum Electrodynamics". *Science* (August 12, 1966) **153** (3737): 699–708. Bibcode:1966Sci...153..699F. doi:10.1126/science.153.3737.699. PMID 17791121.

- Feynman, Richard P. (1974a). "Structure of the proton". *Science* (February 15, 1974) **183** (4125): 601–610. Bibcode:1974Sci...183..601F. doi:10.1126/science.183.4125.601. PMID 17778830.

- Feynman, Richard P. (1974). "Cargo Cult Science" (PDF). *Engineering and Science* **37** (7).

- Feynman, Richard P.; Kleinert, Hagen (1986). "Effective classical partition functions". *Physical Review A* (December 1986) **34** (6): 5080–5084. Bibcode:1986PhRvA..34.5080F. doi:10.1103/PhysRevA.34.5080. PMID 9897894.

Textbooks and lecture notes

The Feynman Lectures on Physics is perhaps his most accessible work for anyone with an interest in physics, compiled from lectures to Caltech undergraduates in 1961–64. As news of the lectures' lucidity grew, professional physicists and graduate students began to drop in to listen. Co-authors Robert B. Leighton and Matthew Sands, colleagues of Feynman, edited and illustrated them into book form. The work has endured and is useful to this day. They were edited and supplemented in 2005 with "Feynman's Tips on Physics: A Problem-Solving Supplement to the Feynman Lectures on Physics" by Michael Gottlieb and Ralph Leighton (Robert Leighton's son), with support from Kip Thorne and other physicists.

- Feynman, Richard P.; Leighton, Robert B.; Sands, Matthew (2005) [1970]. *The Feynman Lectures on Physics: The Definitive and Extended Edition* (2nd ed.). Addison Wesley. ISBN 0-8053-9045-6. Includes *Feynman's Tips on Physics* (with Michael Gottlieb and Ralph Leighton), which includes four previously unreleased lectures on problem solving, exercises by Robert Leighton and Rochus Vogt, and a historical essay by Matthew Sands. Three volumes; originally published as separate volumes in 1964 and 1966.

- Feynman, Richard P. (1961). *Theory of Fundamental Processes*. Addison Wesley. ISBN 0-8053-2507-7.

- Feynman, Richard P. (1962). *Quantum Electrodynamics*. Addison Wesley. ISBN 978-0-8053-2501-0.

- Feynman, Richard P.; Hibbs, Albert (1965). *Quantum Mechanics and Path Integrals*. McGraw Hill. ISBN 0-07-020650-3.

- Feynman, Richard P. (1967). *The Character of Physical Law: The 1964 Messenger Lectures*. MIT Press. ISBN 0-262-56003-8.

- Feynman, Richard P. (1972). *Statistical Mechanics: A Set of Lectures*. Reading, Mass: W. A. Benjamin. ISBN 0-8053-2509-3.

- Feynman, Richard P. (1985b). *QED: The Strange Theory of Light and Matter*. Princeton University Press. ISBN 0-691-02417-0.

- Feynman, Richard P. (1987). *Elementary Particles and the Laws of Physics: The 1986 Dirac Memorial Lectures*. Cambridge University Press. ISBN 0-521-34000-4.

- Feynman, Richard P. (1995). Brian Hatfield, ed. *Lectures on Gravitation*. Addison Wesley Longman. ISBN 0-201-62734-5.

- Feynman, Richard P. (1997). *Feynman's Lost Lecture: The Motion of Planets Around the Sun* (Vintage Press ed.). London: Vintage. ISBN 0-09-973621-7.

- Feynman, Richard P. (2000). Tony Hey and Robin W. Allen, ed. *Feynman Lectures on Computation*. Perseus Books Group. ISBN 0-7382-0296-7.

Popular works

- Feynman, Richard P. (1985). Ralph Leighton, ed. *Surely You're Joking, Mr. Feynman!: Adventures of a Curious Character*. W. W. Norton & Co. ISBN 0-393-01921-7. OCLC 10925248.

- Feynman, Richard P. (1988). Ralph Leighton, ed. *What Do You Care What Other People Think?: Further Adventures of a Curious Character*. W. W. Norton & Co. ISBN 0-393-02659-0.

- *No Ordinary Genius: The Illustrated Richard Feynman*, ed. Christopher Sykes, W. W. Norton & Co, 1996, ISBN 0-393-31393-X.

- *Six Easy Pieces: Essentials of Physics Explained by Its Most Brilliant Teacher*, Perseus Books, 1994, ISBN 0-201-40955-0.

- *Six Not So Easy Pieces: Einstein's Relativity, Symmetry and Space-Time*, Addison Wesley, 1997, ISBN 0-201-15026-3.

- *The Meaning of It All: Thoughts of a Citizen Scientist*, Perseus Publishing, 1999, ISBN 0-7382-0166-9.

- *The Pleasure of Finding Things Out: The Best Short Works of Richard P. Feynman*, edited by Jeffrey Robbins, Perseus Books, 1999, ISBN 0-7382-0108-1.

- *Classic Feynman: All the Adventures of a Curious Character*, edited by Ralph Leighton, W. W. Norton & Co, 2005, ISBN 0-393-06132-9. Chronologically re-ordered omnibus volume of *Surely You're Joking, Mr. Feynman!* and *What Do You Care What Other People Think?*, with a bundled CD containing one of Feynman's signature lectures.

- *Quantum Man*, Atlas books, 2011, Lawrence M. Krauss, ISBN 978-0-393-06471-1.

- "Feynman: The Graphic Novel" Jim Ottaviani and Leland Myrick, ISBN 978-1-59643-259-8.

Audio and video recordings

- *Safecracker Suite* (a collection of drum pieces interspersed with Feynman telling anecdotes)

- *Los Alamos From Below* (audio, talk given by Feynman at Santa Barbara on February 6, 1975)

- *Six Easy Pieces* (original lectures upon which the book is based)

- *Six Not So Easy Pieces* (original lectures upon which the book is based)

- The Feynman Lectures on Physics: The Complete Audio Collection

- Samples of Feynman's drumming, chanting and speech are included in the songs "Tuva Groove (Bolur Daa-Bol, Bolbas Daa-Bol)" and "Kargyraa Rap (Dürgen Chugaa)" on the album *Back Tuva Future, The Adventure Continues* by Kongar-ool Ondar. The hidden track on this album also includes excerpts from lectures without musical background.

- The Messenger Lectures, given at Cornell in 1964, in which he explains basic topics in physics. Available on Project Tuva for free (See also the book *The Character of Physical Law*)

- Take the world from another point of view [video-recording] / with Richard Feynman; Films for the Hu (1972)

- The Douglas Robb Memorial Lectures Four public lectures of which the four chapters of the book QED: The Strange Theory of Light and Matter are transcripts. (1979)

- The Pleasure of Finding Things Out on YouTube (1981) (not to be confused with the later published book of same title)

- Richard Feynman: Fun to Imagine Collection, BBC Archive of 6 short films of Feynman talking in a style that is accessible to all about the physics behind common to all experiences. (1983)

- *Elementary Particles and the Laws of Physics* (1986)

- Tiny Machines: The Feynman Talk on Nanotechnology (video, 1984)

- Computers From the Inside Out (video)

- Quantum Mechanical View of Reality: Workshop at Esalen (video, 1983)

- Idiosyncratic Thinking Workshop (video, 1985)

- Bits and Pieces —From Richard's Life and Times (video, 1988)

- Strangeness Minus Three (video, BBC Horizon 1964)

- No Ordinary Genius (video, Cristopher Sykes Documentary)

- Richard Feynman —The Best Mind Since Einstein (video, Documentary)

- The Motion of Planets Around the Sun (audio, sometimes titled "Feynman's Lost Lecture")

- Nature of Matter (audio)

7.3.12 See also

- List of things named after Richard Feynman
- Feynman diagram
- Feynman checkerboard
- Flexagon
- Foresight Nanotech Institute Feynman Prize
- List of physicists
- List of theoretical physicists
- Morrie's law
- Negative probability
- One-electron universe
- Stückelberg–Feynman interpretation
- Wheeler–Feynman absorber theory

7.3.13 References

[1] Richard Feynman at the Mathematics Genealogy Project

[2] Mehra, J. (2002). "Richard Phillips Feynman 11 May 1918 – 15 February 1988". *Biographical Memoirs of Fellows of the Royal Society* **48**: 97–128. doi:10.1098/rsbm.2002.0007.

[3] Krauss 2011, p. 168

[4] Tindol, Robert (December 2, 1999). "Physics World poll names Richard Feynman one of 10 greatest physicists of all time" (Press release). California Institute of Technology. Archived from the original on March 21, 2012. Retrieved December 1, 2012.

[5] West, Jacob (June 2003). "The Quantum Computer" (PDF). Retrieved September 20, 2009.

[6] Deutsch 1992, pp. 57–61.

[7] "Richard P. Feynman – Biographical". The Nobel Foundation. Retrieved April 23, 2013.

[8] *Guide to Nobel Prize – Richard P. Feynman*. Encyclopaedia Britannica. Retrieved March 31, 2013.

[9] J. J. O'Connor; E. F. Robertson (August 2002). "Richard Phillips Feynman". University of St. Andrews. Retrieved April 23, 2013.

[10] "Richard Phillips Feynman". Nobel-winners.com. Retrieved April 23, 2013.

[11] Feynman 1988, p. 25.

[12] Harrison, John. "Physics, bongos and the art of the nude". *The Daily Telegraph*. Retrieved April 23, 2013.

[13] Surely You're Joking, Mr. Feynman! 2001, p. 284–287.

[14] Chown 1985, p. 34.

[15] Close 2011, p. 58.

[16] Sykes 1994, p. 54.

[17] Friedman 2004, p. 231.

[18] Henderson 2011, p. 8.

[19] Charles Hirshberg (March 23, 2014). "My Mother, the Scientist". Popular Science. Retrieved April 23, 2013.

[20] Sykes 1994.

[21] Gleick 1992, p. 30.

[22] Schweber 1994, p. 374.

[23] Feynman 1992, p. 72

[24] Schwach, Howard (April 15, 2005). "Museum Tracks Down FRHS Nobel Laureates". *The Wave*. Retrieved April 23, 2013.

[25] Ottaviani 2011, p. 28.

[26] "Putnam Competition Individual and Team Winners". MMA: Mathematical Association of America. 2014. Retrieved March 8, 2014.

[27] Gleick 1992, p. 84.

[28] Feynman 1992, p. 134

[29] Feynman 1985

[30] Robert H. March. "Physics at the University of Wisconsin: A History". *Physics in Perspective* **5** (2): 130–149. Bibcode:2003PhP.....5..130M. doi:10.1007/s00016-003-0142-6.

[31] LeVine 2009

[32] Hey & Walters 1987.

[33] Feynman 1987

[34] Bethe 1991, p. 241.

[35] Schwinger 1958

[36] Feynman & Hibbs 1965

[37] Kac, Mark (1949). "On Distributions of Certain Wiener Functionals". *Transactions of the American Mathematical Society* **65** (1): 1–13. doi:10.2307/1990512. JSTOR 1990512.

[38] Gleick 1992

[39] Pines, David (1989). "Richard Feynman and Condensed Matter Physics". *Physics Today* **42** (2): 61. Bibcode:1989PhT....42b..61P. doi:10.1063/1.881194.

[40] Feynman 2005

[41] Sykes 1996

[42] Feynman 1961

[43] Feynman 1995

[44] Feynman 1970

[45] Gribbin & Gribbin 1997, p. 170

[46] Feynman 1974b

[47] Kleinert, Hagen (1999). "Specific heat of liquid helium in zero gravity very near the lambda point". *Physical Review D* **60** (8): 085001. arXiv:hep-th/9812197. Bibcode:1999PhRvD..60h5001K. doi:10.1103/PhysRevD.60.085001.

[48] Lipa, J. A.; Nissen, J.; Stricker, D.; Swanson, D.; Chui, T. (2003). "Specific heat of liquid helium in zero gravity very near the lambda point". *Physical Review B* **68** (17): 174518. arXiv:cond-mat/0310163. Bibcode:2003PhRvB..68q4518L. doi:10.1103/PhysRevB.68.174518.

[49] Hillis 1989

[50] Hooft, Gerard 't Hooft. "On the Foundations of Superstring Theory". http://www.staff.science.uu.nl/~{}hooft101/.

[51] Feynman 1988, p. 151

[52] James Gleick (February 17, 1988). "Richard Feynman Dead at 69; Leading Theoretical Physicist". *The New York Times*. Retrieved April 23, 2013.

[53] Richard Feynman. "Appendix F – Personal observations on the reliability of the Shuttle". Kennedy Space Center.

[54] Bricken, Rob. "Here's William Hurt as the legendary physicist Richard Feynman!". io9. Retrieved April 23, 2013.

[55] Brian 2008, p. 49 Interviewer: Do you call yourself an agnostic or an atheist? Feynman: An atheist. Agnostic for me would be trying to weasel out and sound a little nicer than I am about this.

[56] "I love my wife. My wife is dead". Letters of Note. Retrieved April 23, 2013.

[57] Leighton 2000.

[58] Feynman 2005, p. 173

[59] Feynman 1988

[60] "Anti-clock(1979)". IMDB. Retrieved April 23, 2013.

[61] "Richard Feynman at Find a Grave". Retrieved October 4, 2008.

[62] ibdb.com

[63] "Who is Richard Feynman?". feynmangroup.com. Retrieved December 1, 2012.

[64] "American Scientists Series Slideshow". Beyondthepref.com. Retrieved December 1, 2012.

[65] "Fermilab Open House: Computing Division". fnal.gov. Retrieved December 1, 2012.

[66] "Real Time Opera". rtopera.org. Retrieved December 1, 2012.

[67] Gleick, James (September 20, 1992). "Part Showman, All Genius". *The New York Times*. p. SM38. Retrieved February 23, 2015.

[68] Great Mind Richard Feynman Birthday | Manhattan Project and Challenger Disaster | Quantum Electrodynamics | Biography. Techie-buzz.com (May 10, 2011). Retrieved on May 6, 2012.

[69] Jim Ottaviani; Leland Myrick (2011). *Feynman*. New York: First Second. ISBN 978-1-59643-259-8. OCLC 664838951.

[70] "The Challenger". BBC. Retrieved March 18, 2013.

[71] "The Challenger". BBC Two. Retrieved March 19, 2013.

7.3.14 Bibliography

- Bethe, Hans A. (1991). *The Road from Los Alamos*. Masters of Modern Physics **2**. New York: Simon and Schuster. ISBN 0-671-74012-1. OCLC 24734608.

- Brian, Denis (2008). *The Voice of Genius: Conversations with Nobel Scientists and Other Luminaries*. Basic Books. ISBN 978-0-465-01139-1.

- Chown, Marcus (May 2, 1985). "Strangeness and Charm". *New Scientist*: 34. ISSN 0262-4079.

- Close, Frank (2011). *The Infinity Puzzle: The Personalities, Politics, and Extraordinary Science Behind the Higgs Boson*. Oxford University Press. ISBN 978-0-19-959350-7.

- Deutsch, David (June 1, 1992). "Quantum computation". *Physics World*: 57–61. ISSN 0953-8585.

- Edwards, Steven Alan (2006). *The Nanotech Pioneers*. Wiley. ISBN 978-3-527-31290-0. OCLC 64304124.

- Feynman, Richard P. (1986). *Rogers Commission Report, Volume 2 Appendix F – Personal Observations on Reliability of Shuttle*. NASA.

- Feynman, Richard P. (1987). Ralph Leighton, ed. "Mr. Feynman Goes to Washington". *Engineering and Science* (Caltech) **51** (1): 6–22. ISSN 0013-7812.

- Feynman, Michelle, ed. (2005). *Perfectly Reasonable Deviations from the Beaten Track: The Letters of Richard P. Feynman*. Basic Books. ISBN 0-7382-0636-9. (Published in the UK under the title: *Don't You Have Time to Think?*, with additional commentary by Michelle Feynman, Allen Lane, 2005, ISBN 0-7139-9847-4.)

- Friedman, Jerome (2004). "A Student's View of Fermi". In Cronin, James W. *Fermi Remembered*. Chicago: University of Chicago Press. ISBN 978-0-226-12111-6.

- Gribbin, John; Gribbin, Mary (1997). *Richard Feynman: A Life in Science*. Dutton. ISBN 0-525-94124-X.

- Gleick, James (1992). *Genius: The Life and Science of Richard Feynman*. Pantheon Books. ISBN 0-679-40836-3. OCLC 243743850.

- Henderson, Harry (2011). *Richard Feynman: Quarks, Bombs, and Bongos*. Chelsea House Publishers. ISBN 978-0-8160-6176-1.

- Hey, Tony; Walters, Patrick (1987). *The quantum universe*. Cambridge University Press. ISBN 978-0-521-31845-7.

- Hillis, W. Daniel (1989). "Richard Feynman and The Connection Machine". *Physics Today* (Institute of Physics) **42** (2): 78. Bibcode:1989PhT....42b..78H. doi:10.1063/1.881196.

- Krauss, Lawrence M. (2011). *Quantum Man: Richard Feynman's Life in Science*. W. W. Norton & Company. ISBN 0-393-06471-9. OCLC 601108916.

- Leighton, Ralph (2000). *Tuva Or Bust!:Richard Feynman's last journey*. W. W. Norton & Company. ISBN 0-393-32069-3.

- LeVine, Harry (2009). *The Great Explainer: The Story of Richard Feynman*. Greensboro, North Carolina: Morgan Reynolds. ISBN 978-1-59935-113-1.

- Ottaviani, Jim; Myrick, Leland; Sycamore, Hilary (2011). *Feynman* (1st ed.). New York: First Second. ISBN 978-1-59643-259-8.

- Schweber, Silvan S. (1994). *QED and the Men Who Made It: Dyson, Feynman, Schwinger, and Tomonaga*. Princeton University Press. ISBN 0-691-03327-7.

- Schwinger, Julian, ed. (1958). *Selected Papers on Quantum Electrodynamics*. Dover. ISBN 0-486-60444-6.

- Sykes, Christopher (1994). *No ordinary genius : the illustrated Richard Feynman*. New York: W. W. Norton. ISBN 0-393-03621-9.

7.3.15 Further reading

Articles

- *Physics Today*, American Institute of Physics magazine, February 1989 Issue. (Vol. 42, No. 2.) Special Feynman memorial issue containing non-technical articles on Feynman's life and work in physics.

Books

- Brown, Laurie M. and Rigden, John S. (editors) (1993) *Most of the Good Stuff: Memories of Richard Feynman* Simon and Schuster, New York, ISBN 0-88318-870-8. Commentary by Joan Feynman, John Wheeler, Hans Bethe, Julian Schwinger, Murray Gell-Mann, Daniel Hillis, David Goodstein, Freeman Dyson, and Laurie Brown

- Dyson, Freeman (1979) *Disturbing the Universe*. Harper and Row. ISBN 0-06-011108-9. Dyson's autobiography. The chapters "A Scientific Apprenticeship" and "A Ride to Albuquerque" describe his impressions of Feynman in the period 1947–48 when Dyson was a graduate student at Cornell

- Gleick, James (1992) *Genius: The Life and Science of Richard Feynman*. Pantheon. ISBN 0-679-74704-4

- LeVine, Harry, III (2009) *The Great Explainer: The Story of Richard Feynman* (*Profiles in Science* series) Morgan Reynolds, Greensboro, North Carolina, ISBN 978-1-59935-113-1; for high school readers

- Mehra, Jagdish (1994) *The Beat of a Different Drum: The Life and Science of Richard Feynman*. Oxford University Press. ISBN 0-19-853948-7

- Gribbin, John and Gribbin, Mary (1997) *Richard Feynman: A Life in Science*. Dutton, New York, ISBN 0-525-94124-X

- Milburn, Gerard J. (1998) *The Feynman Processor: Quantum Entanglement and the Computing Revolution* Perseus Books, ISBN 0-7382-0173-1

- Mlodinow, Leonard (2003) *Feynman's Rainbow: A Search For Beauty In Physics And In Life* Warner Books. ISBN 0-446-69251-4 Published in the United Kingdom as *Some Time With Feynman*

- Ottaviani, Jim and Myrick, Leland (2011) *Feynman*. First Second. ISBN 978-1-59643-259-8 OCLC 664838951.

Films and plays

- *Infinity*, a movie directed by Matthew Broderick and starring Matthew Broderick as Feynman, depicting Feynman's love affair with his first wife and ending with the Trinity test. 1996.

- Parnell, Peter (2002) "QED" Applause Books, ISBN 978-1-55783-592-5, (play).

- Whittell, Crispin (2006) "Clever Dick" Oberon Books, (play)

- "The Pleasure of Finding Things Out". Feynman talks about his life in science and his love of exploring nature. 1981, BBC Horizon. See Christopher Sykes Productions.

- "The Quest for Tannu Tuva", with Richard Feynman and Ralph Leighton. 1987, BBC Horizon and PBS Nova (entitled "Last Journey of a Genius").

- "No Ordinary Genius" A two-part documentary about Feynman's life and work, with contributions from colleagues, friends and family. 1993, BBC Horizon and PBS Nova (a one-hour version, under the title "The Best Mind Since Einstein") (2 × 50-minute films)

- The Challenger (2013) A BBC Two factual drama starring William Hurt, tells the story of American Nobel prize-winning physicist Richard Feynman's determination to reveal the truth behind the 1986 space shuttle Challenger disaster.

- The Fantastic Mr Feynman. One hour documentary. 2013, BBC TV.

- *How We Built the Bomb*, a television movie directed by Seth Skundrick, 2015.

7.3.16 External links

- Official website
- The Feynman Lectures on Physics Website by Michael Gottlieb, assisted by Rudolf Pfeiffer and Caltech
- Feynman Online!, a site dedicated to Feynman
- Richard Feynman at the Internet Movie Database
- Feynman and the Connection Machine
- Richard Feynman (Interviews, with and about) – American Institute of Physics

7.4 Murray Gell-Mann

Murray Gell-Mann (/ˈmʌri ˈɡɛl ˈmæn/; born September 15, 1929) is an American physicist who received the 1969 Nobel Prize in physics for his work on the theory of elementary particles. He is the Robert Andrews Millikan Professor of Theoretical Physics Emeritus at the California Institute of Technology, a Distinguished Fellow and co-founder of the Santa Fe Institute, Professor in the Physics and Astronomy Department of the University of New Mexico, and the Presidential Professor of Physics and Medicine at the University of Southern California.[1] Gell-Mann has spent several periods at CERN, among others as a John Simon Guggenheim Memorial Foundation Fellow in 1972.[2]

He introduced, independently of George Zweig, the quark—constituents of all hadrons—having first identified the SU(3) flavor symmetry of hadrons. This symmetry is now understood to underlie the light quarks, extending isospin to include strangeness, a quantum number which he also discovered.

He developed the V–A theory of the weak interaction in collaboration with Richard Feynman. In the 1960s, he introduced current algebra as a method of systematically exploiting symmetries to extract predictions from quark models, in the absence of reliable dynamical theory. This method led to model-independent sum rules confirmed by experiment and provided starting points underpinning the development of the standard theory of elementary particles.

Gell-Mann, along with Maurice Lévy, developed the sigma model of pions, which describes low-energy pion interactions. Modifying the integer-charged quark model of Moo-Young Han and Yoichiro Nambu, Harald Fritzsch and Gell-Mann were the first to write down the modern accepted theory of quantum chromodynamics, although they did not anticipate asymptotic freedom. In 1969 he received the Nobel Prize in physics for his contributions and discoveries concerning the classification of elementary particles and their interactions.[3]

Gell-Mann is responsible, together with Pierre Ramond and Richard Slansky, and independently of Peter Minkowski, Rabindra Mohapatra, Goran Senjanovic, Sheldon Lee Glashow, and Tsutomu Yanagida, for the see-saw theory of neutrino masses, that produces masses at the large scale in any theory with a right-handed neutrino. He is also known to have played a large role in keeping string theory alive through the 1970s and early 1980s, supporting that line of research at a time when it was unpopular.

Gell-Mann is a proponent of the consistent histories approach to understanding quantum mechanics.

7.4.1 Early life and education

Gell-Mann was born in lower Manhattan into a family of Jewish immigrants from the Austro-Hungarian Empire.[4][5] His parents were Pauline (née Reichstein) and Arthur Isidore Gell-Mann, who taught English as a Second Language (ESL).[6]

Propelled by an intense boyhood curiosity and love for nature and mathematics, he graduated valedictorian from the Columbia Grammar & Preparatory School and subsequently entered Yale at the age of 15 as a member of Jonathan Edwards College. At Yale, he participated in the William Lowell Putnam Mathematical Competition and was on the team representing Yale University (along with Murray Gerstenhaber and Henry O. Pollak) that won the second prize in 1947. Gell-Mann earned a bachelor's degree in physics from Yale in 1948, and a PhD in physics from MIT in 1951. His advisor at MIT was Victor Weisskopf.

7.4.2 Physics career

In 1958, Gell-Mann and Richard Feynman, in parallel with the independent team of George Sudarshan and Robert Marshak, discovered the chiral structures of the weak interaction in physics. This work followed the experimental discovery of the violation of parity by Chien-Shiung Wu, as suggested by Chen Ning Yang and Tsung-Dao Lee, theoretically.

Gell-Mann's work in the 1950s involved recently discovered cosmic ray particles that came to be called kaons and hyperons. Classifying these particles led him to propose that a quantum number called strangeness would be conserved by the strong and the electromagnetic interactions, but not by the weak interactions. Another of Gell-Mann's ideas is the Gell-Mann-Okubo formula, which was, initially, a formula based on empirical results, but was later explained by his quark model. Gell-Mann and Abraham Pais were involved in explaining several puzzling aspects of the physics of these particles.

In 1961, this led him (and Kazuhiko Nishijima) to introduce a classification scheme for hadrons, elementary particles that participate in the strong interaction. (This scheme had been independently proposed by Yuval Ne'eman.) This scheme is now explained by the quark model. Gell-Mann referred to the scheme as the *Eightfold Way*, because of the *octets* of particles in the classification. (The term is a reference to the eightfold way of Buddhism.)

In 1964, Gell-Mann and, independently, George Zweig went on to postulate the existence of quarks, particles of which the hadrons of this scheme are composed. The name was coined by Gell-Mann and is a reference to the novel *Finnegans Wake*, by James Joyce ("Three quarks for Muster Mark!" book 2, episode 4.) Zweig had referred to the particles as "aces",*[7] but Gell-Mann's name caught on. Quarks, antiquarks, and gluons were soon established as the underlying elementary objects in the study of the structure of hadrons. He was awarded a Nobel Prize in physics in 1969 for his contributions and discoveries concerning the classification of elementary particles and their interactions.*[8]

In 1972 he and Harald Fritzsch introduced the conserved quantum number "color charge", and later, together with Heinrich Leutwyler, they coined the term quantum chromodynamics (QCD) as the gauge theory of the strong interaction. The quark model is a part of QCD, and it has been robust enough naturally to accommodate the discovery of new "flavors" of quarks, which superseded the eightfold way scheme.

During the 1990s, Gell-Mann's interest turned to the emerging study of complexity. He played a central role in the founding of the Santa Fe Institute, where he continues to work as a distinguished professor.

He wrote a popular science book about these matters, *The Quark and the Jaguar: Adventures in the Simple and the Complex*. The title of the book is taken from a line of a poem by Arthur Sze: "The world of the quark has everything to do with a jaguar circling in the night".*[9]

The author George Johnson has written a biography of Gell-Mann, which is titled *Strange Beauty: Murray Gell-Mann, and the Revolution in 20th-Century Physics*, which Dr. Gell-Mann has criticized as inaccurate. The Nobel Prize–winning physicist Philip Anderson, in his chapter on Gell-Mann,*[10] says that Johnson's biography is excellent.

Both Anderson and Johnson say that Gell-Mann is a perfectionist and that his semibiographical, *The Quark and the Jaguar* is consequently incomplete.

7.4.3 Personal life

Gell-Mann married Marcia Southwick in 1992, after the death of his first wife, J. Margaret Dow (d. 1981), whom he married in 1955. His children are Elizabeth Sarah Gell-Mann (b. 1956) and Nicholas Webster Gell-Mann (b. 1963); and he has a stepson, Nicholas Southwick Levis (b. 1978).

Gell-Mann has interests in birdwatching, collecting antiques, ranching, historical linguistics, archaeology, natural history, the psychology of creative thinking, other subjects connected with biological, and cultural evolution and with learning.*[3]*[11] Along with S. A. Starostin, he established the *Evolution of Human Languages project*[12] at the Santa Fe Institute.

He is currently the Robert Andrews Millikan Professor of Theoretical Physics Emeritus at California Institute of Technology as well as a University Professor in the Physics and Astronomy Department of the University of New Mexico in Albuquerque, New Mexico, and the Presidential Professor of Physics and Medicine at the University of Southern California. He is a member of the editorial board of the *Encyclopædia Britannica*. In 1984 Gell-Mann co-founded the Santa Fe Institute—a non-profit theoretical research institute in Santa Fe, New Mexico—to study complex systems and disseminate the notion of a separate interdisciplinary study of complexity theory.

He was a postdoctoral fellow at the Institute for Advanced Study in 1951, and a visiting research professor at the University of Illinois at Urbana–Champaign from 1952 to 1953. He was a visiting associate professor at Columbia University and an associate professor at the University of Chicago in 1954–55 before moving to the California Institute of Technology, where he taught from 1955 until he retired in 1993.

As a humanist and an agnostic, Gell-Mann is a Humanist Laureate in the International Academy of Humanism.*[13]*[14]

Gell-Mann endorsed Barack Obama for the United States presidency in October 2008.*[15]

7.4.4 Awards and honors

- Nobel Prize in Physics (1969)

- Ernest O. Lawrence Award (1966)

- Academy of Achievement Golden Plate Award (1962)
- Albert Einstein Medal (2005)
- Yale University – D.Sc (h.c.), 1959
- American Physical Society – Dannie Heineman Prize for Mathematical Physics, 1959
- University of Chicago – Sc.D.(h.c.), 1967
- Franklin Medal, 1967
- National Academy of Sciences – John J. Carty Award, 1968[*][16]
- University of Illinois – Sc.D.(h.c.), 1968
- Wesleyan University – Sc.D.(h.c.), 1968
- Research Corporation Award, 1969
- University of Turin, Italy – Honorary Doctorate, 1969
- University of Utah – Sc.D.(h.c.), 1970
- Columbia University – Sc.D.(h.c.), 1977
- University of Cambridge, England – Sc.D.(h.c.), 1980
- United Nations Environment Programme Roll of Honor for Environmental Achievement (The Global 500), 1988
- World Federation of Scientists – Erice Prize, 1990
- University of Oxford, England – D.Sc.(h.c.), 1992
- Southern Illinois University – Sc.D.(h.c.), 1993
- University of Florida – Sc.D.(h.c.), Doctorate of Natural Resources, 1994
- Southern Methodist University – Sc.D.(h.c.), 1999
- American Humanist Association – Humanist of the Year, 2005
- Helmholtz-Medal of the Berlin-Brandenberg Academy of Sciences and Humanities, 2014[*][17]

7.4.5 See also

- List of Jewish Nobel laureates
- Quark
- Gell-Mann matrices

7.4.6 References

[1] "Nobel Prize Winner Appointed Presidential Professor at USC".

[2] "CERN-affiliated article by Gell-Mann". Springer. Retrieved 11 June 2015.

[3] http://www.nobelprize.org/nobel_prizes/physics/laureates/1969/gell-mann-bio.html

[4] M. Gell-Mann (October 1997). "My Father". *Web of Stories*. Retrieved 2010-10-01.

[5] J. Brockman (2003). "The Making of a Physicist: A talk with Murray Gell-Mann". *Edge.org*. Retrieved 2010-10-01.

[6] Profile, imdb.com; accessed April 26, 2015.

[7] G. Zweig (1980) [1964]. "An SU(3) model for strong interaction symmetry and its breaking II". In D. Lichtenberg and S. Rosen. *Developments in the Quark Theory of Hadrons* **1**. Hadronic Press. pp. 22–101.

[8] Nobel Prize in Physics, 1969

[9] "Murray Gell-Mann – Physicist – The decision to write "The Quark and the Jaguar" - Web of Stories".

[10] Anderson, Philip W. (2011). "Ch. V Genius. Search for Polymath's Elementary Particles". *More and Different: Notes from a Thoughtful Curmudgeon*. World Scientific. pp. 241–2. ISBN 978-981-4350-14-3.Philip Anderson, *More and Different*, Chapter V, World Scientific, 2011.

[11] SANTA FE, New Mexico (NM) Political Contributions by Individuals

[12] Peregrine, Peter Neal (2009). *Ancient Human Migrations: A Multidisciplinary Approach*. University of Utah Press. p. ix. ISBN 978-0-87480-942-8. "Sergei+Starostin+and+I+established+the+Evolution+of+Human+Languages+p "Sergei Starostin and I established the Evolution of Human Languages project"

[13] The International Academy of Humanism at the website of the Council for Secular Humanism. Retrieved 18 October 2007. Some of this information is also at the International Humanist and Ethical Union website

[14] Herman Wouk (2010). *The Language God Talks: On Science and Religion*. Hachette Digital, Inc. ISBN 9780316096751. Feynman, Gell-Man, Weinberg, and their peers accept Newton's incomparable stature and shrug off his piety, on the kindly thought that the old man got into the game too early. ...As for Gell-Mann, he seems to see nothing to discuss in this entire God business, and in the index to *The Quark and the Jaguar* God goes unmentioned. Life he called a "complex adaptive system" which produces interesting phenomena such as the jaguar and Murray Gell-Mann, who discovered the quark. Gell-Mann is a Nobel-class tackler of problems, but for him the existence of God is not one of them.

[15] https://www.youtube.com/watch?v=hFaTrOFXrs8

[16] "John J. Carty Award for the Advancement of Science". National Academy of Sciences. Retrieved 7 March 2011.

[17] http://www.bbaw.de/presse/pressemitteilungen/pressemitteilungen/portal_factory/PDFs/bbaw-10-2014

7.4.7 Further reading

- Biography and Bibliographic Resources, from the Office of Scientific and Technical Information, United States Department of Energy

- Encyclopædia Britannica's Biography of Murray Gell-Mann

- Fritzsch, H.; Gell-Mann, M.; Leutwyler, H. (26 November 1973). "Advantages of the color octet gluon picture" (PDF). *Physics Letters B* **47** (4): 365–8. Bibcode:1973PhLB...47..365F. doi:10.1016/0370-2693(73)90625-4.

- Fritzsch, H.; Gell-Mann, M. (1972). "Current algebra- quarks and what else?". In Jackson, J.D.; Roberts, A.; International Union of Pure and Applied Physics. *Proceedings of the XVI International Conference on High Energy Physics* **2**. National Accelerator Laboratory. pp. 135–165. OCLC 57672574.

- Murray Gell-Mann tells his life story at Web of Stories

- Strange Beauty home page

- The Making of a Physicist: A Talk With Murray Gell-Mann

- Berreby, D. (8 May 1994). "The Man Who Knows Everything". *New York Times*.

- The Man With Five Brains

- The many worlds of Murray Gell-Mann

- The Simple and the Complex, Part I: The Quantum and the Quasi-Classical with Murray Gell-Mann, Ph.D.

- Nobel Prize Biography

7.4.8 External links

- Inspire profile of Murray Gell-Mann's publication (nuclear and particle physics)

- Biography and Bibliographic Resources, from the Department of Energy, Office of Scientific & Technical Information

- Gell-Mann's Home Page at SFI
- Murray Gell-Mann at TED
    - "Beauty, truth and ... physics?" (TED2007)
    - "The ancestor of language" (TED2007)
- Murray Gell-Mann Video Interview with the Academy of Achievement in 1990
- Murray Gell-Mann talks quarks (Video)
- Murray Gell-Mann at the Mathematics Genealogy Project
- Scientific publications of M. Gell-Mann on INSPIRE-HEP

7.5 Sheldon Lee Glashow

Sheldon Lee Glashow (born December 5, 1932) is a Nobel Prize winning American theoretical physicist. He is the Metcalf Professor of Mathematics and Physics at Boston University and Higgins Professor of Physics, Emeritus, at Harvard University, and is a member of the Board of Sponsors for the *Bulletin of the Atomic Scientists*.

7.5.1 Birth and education

Sheldon Lee Glashow was born in New York City, to Jewish immigrants from Russia, Bella (Rubin) and Lewis Gluchovsky, a plumber.[1] He graduated from Bronx High School of Science in 1950. Glashow was in the same graduating class as Steven Weinberg, whose own research, independent of Glashow's, would result in the two and Abdus Salam sharing the same 1979 Nobel Prize in Physics (see below).[2] Glashow received a Bachelor of Arts degree from Cornell University in 1954 and a Ph.D. degree in physics from Harvard University in 1959 under Nobel-laureate physicist Julian Schwinger. Afterwards, Glashow became a NSF fellow at NORDITA and joined the University of California, Berkeley where he was an Associate Professor from 1962–66.[3] He joined the Harvard physics department as a professor in 1966, and was named Higgins Professor of Physics in 1979; he became emeritus in 2000. Glashow has been a visiting professor or scientist at CERN, the University of Marseilles, MIT, Brookhaven Laboratory, Texas A&M, the University of Houston, and Boston University.[2]

7.5.2 Research

In 1961, Glashow extended electroweak unification models due to Schwinger by including a short range neutral current, the Z0. The resulting symmetry structure that Glashow proposed, SU(2) × U(1), forms the basis of the accepted theory of the electroweak interactions. For this discovery, Glashow along with Steven Weinberg and Abdus Salam, was awarded the 1979 Nobel Prize in Physics.

In collaboration with James Bjorken, Glashow was the first to predict a fourth quark, the charm quark, in 1964. This was at a time when 4 leptons had been discovered but only 3 quarks proposed. The development of their work in 1970, the GIM mechanism showed that the two quark pairs: (d,s), (u,c), would largely cancel out flavor changing neutral currents, which had been observed experimentally at far lower levels than theoretically predicted on the basis of 3 quarks only. The prediction of the charm quark also removed a technical disaster for any quantum field theory with unequal numbers of quarks and leptons —an anomaly —where classical field theory symmetries fail to carry over into the quantum theory.

In 1973, Glashow and Howard Georgi proposed the first grand unified theory. They discovered how to fit the gauge forces in the standard model into an SU(5) group, and the quarks and leptons into two simple representations. Their theory qualitatively predicted the general pattern of coupling constant running, with plausible assumptions, it gave rough mass ratio values between third generation leptons and quarks, and it was the first indication that the law of Baryon number is inexact, that the proton is unstable. This work was the foundation for all future unifying work.

Glashow shared the 1977 J. Robert Oppenheimer Memorial Prize with Feza Gürsey.*[4]*[5]

7.5.3 Superstring theory

Glashow is a skeptic of Superstring theory due to its lack of experimentally testable predictions. He had campaigned to keep string theorists out of the Harvard physics department, though the campaign failed.*[6] About ten minutes into "String's the Thing", the second episode of *The Elegant Universe* TV series, he describes superstring theory as a discipline distinct from physics, saying "...you may call it a tumor, if you will..." .*[7]

7.5.4 Personal life

Glashow is married to the former Joan Shirley Alexander. They have four children.*[2] Joan's sister was Lynn Margulis, making Carl Sagan his former brother-in-law. Daniel Kleitman, who was also a doctoral student of Julian Schwinger, is his brother-in-law, through Joan's other sister, Sharon.

In 2003 he was one of 22 Nobel Laureates who signed the

Professor Glashow's KHC PY 101 Energy class, at Boston University's Kilachand Honors College (Spring 2011)

Humanist Manifesto.*[8]

7.5.5 Works

- *The charm of physics* (1991) ISBN 0-88318-708-6
- *From alchemy to quarks: the study of physics as a liberal art* (1994) ISBN 0-534-16656-3
- *Interactions: a journey through the mind of a particle physicist and the matter of this world* (1988) ISBN 0-446-51315-6
- *First workshop on grand unification: New England Center, University of New Hampshire, April 10–12, 1980* edited with Paul H. Frampton and Asim Yildiz (1980) ISBN 0-915692-31-7
- *Third Workshop on Grand Unification, University of North Carolina, Chapel Hill, April 15–17, 1982* edited with Paul H. Frampton and Hendrik van Dam (1982) ISBN 3-7643-3105-4
- "Desperately Seeking Superstrings?" with Paul Ginsparg in *Riffing on Strings: Creative Writing Inspired by String Theory* (2008) ISBN 978-0-9802114-0-5

7.5.6 See also

- List of Jewish Nobel laureates
- GIM mechanism

7.5.7 References

Notes

[1] Sheldon Lee Glashow – Britannica Encyclopedia. Britannica.com. Retrieved on 2012-07-27.

[2] Glashow's autobiography. Nobelprize.org. Retrieved on 2012-07-27.

[3] Sheldon Glashow. Jewishvirtuallibrary.org. Retrieved on 2012-07-27.

[4] Walter, Claire (1982). *Winners, the blue ribbon encyclopedia of awards*. Facts on File Inc. p. 438. ISBN 9780871963864.

[5] "Gürsey and Glashow share Oppenheimer memorial". *Physics Today* (American Institute of Physics). May 1977. doi:10.1063/1.3037556. Retrieved 1 March 2015.

[6] Jim Holt (2006-10-02), "Unstrung", The New Yorker. Retrieved on 2012-07-27.

[7] "[T]here ain't no experiment that could be done nor is there any observation that could be made that would say, `You guys are wrong.' The theory is safe, permanently safe." He also said, "Is this a theory of Physics or Philosophy? I ask you" NOVA interview

[8] "Notable Signers". *Humanism and Its Aspirations*. American Humanist Association. Retrieved October 2, 2012.

7.5.8 External links

- Sheldon Lee Glashow at the Mathematics Genealogy Project
- Nobel lecture
- Biography and Bibliographic Resources, from the Office of Scientific and Technical Information, United States Department of Energy
- Sheldon Lee Glashow
- Interview with Glashow on Superstrings
- Contributions to the theory of the unified weak and electromagnetic interaction between elementary particles, including inter alia the prediction of the weak neutral current.
- Sheldon Glashow Boston University Physics Department
- Sheldon Glashow Photos
- Interview with Glashow About Contemporary Physics and Winning the Nobel Prize
- Scientific publications of S. L. Glashow on INSPIRE-HEP

7.6 Haim Harari

Haim Harari (Hebrew: חיים הררי) (born 18 November 1940) is an Israeli theoretical physicist who has made contributions in particle physics, science education, and other fields.

Haim Harari

7.6.1 Biography

Haim Harari was born in Jerusalem. His family has lived in the area which is now Israel for five generations. His parents were Knesset member Yizhar Harari and Dina Neumann. Harari received his M.Sc. and Ph.D. in Physics from Hebrew University of Jerusalem.

7.6.2 Academic career

In 1967, after completing his Ph.D, he became the youngest professor ever at the Weizmann Institute. He is the Chair of the Board of the Davidson Institute of Science Education at the Weizmann Institute and Chair of the Management Committee of the Weizmann Global Endowment Management Trust in New York. He was the President, from 1988 to 2001, of the Weizmann Institute of Science. During his presidency, the Weizmann Institute, entirely dedicated to basic research, became one of the leading royalty earning academic research organizations in the world.

Haim Harari has made major contributions to three different fields: particle physics research on the international scene, science education in the Israeli school system and science administration and policy making.

Harari coined the name of the top and bottom quarks,*[1]*[2] predicted in 1973 by Kobayashi and Maskawa,*[3] and made the first complete statement of the standard six quarks and six leptons model of particle physics (at the Stanford 1975 Lepton-Photon Conference).*[1] He also proposed the Rishon Model,*[4]*[5] a model for a substructure of quarks and leptons, currently believed to be the most fundamental particles in nature. There is no experimental evidence yet for such substructure.

He is the founder of Perach, a national tutoring and mentoring project*[6] in which Israeli undergraduates receive a tuition fellowship in return for devoting four hours per week to a child from an underprivileged socioeconomic background. He also initiated and established a science teaching center in which high school students perform all their physics studies in advanced laboratories and with highly qualified teachers, instead of pursuing the same in their own schools. Harari has been chairman of both projects, since their founding.

7.6.3 Awards and recognition

- Membership in the Israel Academy of Sciences (1978);
- Membership in the American Academy of Arts and Sciences (2010);
- Rothschild Prize in Physics (1976);
- Israel Prize, in the exact sciences (1989);*[7]
- The EMET Prize for Art, Science and Culture in Education (2004);*[8]
- Commander Cross of the Order of Merit presented by the President of Germany;
- Cross of Honor, Science and Art, First Class presented by Austria;
- Golden cross of honor for service to the land of Lower Austria (2011);
- Harnack medal from the Max Planck Institute (2001), acknowledging his contribution to co-operation between the Max Planck Society and the Weizmann Institute.

In 2004 Harari gave a speech entitled "A View from the Eye of the Storm" offering insights into the problems of the Middle East. He eventually turned it into a book of the same name.

7.6.4 Published works

- *A View from the Eye of the Storm: Terror and Reason in the Middle East,* HarperCollins, 2005.

7.6.5 References

[1] H. Harari (1975). "A new quark model for hadrons". *Physics Letters B* **57B** (3): 265. Bibcode:1975PhLB...57..265H. doi:10.1016/0370-2693(75)90072-6.

[2] K.W. Staley (2004). *The Evidence for the Top Quark*. Cambridge University Press. pp. 31–33. ISBN 978-0-521-82710-2.

[3] M. Kobayashi, T. Maskawa (1973). "CP-Violation in the Renormalizable Theory of Weak Interaction". *Progress of Theoretical Physics* **49** (2): 652–657. Bibcode:1973PThPh..49..652K. doi:10.1143/PTP.49.652.

[4] H. Harari (1979). "A schematic model of quarks and leptons". *Physics Letters B* **86** (1): 83–86. Bibcode:1979PhLB...86...83H. doi:10.1016/0370-2693(79)90626-9.

[5] H. Harari, N. Seiberg (1982). "The rishon model". *Physics Letters B* **204** (1): 141–167. Bibcode:1982NuPhB.204..141H. doi:10.1016/0550-3213(82)90426-6.

[6] "Perach Official Site".

[7] "Israel Prize Official Site - Recipients in 1989 (in Hebrew)".

[8] "EMET Prize Official Site - Recipients in 2004".

7.6.6 External links

- Haim Harari's homepage at the Weizmann Institute of Science
- Davidson Institute of Science Education
- Book: A View from the Eye of the Storm - Terror and Reason in the Middle East

7.6.7 See also

- List of Israel Prize recipients
- Harari

7.7 John Iliopoulos

John Iliopoulos (Greek language: Ιωάννης Ηλιόπουλος; 1940, Kalamata) is a Greek physicist and the first person to present the Standard Model of particle physics in a single report. He is best known for his prediction of the charm quark with Sheldon Lee Glashow and Luciano Maiani (the "GIM mechanism").*[1] Iliopoulos is also known for the Fayet-Iliopoulos D-term formula, which was introduced in 1974. He is currently an honorary member of Laboratory of theoretical physics of École Normale Supérieure, Paris.

Iliopoulos graduated from National Technical University of Athens (NTUA) in 1962 as an Mechanical-Electrical Engineer. He continued his studies in the field of Theoretical Physics in Paris University, and in 1963 he obtained the D.E.A, in 1965 the Doctorat 3e Cycle, and in 1968 the Doctorat d' Etat titles. Between the years 1966 and 1968 he was a scholar at CERN, Geneva. From 1969 till 1971 he was a Research Assosiate in Harvard University. In 1971 he returned in Paris and began working at CNRS. He also held the director position of the Laboratory of Theoretical Physics of Ecole Normale Superieure between the years 1991-1995 and 1998-2002. In 2002, Iliopoulos was the first recipient of the Aristeio prize, which has been instituted to recognize Greeks who have made significant contributions towards furthering their chosen fields of science. Iliopoulos and Maiani were jointly awarded the 1987 Sakurai Prize for theoretical particle physics. In 2007 Iliopoulos and Maiani received the Dirac Medal of the ICTP "(f)or their work on the physics of the charm quark, a major contribution to the birth of the Standard Model, the modern theory of Elementary Particles."

7.7.1 Awards

- 1987 Sakurai Prize of the American Physical Society
- 1996 Doctor honoris causa, Université de la Méditerranée, Aix-Marseille
- 2005 Matteucci Medal, Accademia Nazionale dei XL
- 2007 Dirac Medal, Abdus Salam International Centre for Theoretical Physics, Trieste, Italy

7.7.2 See also

- GIM mechanism

7.7.3 References

[1] S. L. Glashow, J. Iliopoulos, and L. Maiani (1970). "Weak Interactions with Lepton-Hadron Symmetry". *Phys. Rev.* **D2** (7): 1285. Bibcode:1970PhRvD...2.1285G. doi:10.1103/PhysRevD.2.1285.

7.7.4 External links

- *Science World*
- Matteucci Medal National Academy of Science, Italy

7.8 Makoto Kobayashi

Makoto Kobayashi (小林誠 *Kobayashi Makoto*) (born April 7, 1944 in Nagoya, Japan) is a Japanese physicist known for his work on CP-violation who was awarded one fourth of the 2008 Nobel Prize in Physics "for the discovery of the origin of the broken symmetry which predicts the existence of at least three families of quarks in nature."[*][3]

7.8.1 Biography

After completing his PhD at Nagoya University in 1972, Kobayashi worked as a research associate on particle physics at Kyoto University. Together, with his colleague Toshihide Maskawa, he worked on explaining CP-violation within the Standard Model of particle physics. Kobayashi and Maskawa's theory required that there were at least three generations of quarks, a prediction that was confirmed experimentally four years later by the discovery of the bottom quark.

Kobayashi and Maskawa's article, "CP Violation in the Renormalizable Theory of Weak Interaction",[*][4] published in 1973, is the fourth most cited high energy physics paper of all time as of 2010.[*][5] The Cabibbo–Kobayashi–Maskawa matrix, which defines the mixing parameters between quarks was the result of this work. Kobayashi and Maskawa were jointly awarded half of the 2008 Nobel Prize in Physics for this work, with the other half going to Yoichiro Nambu.[*][3]

Academic career

- April, 1972 : Research Associate of Kyoto University
- July, 1979 : Assistant Professor of the National Laboratory of High Energy Physics
- April, 1989 : Professor of the National Laboratory of High Energy Physics, Head of Physics Division II
- April, 1997 : Professor of the Institute of Particle and Nuclear Science, KEK Head of Physics Division II
- April, 2003 : Director, Institute of Particle and Nuclear Studies, KEK
- April, 2004 : Trustee (Director, Institute of Particle and Nuclear Studies), KEK (Inter-University Research Institute Corporation)
- June, 2006 : Professor emeritus of KEK.

Honors

- In October 2008, Kobayashi was honored with Japan's Order of Culture; and an awards ceremony for the Order of Culture was held at the Tokyo Imperial Palace.

7.8.2 Personal life

Kobayashi was born and educated in Nagoya, Japan. He married Sachiko Enomoto in 1975; they had one son, Junichiro. After his first wife died, Kobayashi married Emiko Nakayama in 1990, they had a daughter, Yuka.[*][6]

7.8.3 See also

- *Progress of Theoretical Physics*

7.8.4 References

[1] "Makoto Kobayashi" (Press release). High Energy Accelerator Research Organization. 6 July 2007. Retrieved 2008-10-04.

[2] L. Hoddeson (1977). "Flavor Mixing and *CP* Violation". *The Rise of the Standard Model*. Cambridge University Press. p. 137. ISBN 0-521-57816-7.

[3] *The Nobel Prize in Physics 2008*, The Nobel Foundation, retrieved 2009-10-17

[4] M. Kobayashi, T. Maskawa (1973). "CP-Violation in the Renormalizable Theory of Weak Interaction". *Progress of Theoretical Physics* **49** (2): 652–657. Bibcode:1973PThPh..49..652K. doi:10.1143/PTP.49.652.

[5] "Top Cited Articles of All Time (2010 edition)". SPIRES database. Retrieved 2014-06-21.

[6] "Makoto Kobayashi (Autobiography)". The Nobel Foundation.

7.8.5 External links

- Progress of Theoretical Physics
- Makoto Kobayashi, Professor emeritus of KEK

7.9 Leon M. Lederman

Leon Max Lederman (born July 15, 1922) is an American experimental physicist who received, along with Martin Lewis Perl, the Wolf Prize in Physics in 1982, for their research on quarks and leptons, and the Nobel Prize for Physics in 1988, along with Melvin Schwartz and Jack Steinberger, for their research on neutrinos. He is Director Emeritus of Fermi National Accelerator Laboratory (Fermilab) in Batavia, Illinois, USA. He founded the Illinois Mathematics and Science Academy, in Aurora, Illinois in 1986, and has served in the capacity of Resident Scholar since 1998.*[2] In 2012, he was awarded the Vannevar Bush Award for his extraordinary contributions to understanding the basic forces and particles of nature.*[3]

7.9.1 Early life and career

Lederman was born in New York City, New York, the son of Minna (née Rosenberg) and Morris Lederman, a laundryman.*[4] Lederman graduated from the James Monroe High School in the South Bronx. He received his bachelor's degree from the City College of New York in 1943, and received a Ph.D. from Columbia University in 1951. He then joined the Columbia faculty and eventually became Eugene Higgins Professor of Physics. In 1960, on leave from Columbia, he spent some time at CERN in Geneva as a Ford Foundation Fellow.*[5] He took an extended leave of absence from Columbia in 1979 to become director of Fermilab. Resigning from Columbia (and retiring from Fermilab) in 1989 to teach briefly at the University of Chicago, he then moved to the physics department of the Illinois Institute of Technology, where he currently serves as the Pritzker Professor of Science. In 1991, Lederman became President of the American Association for the Advancement of Science.

Lederman is also one of the main proponents of the "Physics First" movement. Also known as "Right-side Up Science" and "Biology Last," this movement seeks to rearrange the current high school science curriculum so that physics precedes chemistry and biology.

A former president of the American Physical Society, Lederman also received the National Medal of Science, the Wolf Prize and the Ernest O. Lawrence Medal. Lederman serves as President of the Board of Sponsors of The Bulletin of the Atomic Scientists. He also served on the board of trustees for Science Service, now known as Society for Science & the Public, from 1989 to 1992, and is a member of the JASON defense advisory group.*[6]

Among his achievements are the discovery of the muon neutrino in 1962 and the bottom quark in 1977. These helped establish his reputation as among the top particle physicists.

In 1977, a group of physicists led by Leon Lederman announced that a particle with a mass of about 6.0 GeV was being produced by the Fermilab particle accelerator. The particle's initial name was the greek letter Upsilon (Υ). After taking further data, the group discovered that this particle did not actually exist, and the "discovery" was named "Oops-Leon" as a pun on the original name (mispronounced /ˈjuːpsilɒn/) and Lederman's first name.

As the director of Fermilab and subsequent Nobel physics prizewinner, Leon Lederman was a very prominent early supporter – some sources say the architect*[7] or proposer*[8] – of the Superconducting Super Collider project, which was endorsed around 1983, and was a major proponent and advocate throughout its lifetime.*[9]*[10] Lederman later wrote his 1993 popular science book *The God Particle: If the Universe Is the Answer, What Is the Question?* – which sought to promote awareness of the significance of such a project – in the context of the project's last years and the changing political climate of the 1990s.*[11] The increasingly moribund project was finally shelved that same year after some $2 billion of expenditure.*[7]

In 1988, Lederman received the Nobel Prize for Physics along with Melvin Schwartz and Jack Steinberger "for the neutrino beam method and the demonstration of the doublet structure of the leptons through the discovery of the muon neutrino".*[2] Lederman also received the National Medal of Science (1965), the Elliott Cresson Medal for Physics (1976), the Wolf Prize for Physics (1982) and the Enrico Fermi Award (1992).

In 1995, he received the Chicago History Museum "Making History Award" for Distinction in Science Medicine and Technology.

Lederman was an early supporter of Science Debate 2008, an initiative to get the then-candidates for president, Barack Obama and John McCain, to debate the nation's top science policy challenges. In October 2010, Lederman participated in the USA Science and Engineering Festival's Lunch with a Laureate program where middle and high school students got to engage in an informal conversation with a Nobel Prize-winning scientist over a brown-bag lunch.*[12] Lederman was also a member of the USA Science and Engineering Festival's Advisory Board *[13] and CRDF Global.

7.9.2 Personal life

Lederman was born in New York to a family of Jewish immigrants from Russia.*[14] His father operated a hand laundry while encouraging Leon to pursue his education. He went to elementary school in New York City, continuing on to college and his doctorate in the city.*[15]

In his book, *The God Particle: If the Universe Is the Answer, What Is the Question?*, Lederman writes that, although he was a chemistry major, he became fascinated with physics, because of the clarity of the logic and the unambiguous results from experimentation. His best friend during his college years, Martin Klein, convinced him of "the splendors of physics during a long evening over many beers." After that conversation he became resolute and unwavering regarding his desire to pursue physics. When he joined the Army with a B.S. in Chemistry, he was determined to become a physicist following his service.*[16]

After three years in the U.S. Army during World War II, he took up physics at Columbia University, and received his Masters in 1948. Lederman began his Ph.D research working with Columbia's Nevis synchro-cyclotron,*[17] which was the most powerful particle accelerator in the world at that time.*[16] Dwight D. Eisenhower, then the president of Columbia University, and future president of the United States, cut the ribbon dedicating the synchro-cyclotron in June 1950.*[18] These atom smashers were just coming of age at this time and created the new discipline of particle physics.*[16]

After receiving his Ph.D and then becoming a faculty member at Columbia University he was promoted to full professor in 1958.*[19]

In "The God Particle" he once wrote "The history of atomism is one of reductionism – the effort to reduce all the operations of nature to a small number of laws governing a small number of primordial objects."*[20] And this was the quest he undertook. This book shows that he pursued the quark, and hopes to find the Higgs boson. The top quark, which he and other physicists realized must exist according to the standard model, was, in fact, produced at Fermilab not long after this book was published.*[21]

He is known for his sense of humor in the physics community.*[22] On August 26, 2008 Lederman was video-recorded by a science focused organization called ScienCentral, on the street in a major U.S. city, answering questions from passersby.*[23] He answered questions such as "What is the strong force?" and "What happened before the Big Bang?".

He has three children with his first wife, Florence Gordon, and now lives with his second wife, Ellen (Carr), in Driggs, Idaho.*[24]

He is an atheist.*[25]*[26]

7.9.3 Publications

- *The God Particle: If the Universe Is the Answer, What Is the Question?* by Leon M. Lederman, Dick Teresi (ISBN 0-385-31211-3)

- *From Quarks to the Cosmos* by Leon Lederman and David N. Schramm (ISBN 0-7167-6012-6)

- *Portraits of Great American Scientists* Leon M. Lederman, et al. (ISBN 1-57392-932-8)

- *Symmetry and the Beautiful Universe* Leon M. Lederman and Christopher T. Hill (ISBN 1-59102-242-8)

- *What We'll Find Inside the Atom* by Leon Lederman is an essay he wrote for the September 15, 2008 issue of Newsweek

- "Quantum Physics for Poets" Leon M. Lederman and Christopher T. Hill (ISBN 978-1616142339)

7.9.4 Honorary degrees and awards

- Election to the National Academy of Sciences, 1965.

-

7.9. LEON M. LEDERMAN

- U.S. National Medal of Science, 1965.

- Member, American Academy of Arts and Sciences, 13 May 1970.

- Townsend Harris Medal, Alumni Association of the City College of New York, 1973.

- Elliot Cresson Prize of the Franklin Institute, 1976.

- President Jimmy Carter's Committee on the National Medal of Science, 21 March 1979.

- Wolf Foundation Prize in Physics, Israel, 1982.

- Doctor of Science, University of Chicago, Chicago, Illinois, 10 June 1983.

- Doctor of Humane Letters and Science, IIT, Chicago, Illinois, 17 May 1987.

- Doctor of Science, Lake Forest College, Lake Forest, Illinois, 7 May 1988.

- Honorary Degree of Doctor of Science, Carnegie Mellon University, Pittsburgh, Pennsylvania, 15 May 1988.

- Department of Energy Distinguished Associate Award, May 1988.

- Doctor of Science, City College of New York, New York, New York, 8 June 1988.

- Nobel Prize in Physics, December 1988.

- La laurea honoris causa in Fisica, Universita' degli Studi di Pisa, 21 March 1989.

- Member, American Philosophical Society, 21 April 1989.

- Doctor of Science, honoris causae, Aurora University, Aurora, Illinois, 20 May 1989.

- Doctor of Humane Letters, Columbia College, Chicago, Illinois, 3 June 1989.

- Citation on the occasion of the dedication of IMSA (founded 1985) to the State of Illinois in honor of Leon Lederman and Gov. James Thompson, IMSA, Aurora, Illinois, 10 June 1989.

- Doctor of Humane Letters, Rush University, Chicago, Illinois, 10 June 1989.

- Doctor of Science, University of Illinois, Chicago, Illinois, 11 June 1989.

- Doctor en Filosofia – Fisica, Honoris Causa, Universidad de Guanajuato, Guanajuato, Mexico, 2 August 1989.

- Honorary Degree, Academia Nacional de Ciencias Exactas, Fisicas y Naturales, Buenos Aires, Argentina, 3 November 1989.

- Doctor of Philosophy of Physics, honoris causa, University of Guanajuato, Guanajuato, Mexico, 1989.

- Appointment as member of the Secretary of Energy Advisory Board, U.S. Department of Energy, Washington, DC, 15 February 1990.

- Laureate of the Lincoln Academy, State of Illinois, Springfield, Illinois, 21 April 1990.

- Doctor of Science, Bradley University, Peoria, Illinois, 19 May 1990.

- Doctor of Science, Columbia University, New York, New York, 26 September 1990.

- Public Affairs Committee Award, American Chemical Society, 22 March 1991.

- 1991 William Proctor Prize, Sigma Xi, The Scientific Research Society, 1991.

- Doctor of Humane Letters, honoris causa, Mount Sinai School of Medicine, The City University of New York, New York City, New York, 11 May 1992.

- Doctor of Science, Adelphi University, Long Island, New York, 17 May 1992.

- Doctor of Science, honoris causa, Saint Xavier University, Chicago, Illinois, 23 May 1992.

- Doctor of Science – honoris causa, University of Southampton, Southampton, United Kingdom, 10 July 1992.

- Doctor of Science, honoris causa, Drury College, Springfield, Missouri, 14 October 1992.

- Enrico Fermi Prize of the U.S. Department of Energy, 1992.

- Doctor of Humane Letters, State University of New York, Geneseo, New York, 15 May 1993.

- Doctor of Science, honoris causa, Case-Western Reserve University, Cleveland, Ohio, 23 May 1993.

- President's Medal, The City College, The City University of New York, New York, New York, 28 May 1993.

- Doctor of Science, honoris causa, Marywood College, Scranton, Pennsylvania, 15 October 1993.

- Doctor of Humane Letters, honoris causa, Illinois Benedictine College, Lisle, Illinois, 21 May 1994.

- Appointment as a Tetelman Fellow at Jonathan Edwards College, Yale University, New Haven, Connecticut, 7 November 1994.

- Doctor of Humane Letters, University of Dallas, Dallas, Texas, 15 November 1994.

- The first Enrico Fermi History Maker Award, for distinction in Science, Medicine and Technology, Chicago Historical Society, Chicago, Illinois, 8 June 1995.

- Doctor of Humane Letters, DePaul University, Chicago, Illinois, 11 June 1995.

- Ordem Nacional do Merito Cientifico, Brasilia, Brazil, 13 June 1995.

- Universitario de la Universidad Nacional de San Antonio Abad Del Cusco, Cusco, Peru, Doctor, honoris causa, 16 August 1995.

- University of Notre Dame, Notre Dame, Indiana, Doctor of Science, honoris causa, 18 May 1997.

- Doctor of Science, Bethany College, Bethany, Virginia, 24 May 1997.

- Doctor of Science, Illinois State University, Normal, Illinois, 17 May 1998.

- Doctor of Science, honoris causa, Clark University, Worcester, Massachusetts, 17 May 1998.

- Doctor of Science, Pennsylvania State University, University Park, Pennsylvania, December 1998.

- Doctor, honoris causa, Institute for High Energy Physics, Protvino, Russia, 15 July 1998.

- Member, President Bill Clinton's Commission on White House Fellowships, 13 April 1999.

- Diploma, Miembro Correspondiente, La Academia Mexicana de Ciencias, October 1999.

- 1999 Medallion, Division of Particles and Fields, Mexican Physical Society, Mérida, Yucatán, Mexico, 11 November 1999.

- Honorary Professor, Beijing Normal University, Beijing, China, 5 November 2000.

- Doctor of Humane Letters, Roosevelt University, Chicago, Illinois, 21 January 2001.

- Doctor of Science, Florida Institute of Technology, Melbourne, Florida, May 2004.

- Doctor of Public Service, The George Washington University, Washington, D.C., 16 May 2004.

- Doctor of Science Education, honoris causa, The Ohio State University, Columbus, Ohio, 12 December 2004.

- Doctor of Humane Letters, Honoris Causa, Dominican University, River Forest, Illinois, 6 May 2006.

- Doctor of Science, Honoris Causa, Gustavus Adolphus College, St. Peter, Minnesota, 24 October 2008.

- Vannevar Bush Prize, 2012.*[27]

7.9.5 See also

- List of Jewish Nobel laureates

7.9.6 References

[1] American Scientists - Charles W. Carey - Google Books

[2] Lederman, Leon M. (1988). Frängsmyr, Tore; Ekspång, Gösta, eds. "The Nobel Prize in Physics 1988: Leon M. Lederman, Melvin Schwartz, Jack Steinberger". *Nobel Lectures, Physics 1981–1990* (Singapore: World Scientific Publishing Co.). Retrieved 22 May 2012.

[3] National Science Board - Honorary Awards - Vannevar Bush Award Recipients

[4] Fermilab: Physics, the Frontier, and Megascience - Lillian Hoddeson, Adrienne W. Kolb, Catherine Westfall - Google Books

[5] "CERN affiliated article by Lederman". Springer. Retrieved 11 June 2015.

[6] Horgan, John (April 16, 2006). "Rent-a-Genius". The New York Times.

[7] ASCHENBACH, JOY (1993-12-05). "No Resurrection in Sight for Moribund Super Collider : Science: Global financial partnerships could be the only way to salvage such a project. But some feel that Congress delivered a fatal blow.". *Los Angeles Times*. Retrieved 16 January 2013. Disappointed American physicists are anxiously searching for a way to salvage some science from the ill-fated superconducting super collider ... "We have to keep the momentum and optimism and start thinking about international collaboration," said Leon M. Lederman, the Nobel Prize-winning physicist who was the architect of the super collider plan

[8] Lillian Hoddeson; Adrienne Kolb. "Vision to reality: From Robert R. Wilson's frontier to Leon M. Lederman's Fermilab". arXiv:1110.0486. Lederman also planned what he saw as Fermilab's next machine, the Superconducting SuperCollider (SSC)

[9] Abbott, Charles (June 1987). "Illinois Issues journal, June 1987". p. 18. Lederman, who considers himself an unofficial propagandist for the super collider, said the SSC could reverse the physics brain drain in which bright young physicists have left America to work in Europe and elsewhere. (direct link to article:)

[10] Kevles, Dan. "Good-bye to the SSC" (PDF). *California Institute of Technology "Engineering & Science"*. 58 no. 2 (Winter 1995): 16–25. Retrieved 16 January 2013. Lederman, one of the principal spokesmen for the SSC, was an accomplished high-energy experimentalist who had made Nobel Prize-winning contributions to the development of the Standard Model during the 1960s (although the prize itself did not come until 1988). He was a fixture at congressional hearings on the collider, an unbridled advocate of its merits []

[11] Calder, Nigel (2005). *Magic Universe:A Grand Tour of Modern Science*. pp. 369–370. The possibility that the next big machine would create the Higgs became a carrot to dangle in front of funding agencies and politicians. A prominent American physicist, Leon lederman, advertised the Higgs as The God Particle in the title of a book published in 1993 ...Lederman was involved in a campaign to persuade the US government to continue funding the Superconducting Super Collider... the ink was not dry on Lederman's book before the US Congress decided to write off the billions of dollars already spent

[12] Archived December 30, 2010 at the Wayback Machine

[13] USA Science and Engineering Festival - Advisors

[14] Humes, Edward (2006), *Over Here: How the G.I. Bill Transformed the American Dream*, Houghton Mifflin Harcourt, p. 275, ISBN 9780151007103.

[15] Lederman, Leon M. "Leon M. Lederman – Autobiography." http://nobelprize.org/nobel_prizes/physics/laureates/1988/lederman-autobio.html Retrieved 06 April 2009. Published 1988, The Nobel Foundation.

[16] The God Particle: If the Universe is the Answer, What is the Question – page 5
by Leon Lederman with Dick Teresi (copyright 1993) Houghton Mifflin Company

[17] see **High Energy Physics** section on this page: http://www.columbia.edu/cu/physics/news/DeptHistory/index/

[18] Eisenhower at ribbon cutting for the dedication of the Nevis synchro-cyclotron http://www.columbia.edu/cu/physics/images/columbia-cyclotron.jpeg

[19] The God Particle: If the Universe is the Answer, What is the Question – page 296
by Leon Lederman with Dick Teresi (copyright 1993) Houghton Mifflin Company

[20] The God Particle: If the Universe is the Answer, What is the Question – page 87
by Leon Lederman with Dick Teresi (copyright 1993) Houghton Mifflin Company

[21] Observation of Top Quark Production in [anti-p][and][p] Collisions with the Collider Detector at Fermilab http://prola.aps.org/abstract/PRL/v74/i14/p2626_1

[22] The God Particle: If the Universe is the Answer, What is the Question – page 17 – by Leon Lederman with Dick Teresi (copyright 1993) Houghton Mifflin Company

[23] Street Corner Science with Leon Lederman, http://blogs.discovermagazine.com/cosmicvariance/2008/08/26/street-corner-science-with-leon-lederman

[24] http://411.info/people/Idaho/Driggs/Lederman-Leon/132837548.html." http://nobelprize.org/nobel_prizes/physics/laureates/1988/lederman-autobio.html Retrieved 29.7.2008. Published 1988, The Nobel Foundation.

[25] Dan Falk (2005). "What About God?". *Universe on a T-Shirt: The Quest for the Theory of Everything*. Arcade Publishing. p. 195. ISBN 9781559707336. "Physics isn't a religion. If it were, we'd have a much easier time raising money." - Leon Lederman

[26] Babu Gogineni (July 10, 2012). "It's the Atheist Particle, actually". Postnoon News. Retrieved 10 July 2012. Leon Lederman is himself an atheist and he regrets the term, and Peter Higgs who is an atheist too, has expressed his displeasure, but the damage has been done!

[27] Fermilab History and Archives Project | Leon M. Lederman Honorary Degrees and Awards

7.9.7 External links

- Education, Politics, Einstein and Charm *The Science Network* interview with Leon Lederman

- Biography and Bibliographic Resources, from the Office of Scientific and Technical Information, United States Department of Energy

- Fermilab's Leon M. Lederman webpage

- The Nobel Prize in Physics 1988

- Video Interview with Lederman from the Nobel Foundation

- Leon M. Lederman – Autobiography

- Timeline of Nobel Prize Winners in Physics webpage for Leon Max Lederman

- *Story of Leon* by Leon Lederman

- Honeywell – Nobel Interactive Studio

- 1976 Cresson Medal recipient from The Franklin Institute.

- Scientific publications of Leon M. Lederman on INSPIRE-HEP

Professor Luciano Maiani : president of the CERN Council from January 1997 until December 1997, and director general of the organisation from 1 January 1999

7.10 Luciano Maiani

Luciano Maiani (born 16 July 1941[*][1] in Rome) is a San Marino citizen physicist best known for his prediction of the charm quark with Sheldon Lee Glashow and John Iliopoulos (the "GIM mechanism"[*][2]).

7.10.1 Academic history

In 1964 Luciano Maiani received his degree in physics and he became a research associate at the Istituto Superiore di Sanità in Italy.[*][3] During that same year he collaborated with R. Gatto's theoretical physics group at the University of Florence.[*][3] He crossed the Atlantic in 1969 to do a post-doctoral fellowship at Harvard's Lyman Laboratory of Physics.[*][3] In 1976 Maiani became a professor of theoretical physics at the University of Rome,[*][3] however he traveled widely during this period, holding visiting professorships at the Ecole Normale Supérieure of Paris (1977)[*][3] and CERN (1979–1980 and 1985–1986).[*][3]

Maiani also took an interest in the direction of particle physics research start on CERN's Scientific Policy Committee from 1984 to 1991.[*][3] Then, in 1993, he became president of Italy's Istituto Nazionale di Fisica Nucleare (INFN).[*][3] From 1993 to 1996 Maiani served as a scientific delegate in CERN council and then as that council's president in 1997.[*][3] Thereafter he became director general of CERN, serving from 1 January 1999[*][4] through the end of 2003.[*][5] From 1995-1997 Maiani chaired the Italian Comitato Tecnico Scientifico, Fondo Ricerca Applicata. At the end of 2007 he was proposed as president of Consiglio Nazionale delle Ricerche, but his nomination was suspended temporally after he signed a letter criticizing the rector of 'La Sapienza' University in Rome, who invited Pope Benedict XVI to give a *lectio magistralis* in 2008.[*][6] However he became the President of CNR since 2008.

Luciano Maiani has authored over 100 scientific publications on the theory of elementary particles often with several co-authors. In 1970 he predicted the charmed quark in a paper with Glashow and Iliopoulos which was later discovered at SLAC and Brookhaven in 1974 and led to a Nobel Prize in Physics for the discoverers. Working with Guido Altarelli in 1974 they explained that the observed octet enhancement in weak non-leptonic decays was due to a leading gluon exchange effect in quantum chromodynamics. They later extended this effect to describe the weak non-leptonic decays of charm and bottom quarks as well and also produced a parton model description of heavy flavor weak decays. In 1976 Maiani analyzed the of CP violation in the six-quark theory and predicted the very small electric dipole moment of the neutron. In the 1980s he started using the numerical simulation of lattice QCD and this led to the first prediction of the decay constant of pseudoscalar charmed mesons and of B mesons. A proponent of supersymmetry, Maiani once said that the search for it was "primary goal of modern particle physics" .[*][7] He has not confined his interest to the theoretical side of physics either, with involvement in ALPI, EUROBALL, DAFNE, VIRGO and the LHC.

7.10.2 Awards

- 1979 Matteucci Medal, Accademia Nazionale dei XL

- 1987 Sakurai Prize of the American Physical Society

- 1996 Doctor honoris causa, Université de la Méditerranée, Aix-Marseille

- 2007 Dirac Medal, Abdus Salam International Centre for Theoretical Physics, Trieste, Italy

- 2010 Doctor honoris causa, Benemérita Universidad Autónoma de Puebla, Puebla, México

7.10.3 See also

- GIM mechanism

7.10.4 External links

- Scientific publications of Luciano Maiani on INSPIRE-HEP

7.10.5 References

[1] "Archives of Luciano Maiani (1941-), Director-General of CERN from 1999 to 2003". CERN. Retrieved 1 September 2015.

[2] "Weak Interactions with Lepton-Hadron Symmetry". Phys.Rev. D (APS) **2**: 1285–1292. 1970. doi:10.1103/PhysRevD.2.1285.

[3] "Professor Luciano Maiani is new President of CERN Council". CERN. Retrieved 1 September 2015.

[4] "Professor Luciano Maiani chosen as next Director General of CERN". CERN. Retrieved 1 September 2015.

[5] "CERN Council looks to bright future". CERN. Retrieved 1 September 2015.

[6] Aprileonline.info: CNR, Maiani e il giudizio divino

[7] CERN Press Release – Professor Luciano Maiani is chosen as next Director General of CERN

7.11 Toshihide Maskawa

Toshihide Maskawa (or **Masukawa**) (益川敏英 *Masukawa Toshihide*, born February 7, 1940 in Nagoya, Japan) is a Japanese theoretical physicist known for his work on CP-violation who was awarded one quarter of the 2008 Nobel Prize in Physics "for the discovery of the origin of the broken symmetry which predicts the existence of at least three families of quarks in nature." *[1]

7.11.1 Biography

A native of Aichi Prefecture, Maskawa graduated from Nagoya University in 1962 and received a Ph.D in particle physics from the same university in 1967. At Kyoto University in the early 1970s, he collaborated with Makoto Kobayashi on explaining broken symmetry (the CP violation) within the Standard Model of particle physics. Maskawa and Kobayashi's theory required that there be at least three generations of quarks, a prediction that was confirmed experimentally four years later by the discovery of the bottom quark.

Maskawa and Kobayashi's 1973 article, "CP Violation in the Renormalizable Theory of Weak Interaction",*[2] is the fourth most cited high energy physics paper of all time as of 2010.*[3] The Cabibbo–Kobayashi–Maskawa matrix, which defines the mixing parameters between quarks was the result of this work. Kobayashi and Maskawa were jointly awarded half of the 2008 Nobel Prize in Physics for this work, with the other half going to Yoichiro Nambu.*[1]

Maskawa was Director of the Yukawa Institute for Theoretical Physics from 1997 to 2003.*[4] He is now special professor and director general of Kobayashi-Maskawa Institute for the Origin of Particles and the Universe at Nagoya University, director of Maskawa Institute for Science and Culture at Kyoto Sangyo University and professor emeritus of Kyoto University .

7.11.2 Awards and honors

- Nobel Prize in Physics (2008)
- Japan Order of Culture (2008)
- Asahi Prize (1994)
- Japan Academy Prize (1985)
- Sakurai Prize (1985)

7.11.3 Notes

[1] "The Nobel Prize in Physics 2008". The Nobel Foundation. Retrieved 2009-10-17.

[2] M. Kobayashi, T. Maskawa (1973). "CP-Violation in the Renormalizable Theory of Weak Interaction". *Progress of Theoretical Physics* **49** (2): 652–657. Bibcode:1973PThPh..49..652K. doi:10.1143/PTP.49.652.

[3] "Top Cited Articles of All Time (2010 edition)". SLAC. 2009. Retrieved 2014-06-21.

[4] "History of YITP". Yukawa Institute for Theoretical Physics.

7.12 Yuval Ne'eman

Yuval Ne'eman (Hebrew: יובל נאמן, 14 May 1925 – 26 April 2006) was an Israeli theoretical physicist, military scientist, and politician. He was a minister in the Israeli government in the 1980s and early 1990s.*[1]

7.12.1 Biography

Yuval Ne'eman was born in Tel Aviv*[2] during the Mandate era, graduated from high school at the age of 15, and studied mechanical engineering at the Technion.

At the age of 15, Ne'eman also joined the Haganah. During the 1948 Arab-Israeli War Ne'eman served in the Israeli Defense Forces (IDF) as battalion deputy commander, then as Operations Officer of Tel Aviv, and commander of Givati Brigade.

Later (1952–54) he served as Deputy Commander of Operations Department of General Staff, Commander of Planning Department of IDF. In this role, he helped organize the IDF into a reservist-based army, developed the mobilization system, and wrote the first draft of the Israel defense doctrine.

Between 1958 and 1960 Ne'eman was IDF Attaché in Great Britain, where he also studied for a PhD in physics under the supervision of 1979 Nobel Prize in Physics winner Abdus Salam at Imperial College London. In 1961, he was demobilized from the IDF with a rank of Colonel.

Between 1998-2002 Ne'eman was the head of the Israeli Engineer association *[3]*[4]

7.12.2 Scientific career

One of his greatest achievements in physics was his 1961 discovery of the classification of hadrons through the SU(3) flavour symmetry, now named the *Eightfold Way*, which was also proposed independently by Murray Gell-Mann. This SU(3) symmetry laid the foundation of the quark model, proposed by Gell-Mann and George Zweig in 1964 (independently of each other).

Discussion in the main lecture hall at the École de Physique des Houches (Les Houches Physics School), 1972. From left, Yuval Ne'eman, Bryce DeWitt, Kip Thorne.

Ne'eman was founder and director of the School of Physics and Astronomy at Tel Aviv University from 1965 to 1972, president of Tel Aviv University from 1971 to 1975, and director of its Sackler Institute of Advanced Studies from 1979 to 1997. He was also the co-director (along with Sudarshan) of the Center for Particle Theory at the University of Texas, Austin from 1968 to 1990. He was a strong believer in the importance of space research and satellites to Israel's economic future and security, and thus founded the Israeli Space Agency in 1983, which he chaired almost until his death. He also served on the Atomic Energy Commission from 1965 to 1984 and held the position of scientific director in its Soreq facility. Neeman was chief scientist of the Defense Ministry from 1974 to 1976.

He was described as "one of the most colorful figures of modern science"*[5] and co-authored *The Particle Hunters*, which was published in English in 1986. *The Times Literary Supplement* hailed this book as "the best guide to quantum physics at present available".*[6]

7.12.3 Awards and honours

- In 1969, Ne'eman received the Israel Prize*[7] in the field of exact sciences (which he returned in 1992 in protest of the award of the Israel Prize to Emile Habibi).

- In 1970, he received the Albert Einstein Award*[7] for his unique contribution in the field of physics.

- In 1972, he was elected to the National Academy of Sciences.

- In 1984, he received the Wigner Medal, which is awarded every 2 years for "outstanding contributions to the understanding of physics through group theory."

- In 2003, he received the EMET Prize for Arts, Sciences and Culture for his pioneering contribution in the deciphering of the atomic nucleus and its components, and for his enormous scientific contribution to the development of sub-atomic physics in Israel.

He was also awarded with the College de France Medal and the Officer's Cross of the French Order of Merit (Paris, 1972), the Wigner Medal (Istanbul-Austin, 1982), Birla Science Award (Hyderabad, 1998) and additional prizes and honorary doctorates from universities in Europe and USA.*[8]

7.12.4 Political career

In the late 1970s, Ne'eman founded Tehiya, a right-wing breakaway from Likud, formed in opposition to Menachem

Begin's support for the Camp David talks that paved the way for peace with Egypt and the evacuation of Yamit. He was elected to the Knesset in the 1981 elections in which Tehiya won three seats. The party joined Begin's coalition about a year after the elections and Ne'eman was appointed Minister of Science and Development, the role later changed to Minister of Science and Technology.

He retained his seat in the 1984 elections, but Tehiya were not included in the grand coalition formed by the Alignment and Likud. After the 1988 elections, Tehiya were again excluded from the governing coalition. Ne'eman resigned from the Knesset on 31 January 1990 and was replaced by Gershon Shafat. However, Tehiya joined the government in June after the Alignment had left, and he was appointed Minister of Energy and Infrastructure and Minister of Science and Technology despite not retaking his seat in the Knesset. He lost his ministerial position following the 1992 elections and did not return to politics.

7.12.5 Death

He died at age 80,*[6] on 26 April 2006 in the Ichilov Hospital, Tel Aviv, from a stroke.*[9] He left a wife, Dvora; a son and daughter; and a sister, Ruth Ben-Yisrael.

7.12.6 See also

- List of Israel Prize recipients

7.12.7 References

[1] "In Remembrance of Yuval Ne'eman", Teddy Ne'eman (son of Yuval Ne'eman), PhysicaPlus (פיזיקהפלוס), online magazine of the Israel Physical Society, Issue No. 7

[2] Watson, Andrew. "Yuval Ne'eman Dies at 80 - ScienceNOW". sciencemag.org. Retrieved 27 August 2011.

[3]

[4]

[5] Yuval Ne'eman Dies at 80 – Watson 2006 (426): 1. ScienceNOW

[6] Lawrence Joffe (14 May 2006). "Obituary: Yuval Ne'eman | Science". *The Guardian* (London). Retrieved 27 August 2011.

[7] "Yuval Ne'eman". Utexas.edu. Retrieved 27 August 2011.

[8] Yival Neeman Israel Science and Technology

[9] Nadav Shragai (26 April 2006). "Professor, veteran politician Yuval Ne'eman dies at 81". *Haaretz*. Retrieved 27 August 2011.

7.12.8 External links

- Yuval Ne'eman on the Knesset website
- Jerusalem Post obituary
- Yuval Ne'eman's papers in the INSPIRE-HEP database
- Jewish Physicists list

7.13 Kazuhiko Nishijima

Kazuhiko Nishijima (西島和彦 *Nishijima Kazuhiko*) (4 October 1926 – 15 February 2009) was a Japanese physicist who made significant contributions to particle physics. He was professor emeritus at the University of Tokyo and Kyoto University until his death in 2009.*[1]

He was born in Tsuchiura, Japan. He is most well known for his work on the Gell-Mann–Nishijima formula, and the concept of strangeness, which he called the "eta-charge" or "η-charge", after the eta meson (η).*[1]*[2]*[3]

He was nominated for the Nobel Prize in Physics in 1960 and 1961.*[4]

7.13.1 Life

Nishijima was born in Tsuchiura, Japan on 4 October 1926.*[3] He obtained his diploma in physics at the University of Tokyo in 1948, and his PhD from Osaka University in 1955 for his thesis on the nuclear potential.*[3]

In 1950, while at Osaka University, Nishijima was hired by Yoichiro Nambu to work on the theory of strong interactions and of strange particles (then called V particles).*[3] While studying the decay of these particles, Nishijima developed, with Tadao Nakano, and independently of Murray Gell-Mann, a formula that would relate the quantum numbers of these particles, the Gell-Mann–Nishijima formula (or sometimes the *NNG formula*, for Nishijima, Nakano, and Gell-Mann).*[3]

$$Q = I_3 + \frac{B+S}{2}$$

where Q is the electric charge, I_3 is the isospin projection, B is the baryon number, and S is the strangeness quantum number of the particle. This formula was pivotal for the later development of the quark model by Gell-Mann*[5] and George Zweig*[6]*[7] in 1964 (independently of each other).

From 1956 to 1958, Nishijima worked in Göttingen, Germany, upon the invitation of Werner Heisenberg.*[3] In 1958, he moved to the United States and joined the Institute for Advanced Study in Princeton, New Jersey.*[3] A year and a half later, he became a professor at the University of Illinois at Urbana–Champaign.*[3] In 1966, he returned to the University of Tokyo, where he founded a theoretical physics research group and served in some administrative positions.*[3] From 1986 until 1989, he served as the director of the Research Institute for Fundamental Physics at Kyoto University, and from 1995 until 2005, he was the president of the Nishina Memorial Foundation, a foundation that promotes physics in Japan.*[3]

Nishijima kept active in research until near the end of his life. His last subjects of research were color confinement and noncommutative quantum field theory.*[3] He died of leukemia on 15 February 2009 at the age of 82.*[1]*[3]

7.13.2 Books

- Nishijima, K (1963). *Fundamental Particles*. W. A. Benjamin. OCLC 536472.
- Nishijima, K (1998) [1974]. *Fields and Particles: Field Theory and Dispersion Relations* (4th ed.). Benjamin Cummings. ISBN 0-8053-7399-3.

7.13.3 Awards

- Nishina Memorial Prize*[3]
- Japan Academy Prize*[3]
- Order of Culture of Japan*[3]*[8]
- Guggenheim Fellowship*[9]

7.13.4 References

[1] "Particle Physicist Kazuhiko Nishijima dies at 82". The Japan Times. 18 February 2009. Retrieved 2010-07-16.

[2] Nishijima, K (1955). "Charge Independence Theory of V Particles". *Progress of Theoretical Physics* **13** (3): 285. Bibcode:1955PThPh..13..285N. doi:10.1143/PTP.13.285.

[3] Nambu, Y. (2009). "Kazuhiko Nishijima". *Physics Today* **62** (8): 58. Bibcode:2009PhT....62h..58N. doi:10.1063/1.3206100.

[4] 東京新聞: 朝永氏、受賞前に７回「候補」ノーベル賞選考資料: 国際 東京新聞、2014 年 8 月 14 日夕刊

[5] Gell-Mann, M (1964). "A Schematic Model of Baryons and Mesons". *Physics Letters* **8** (3): 214–215. Bibcode:1964PhL.....8..214G. doi:10.1016/S0031-9163(64)92001-3.

[6] Zweig, G (1964). "An SU(3) Model for Strong Interaction Symmetry and its Breaking". *CERN Report No.8181/Th 8419*.

[7] Zweig, G (1964). "An SU(3) Model for Strong Interaction Symmetry and its Breaking: II". *CERN Report No.8419/Th 8412*.

[8] 本学名誉教授の西島和彦先生が平成 15 年度の文化勲章を受章 (in Japanese). University of Tokyo. Retrieved 2010-07-16.

[9] "Kazuhiko Nishijima, Guggenheim Fellow 1965". John Simon Guggenheim Memorial Foundation. Retrieved 2009-08-18.

7.13.5 Further reading

- Kawarabayashi, K (1987). A, Ukawa, ed. *Wandering in the Fields: Festschrift for Professor Kazuhiko Nishijima on the Occasion of His Sixtieth Birthday*. World Scientific. ISBN 9971-5-0363-8.

7.14 Burton Richter

Burton Richter (born March 22, 1931) is a Nobel Prize-winning American physicist. He led the Stanford Linear Accelerator Center (SLAC) team which co-discovered the J/ψ meson in 1974, alongside the Brookhaven National Laboratory (BNL) team led by Samuel Ting. This discovery was part of the so-called November Revolution of particle physics. He was the SLAC director from 1984 to 1999.

7.14.1 Life

A native of New York City, Richter was born into a Jewish*[3] family in Brooklyn, and was raised in the Queens neighborhood of Far Rockaway.*[4] His parents were Fanny (Pollack) and Abraham Richter, a textile worker.*[5] He graduated from Far Rockaway High School, a school that also produced fellow laureates Baruch Samuel Blumberg and Richard Feynman.*[6] He attended Mercersburg Academy in Pennsylvania, then continued on to study at the Massachusetts Institute of Technology, where he received his bachelor's degree in 1952 and his PhD in 1956. He was director of the Stanford Linear Accelerator Center (SLAC) from 1984 to 1999.

As a professor at Stanford University, Richter built a particle accelerator called SPEAR (Stanford Positron-Electron Asymmetric Ring) with the help of David Ritson and the support of the U.S. Atomic Energy Commission. With it he led a team that discovered a new subatomic particle he called a ψ (psi). This discovery was also made by the team led by Samuel Ting at Brookhaven National Laboratory, but he called the particle J. The particle thus became known as the J/ψ meson. Richter and Ting were jointly awarded the 1976 Nobel Prize in Physics for their work.

Richter was a member of the JASON advisory group and serves on the board of directors of Scientists and Engineers for America, an organization focused on promoting sound science in American government.*[7]

In May 2007, he visited Iran and Sharif University of Technology.*[8]

In 2012, President Barack Obama announced that Burton Richter was a co-recipient of the Enrico Fermi Award, along with Mildred Dresselhaus.*[7]

In 2013, Richter commented on an open letter from Tom Wigley, Kerry Emanuel, Ken Caldeira, and James Hansen, that Angela Merkel was "wrong to shut down nuclear".*[9]

In 2014, Richter was among the residents of a continuing care retirement center filing a lawsuit alleging refundable entrance fees were sent out of state. This may be the first legal complaint challenging a continuing care retirement home's financial practices.*[10]*[11]*[12] At a hearing on September 9, 2014 in Federal Court, attorneys allege Richter read the contracts, saw significant problems, and is entitled to pursue a legal judgement concerning the use of his money.*[13]

7.14.2 See also

- List of Jewish Nobel laureates
- List of independent discoveries

7.14.3 References

[1] Burton Richter (1956). *Photoproduction of Positive Pions from Hydrogen by 265 MEV Gamma Rays* (PDF) (Thesis). Retrieved 2014-02-20.

[2] MIT libraries Ph.D. Thesis record

[3] Shalev, Baruch A. (2002). *100 Years of Nobel Prizes*. The Americas Group. p. 61. ISBN 978-0-935047-37-0

[4] Crease, Robert P.; Mann, Charles C. (October 26, 1986). "In Search of the Z Particle". *The New York Times*. Retrieved 2007-10-02. Burton Richter was born in Brooklyn 55 years ago, but grew up in Far Rockaway, Queens.

[5] http://www.encyclopedia.com/topic/Burton_Richter.aspx

[6] Schwach, Howard (April 15, 2005). "Museum tracks down FRHS Nobel laureates". *The Wave*. Retrieved 2007-10-02. Burton Richter graduated from Far Rockaway High School in 1948.

[7] "President Obama Names Scientists Mildred Dresselhaus and Burton Richter as the Enrico Fermi Award Winners". January 11, 2012.

[8] Erdbrink, Thomas (June 6, 2008). "Iran makes the sciences a part of its revolution". *The Washington Post*. Retrieved April 27, 2010.

[9] "Environmental scientists tout nuclear power to avert climate change - CNN.com". *CNN*. November 3, 2013.

[10] *Burton Richter, Linda Collins Cork, Georgia L. May, Thomas Merigan, Alfred Spivack, Janice R. Anderson v. CC-Palo Alto, Inc.* (United States District Court for the Northern District of California). Text

[11] Jason Green (2014-02-20). "Residents sue Palo Alto retirement community for 'up-streaming' $190 million". San Jose Mercury News. Retrieved 2014-02-20.

[12] KTVU (Feb 19, 2014). "Residents file suit against high-end Palo Alto seniors home". Cox Media Group Television. Retrieved 2014-02-20.

[13] Sue Dreman (Sep 12, 2014). "Vi seniors take case to federal court". Retrieved 2014-11-08.

7.14.4 Publications

- Barber, W. C.; Richter, B.; Panofsky, W. K. H.; O'Neill, G. K. & B. Gittelman. "An Experiment on the Limits of Quantum Electro-dynamics", High-Energy Physics Laboratory at Stanford University, Princeton University, United States Department of Energy (through predecessor agency the Atomic Energy Commission), Office of Naval Research, (June 1959).

- Richter, B. "Design Considerations for High Energy Electron – Positron Storage Rings", Stanford Linear Accelerator Center, Stanford University, United States Department of Energy (through predecessor agency the Atomic Energy Commission), (November 1966).

- Boyarski, A. M.; Coward, D.; Ecklund, S.; Richter, B.; Sherden, D.; Siemann, R. & C. Sinclair. "Inclusive Yields of $pi\{sup +\}$, $pi\{sup -\}$, $K\{sup +\}$, and $K\{sup -\}$ from $H\{sub 2\}$ Photoproduced at 18 GeV at Forward Angles", Stanford Linear Accelerator Center, Stanford University, United States Department of Energy (through predecessor agency the Atomic Energy Commission), (1971).

- Richter, B. "Total Hadron Cross Section, New Particles, and Muon Electron Events in $e\{sup +\}e\{sup -\}$ Annihilation at SPEAR", Stanford Linear Accelerator Center, Stanford University, United States Department of Energy (through predecessor agency the

- U.S. Energy Research and Development Administration (ERDA)), (January 1976).

- Richter, B. "Forty-five Years of e{sup+}e{sup-} Annihilation Physics: 1956 to 2001", Stanford Linear Accelerator Center, United States Department of Energy, (August 1984).

- Richter, B. "Charting the Course for Elementary Particle Physics", Stanford Linear Accelerator Center, United States Department of Energy, (February 16, 2007).

7.14.5 External links

- Nobelprize.org autobiography
- Nobel Lecture (PDF format)
- The Nobel Prize in Physics 1976
- SLAC press image
- Biography and Bibliographic Resources, from the Office of Scientific and Technical Information, United States Department of Energy
- NIF Secretary of Energy Board
- A Celebration Honoring Burton Richter SLAC image gallery

7.15 Samuel C. C. Ting

This is a Chinese name; the family name is Ting.

Samuel Chao Chung Ting (Chinese: 丁肇中; pinyin: Dīng Zhàozhōng; Wade-Giles: Ting Chao-chung) (born January 27, 1936) is an American physicist who received the Nobel Prize in 1976, with Burton Richter, for discovering the subatomic J/ψ particle. He is the principal investigator for the international $1.5 billion Alpha Magnetic Spectrometer experiment which was installed on the International Space Station on 19 May 2011.

7.15.1 Biography

Samuel Ting was born on January 27, 1936, in Ann Arbor, Michigan. His parents, Kuan-hai Ting (丁觀海) and Tsun-ying Jeanne Wang (王雋英), met and married as graduate students at the University of Michigan. His parents were from Rizhao County (日照縣) in the Shandong province of China.*[1]

Ting's parents returned to Rizhao two months after his birth.*[1] Due to the Japanese invasion, his education was disrupted, and he was mostly home-schooled by his parents. Because of the Chinese Civil War, his parents escaped to Taiwan and started to teach engineering in local academic institution. From 1948, Ting attended high school and college in Taiwan, but he soon dropped out the college at the freshman year.*[2]*[3]

In 1956, Ting was invited to attend the University of Michigan. There, he studied engineering, mathematics, and physics. In 1959, he was awarded BAs in both mathematics and physics, and in 1962, he earned a doctorate in physics. In 1963, he worked at the European Organization for Nuclear Research, which would later become CERN. From 1965, he taught at Columbia University and worked at the Deutsches Elektronen-Synchrotron (DESY) in Germany. Since 1969, Ting has been a professor at the Massachusetts Institute of Technology (MIT). Ting is a member of the United States National Academy of Sciences, an academician of the Chinese Academy of Sciences, and a foreign academician of Academia Sinica.*[2]

7.15.2 Nobel Prize

Main article: J/ψ meson

In 1976, Ting was awarded the Nobel Prize in Physics, which he shared with Burton Richter of the Stanford Linear Accelerator Center, for the discovery of the J/ψ meson nuclear particle. They were chosen for the award, in the words of the Nobel committee, "for their pioneering work in the discovery of a heavy elementary particle of a new kind." *[4] The discovery was made in 1974 when Ting was heading a research team at MIT exploring new regimes of high energy particle physics.*[5]

Ting gave his Nobel Prize acceptance speech in Mandarin. Although there had been Chinese recipients before (Tsung-Dao Lee and Chen Ning Yang), none had previously delivered the acceptance speech in Chinese. In his Nobel banquet speech, Ting emphasized the importance of experimental work:

> *In reality, a theory in natural science cannot be without experimental foundations; physics, in particular, comes from experimental work. I hope that awarding the Nobel Prize to me will awaken the interest of students from the developing nations so that they will realize the importance of experimental work.*[6]

7.15.3 Alpha Magnetic Spectrometer

Main article: Alpha Magnetic Spectrometer

In 1995, not long after the cancellation of the Superconducting Super Collider project had severely reduced the possibilities for experimental high-energy physics on Earth, Ting proposed the Alpha Magnetic Spectrometer, a space-borne cosmic-ray detector. The proposal was accepted and he became the principal investigator and has been directing the development since then. A prototype, *AMS-01*, was flown and tested on Space Shuttle mission STS-91 in 1998. The main mission, *AMS-02*, was then planned for launch by the Shuttle and mounting on the International Space Station.[*][7]

This project is a massive $1.5 billion undertaking involving 500 scientists from 56 institutions and 16 countries. After the 2003 Space Shuttle Columbia disaster, NASA announced that the Shuttle was to be retired by 2010 and that *AMS-02* was not on the manifest of any of the remaining Shuttle flights. Dr. Ting was forced to (successfully) lobby the United States Congress and the public to secure an additional Shuttle flight dedicated to this project. Also during this time, Ting had to deal with numerous technical problems in fabricating and qualifying the large, extremely sensitive and delicate detector module for space. *AMS-02* was successfully launched on Shuttle mission STS-134 on 16 May 2011 and was installed on the International Space Station on 19 May 2011.[*][8] [*][9]

7.15.4 Personal life

In 1960 Ting married Kay Kuhne, and together they had two daughters, Jeanne Ting Chowning and Amy Ting. In 1985 he married Dr. Susan Carol Marks, and they had one son, Christopher.[*][3]

7.15.5 Publications

- (with S. J. Brodsky) "Timelike Momenta In Quantum Electrodynamics," Columbia University, United States Department of Energy (through predecessor agency the Atomic Energy Commission), (December 1965).

- "Hadron and Photon Production of J Particles and the Origin of J Particles," Massachusetts Institute of Technology, International Conference on High Energy Particle Physics, Palermo, Sicily, Italy, (June 23, 1975).

7.15.6 See also

- MIT Physics Department

- List of multiple discoveries

- J/ψ meson

7.15.7 References

[1] Ng, Franklin (1995). *The Asian American encyclopedia*. Marshall Cavendish. pp. 1, 490. ISBN 978-1-85435-684-0.

[2] "About The Programs - Personal Journeys: Samuel C.C. Ting". *A Bill Moyers Special - Becoming American - The Chinese Experience*. 2003. Retrieved June 2, 2014.

[3] Samuel C.C. Ting. "Samuel C.C. Ting - Biographical". The Nobel Foundation. Retrieved 20 Sep 2014.

[4] "The Nobel Prize in Physics 1976". nobelprize.org. Retrieved 2009-10-09.

[5] "Experimental Observation of a Heavy Particle J". *Physical Review Letters* **33** (23): 1404–1406. 1974. Bibcode:1974PhRvL..33.1404A. doi:10.1103/PhysRevLett.33.1404.

[6] "Samuel C.C.Ting - Banquet Speech". *Nobelprize.org. Nobel Media AB 2013*. Dec 10, 1976. Retrieved June 1, 2014.

[7] "Alpha Magnetic Spectrometer - 02 (AMS-02)". NASA. 2009-08-21. Retrieved 2009-09-03.

[8] Jeremy Hsu (2009-09-02). "Space Station Experiment to Hunt Antimatter Galaxies". Space.com. Retrieved 2009-09-02.

[9] A Costly Quest for the Dark Heart of the Cosmos (New York Times, November 16, 2010)

7.15.8 External links

- Autobiography

- Biography and Bibliographic Resources, from the Office of Scientific and Technical Information, United States Department of Energy

- Nobel speech in translation

- Faculty page at MIT

- Nobel-Winners.com Bio

- PBS bio

- Scientific publications of Samuel C. C. Ting on INSPIRE-HEP

7.16 George Zweig

George Zweig (born May 30, 1937) is an American physicist. He was trained as a particle physicist under Richard Feynman.*[1] He introduced, independently of Murray Gell-Mann, the quark model (although he named it "aces"). He later turned his attention to neurobiology. He has worked as a Research Scientist at Los Alamos National Laboratory and MIT, and in the financial services industry.

7.16.1 Early life

Zweig was born in Moscow, Russia into a Jewish family.*[2] His father was a type of civil engineer known as a structural engineer. He graduated from the University of Michigan in 1959, with a bachelor's degree in mathematics, having taken numerous physics courses as electives. He earned a PhD degree in theoretical physics at the California Institute of Technology in 1964 .

7.16.2 Career

Zweig proposed the existence of quarks at CERN, independently of Murray Gell-Mann, right after defending his PhD dissertation. Zweig dubbed them "aces", after the four playing cards, because he speculated there were four of them (on the basis of the four extant leptons known at the time).*[3]*[4] The introduction of quarks provided a cornerstone for particle physics.

Like Gell-Mann, he realized that several important properties of particles such as baryons (e.g., protons and neutrons) could be explained by treating them as triplets of other constituent particles (which he called aces and Gell-Mann called quarks), with fractional baryon number and electric charge. Unlike Gell-Mann, Zweig was partly led to his picture of the quark model*[5]*[6] by the peculiarly attenuated decays of the φ meson to $\varrho\,\pi$, a feature codified by what is now known as the OZI Rule, the "Z" in which stands for "Zweig". In subsequent technical terminology, ultimately Gell-Mann's quarks were closer to "current quarks", while Zweig's to "constituent quarks" .*[7]

As pointed out by astrophysicist John Gribbin, Gell-Mann deservedly received the Nobel Prize for physics in 1969, for his overall contributions and discoveries concerning the classification of elementary particles and their interactions; at that time, quark theory had not become fully accepted,*[8] and was not specifically mentioned in the official citation of the prize. In later years, when quark theory became established as the standard model of particle physics, the Nobel committee presumably felt they couldn't recognize Zweig as the scientist who first spelled out the theory's implications in detail and suggested that they might be real, without including Gell-Mann again. Nevertheless, in 1977 Richard Feynman nominated both Gell-Mann and Zweig for the Nobel prize, presumed to be his only nomination for such.*[9] Whatever the reason, despite Zweig's contributions to a theory central to modern physics, he has not yet been awarded a Nobel prize.*[10]

Zweig later turned to hearing research and neurobiology, and studied the transduction of sound into nerve impulses in the cochlea of the human ear,*[11] and how the brain maps sound onto the spatial dimensions of the cerebral cortex. In 1975, while studying the ear,*[12] he discovered a version of the continuous wavelet transform, the cochlear transform.

In 2003, Zweig joined the quantitative hedge fund Renaissance Technologies, founded by the former Cold War code breaker James Simons. He left the firm in 2010.

Once his four year confidentiality agreement with Renaissance Technologies expired, the 78 year old Zweig returned to Wall Street and co-founded a quantitative hedge fund, called Signition, with two younger partners. They hope to begin trading in 2015. Zweig was quoted as saying "life can be very boring" without work.*[13]

7.16.3 Awards and honors

- MacArthur Prize Fellowship (1981)
- National Academy of Sciences (1996)
- Sakurai Prize (2015)

7.16.4 References

[1] "George Zweig" . Mathematics Genealogy Project (North Dakota State University). Retrieved 2010-03-18.

[2] Panos Charitos interviews George Zweig (2013) CERN Interview.

[3] G. Zweig (1964), "An SU(3) model for strong interaction symmetry and its breaking", In *Lichtenberg, D. B. (Ed.), Rosen, S. P. (Ed.): Developments In The Quark Theory Of Hadrons, Vol. 1*, 22-101 and CERN Geneva - TH. 401 (REC.JAN. 64) 24p.

[4] G. Zweig (1964), "An SU(3) model for strong interaction symmetry and its breaking II", Published in 'Developments in the Quark Theory of Hadrons'. Volume 1. Edited by D. Lichtenberg and S. Rosen. Nonantum, Mass., Hadronic Press, 1980. pp. 22-101.

[5] G. Zweig (1980), "Origins of the Quark Model", CALT-68-805

[6] G. Zweig (2013), "Concrete Quarks: The Beginning of the End" (PDF), *CERN colloquium*

[7] Concrete quarks: CERN 2013 colloquium, ditto, FNAL 2014

[8] Missing hadronic resonance states predicted by the quark model were only established in the early 1970s. Appreciation that flavor SU(3) reflects nothing beyond the symmetries of the three lightest quarks had to wait until the late 1970s. Understanding of the reason free quark searches were turning up negative was lacking until 1974.

[9] G.Zweig (2010), "Memories of Murray and the Quark Model", *International Journal of Modern Physics A* **25**: 3863, arXiv:1007.0494, Bibcode:2010IJMPA..25.3863Z, doi:10.1142/S0217751X10050494

[10] J. Gribbin (1995), *Schrödinger's Kittens and the Search For Reality*, Phoenix, p. ix, 261 p. : ill. ; 29 cm., ISBN 978-1-85799-402-5

[11] Zweig, G. (1995). "The origin of periodicity in the spectrum of evoked otoacoustic emissions". *The Journal of the Acoustical Society of America* **98** (4): 2018. doi:10.1121/1.413320.

[12] Zweig, G. (1976). "Basilar Membrane Motion". *Cold Spring Harbor Symposia on Quantitative Biology* **40**: 619–33. doi:10.1101/SQB.1976.040.01.058. PMID 820509., Zweig, G.; Lipes, R.; Pierce, J. R. (1976). "The cochlear compromise". *The Journal of the Acoustical Society of America* **59** (4): 975–82. doi:10.1121/1.380956. PMID 1262596.

[13] http://www.wsj.com/articles/at-78-scientist-is-starting-a-hedge-fund-1437693849

Chapter 8

Appendix C – Selected facilities and experiments

8.1 Brookhaven National Laboratory

Brookhaven National Laboratory (**BNL**) is a United States national laboratory located in Upton, New York, on Long Island, and was formally established in 1947 at the site of Camp Upton, a former U.S. Army base. Its name stems from its location in the greater area of the Town of Brookhaven.

8.1.1 Operation

Brookhaven, which originally was owned by the Atomic Energy Commission, is now owned by the Commission's successor, the United States Department of Energy, which subcontracts the actual research and operation to universities and research organizations. It is currently operated by Brookhaven Science Associates LLC, which is an equal partnership of Stony Brook University and Battelle Memorial Institute. It was operated by Associated Universities, Inc. (AUI), from 1947 until 1998. Associated lost the contract to manage the BNL in 1998 in the wake of a 1994 fire at the facility's high-beam flux reactor that exposed several workers to radiation and reports in 1997 of a leak of tritium into the groundwater of the Long Island Central Pine Barrens, on which the facility sits.*[1]*[2]

Co-located with the laboratory is the Upton, New York forecast office of the National Weather Service.

BNL is staffed by approximately 3,000 scientists, engineers, technicians, and support personnel, and hosts 4,000 guest investigators every year.*[3] Discoveries made at the lab have won seven Nobel Prizes.*[4]

The laboratory has its own police station, fire department, and ZIP code (11973). In total, the lab spans a 5,265-acre (21 km²) area. BNL is served by a rail spur operated as needed by the New York and Atlantic Railway.

8.1.2 History

Military conscripts entering the Camp Upton site, which would in 1947 be repurposed as BNL, in 1917

Brookhaven National Laboratory was established in 1947 on the site of Camp Upton, a training center during both World War I and World War II. After the latter war, the camp was deemed no longer necessary. Meanwhile, discussions were underway for the creation of a new science facility that would focus on the peaceful uses of atomic energy. For this task a nonprofit corporation was established that consisted of representatives from nine major research universities —Columbia, Cornell, Harvard, Johns Hopkins, MIT, Princeton, Penn, Rochester, and Yale University. With the corporation finding the Camp Upton site ideal in terms of space and transportation, a plan was conceived to convert the military camp into a research facility.

On March 21, 1947, the Camp Upton site was officially transferred from the U.S. War Department to the new U.S.

Atomic Energy Commission (AEC), predecessor to the U.S. Department of Energy (DOE).

8.1.3 Major programs

Although originally conceived as a nuclear research facility, its mission has greatly expanded. Its foci are now:

The Center for Functional Nanomaterials (CFN)

Satoshi Ozaki posed with a magnet for the Relativistic Heavy Ion Collider in 1991

- Nuclear and high-energy physics[*][5]
- Physics and chemistry of materials[*][6]
- Environmental[*][7] and energy research
- Nonproliferation[*][8]
- Neurosciences and medical imaging[*][9]
- Structural biology[*][10]

8.1.4 Major facilities

- Relativistic Heavy Ion Collider (RHIC), which was designed to research quark–gluon plasma.[*][11] Until 2008 it was the world's most powerful particle accelerator.
- Center for Functional Nanomaterials (CFN), it is used for the study of nanoscale materials.[*][12]

- National Synchrotron Light Source (NSLS), the lab's most popular machine which is said to have attracted more researchers in the world than any other facility.[*][13] It was involved in the work that won the 2003 Nobel Prize in Chemistry.[*][14]

- Alternating Gradient Synchrotron, a particle accelerator that was used in three of the lab's Nobel prizes.[*][15]

- Accelerator Test Facility, generates, accelerates and monitors particle beams.[*][16]

- Tandem Van de Graaff, once the world's largest electrostatic accelerator.[*][17]

- New York Blue Gene supercomputer, an 18 rack Blue Gene/L and a 2 rack Blue Gene/P massively parallel supercomputer that involves a cooperative effort between Brookhaven National Laboratory and Stony Brook University.[*][18] It is the world's 5th fastest supercomputer and the world's 2nd most powerful for open access research as of 2008.[*][19]

8.1.5 Plans

The lab is building NSLS-II, which in 2015 will replace the NSLS after more than 30 years of operation.[*][20]

8.1.6 Off-site contributions

It is a contributing partner to ATLAS experiment, one of the four detectors located at the Large Hadron Collider (LHC). It is currently operating at CERN near Geneva, Switzerland.

Brookhaven is also responsible for the design of the SNS accumulator ring in partnership with Spallation Neutron Source in Oak Ridge, Tennessee.

Exterior of National Synchrotron Light Source II facility, taken 22 July 2012 during Brookhaven National Laboratory "Summer Sundays" public tour.

8.1.7 Public access

For other than approved Public Events the Laboratory is closed to the general public. The lab is open to the public on Sundays during the summer for tours and special programs. The public access program is referred to as 'Summer Sundays' and takes place from mid-July to mid-August, and features a science show and a tour of the facilities. The laboratory also hosts science fairs, science bowls, and robotics competitions for local schools. The Lab estimates that each year it enhances the science education of roughly 24,000 kindergarten to 12th grade LI students, more than 100 undergraduates, and 550 teachers from across the United States.

8.1.8 Controversy

In January 1997, ground water samples taken by BNL staff revealed concentrations of tritium that were twice the allowable federal drinking water standards—some samples taken later were 32 times the standard. The tritium was found to be leaking from the laboratory's High Flux Beam Reactor's spent-fuel pool into the aquifer that provides drinking water for nearby Suffolk County residents.

DOE's and BNL's investigation of this incident concluded that the tritium had been leaking for as long as 12 years without DOE's or BNL's knowledge. Installing wells that could have detected the leak was first discussed by BNL engineers in 1993, but the wells were not completed until 1996. The resulting controversy about both BNL's handling of the tritium leak and perceived lapses in DOE's oversight led to the termination of AUI as the BNL contractor in May 1997.

The responsibility for failing to discover Brookhaven's tritium leak has been acknowledged by laboratory managers, and DOE admits it failed to properly oversee the laboratory's operations. Brookhaven officials repeatedly treated the need for installing monitoring wells that would have detected the tritium leak as a low priority despite public concern and the laboratory's agreement to follow local environmental regulations. DOE's on-site oversight office, the Brookhaven Group, was directly responsible for Brookhaven's performance, but it failed to hold the laboratory accountable for meeting all of its regulatory commitments, especially its agreement to install monitoring wells. Senior DOE leadership also shared responsibility because they failed to put in place an effective system that encourages all parts of DOE to work together to ensure that contractors meet their responsibilities on environmental, safety and health issues. Unclear responsibilities for environment, safety and health matters has been a recurring problem for DOE management.

8.1.9 Nobel Prizes

Nobel Prize in Physics

- 1957 – Chen Ning Yang and Tsung-Dao Lee – parity laws*[21]

- 1976 – Samuel C. C. Ting – Psi particle*[22]

- 1980 – James Cronin and Val Logsdon Fitch – CP-violation*[23]

- 1988 – Leon M. Lederman, Melvin Schwartz, Jack Steinberger – Muon neutrino*[24]

- 2002 – Raymond Davis, Jr. – Neutrino oscillation*[25]

Nobel Prize in Chemistry

- 2003 – Roderick MacKinnon – Ion channel*[14]

- 2009 – Venkatraman Ramakrishnan and Thomas A. Steitz – Ribosome*[26]

8.1.10 See also

- Center for the Advancement of Science in Space—operates the US National Laboratory on the ISS.

- Tennis for Two*[27]*[28]

- Goldhaber fellows

8.1.11 References

[1] Atomic Laboratory on Long Island to Be Mighty Research Center – New York Times – March 1, 1947

[2]

[3] "About BNL". Bnl.gov. Retrieved 2012-07-25.

[4] "Nobel Prizes at BNL". Bnl.gov. Retrieved 2012-07-25.

[5] "Physics Department". Bnl.gov. 2008-05-12. Retrieved 2010-03-17.

[6] "Homepage, Basic Energy Sciences Directorate". Bnl.gov. Retrieved 2010-03-17.

[7] "Environmental Sciences Department". Bnl.gov. 2009-02-04. Retrieved 2010-03-17.

[8] "Brookhaven National Laboratory Nonproliferation and National Security Programs". Bnl.gov. 2010-02-02. Retrieved 2010-03-17.

[9] "Radiotracer Chemistry and Instrumentation for Biological Imaging". Bnl.gov. Retrieved 2010-03-17.

[10] "Biology Department – Brookhaven National Laboratory". Biology.bnl.gov. Retrieved 2010-03-17.

[11] "RHIC | Relativistic Heavy Ion Collider". Bnl.gov. Retrieved 2010-03-17.

[12] "Center for Functional Nanomaterials, Brookhaven National Laboratory". Bnl.gov. Retrieved 2010-03-17.

[13] "National Synchrotron Light Source". Nsls.bnl.gov. Retrieved 2010-03-17.

[14] "Nobel Prize | 2003 Chemistry Prize, Roderick MacKinnon". Bnl.gov. Retrieved 2010-03-17.

[15] "Alternating Gradient Synchrotron". Bnl.gov. 2008-01-31. Retrieved 2010-03-17.

[16] "Accelerator Test Facility". Bnl.gov. 2008-01-31. Retrieved 2010-03-17.

[17] "Tandem Van de Graaff". Bnl.gov. 2008-02-28. Retrieved 2010-03-17.

[18] New York Blue Gene supercomputer

[19] "Computing Power for Scientific Discovery" (PDF). BNL.gov. Retrieved August 4, 2010.

[20] "NSLS-II: The Future National Synchrotron Light Source". bnl.gov. Retrieved August 4, 2010.

[21] "Nobel Prize | 1957 Physics Prize, Lee and Yang". Bnl.gov. Retrieved 2010-03-17.

[22] "Nobel Prize | 1976 Prize in Physics, Samuel Ting". Bnl.gov. Retrieved 2010-03-17.

[23] "Nobel Prize | 1980 Physics Prize, Cronin and Fitch". Bnl.gov. Retrieved 2010-03-17.

[24] "Nobel Prize | 1988 Prize in Physics, Lederman, Schwartz and Steinberger". Bnl.gov. Retrieved 2010-03-17.

[25] "Nobel Prize | 2002 Physics Prize, Raymond Davis jr". Bnl.gov. Retrieved 2010-03-17.

[26] "Nobel Prize | 2009 Chemistry Prize, Venkatraman Ramakrishnan and Thomas A. Steitz". Bnl.gov. Retrieved 2010-05-20.

[27] "The anatomy of the first video game - On the Level". MSNBC. 2008-10-23. Retrieved 2010-03-17.

[28] "'+alt+'". Bnl.gov. Retrieved 2010-03-17.

- "Dr. Strangelet or: How I Learned to Stop Worrying and Love the Big Bang"

8.1.12 External links

- Brookhaven National Lab Official Website
- Physics Today: DOE Shuts Brookhaven Lab's HFBR in a Triumph of Politics Over Science 404
- Summer Sundays at Brookhaven National Laboratory
- Annotated bibliography for Brookhaven Laboratory from the Alsos Digital Library for Nuclear Issues
- Headlines

Coordinates: 40°52′24″N 72°52′19″W / 40.873346°N 72.872057°W

8.2 Belle experiment

The **Belle experiment** is a particle physics experiment conducted by the **Belle Collaboration**, an international collaboration of more than 400 physicists and engineers investigating CP-violation effects at the High Energy Accelerator Research Organisation (KEK) in Tsukuba, Ibaraki Prefecture, Japan.

The Belle detector, located at the collision point of the e−e+ asymmetric-energy collider (KEKB), is a multilayer particle detector. Its large solid angle coverage, vertex location with precision on the order of tens of micrometres (provided by a silicon vertex detector), good pion–kaon separation at the momenta range from 100 MeV/c till few GeV/c (provided by a novel Cherenkov detector), and few-percent precision electromagnetic calorimetry (CsI(Tl) scintillating crystals) allow for many other scientific searches apart from

CP-violation. Extensive studies of rare decays, searches for exotic particles and precision measurements of B mesons, D mesons, and tau particles have been carried out and have resulted in almost 300 publications in physics journals.

Highlights of the Belle experiment so far include

- the first observation of CP-violation outside of the kaon system (2001)
- observation of: $B \to K^* l^+ l^-$ and $b \to s l^+ l^-$
- measurement of ϕ_3 using the $B \to DK, D \to K_S \pi^+ \pi^-$ Dalitz plot
- measurement of the CKM quark mixing matrix elements $|V_{ub}|$ and $|V_{cb}|$
- observation of direct CP-violation in $B^0 \to \pi^+ \pi^-$ and $B^0 \to K^- \pi^+$
- observation of $b \to d$ transitions
- evidence for $B \to \tau \nu$
- observations of a number of new particles including the X(3872)

The Belle experiment operated at the KEKB accelerator, the world's highest luminosity machine. The instantaneous luminosity exceeded 2.11×10^{34} cm*−2·s*−1. The integrated luminosity collected at the $\Upsilon(4S)$ resonance mass is ~710 fb*−1 (corresponds to 771 million BB meson pairs). Most data is recorded on the $\Upsilon(4S)$ resonance, which decays to pairs of B mesons. About 10% of the data is recorded below the $\Upsilon(4S)$ resonance in order to study backgrounds. In addition, Belle has carried out special runs at the $\Upsilon(5S)$ resonance to study B

s mesons as well as on the $\Upsilon(1S)$, $\Upsilon(2S)$ and $\Upsilon(3S)$ resonances to search for evidence of Dark Matter and the Higgs Boson. The samples of $\Upsilon(1S)$, $\Upsilon(2S)$ and $\Upsilon(5S)$ collected by Belle are the world largest samples available.

The Belle II B-factory, an upgraded facility with two orders of magnitude more luminosity, has been approved in June 2010.*[1] The design and construction work is ongoing.

8.2.1 See also

- B-factory
- B–B oscillation

8.2.2 References

[1] KEK press release

8.2.3 External links

- Official Belle Website
- Official Belle 2 Website

Coordinates: 36°09′28″N 140°04′31″E / 36.15778°N 140.07528°E

8.3 BaBar experiment

The **BaBar experiment**, or simply **BaBar**, is an international collaboration of more than 500 physicists and engineers studying the subatomic world at energies of approximately ten times the rest mass of a proton (~10 GeV). Its design was motivated by the investigation of Charge Parity violation. BaBar is located at the SLAC National Accelerator Laboratory, which is operated by Stanford University for the Department of Energy in California.

8.3.1 Physics

BaBar was set up to understand the disparity between the matter and antimatter content of the universe by measuring Charge Parity violation. CP symmetry is a combination of **C**harge-conjugation symmetry (C symmetry) and **P**arity symmetry (P symmetry), each of which are conserved separately except in weak interactions. BaBar focuses on the study of CP violation in the B meson system. The name of the experiment is derived from the nomenclature for the B meson (symbol **B**) and its antiparticle (symbol **B**, pronounced **B bar**). The experiment's mascot was accordingly chosen to be Babar the Elephant.

If CP symmetry holds, the decay rate of B mesons and their antiparticles should be equal. Analysis of secondary particles produced in the BaBar detector showed this was not the case – in the summer of 2002, definitive results were published based on the analysis of 87 million B/B meson-pair events, clearly showing the decay rates were not equal. Consistent results were found by the Belle experiment at the KEK laboratory in Japan.

CP violation was already predicted by the Standard Model of particle physics, and well established in the neutral kaon system (K/K meson pairs). The BaBar experiment has increased the accuracy to which this effect has been experimentally measured. Currently, results are consistent with the standard model, but further investigation of a greater variety of decay modes may reveal discrepancies in the future.

The BaBar detector is a multilayer particle detector. Its large solid angle coverage (near hermetic), vertex location with precision on the order of 10 μm (provided by a silicon vertex detector), good pion–kaon separation at multi-GeV momenta (provided by a novel Cherenkov detector), and few-percent precision electromagnetic calorimetry (CsI(Tl) scintillating crystals) allow a list of other scientific searches apart from CP violation in the B meson system.*[1] Studies of rare decays and searches for exotic particles and precision measurements of phenomena associated with mesons containing bottom and charm quarks, as well as phenomena associated with tau leptons are possible.

The BaBar detector ceased operation on 7 April 2008, but data analysis is ongoing.

8.3.2 Detector description

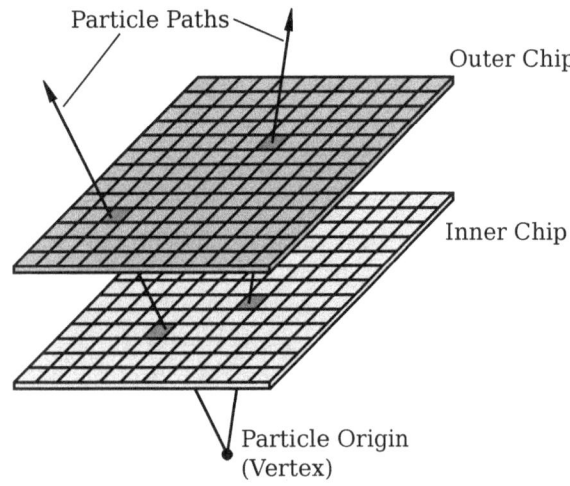

Principle of silicon vertex detectors: the particles' origin, where the event that created them occurred, can be found by extrapolating backwards from the charged regions (red) left on the sensors.

The BaBar detector is cylindrical with the interaction region at the center. In the interaction region, 9 GeV electrons collide with 3.1 GeV antielectrons (sometimes called positrons) to produce a center-of-mass collision energy of 10.58 GeV, corresponding to the $\Upsilon(4S)$ resonance. The $\Upsilon(4S)$ decays immediately into a pair of B mesons – half the time B+B− and half the time B0B0. To detect the particles there are a series of subsystems arranged cylindrically around the interaction region. These subsystems are as follows, in order from inside to outside:

- **Silicon Vertex Tracker (SVT)**

  Made from 5 layers of double-sided silicon strips, the SVT records charged particle tracks very close to the interaction region inside BaBar.

- **Drift Chamber (DCH)**

  Less expensive than silicon, the 40 layers of wires in this gas chamber detect charged particle tracks out to a much larger radius, providing a measurement of their momenta. In addition, the DCH also measures the energy loss of the particles as they pass through matter. See Bethe-Bloch formula.

- **Detector of Internally Reflected Cherenkov Light (DIRC)**

  The DIRC is composed of 144 quartz bars which radiate and focus Cherenkov radiation to differentiate between kaons and pions.

- **Electromagnetic Calorimeter (EMC)**

  Made from 6580 CsI crystals, the EMC identifies electrons and antielectrons, which allows for the reconstruction of the particle tracks of photons (and thus of neutral pions ($\pi 0$)) and of "long Kaons" (K L), which are also electrically neutral.

- **Magnet**

  The Magnet produces a 1.5 T field inside the detector, which bends the tracks of charged particles allowing deduction of their momentum.

- **Instrumented Flux Return (IFR)**

  The IFR is designed to return the flux of the 1.5 T magnet, so it is mostly iron but there is also instrumentation to detect muons and long kaons. The IFR is broken into 6 sextants and two endcaps. Each of the sextants has empty spaces which held the 19 layers of Resistive Plate Chambers (RPC), which were replaced in 2004 and 2006 with Limited Streamer Tubes (LST) interleaved with brass. The brass is there to add mass for the interaction length since the LST modules are so much less massive than the RPCs. The LST system is designed to measure all three cylindrical coordinates of a track: which individual tube was hit gives the φ coordinate, which layer the hit was in gives the ϱ coordinate, and finally the z-planes atop the LSTs measure the z coordinate.

8.3.3 Notable events

On 9 October 2005, BaBar recorded a record luminosity just over 1×10^{34} cm*−2s*−1 delivered by the PEP-II positron-electron collider.*[2] This represents 330% of the luminosity that PEP-II was designed to deliver, and was produced along with a world record for stored current in an electron storage ring at 1,732 mA, paired with a record 2,940 mA of positrons. "For the BaBar experiment, higher luminosity means generating more collisions per second, which translates into more accurate results and the ability to find physics effects they otherwise couldn't see." *[3]

In 2008, BaBar physicists detected the lowest energy particle in the bottomonium quark family. Spokesman Hassan Jawahery said: "These results were highly sought after for over 30 years and will have an important impact on our understanding of the strong interactions." *[4]

In May 2012 BaBar reported*[5]*[6] that their recently analyzed data may suggest possible flaws in the Standard Model of particle physics. These data show that a particular type of particle decay called "B to D-star-tau-nu" happens more often than the Standard Model says it should. In this type of decay, a particle called the B-bar meson decays into a D meson, an antineutrino and a tau-lepton.*[7] While the level of certainty of the excess (3.4 sigma) is not enough to claim a break from the Standard Model, the results are a potential sign of something amiss and are likely to impact existing theories, including those attempting to deduce the properties of Higgs bosons. However, results at LHCb have demonstrated no significant deviation from the Standard Model prediction of very nearly zero asymmetry.*[8]*[9]

8.3.4 Data record

8.3.5 See also

- B-Factory
- B-Bbar oscillation

8.3.6 Notes

[1] Aubert, B.; Bazan, A.; Boucham, A.; Boutigny, D.; De Bonis, I.; Favier, J.; Gaillard, J. -M.; Jeremie, A.; Karyotakis, Y.; Le Flour, T.; Lees, J. P.; Lieunard, S.; Petitpas, P.; Robbe, P.; Tisserand, V.; Zachariadou, K.; Palano, A.; Chen, G. P.; Chen, J. C.; Qi, N. D.; Rong, G.; Wang, P.; Zhu, Y. S.; Eigen, G.; Reinertsen, P. L.; Stugu, B.; Abbott, B.; Abrams, G. S.; Amerman, L.; Borgland, A. W. (2002). "The BABAR detector". *Nuclear Instruments and Methods in Physics Research Section A: Accelerators, Spectrometers, Detectors and Associated Equipment* **479**: 1. doi:10.1016/S0168-9002(01)02012-5.

[2] Daily PEP-II-delivered and BaBar-recorded luminosities (bar chart). Accessed 11 October 2005.

[3] Dynamic Performance from SLAC B-Factory. Accessed 11 October 2005.

[4] Physicists Discover New Particle: the Bottom-most 'Bottomonium' 2008-07-10, Accessed 2009-08-02

[5] BABAR Data in Tension with the Standard Model (SLAC press-release).

[6] BaBar Collaboration, *Evidence for an excess of B -> D(*) Tau Nu decays*, arXiv:1205.5442.

[7] BaBar data hint at cracks in the Standard Model (EScience-News.com).

[8] Article on LHCb results

[9] 2012 LHCb paper

[10] BaBar Accelerator and Detector Performance Data, accessed 11 July 2009

8.3.7 External links

- "BaBar Experiment Confirms Time Asymmetry".
- Official BaBar Website
- BaBar Public Home Page
- Report of 2001 announcement about detection of CP violation
- YouTube Video of the BaBar Control Room April 30, 2007

8.4 CDF experiment

For other uses of "CDF", see CDF (disambiguation).

The **Collider Detector at Fermilab** (CDF) experimental collaboration studies high energy particle collisions at the Tevatron, the world's former highest-energy particle accelerator. The goal is to discover the identity and properties of the particles that make up the universe and to understand the forces and interactions between those particles.

CDF is an international collaboration of about 600 physicists (from about 30 American universities and National laboratories and about 30 groups from universities and national laboratories from Italy, Japan, UK, Canada, Germany, Spain, Russia, Finland, France, Taiwan, Korea,

Wilson Hall at Fermilab

Part of the CDF detector

and Switzerland). The CDF detector itself weighs 5000 tons [*], is about 12 meters in all three dimensions. The goal of the experiment is to measure exceptional events out of the billions of collisions in order to:

- Look for evidence for phenomena beyond the Standard Model of particle physics
- Measure and study the production and decay of heavy particles such as the Top and Bottom Quarks, and the W and Z bosons
- Measure and study the production of high-energy particle jets and photons
- Study other phenomena such as diffraction

The Tevatron collides protons and antiprotons at a center-of-mass energy of about 2 TeV. The very high energy available for these collisions makes it possible to produce heavy particles such as the Top quark and the W and Z bosons, which weigh much more than a proton (or antiproton).

These heavier particles are identified through their characteristic decays. The CDF apparatus records the trajectories and energies of electrons, photons and light hadrons. Neutrinos do not register in the apparatus leading to an apparent *missing energy*. Other hypothetical particles might leave a missing energy signature, and some searches for new phenomena are based on that.

There is another experiment similar to CDF called D0 located at another point on the Tevatron ring.

8.4.1 History of CDF

There are currently two particle detectors located on the Tevatron at Fermilab: CDF and D0. CDF predates D0 as the first detector on the Tevatron. Construction of CDF began in 1982 under the leadership of John Peoples. The Tevatron was completed in 1983 and CDF began to take data in 1985.[*][1] Over the years, two major updates have been made to CDF. The first upgrade began in 1989 and the second upgrade began in 2001. Each upgrade is considered a "run." Run 0 was the run before any upgrades, Run I was after the first upgrade and Run II was after the second upgrade. Run II includes upgrades on the central tracking system, preshower detectors and extension on muon coverage.[*][2]

8.4.2 Discovery of the top quark

One of CDF's most famous observations is the observation of the top quark in February 1995.[*][3] The existence of the top quark was hypothesized after the observation of the Upsilon in 1977, which was found to consist of a bottom quark and an anti-bottom quark. The Standard Model, which today is the most widely accepted theory describing the particles and interactions, predicted the existence of three generations of quarks. The first generation quarks are the up and down quarks, second generation quarks are strange and charm, and third generation are top and bottom. The existence of the bottom quark solidified physicists' conviction that the top quark existed.[*][4] The top quark was the very last quark to be observed, mostly due to its comparatively high mass. Whereas, the masses of the other quarks range from .005 GeV (up quark) to 4.7GeV (bottom quark), the top quark has a mass of 175 GeV.[*][5] Only Fermilab's Tevatron had the energy capability to produce and detect top anti-top pairs. The large mass of the top quark caused the top quark to decay almost instantaneously, within the order of 10^{-25} seconds, making it extremely difficult to observe. The Standard Model predicts that the top quark may decay leptonically into a bottom quark and a W boson. This W boson may then decay into a lepton and neutrino ($t \rightarrow Wb \rightarrow \nu lb$). Therefore, CDF worked to reconstruct top events, looking specifically for evidence of bottom quarks, W bosons neutrinos. Finally in February 1995, CDF had enough evidence to say that they had "discovered" the top quark.*[6]

8.4.3 How CDF works

In order for physicists to understand the data corresponding to each event, they must understand the components of the CDF detector and how the detector works. Each component affects what the data will look like. Today, the 5000-ton detector sits in B0 and analyzes millions of beam collisions per second.*[7] The detector is designed in many different layers. Each of these layers work simultaneously with the other components of the detector in an effort to interact with the different particles, thereby giving physicists the opportunity to "see" and study the individual particles.

CDF can be divided into layers as follows:

- Layer 1: Beam Pipe
- Layer 2: Silicon Detector
- Layer 3: Central Outer Tracker
- Layer 4: Solenoid Magnet
- Layer 5: Electromagnetic Calorimeters
- Layer 6: Hadronic Calorimeters
- Layer 7: Muon Detectors

CDF silicon vertex detector

Cross section of the silicon detector

8.4.4 Layer 1: the beam pipe

The beam pipe is the innermost layer of CDF. The beam pipe is where the protons and anti-protons, traveling at approximately 0.99996 c, collide head on. Each of the protons is moving extremely close to the speed of light with extremely high energies. Therefore, in a collision, much of the energy is converted into mass. This allows proton- anti-proton annihilation to produce daughter particles, such as top quarks with a mass of 175 GeV, much heavier than the original protons.*[8]

8.4.5 Layer 2: silicon detector

Surrounding the beam pipe is the silicon detector. This detector is used to track the path of charged particles as they travel through the detector. The silicon detector begins at a radius of $r = 1.5$ cm from the beam line and extends to a radius of $r = 28$ cm from the beam line.*[2] The silicon detector is composed of seven layers of silicon arranged in a barrel shape around the beam pipe. Silicon is often used in charged particle detectors because of its high sensitivity, allowing for high-resolution vertex and tracking.*[9] The first layer of silicon, known as Layer 00, is a single sided detector designed to separate signal from background even under extreme radiation. The remaining layers are double sided and radiation-hard, meaning that the layers are protected from damage from radioactivity.*[2] The silicon works to track the paths of charged particles as they pass through the detector by ionizing the silicon. The density of the silicon, coupled with the low ionization energy of silicon, allow ionization signals to travel quickly.*[9] As a particle travels through the silicon, its position will be recorded in 3 dimensions. The silicon detector has a track hit resolution of 10 μm, and impact parameter resolution of 30 μm.*[2] Physicists can look at this trail of ions and determine the path that the particle took.*[8] As the silicon detector is located within a magnetic field, the curvature of the path through the silicon allows physicists to calculate the momentum of the particle. More curvature means less momentum and vice versa.

8.4.6 Layer 3: central outer tracker (COT)

Outside of the silicon detector, the central outer tracker works in much the manner as the silicon detector as it is also used to track the paths of charged particles and is also located within a magnetic field. The COT, however, is not made of silicon. Silicon is tremendously expensive and is not practical to purchase in extreme quantities. COT is a gas chamber filled with tens of thousands of gold wires arranged in layers and argon gas. Two types of wires are used in the COT: sense wires and field wires. Sense wires are thinner and attract the electrons that are released by the argon gas as it is ionized. The field wires are thicker than the sense wires and attract the positive ions formed from the release of electrons.*[8] There are 96 layers of wire and each wire is placed approximately 3.86 mm apart from one another.*[2] As in the silicon detector, when a charged particle passes through the chamber it ionizes the gas. This signal is then carried to a nearby wire, which is then carried to the computers for read-out. The COT is approximately 3.1 m long and extends from $r = 40$ cm to $r = 137$ cm. Although the COT is not nearly as precise as the silicon detector, the COT has a hit position resolution of 140 μm and a momentum resolution of 0.0015 $(GeV/c)^*-1$.*[2]

8.4.7 Layer 4: solenoid magnet

The solenoid magnet surrounds both the COT and the silicon detector. The purpose of the solenoid is to bend the trajectory of charged particles in the COT and silicon detector by creating a magnetic field parallel to the beam.*[2] The solenoid has a radius of r=1.5 m and is 4.8 m in length. The curvature of the trajectory of the particles in the magnet field allows physicists to calculate the momentum of each of the particles. The higher the curvature, the lower the momentum and vice versa. Because the particles have such a high energy, a very strong magnet is needed to bend the paths of the particles. The solenoid is a superconducting magnet cooled by liquid helium. The helium lowers the temperature of the magnet to 4.7 K or −268.45 °C which reduces the resistance to almost zero, allowing the magnet to conduct high currents with minimal heating and very high efficiency, and creating a powerful magnetic field.*[8]

8.4.8 Layers 5 and 6: electromagnetic and hadronic calorimeters

Calorimeters quantify the total energy of the particles by converting the energy of particles to visible light though polystyrene scintillators. CDF uses two types of calorimeters: electromagnetic calorimeters and hadronic calorimeters. The electromagnetic calorimeter measures the energy of light particles and the hadronic calorimeter measures the energy of hadrons.*[8] The central electromagnetic calorimeter uses alternating sheets of lead and scintillator. Each layer of lead is approximately 20 mm ($^3/_4$ in) wide. The lead is used to stop the particles as they pass through the calorimeter and the scintillator is used to quantify the energy of the particles. The hadronic calorimeter works in much the same way except the hadronic calorimeter uses steel in place of lead.*[2] Each calorimeter forms a wedge, which consists of both an electromagnetic calorimeter and a hadronic calorimeter. These wedges are about 2.4 m (8 ft) in length and are arranged around the solenoid.*[8]

8.4.9 Layer 7: muon detectors

The final "layer" of the detector consists of the muon detectors. Muons are charged particles that may be produced when heavy particles decay. These high-energy particles hardly interact so the muon detectors are strategically placed at the farthest layer from the beam pipe behind large walls of steel. The steel ensures that only extremely high-energy particles, such as neutrinos and muons, pass through to the muon chambers.*[8] There are two aspects of the muon detectors: the planar drift chambers and scintillators. There are four layers of planar drift chambers, each with the capability of detecting muons with a transverse momentum $p_T > 1.4$ GeV/c.*[2] These drift chambers work in the same way as the COT. They are filled with gas and wire. The charged muons ionize the gas and the signal is carried to readout by the wires.*[8]

8.4.10 Conclusion

Understanding the different components of the detector is important because the detector determines what data will look like and what signal one can expect to see for each of your particles. It is important to remember that a detector is basically a set of obstacles used to force particles to interact, allowing physicists to "see" the presence of a certain particle. If a charged quark is passing through the detector, the evidence of this quark will be a curved trajectory in the silicon detector and COT deposited energy in the calorimeter. If a neutral particle, such as a neutron, passes through the detector, there will be no track in the COT and silicon detector but deposited energy in the hadronic calorimeter. Muons may appear in the COT and silicon detector and as deposited energy in the muon detectors. Likewise, a neutrino, which rarely if ever interacts, will express itself only in the form of missing energy.

8.4.11 References

[1] Jean, Reising. "History and Archives Project." About Fermilab - History and Archives Project - Main Page. 2006. Fermi National Accelerator Laboratory. 10 May 2009 http://history.fnal.gov/

[2] "Brief Description of the CDF Detector in Run II." (2004): 1-2.

[3] Kilminster, Ben. "CDF "Results of the Week" in Fermilab Today." The Collider Detector at Fermilab. Collider Detector at Fermilab. 28 Apr. 2009 <http://www-cdf.fnal.gov/rotw/CDF_ROW_descriptions.html>.

[4] Lankford, Andy. "Discovery of the Top Quark." Collider Detector at Fermilab. 25 Apr. 2009 <http://www.ps.uci.edu/physics/news/lankford.html>.

[5] "Quark Chart." The Particle Adventure. Particle Data Group. 5 May 2009 <http://www3.fi.mdp.edu.ar/fc3/particle/quark_chart.html>.

[6] Quigg, Chris. "Discovery of the Top Quark." 1996. Fermi National Accelerator Laboratory. 8 May 2009 <http://lutece.fnal.gov/Papers/PhysNews95.html>.

[7] Yoh, John (2005). Brief Introduction to the CDF Experiment. Retrieved April 28, 2008, Web site: http://www-cdf.fnal.gov/events/cdfintro.html <http://www-cdf.fnal.gov/upgrades/tdr/tdr.html>

[8] Lee, Jenny (2008). The Collider Detector at Fermilab. Retrieved September 26, 2008, from CDF Virtual Tour Web site: http://www-cdf.fnal.gov/

[9] "Particle Detectors." Particle Data Group. 24 July 2008. Fermi National Accelerator Laboratory. 11 May 2009 <http://pdg.lbl.gov/2008/reviews/rpp2008-rev-particle-detectors.pdf>.

8.4.12 Further reading

* Worlds within the atom, National Geographic article, May, 1985

8.4.13 External links

* Fermilab news page
* The Collider Detector At Fermilab (CDF)

8.5 DØ experiment

The **DØ experiment** (sometimes written **D0 experiment**, or **DZero experiment**) consists of a worldwide collaboration of scientists conducting research on the fundamental

DØ under construction, the installation of the central tracking system

DØ's control room

nature of matter. DØ was one of two major experiments (the other is the CDF experiment) located at the world's second highest-energy accelerator,[*][1] the Tevatron Collider at the Fermilab in Batavia, Illinois, USA.

The research is focused on precise studies of interactions of protons and antiprotons at the highest available energies. It involves an intense search for subatomic clues that reveal

DØ Detector with large liquid argon calorimeter

the character of the building blocks of the universe.

8.5.1 Overview

The DØ experiment is located at one of the interaction regions, where proton and antiproton beams intersect, on the Tevatron synchrotron ring, labelled 'DØ'. It is expected to record data until the end of 2011. DØ is an international collaboration of about 550 physicists from 89 universities and national laboratories from 18 countries.

The experiment is a test of the Standard Model of particle physics. It is sensitive in a general way to the effects of high energy collisions and so is meant to be a highly model-independent probe of the theory. This is accomplished by constructing and upgrading a large volume elementary particle detector.

The detector is designed to stop as many as possible of the subatomic particles created from energy released by colliding proton/antiproton beams. The interaction region where the matter–antimatter annihilation takes place is close to the geometric center of the detector. The beam collision area is surrounded by tracking chambers in a strong magnetic field parallel to the direction of the beam(s). Outside the tracking chamber are the pre-shower detectors and the calorimeter. Muon chambers form the last layer in the detector. The whole detector is encased in concrete blocks which act as radiation shields. About 1.7 million collisions of the proton and antiproton beams are inspected every second and about 100 collisions per second are recorded for further studies.

8.5.2 Physics research

Higgs boson

One of the main physics goal of the DØ experiment is the search for the Higgs boson predicted by the Standard Model of particle physics. The LEP experiments at CERN have excluded the existence of such a Higgs boson with a mass smaller than 114.4 GeV/c^2. The combined measurements of the DØ and CDF experiments reported in January 2010 exclude a Higgs boson with a mass between 162 and 166 [[GeV/c²]].*[2]

On December 22, 2011, The DØ collaboration reported about the most stringent constraints on MSSM Higgs boson production in p-p collisions at sqrt(s)=1.96 TeV: "Upper limits on MSSM Higgs boson production are set for Higgs boson masses ranging from 90 to 300 GeV, and excludes tanβ>20-30 for Higgs boson masses below 180 GeV." *[3]

Top quark

On March 4, 2009, the DØ and CDF collaborations both announced the discovery of the production of single top quarks in proton-antiproton collisions. This process occurs at about half the rate as the production of top quark pairs but is much more difficult to observe since it is more difficult to distinguish from other processes that happen at much higher rate. The observation of single top quarks is used to measure the element V_{tb} of the CKM matrix.*[4]

- DZero's top-quark physics group's home page

New particle

From a press release dated June 13, 2007:

> Physicists of the DZero experiment at the Department of Energy's Fermi National Accelerator Laboratory have discovered a new heavy particle, the Ξ_b (pronounced "zigh sub b") baryon, with a mass of 5.774±0.019 GeV/c^2, approximately six times the proton mass. The newly discovered electrically charged Ξ_b baryon, also known as the "cascade b," is made of a down, a strange and a bottom quark. It is the first observed baryon formed of quarks from all three

families of matter. Its discovery and the measurement of its mass provide new understanding of how the strong nuclear force acts upon the quarks, the basic building blocks of matter.

B mesons

The DØ collaboration has published results which may explain the matter-antimatter asymmetry responsible for the abundance of matter in the universe.*[5] B mesons, which oscillate between their matter and antimatter state trillions of times each second, may take longer to decay into antimatter than matter. This would eventually lead to a slightly greater abundance of matter than antimatter, explaining why some matter remains after annihilation in the early universe. Experimental results from physicists at the Large Hadron Collider, however, have suggested that "the difference from the Standard Model is insignificant." *[6]

8.5.3 Detector

Silicon Microstrip Tracker

The point where the beams collide is surrounded by "tracking detectors" to record the tracks (trajectories) of the high energy particles produced in the collision. The measurements closest to the collision are made using silicon detectors. These are flat wafers of silicon chip material. They give very precise information, but they are expensive, so they are concentrated closest to the beam where they do not have to cover as much area. The information from the silicon detector can be used to identify b-quarks (like the ones produced from the decay of a Higgs particle).

Central Fiber Tracker

Outside the silicon, DØ has an outer tracker made using scintillating fibers, which produce photons of light when a particle passes through. The whole tracker is immersed in a powerful magnetic field so the particle tracks are curved; from the curvature, the momentum can be deduced.

Calorimeter

Outside the tracker is a dense absorber to capture particles and measure their energies. This is called a calorimeter. It uses uranium metal bathed in liquefied argon; the uranium causes particles to interact and lose energy, and the argon detects the interactions and gives an electrical signal that can be measured.

Muon Detector

The outermost layer of the detector detects muons. Muons are unstable particles but they live long enough to leave the detector. High energy muons are quite rare and a good sign of interesting collisions. Unlike most common particles they don't get absorbed in the calorimeter so by putting particle detectors outside it, muons can be identified. The muon system is very large because it has to surround all of the rest of the detector, and it is the first thing that you see when looking at DØ.

8.5.4 Trigger and DAQ

2.5 million proton-antiproton collisions happen every second in the detector. Because this exceeds current computing capabilities, only 20-50 events can be stored on tape per second. Therefore, an intricate Data Acquisition (DAQ) system is implemented at D0 that determines which events are "interesting" enough to written to tape and which can be thrown out. DAQ takes place in three stages, somewhat analogous to a digital camera. The stages are set up such that the first is the fastest, but least exclusive and the third is slowest, but most exclusive. The first stage is a hardware stage and operates at 2.5 MHz. It is like the CMOS sensor in a digital camera. It detects the events and converts raw data into something useful. It then very quickly determines if the event is worth keeping and if it is, it sends it to the second stage. The second stage is both hardware- and software-based, and operates at about 1000 Hz. It further determines whether the event is "interesting". It is similar to the RAM storage in a digital camera, temporarily storing the data until it can be sent to the third stage. Finally, the third stage is entirely software based. It reads through each event to see if it is worth storing and writes those worthy to tape. It is similar to the SD card in a digital camera, writing the events to permanent storage.

8.5.5 References

[1] after the Large Hadron Collider

[2] T. Aaltonen *et al.* (CDF and DØ Collaborations) (2010). "Combination of Tevatron searches for the standard model Higgs boson in the W^*+W^*- decay mode". *Physical Review Letters* **104** (6). arXiv:1001.4162. Bibcode:2010PhRvL.104f1802A. doi:10.1103/PhysRevLett.104.061802.

[3] "Search for Higgs bosons of the minimal supersymmetric standard model in p-p collisions at sqrt(s)=1.96 TeV" (PDF), *Phys. Lett. B* (DØ Collaboration), 22 December 2011, arXiv:1112.5431, Bibcode:2012PhLB..710..569D, doi:10.1016/j.physletb.2012.03.021

[4] V.M. Abazov *et al.* (DØ Collaboration) (2009). "Observation of Single Top Quark Production". *Physical Review Letters* **103** (9): 092001. arXiv:0903.0850. Bibcode:2009PhRvL.103i2001A. doi:10.1103/PhysRevLett.103.092001.

[5] Overbye, Dennis. (May 17, 2010), "A New Clue to Explain Existence", *New York Times*

[6] Timmer, John. (September 2011), "LHCb detector causes trouble for supersymmetry theory", *Ars Technica*

8.5.6 External links

- The DØ Experiment

8.6 Fermilab

Fermi National Accelerator Laboratory (**Fermilab**), located just outside Batavia, Illinois, near Chicago, is a United States Department of Energy national laboratory specializing in high-energy particle physics. Since 2007, Fermilab has been operated by the Fermi Research Alliance, a joint venture of the University of Chicago, Illinois Institute of Technology and the Universities Research Association (URA). Fermilab is a part of the Illinois Technology and Research Corridor.

Fermilab's Tevatron was a landmark particle accelerator; at 3.9 miles (6.3 km) in circumference, it was the world's second-largest energy particle accelerator (after CERN's Large Hadron Collider, which is 27 km in circumference), until it was shut down in 2011. In 1995, the discovery of the top quark was announced by researchers who used the Tevatron's CDF and DØ detectors.

In addition to high-energy collider physics, Fermilab hosts smaller fixed-target and neutrino experiments, such as MiniBooNE and MicroBooNE (Mini Booster Neutrino Experiment and Micro Booster Neutrino Experiment), SciBooNE (SciBar Booster Neutrino Experiment) and MINOS (Main Injector Neutrino Oscillation Search). The MiniBooNE detector is a 40-foot (12 m) diameter sphere that contains 800 tons of mineral oil lined with 1,520 phototube detectors. An estimated 1 million neutrino events are recorded each year. SciBooNE is the newest neutrino experiment at Fermilab; it sits in the same neutrino beam as MiniBooNE but has fine-grained tracking capabilities. The MINOS experiment uses Fermilab's NuMI (Neutrinos at the Main Injector) beam, which is an intense beam of neutrinos that travels 455 miles (732 km) through the Earth to the Soudan Mine in Minnesota.

In the public realm, Fermilab hosts many cultural events: not only public science lectures and symposia, but also classical and contemporary music concerts, folk dancing and arts galleries. The site is open from dawn to dusk to visitors who present valid photo identification.

Asteroid 11998 Fermilab is named in honor of the laboratory.

8.6.1 History

Robert Rathbun Wilson Hall

Weston, Illinois, was a community next to Batavia voted out of existence by its village board in 1966 to provide a site for Fermilab.*[2]

The laboratory was founded in 1967 as the **National Accelerator Laboratory**; it was renamed in honor of Enrico Fermi in 1974. The laboratory's first director was Robert Rathbun Wilson, under whom the laboratory opened ahead of time and under budget. Many of the sculptures on the site are of his creation. He is the namesake of the site's high-rise laboratory building, whose unique shape has become the symbol for Fermilab and which is the center of activity on the campus.

After Wilson stepped down in 1978 to protest the lack of funding for the lab, Leon M. Lederman took on the job. It was under his guidance that the original accelerator was replaced with the Tevatron, an accelerator capable of colliding proton and an antiproton at a combined energy of 1.96 TeV. Lederman stepped down in 1989 and remains Director Emeritus. The science education center at the site was named in his honor.

The later directors include:

- John Peoples, 1989 to 1999
- Michael S. Witherell, July 1999 to June 2005
- Piermaria Oddone, July 2005 to July 2013*[3]

- Nigel Lockyer, September 2013 to the present*[4]

Fermilab continues to participate in the work in the LHC; it serves as a Tier 1 site in the Worldwide LHC Computing Grid.*[5]

8.6.2 Accelerators

Current state

As of 2014, the first stage in the acceleration process (pre-accelerator injector) takes place in two ion sources which turn hydrogen gas into H*− ions. The gas is introduced into a container lined with molybdenum electrodes, each a matchbox-sized, oval-shaped cathode and a surrounding anode, separated by 1 mm and held in place by glass ceramic insulators. A magnetron generates a plasma to form the ions near the metal surface. The ions are accelerated by the source to 35 keV and matched by low energy beam transport (LEBT) into the radio-frequency quadrupole (RFQ) which applies a 750 keV electrostatic field giving the ions their second acceleration. At the exit of RFQ, the beam is matched by medium energy beam transport (MEBT) into the entrance of the linear accelerator (linac).*[6]

The next stage of acceleration is linear particle accelerator (linac). This stage consists of two segments. The first segment has 5 vacuum vessel for drift tubes, operating at 201 MHz. The second stage has 7 side-coupled cavities, operating at 805 MHz. At the end of linac, the particles are accelerated to 400 MeV, or about 70% of the speed of light.*[7]*[8] Immediately before entering the next accelerator, the H*− ions pass through a carbon foil, becoming H*+ ions (protons).*[9]

The resulting protons then enter the booster ring, a 468 m-circumference circular accelerator whose magnets bend beams of protons around a circular path. The protons travel around the Booster about 20,000 times in 33 milliseconds, adding energy with each revolution until they leave the Booster accelerated to 8 GeV.*[9]

The final acceleration is applied by the Main Injector, which is the smaller of the two rings in the last picture below (foreground). Completed in 1999, it has become Fermilab's "particle switchyard" in that it can route protons to any of the experiments installed along the beam lines after accelerating them to 120 GeV. Until 2011, the Main Injector provided protons to the antiproton ring and the Tevatron for further acceleration but now provides the last push before the particles reach the beam line experiments.

- Two ion sources at the center with two high-voltage electronics cabinets next to them*[1]

- Beam direction right to left: RFQ (silver), MEBT (green), first drift tube linac (blue)*[1]
- A 7835 power amplifier that is used at the first stage of linac*[2]
- A 12 MW klystron used at the second stage of linac*[2]
- A cutaway view of the 805 MHz side-couple cavities*[3]
- Booster ring*[4]
- Fermilab's accelerator rings

1. ^ *a* *b* "35 years of H- ions at Fermilab" (PDF). *Fermilab*. Retrieved 12 August 2015.

2. ^ Cite error: The named reference slideshow was invoked but never defined (see the help page).

3. ^ May, Michael P.; Fritz, James R.; Jurgens, Thomas G.; Miller, Harold W.; Olson, James; Snee, Daniel (1990). "Mechanical Construction of the 805 MHz Side Couple Cavities for the Fermilab Linac Upgrade" (PDF). *Proceedings of the Linear Accelerator Conference 1990, Albuquerque, New Mexico, USA*. Retrieved 13 August 2015.

4. ^ "Wilson Hall & vicinity". *Fermilab*. Retrieved 12 August 2015.

Proton improvement plan

In recognizing higher demands of proton beams to support new experiments, Fermilab started an initiative to enhance their accelerators. The project started in 2011 and will continue for many years.*[10] The project has two phases called Proton Improvement Plan (PIP) and Proton Improvement Plan-II (PIP-II).*[11]

PIP (2011–2018) The overall goals of PIP are to increase the repetition rate of the Booster beam from 7 Hz to 15 Hz and replace old hardware to increase reliability of the operation.*[11] Before the start of the PIP project, a replacement of the pre-accelerator injector was underway. The replacement of almost 40-year-old Cockcroft–Walton generators to RFQ started in 2009 and completed in 2012. At the linac stage, the analog beam position monitor (BPM) modules were replaced with digital boards in 2013. A replacement of Linac vacuum pumps and related hardware is expected to be completed in 2015. A study on the replacement of 201-MHz drift tubes is still ongoing. At the boosting stage, a major component of the PIP is to upgrade the Booster ring to

15-Hz operation. The Booster has 19 radio frequency stations. Originally, the Booster stations were operating without solid-state drive system which was acceptable for 7-Hz, but not for 15-Hz operation. A demonstration project in 2004 converted one of the stations to solid state drive prior to the PIP project. As part of the project, the remaining stations were successfully converted to solid state in 2013. Another major part of the PIP project is to refurbish and replace 40-year-old Booster cavities. Many cavities have been refurbished and tested to operate at 15 Hz repetition rate. The completion of cavity refurbishment is expected to be completed in 2015 and the repetition rate can be gradually increased to 15-Hz operation from that point. A longer term upgrade task is to replace the Booster cavities with a new design. The research and development of the new cavities is underway. The completion of the Booster cavity replacement is expected to be in 2018.*[10]

Prototypes of SRF cavities to be used in the last segment of PIP-II linac[12]*

PIP-II The goals of PIP-II include a plan to delivery 1.2 MW of proton beam power from the Main Injector to the Deep Underground Neutrino Experiment target at 120 GeV and the power near 1 MW at 60 GeV with a possibility to extend the power to 2 MW in the future. The plan should also support the current 8 GeV experiments including Mu2e, g-2, and other short-baseline neutrino experiments. These require an upgrade to the linac to inject to the Booster with 800 MeV. The first option is to add 400 MeV "afterburner" superconducting linac at the tail end of the existing 400 MeV. This requires moving the existing linac up 50 metres (160 ft). However, there are many technical issues with this approach. The preferred option is to build a new 800 MeV superconducting linac to inject to the Booster ring. The new linac site will be located on top of a small portion of Tevatron near the Booster ring in order to take advantage of existing electrical and water, and cryogenic infrastructure. The PIP-II linac will have low energy beam transport line (LEBT), radio frequency quadrupole (RFQ), and medium energy beam transport line (MEBT) operated at the room temperature at with a 162.5 MHz and energy increasing from 0.03 MeV. The first segment of linac will be operated at 162.5 MHz and energy increased up to 11 MeV. The second segment of linac will be operated at 325 MHz and energy increased up to 177 MeV. The last segment of linac will be operated at 650 MHz and will have the final energy level of 800 MeV.*[13]

Project X

Main article: Project X (Accelerator)

Project X is a long range plan to bring accelerators at Fermilab campus to new frontiers. The plan for accelerators focuses on two of the three frontiers that are long-term plan of Fermilab. In the intensity frontier, the new high-intensity accelerators will support experiments that require intense particle beam to understand particles such as neutrinos, muons, kaons and nuclei. In the energy frontier, the accelerators will support detection of new particles and forces with potential future projects such as multi-TeV Muon Collider. The immediate plan of Project X is to focus on the intensity frontier. The project is broken down into 3 stages. Stage one includes upgrade to existing facilities to support immediate experiments. This stage has translated into work done in the Proton Improvement Plan. Stage two includes delivery of three concurrent beam levels: 2.9 MW at 3 GeV; 50–200 kW at 8 GeV and 2.3 MW at 60–120 GeV. Stage three is to build next generation accelerators as the front end to the energy frontier based on international collaboration in projects such as Neutrino Factory and Muon Collider.*[14]

8.6.3 Experiments

- Cryogenic Dark Matter Search (CDMS)
- COUPP: Chicagoland Observatory for Underground Particle Physics
- Dark Energy Survey (DES)
- Deep Underground Neutrino Experiment (DUNE), formerly known as Long Baseline Neutrino Experiment (LBNE)
- Holometer interferometer
- MiniBooNE: Mini Booster Neutrino Experiment
- MicroBooNE: Micro Booster Neutrino Experiment
- MINOS: Main Injector Neutrino Oscillation Search
- MINERvA: Main INjector ExpeRiment with vs on As

8.6. FERMILAB

- MIPP: Main Injector Particle Production
- Mu2e: Muon-to-Electron Conversion Experiment
- Muon g-2
- NOvA: NuMI Off-axis ν_e Appearance
- SELEX: SEgmented Large-X baryon spectrometer EXperiment, run to study charmed baryons
- Sciboone: SciBar Booster Neutrino Experiment
- SeaQuest

Interior of Wilson Hall

8.6.4 Architecture

Fermilab's first director, Wilson, insisted that the site's aesthetic complexion not be marred by a collection of concrete block buildings. The design of the administrative building (Wilson Hall) harkens back to St. Pierre's Cathedral in Beauvais, France. Several of the buildings and sculptures within the Fermilab reservation represent various mathematical constructs as part of their structure.

The Archimedean Spiral is the defining shape of several pumping stations as well as the building housing the MINOS experiment. The reflecting pond at Wilson Hall also showcases a 32-foot-tall (9.8 m) hyperbolic obelisk, designed by Wilson. Some of the high-voltage transmission lines carrying power through the laboratory's land are built to echo the Greek letter π. One can also find structural examples of the DNA double-helix spiral and a nod to the geodesic sphere.

Wilson's sculptures on the site include *Tractricious*, a free-standing arrangement of steel tubes near the Industrial Complex constructed from parts and materials recycled from the Tevatron collider, and the soaring *Broken Symmetry*, which greets those entering the campus via the Pine Street entrance.*[15] Crowning the Ramsey Auditorium is a representation of the Möbius strip with a diameter of more than 8 feet (2.4 m). Also scattered about the access roads and village are a massive hydraulic press and old magnetic containment channels, all painted blue.

8.6.5 Current developments

Fermilab is dismantling the CDF (Collider Detector at Fermilab) and DØ (D0 experiment) facilities, and has been approved to continue moving forward with MINOS, NOvA, G-2, and Liquid Argon Test Facility.

LBNE

Fermilab has been approved and currently stands to become the world leader in Neutrino physics through its Long Baseline Neutrino Experiment (LBNE). Other leaders are CERN, which leads in Accelerator physics with the Large Hadron Collider (LHC), and Japan, which has been approved to build and lead the International Linear Collider (ILC).

"Over 350 people from over 60 institutions participate in the Long-Baseline Neutrino Experiment (LBNE), working together to plan and develop both the experimental facilities and the physics program. LBNE is expected to be fully constructed and ready for operations in 2022.

LBNE plans a world-class program in neutrino physics that will measure fundamental physical parameters to high precision and explore physics beyond the Standard Model. The measurements LBNE makes will greatly increase our understanding of neutrinos and their role in the universe, thereby better elucidating the nature of matter and antimatter.

LBNE will send the world's highest-intensity neutrino beam 800 miles through the Earth's mantle to a large detector, a multi-kiloton volume of target material instrumented such that it can record interactions between neutrinos and the target material. Neutrinos are harmless and can pass right through matter, only very rarely colliding with other matter particles. Therefore, no tunnel is needed; the vast majority of the neutrinos will pass through the mantle's material, and in turn, right through the detector. The experiment will thus need to collect data for a decade or two since neutrinos interact so rarely.

Fermilab, in Batavia, IL, is the host laboratory and the site of LBNE's future beamline, and the Sanford Underground Research Facility (SURF), in Lead, SD, is the site selected to house the massive far detector. The term "baseline" refers to the distance between the neutrino source and the detector.

Why neutrinos: Neutrinos, astonishingly abundant yet not well understood, may provide the key to answering some of the most fundamental questions about the nature of our universe. The discovery that neutrinos are not massless, as previously thought, has opened a first crack in the highly successful Standard Model of Particle Physics. Neutrinos may play a key role in solving the mystery of how the universe came to consist only of matter rather than antimatter."

g−2

Transportation of the 600-ton magnet to Fermilab

"In the summer of 2013, the Muon g−2 team successfully transported a 50-foot-wide electromagnet from Brookhaven National Laboratory in Long Island, New York, to Fermilab in one piece. The move took 35 days and traversed 3,200 miles over land and sea."

"Muon g−2 (pronounced gee minus two) will use Fermilab's powerful accelerators to explore the interactions of short-lived particles known as muons with a strong magnetic field in "empty" space. Scientists know that even in a vacuum, space is never empty. Instead, it is filled with an invisible sea of virtual particles that—in accordance with the laws of quantum physics—pop in and out of existence for incredibly short moments of time. Scientists can test the presence and nature of these virtual particles with particle beams traveling in a magnetic field."

Muon g-2 building (white and orange) that hosts the magnet

Particle discovery

On September 3, 2008, the discovery of a new particle, the bottom Omega baryon (Ω−
b) was announced at the DØ experiment of Fermilab. It is made up of two strange quarks and a bottom quark. This discovery helps to complete the "periodic table of the baryons" and offers insight into how quarks form matter."[16]

8.6.6 Wildlife at Fermilab

In 1967, Wilson brought five American Bison to the site, a bull and four cows, and an additional 21 were provided by the Illinois Department of Conservation. Some fearful locals believed at first that the bison were introduced in order to serve as an alarm if and when radiation at the laboratory reached dangerous levels, but they were assured by Fermilab that this claim had no merit. Today, the herd is a popular attraction that draws many visitors*[17] and the grounds are also a sanctuary for other local wildlife populations.*[18]

Working with the Forest Preserve District of DuPage County, Fermilab has introduced Barn owls to selected structures around the grounds.

8.6.7 See also

- Big Science
- Center for the Advancement of Science in Space—operates the US National Laboratory on the ISS.
- CERN
- Fermi Linux LTS
- Scientific Linux

- Stanford Linear Accelerator Center

8.6.8 References

[1] "DOE Budget Report" (PDF). Retrieved 2014-12-27.

[2] Fermilab. "Before Weston". Retrieved 2009-11-25.

[3] "Fermilab director Oddone announces plan to retire next year". *The Beacon-News.* August 2, 2012. Retrieved 10 July 2013.

[4] "New Fermilab director named". *Crain's Chicago Business.* June 21, 2013. Retrieved 10 July 2013.

[5] National Science Foundation. "The US and LHC Computing". Retrieved 2011-01-11.

[6] Carneiro, J.P. (13 Nov 2014). "Transmission efficiency measurement at the FNAL 4-rod RFQ (FERMILAB-CONF-14-452-APC)" (PDF). *27th International Linear Accelerator Conference (LINAC14).* Retrieved 12 August 2015.

[7] "Fermilab Linac Slide Show Description". *Fermilab.* Retrieved 12 August 2015.

[8] Kubik, Donna (2005). *Fermilab* (PDF). Retrieved 12 August 2015.

[9] "Accelerator". *Fermilab.* Retrieved 12 August 2015.

[10] "FNAL – The Proton Improvement Plan (PIP)". *Proceedings of IPAC2014, Dresden, Germany* (PDF). p. 3409–3411. ISBN 978-3-95450-132-8. Retrieved 15 August 2015.

[11] Holmes, Steve (16 December 2013). *MegaWatt Proton Beams for Particle Physics at Fermilab* (PDF). Fermilab. Retrieved 15 August 2015.

[12] "02 Proton and Ion Accelerators and Applications". *Proceedings of LINAC2014, Geneva, Switzerland* (PDF). September 2014. pp. 171–173. ISBN 978-3-95450-142-7. Retrieved 16 August 2015.

[13] *Proton Improvement Plan-II* (PDF). Fermilab. 12 December 2013. Retrieved 15 August 2015.

[14] *A Fermilab Plan for Discovery* (PDF). 2011. Retrieved 18 August 2015.

[15] "About Fermilab - The Fermilab Campus". 2005-12-01. Retrieved 2007-02-27.

[16] "Fermilab physicists discover "doubly strange" particle". Fermilab. 9 September 2008.

[17] Fermilab (30 December 2005). "Safety and the Environment at Fermilab". Retrieved 2006-01-06.

[18] http://www.fnal.gov/pub/about/campus/ecology/wildlife/ retrieved 3/30/2013

8.6.9 External links

- Fermi National Accelerator Laboratory
  - *Fermilab Today* Daily newsletter
  - Other Fermilab online publications
  - Fermilab Virtual Tour
  - Architecture at the Fermilab campus

Coordinates: 41°49′55″N 88°15′26″W / 41.83194°N 88.25722°W

8.7 Large Hadron Collider

"LHC" redirects here. For other uses, see LHC (disambiguation).

The **Large Hadron Collider** (**LHC**) is the world's largest

A section of the LHC

and most powerful particle collider, the largest, most complex experimental facility ever built, and the largest single machine in the world.[1] It was built by the European Organization for Nuclear Research (CERN) between 1998 and 2008 in collaboration with over 10,000 scientists and engineers from over 100 countries, as well as hundreds of universities and laboratories.[2] It lies in a tunnel 27 kilometres (17 mi) in circumference, as deep as 175 metres (574 ft) beneath the France–Switzerland border near Geneva, Switzerland. Its first research run took place from 30 March 2010 to 13 February 2013 at an initial energy of 3.5 teraelectronvolts (TeV) per beam (7 TeV total), almost 4 times more than the previous world record for a collider,[3] rising to 4 TeV per beam (8 TeV total) from 2012.[4][5] On 13 February 2013 the LHC's first run officially ended, and it was shut down for planned upgrades. 'Test' collisions restarted in the upgraded collider on 5 April

2015,*[6]*[7] reaching 6.5 TeV per beam on 20 May 2015 (13 TeV total, the current world record for particle collisions). Its second research run commenced on schedule, on 3 June 2015.*[8]

The LHC's aim is to allow physicists to test the predictions of different theories of particle physics, high-energy physics and in particular, to prove or disprove the existence of the theorized Higgs boson*[9] and the large family of new particles predicted by supersymmetric theories,*[10] and other unsolved questions of physics, advancing human understanding of physical laws. It contains seven detectors, each designed for certain kinds of research. The proton-proton collision is the primary operation method, but the LHC has also collided protons with lead nuclei for two months in 2013 and used lead–lead collisions for about one month each in 2010, 2011, and 2013 for other investigations.

The LHC's computing grid was (and currently is) a world record holder. Data from collisions was anticipated to be produced at an unprecedented rate for the time, of tens of petabytes per year, a major challenge at the time, to be analysed by a grid-based computer network infrastructure connecting 140 computing centers in 35 countries*[11]*[12] – by 2012 the Worldwide LHC Computing Grid was also the world's largest distributed computing grid, comprising over 170 computing facilities in a worldwide network across 36 countries.*[13]*[14]*[15]

8.7.1 Background

The term *hadron* refers to composite particles composed of quarks held together by the strong force (as atoms and molecules are held together by the electromagnetic force). The best-known hadrons are the baryons protons and neutrons; hadrons also include mesons such as the pion and kaon, which were discovered during cosmic ray experiments in the late 1940s and early 1950s.

A *collider* is a type of a particle accelerator with two directed beams of particles. In particle physics colliders are used as a research tool: they accelerate particles to very high kinetic energies and let them impact other particles. Analysis of the byproducts of these collisions gives scientists good evidence of the structure of the subatomic world and the laws of nature governing it. Many of these byproducts are produced only by high energy collisions, and they decay after very short periods of time. Thus many of them are hard or near impossible to study in other ways.

8.7.2 Purpose

Physicists hope that the LHC will help answer some of the fundamental open questions in physics, concerning the basic laws governing the interactions and forces among the elementary objects, the deep structure of space and time, and in particular the interrelation between quantum mechanics and general relativity, where current theories and knowledge are unclear or break down altogether. Data is also needed from high energy particle experiments to suggest which versions of current scientific models are more likely to be correct – in particular to choose between the Standard Model and Higgsless models and to validate their predictions and allow further theoretical development. Many theorists expect new physics beyond the Standard Model to emerge at the TeV energy level, as the Standard Model appears to be unsatisfactory. Issues possibly to be explored by LHC collisions include:*[16]*[17]

- Are the masses of elementary particles actually generated by the Higgs mechanism via electroweak symmetry breaking?*[18] It is expected that the collider will either demonstrate or rule out the existence of the elusive Higgs boson, thereby allowing physicists to consider whether the Standard Model or its Higgsless alternatives are more likely to be correct.*[19]*[20]*[21]

- Is supersymmetry, an extension of the Standard Model and Poincaré symmetry, realized in nature, implying that all known particles have supersymmetric partners?*[22]*[23]*[24]

- Are there extra dimensions,*[25] as predicted by various models based on string theory, and can we detect them?*[26]

- What is the nature of the dark matter that appears to account for 27% of the mass-energy of the universe?

Other open questions that may be explored using high energy particle collisions:

- It is already known that electromagnetism and the weak nuclear force are different manifestations of a single force called the electroweak force. The LHC may clarify whether the electroweak force and the strong nuclear force are similarly just different manifestations of one universal unified force, as predicted by various Grand Unification Theories.

- Why is the fourth fundamental force (gravity) so many orders of magnitude weaker than the other three fundamental forces? See also Hierarchy problem.

- Are there additional sources of quark flavour mixing, beyond those already present within the Standard Model?

8.7. LARGE HADRON COLLIDER

- Why are there apparent violations of the symmetry between matter and antimatter? See also CP violation.

- What are the nature and properties of quark–gluon plasma, thought to have existed in the early universe and in certain compact and strange astronomical objects today? This will be investigated by *heavy ion collisions*, mainly in ALICE, but also in CMS and ATLAS. Findings published in 2012 confirmed the phenomenon of jet quenching in heavy-ion collisions, and was first observed in 2010.*[27]*[28]*[29]

8.7.3 Design

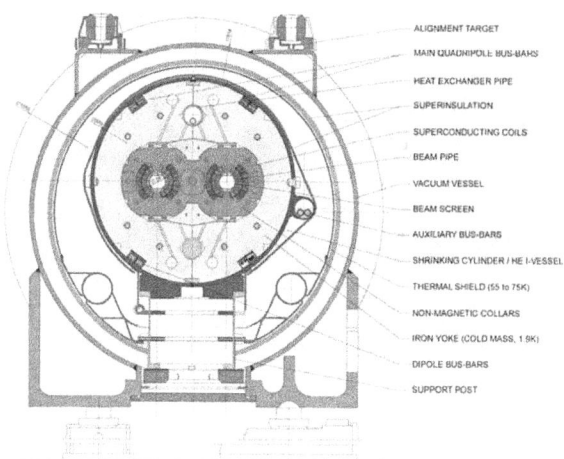

The 2-in-1 structure of the LHC dipole magnets

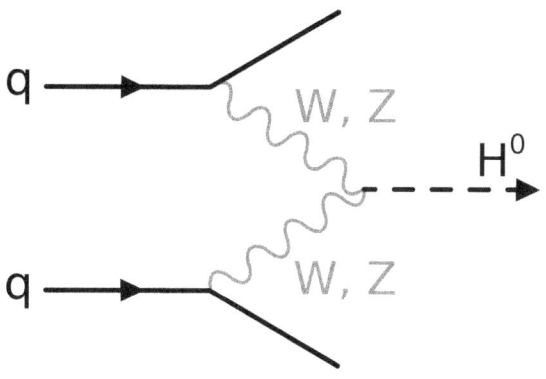

A Feynman diagram of one way the Higgs boson may be produced at the LHC. Here, two quarks each emit a W or Z boson, which combine to make a neutral Higgs.

Map of the Large Hadron Collider at CERN

The LHC is the world's largest and highest-energy particle accelerator.*[30]*[31] The collider is contained in a circular tunnel, with a circumference of 27 kilometres (17 mi), at a depth ranging from 50 to 175 metres (164 to 574 ft) underground.

The 3.8-metre (12 ft) wide concrete-lined tunnel, constructed between 1983 and 1988, was formerly used to house the Large Electron–Positron Collider.*[32] It crosses the border between Switzerland and France at four points, with most of it in France. Surface buildings hold ancillary equipment such as compressors, ventilation equipment, control electronics and refrigeration plants.

The collider tunnel contains two adjacent parallel beamlines (or *beam pipes*) that intersect at four points, each containing a beam, which travel in opposite directions around the ring. Some 1,232 dipole magnets keep the beams on their circular path (see image*[33]), while an additional 392 quadrupole magnets are used to keep the beams focused, in order to maximize the chances of interaction between the particles in the four intersection points, where the two beams cross. In total, over 1,600 superconducting magnets are installed, with most weighing over 27 tonnes.*[34] Approximately 96 tonnes of superfluid helium 4 is needed to keep the magnets, made of copper-clad niobium-titanium, at their operating temperature of 1.9 K (−271.25 °C), making the LHC the largest cryogenic facility in the world at liquid helium temperature.

When running at full design energy of 7 TeV per beam, once or twice a day, as the protons are accelerated from 450 GeV to 7 TeV, the field of the superconducting dipole magnets will be increased from 0.54 to 8.3 teslas (T). The protons will each have an energy of 7 TeV, giving a total collision energy of 14 TeV. At this energy the protons have a Lorentz factor of about 7,500 and move at about 0.999999991 c, or

Superconducting quadrupole electromagnets are used to direct the beams to four intersection points, where interactions between accelerated protons will take place.

about 3 metres per second slower than the speed of light (c).*[35] It will take less than 90 microseconds (μs) for a proton to travel once around the main ring – a speed of about 11,000 revolutions per second. Rather than continuous beams, the protons will be bunched together, into up to 2,808 bunches, with 115 billion protons in each bunch so that interactions between the two beams will take place at discrete intervals, mainly 25 nanoseconds (ns) apart, providing a bunch collision rate of 40 MHz. However it will be operated with fewer bunches when it is first commissioned, giving it a bunch crossing interval of 75 ns.*[36] The design luminosity of the LHC is 10^{34} cm*$^{-2}$s*$^{-1}$.*[37]

Prior to being injected into the main accelerator, the particles are prepared by a series of systems that successively increase their energy. The first system is the linear particle accelerator LINAC 2 generating 50-MeV protons, which feeds the Proton Synchrotron Booster (PSB). There the protons are accelerated to 1.4 GeV and injected into the Proton Synchrotron (PS), where they are accelerated to 26 GeV. Finally the Super Proton Synchrotron (SPS) is used to further increase their energy to 450 GeV before they are at last injected (over a period of several minutes) into the main ring. Here the proton bunches are accumulated, accelerated (over a period of 20 minutes) to their peak energy, and finally circulated for 5 to 24 hours while collisions occur at the four intersection points.*[38]

The LHC physics program is mainly based on proton–proton collisions. However, shorter running periods, typically one month per year, with heavy-ion collisions are included in the program. While lighter ions are considered as well, the baseline scheme deals with lead ions*[39] (see A Large Ion Collider Experiment). The lead ions are first accelerated by the linear accelerator LINAC 3, and the Low-Energy Ion Ring (LEIR) is used as an ion storage and

CMS detector for LHC

cooler unit. The ions are then further accelerated by the PS and SPS before being injected into LHC ring, where they reached an energy of 1.58 TeV per nucleon (or 328 TeV per ion), higher than the energies reached by the Relativistic Heavy Ion Collider. The aim of the heavy-ion program is to investigate quark–gluon plasma, which existed in the early universe.

Detectors

See also: List of Large Hadron Collider experiments

Seven detectors have been constructed at the LHC, located underground in large caverns excavated at the LHC's intersection points. Two of them, the ATLAS experiment and the Compact Muon Solenoid (CMS), are large, general purpose particle detectors.*[31] ALICE and LHCb have more specific roles and the last three, TOTEM, MoEDAL and LHCf, are very much smaller and are for very specialized research. The BBC's summary of the main detectors is:*[40]

Computing and analysis facilities

Main article: LHC Computing Grid

Data produced by LHC, as well as LHC-related simulation, was estimated at approximately 15 petabytes per year (max throughput while running not stated)[*][41] - a major challenge in its own right at the time.

The LHC Computing Grid[*][42] was constructed as part of the LHC design, to handle the massive amounts of data expected for its collisions. It is an international collaborative project that consists of a grid-based computer network infrastructure initially connecting 140 computing centers in 35 countries (over 170 in 36 countries as of 2012). It was designed by CERN to handle the significant volume of data produced by LHC experiments,[*][11][*][12] incorporating both private fiber optic cable links and existing high-speed portions of the public Internet to enable data transfer from CERN to academic institutions around the world.[*][43] The Open Science Grid is used as the primary infrastructure in the United States, and also as part of an interoperable federation with the LHC Computing Grid.

The distributed computing project LHC@home was started to support the construction and calibration of the LHC. The project uses the BOINC platform, enabling anybody with an Internet connection and a computer running Mac OS X, Windows or Linux, to use their computer's idle time to simulate how particles will travel in the tunnel. With this information, the scientists will be able to determine how the magnets should be calibrated to gain the most stable "orbit" of the beams in the ring.[*][44] In August 2011, a second application went live (Test4Theory) which performs simulations against which to compare actual test data, to determine confidence levels of the results.

By 2012 data from over 6 quadrillion (6 x 10^{15}) LHC proton-proton collisions had been analyzed,[*][45] LHC collision data was being produced at approximately 25 petabytes per year, and the LHC Computing Grid had become the world's largest computing grid (as of 2012), comprising over 170 computing facilities in a worldwide network across 36 countries.[*][13][*][14][*][15]

8.7.4 Operational history

The LHC first went live on 10 September 2008, but initial testing was delayed for 14 months from 19 September 2008 to 20 November 2009, following a magnet quench incident that caused extensive damage to over 50 superconducting magnets, their mountings, and the vacuum pipe.[*][46][*][47][*][48][*][49][*][50][*][51]

During its first run (2010 - 2013) the LHC collided two opposing particle beams of either protons at up to 4 teraelectronvolts (4 TeV or 0.64 microjoules), or lead nuclei (574 TeV per nucleus, or 2.76 TeV per nucleon).[*][30][*][52] Its first run discoveries included a particle thought to be the long sought Higgs boson, several composite particles (hadrons) like the χ_b (3P) bottomonium state, the first creation of a quark–gluon plasma, and the first observations of the very rare decay of the B_s meson into two muons ($B_s^0 \rightarrow \mu^+\mu^-$), which challenged the validity of existing models of supersymmetry.[*][53]

Construction

Operational challenges The size of the LHC constitutes an exceptional engineering challenge with unique operational issues on account of the amount of energy stored in the magnets and the beams.[*][38][*][54] While operating, the total energy stored in the magnets is 10 GJ (2,400 kilograms of TNT) and the total energy carried by the two beams reaches 724 MJ (173 kilograms of TNT).[*][55]

Loss of only one ten-millionth part (10^{-7}) of the beam is sufficient to quench a superconducting magnet, while the beam dump must absorb 362 MJ (87 kilograms of TNT) for each of the two beams. These energies are carried by very little matter: under nominal operating conditions (2,808 bunches per beam, 1.15×10^{11} protons per bunch), the beam pipes contain 1.0×10^{-9} gram of hydrogen, which, in standard conditions for temperature and pressure, would fill the volume of one grain of fine sand.

Cost See also: List of megaprojects

With a budget of 7.5 billion euros (approx. $9bn or £6.19bn as of June 2010), the LHC is one of the most expensive scientific instruments[*][56] ever built.[*][57] The total cost of the project is expected to be of the order of 4.6bn Swiss francs (SFr) (approx. $4.4bn, €3.1bn, or £2.8bn as of Jan 2010) for the accelerator and 1.16bn (SFr) (approx. $1.1bn, €0.8bn, or £0.7bn as of Jan 2010) for the CERN contribution to the experiments.[*][58]

The construction of LHC was approved in 1995 with a budget of SFr 2.6bn, with another SFr 210M towards the experiments. However, cost overruns, estimated in a major review in 2001 at around SFr 480M for the accelerator, and SFr 50M for the experiments, along with a reduction in CERN's budget, pushed the completion date from 2005 to April 2007.[*][59] The superconducting magnets were responsible for SFr 180M of the cost increase. There were also further costs and delays due to engineering difficulties encountered while building the underground cavern for the Compact Muon Solenoid,[*][60] and also due to magnet supports which were insufficiently strongly designed and failed their initial testing (2007) and damage from a magnet quench and liquid helium escape (inaugural testing, 2008) *(see: Construction accidents and delays).*[61] Due to lower electricity costs during the summer, the LHC normally does not operate over the winter months,*[62] although an exception over the 2009/10 winter was made to make up for the 2008 start-up delays.

Construction accidents and delays

- On 25 October 2005, José Pereira Lages, a technician, was killed in the LHC when a switchgear that was being transported fell on him.*[63]

- On 27 March 2007 a cryogenic magnet support designed and provided by Fermilab and KEK broke during an initial pressure test involving one of the LHC's inner triplet (focusing quadrupole) magnet assemblies. No one was injured. Fermilab director Pier Oddone stated "In this case we are dumbfounded that we missed some very simple balance of forces" . This fault had been present in the original design, and remained during four engineering reviews over the following years.*[64] Analysis revealed that its design, made as thin as possible for better insulation, was not strong enough to withstand the forces generated during pressure testing. Details are available in a statement from Fermilab, with which CERN is in agreement.*[65]*[66] Repairing the broken magnet and reinforcing the eight identical assemblies used by LHC delayed the startup date, then planned for November 2007.

- On 19 September 2008, during initial testing, a faulty electrical connection led to a magnet quench (the sudden loss of a superconducting magnet's superconducting ability due to warming or electric field effects). Six tonnes of supercooled liquid helium - used to cool the magnets - escaped, with sufficient force to break 10-ton magnets nearby from their mountings, and caused considerable damage and contamination of the vacuum tube *(see 2008 quench incident)*; repairs and safety checks caused a delay of around 14 months.*[67]*[68]*[69]

- Two vacuum leaks were identified in July 2009, and the start of operations was further postponed to mid-November 2009.*[70]

Initial lower currents   Main article: Superconducting magnet § Magnet "training"

In both of its runs (2010 and 2015), the LHC was initially run at energies below its planned operating energy, and ramped up to just 2 x 3.5 TeV energy on its first run (final 2 x 4 TeV), and 2 x 6.5 TeV on its second run (intended: 2 x 7 TeV). This is because massive superconducting magnets require considerable magnet training to handle the high currents involved without losing their superconducting ability. The "training" process involves repeatedly running the magnets with lower currents to provoke any quenches or minute movements that may result. It also takes time to cool down magnets to their operating temperature of around 1.9 K (degrees above absolute zero). Over time the magnet "beds in" and ceases to quench at these lesser currents and can handle the full design current without quenching; CERN media describe the magnets as "shaking out" the unavoidable tiny manufacturing imperfections in their crystals and positions that had initially impaired their ability to handle their planned currents. The magnets over time and with training, gradually become able to handle their full planned currents without quenching.*[71]*[72]

Inaugural tests (2008)

The first beam was circulated through the collider on the morning of 10 September 2008.*[40] CERN successfully fired the protons around the tunnel in stages, three kilometres at a time. The particles were fired in a clockwise direction into the accelerator and successfully steered around it at 10:28 local time.*[73] The LHC successfully completed its major test: after a series of trial runs, two white dots flashed on a computer screen showing the protons travelled the full length of the collider. It took less than one hour to guide the stream of particles around its inaugural circuit.*[74] CERN next successfully sent a beam of protons in a counterclockwise direction, taking slightly longer at one and a half hours due to a problem with the cryogenics, with the full circuit being completed at 14:59.

Quench incident   On 19 September 2008, a magnet quench occurred in about 100 bending magnets in sectors 3 and 4, where an electrical fault led to a loss of approximately six tonnes of liquid helium (the magnets' cryogenic coolant), which was vented into the tunnel. The escaping vapor expanded with explosive force, damaging over 50 superconducting magnets and their mountings, and contaminating the vacuum pipe, which also lost vacuum conditions.*[47]*[48]*[75]

Shortly after the incident CERN reported that the most likely cause of the problem was a faulty electrical connection between two magnets, and that – due to the time needed to warm up the affected sectors and then cool them back down to operating temperature – it would take at least two months to fix.*[76] CERN released an interim technical report*[75] and preliminary analysis of the incident on 15 and

16 October 2008 respectively,*[77] and a more detailed report on 5 December 2008.*[68] The analysis of the incident by CERN confirmed that an electrical fault had indeed been the cause. The faulty electrical connection had led (correctly) to a failsafe power abort of the electrical systems powering the superconducting magnets, but had also caused an electric arc (or discharge) which damaged the integrity of the supercooled helium's enclosure and vacuum insulation, causing the coolant's temperature and pressure to rapidly rise beyond the ability of the safety systems to contain it,*[75] and leading to a temperature rise of about 100 degrees Celsius in some of the affected magnets. Energy stored in the superconducting magnets and electrical noise induced in other quench detectors also played a role in the rapid heating. Around two tonnes of liquid helium escaped explosively before detectors triggered an emergency stop, and a further four tonnes leaked at lower pressure in the aftermath.*[75] A total of 53 magnets were damaged in the incident and were repaired or replaced during the winter shutdown.*[78] This accident was thoroughly discussed in a 22 February 2010 *Superconductor Science and Technology* article by CERN physicist Lucio Rossi.*[79]

In the original timeline of the LHC commissioning, the first "modest" high-energy collisions at a center-of-mass energy of 900 GeV were expected to take place before the end of September 2008, and the LHC was expected to be operating at 10 TeV by the end of 2008.*[80] However, due to the delay caused by the above-mentioned incident, the collider was not operational until November 2009.*[81] Despite the delay, LHC was officially inaugurated on 21 October 2008, in the presence of political leaders, science ministers from CERN's 20 Member States, CERN officials, and members of the worldwide scientific community.*[82]

Most of 2009 was spent on repairs and reviews from the damage caused by the quench incident, along with two further vacuum leaks identified in July 2009 which pushed the start of operations to November of that year.*[70]

First operational run (2010–2013)

On 20 November 2009, low-energy beams circulated in the tunnel for the first time since the incident, and shortly after, on 30 November, the LHC achieved 1.18 TeV per beam to become the world's highest-energy particle accelerator, beating the Tevatron's previous record of 0.98 TeV per beam held for eight years.*[83]

The early part of 2010 saw the continued ramp-up of beam in energies and early physics experiments towards 3.5 TeV per beam and on 30 March 2010, LHC set a new record for high-energy collisions by colliding proton beams at a combined energy level of 7 TeV. The attempt was the third that day, after two unsuccessful attempts in which the protons had to be "dumped" from the collider and new beams had to be injected.*[84] This also marked the start of its main research program.

The first proton run ended on 4 November 2010. A run with lead ions started on 8 November 2010, and ended on 6 December 2010,*[85] allowing the ALICE experiment to study matter under extreme conditions similar to those shortly after the Big Bang.*[86]

CERN originally planned that the LHC would run through to the end of 2012, with a short break at the end of 2011 to allow for an increase in beam energy from 3.5 to 4 TeV per beam.*[4] At the end of 2012 the LHC was shut down until around 2015 to allow upgrade to a planned beam energy of 7 TeV per beam.*[87] In late 2012, in light of the July 2012 discovery of the Higgs boson, the shutdown was postponed for some weeks into early 2013, to allow additional data to be obtained prior to shutdown.

Upgrade (2013–2015)

The LHC was shut down on 13 February 2013 for its planned 2 year upgrade, which would touch on many aspects of the LHC: enabling collisions at 14 TeV, enhancing its detectors and pre-accelerators (the Proton Synchrotron and Super Proton Synchrotron), as well as replacing its ventilation system and 100km of cabling impaired by high-energy collisions from its first run.*[88] The upgraded collider began its long start-up and testing process in June 2014, with the Proton Synchrotron Booster starting on 2 June 2014, the final interconnection between magnets completing and the Proton Sychrotron circulating particles on 18 June 2014, and the first section of the main LHC supermagnet system reaching operating temperature of 1.9 K (−271.25 °C), a few days later.*[89] The first of the main LHC supermagnets were reported to have been successfully "trained" to run at 11,000 amperes (the current required for 6.5 TeV collisions) by 9 December 2014, with the other magnet sectors planned to complete training by Spring 2015.*[90]

Second operational run (2015 onward)

On 5 April 2015 the LHC restarted after a two-year break during which it was extensively upgraded to run at its full specified operating energies of 7 TeV per beam (14 TeV), although initially run at 6.5 TeV per beam (13 TeV total) while the magnets bedded in (known as "training").*[6] Although particle beams travelled in both directions, inside parallel pipes, actual collisions were expected to begin in June.*[6] The first ramp on 10 April 2015 gave promising results as it reached 6.5 TeV.*[91] The upgrades culminated in colliding protons together with a combined energy of 13 TeV.*[92] On 3 June 2015 the LHC started delivering physics data after almost two years offline.*[93]

8.7.5 Timeline of operations

8.7.6 Findings and discoveries

An initial focus of research was to investigate the possible existence of the Higgs boson, a key part of the Standard Model of physics which was predicted by theory but had not yet been observed or proven to exist due to its elusive nature and exceedingly brief lifespan. CERN scientists estimated that, if the Standard Model were correct, the LHC would produce several Higgs bosons every minute, allowing physicists to finally confirm or disprove the Higgs bosons existence. In addition, the LHC allowed to search for supersymmetric particles and other hypothetical particles as possible unknown areas of physics.*[30] Some extensions of the Standard Model predict additional particles, such as the heavy W' and Z' gauge bosons, which were also estimated to be within reach of the LHC to discover.*[108]

First run (data taken 2009–2013)

The first physics results from the LHC, involving 284 collisions which took place in the ALICE detector, were reported on 15 December 2009.*[95] The results of the first proton–proton collisions at energies higher than Fermilab's Tevatron proton–antiproton collisions were published by the CMS collaboration in early February 2010, yielding greater-than-predicted charged-hadron production.*[109]

After the first year of data collection, the LHC experimental collaborations started to release their preliminary results concerning searches for new physics beyond the Standard Model in proton-proton collisions.*[110]*[111]*[112]*[113] No evidence of new particles was detected in the 2010 data. As a result, bounds were set on the allowed parameter space of various extensions of the Standard Model, such as models with large extra dimensions, constrained versions of the Minimal Supersymmetric Standard Model, and others.*[114]*[115]*[116]

On 24 May 2011, it was reported that quark–gluon plasma (the densest matter thought to exist besides black holes) had been created in the LHC.*[99]

Between July and August 2011, results of searches for the Higgs boson and for exotic particles, based on the data collected during the first half of the 2011 run, were presented in conferences in Grenoble*[117] and Mumbai.*[118] In the latter conference it was reported that, despite hints of a Higgs signal in earlier data, ATLAS and CMS exclude with 95% confidence level (using the CLs method) the existence of a Higgs boson with the properties predicted by the Standard Model over most of the mass region between 145 and 466 GeV.*[119] The searches for new particles did not yield signals either, allowing to further constrain the parameter space of various extensions of the Standard Model, including its supersymmetric extensions.*[120]*[121]

On 13 December 2011, CERN reported that the Standard Model Higgs boson, if it exists, is most likely to have a mass constrained to the range 115–130 GeV. Both the CMS and ATLAS detectors have also shown intensity peaks in the 124–125 GeV range, consistent with either background noise or the observation of the Higgs boson.*[122]

On 22 December 2011, it was reported that a new particle had been observed, the χ_b (3P) bottomonium state.*[102]

On 4 July 2012, both the CMS and ATLAS teams announced the discovery of a boson in the mass region around 125–126 GeV, with a statistical significance at the level of 5 sigma. This meets the formal level required to announce a new particle which is consistent with the Higgs boson, but scientists were cautious as to whether it is formally identified as actually being the Higgs boson, pending further analysis.*[123]

On 8 November 2012, the LHCb team reported on an experiment seen as a "golden" test of supersymmetry theories in physics,*[105] by measuring the very rare decay of the B_s meson into two muons ($B_s^0 \rightarrow \mu^+\mu^-$). The results, which match those predicted by the non-supersymmetrical Standard Model rather than the predictions of many branches of supersymmetry, show the decays are less common than some forms of supersymmetry predict, though could still match the predictions of other versions of supersymmetry theory. The results as initially drafted are stated to be short of proof but at a relatively high 3.5 sigma level of significance.*[124] The result was later confirmed by the CMS collaboration.*[125]

In August 2013 the LHCb team revealed an anomaly in the angular distribution of B meson decay products which could not be predicted by the Standard Model; this anomaly had a statistical certainty of 4.5 sigma, just short of the 5 sigma needed to be officially recognized as a discovery. It is unknown what the cause of this anomaly would be, although the Z' boson has been suggested as a possible candidate.*[126]

On 19 November 2014, the LHCb experiment announced the discovery of two new heavy subatomic particles, Ξ'^-_b and Ξ^{*-}_b. Both of them are baryons that are composed of one bottom, one down, and one strange quark. They are excited states of the bottom Xi baryon.*[127]*[128]

On 13 July 2015, the LHCb collaboration at CERN reported results consistent with pentaquark states in the decay of bottom Lambda baryons (Λ0 b).*[129]*[130]*[131]

Second run (2015 onward)

At the conference EPS-HEP 2015 in July, the collaborations presented first cross-section measurements of several particles at the higher collision energy.

8.7.7 Planned "high luminosity" upgrade

Main article: High Luminosity Large Hadron Collider

After some years of running, any particle physics experiment typically begins to suffer from diminishing returns: as the key results reachable by the device begin to be completed, later years of operation discover proportionately less than earlier years. A common outcome is to upgrade the devices involved, typically in energy, in luminosity, or in terms of improved detectors. As well as the planned 2013–2015 increase to its intended 14 TeV collision energy, a luminosity upgrade of the LHC, called the High Luminosity LHC, has also been proposed,*[132] to be made in 2022.

The optimal path for the LHC luminosity upgrade includes an increase in the beam current (i.e. the number of particles in the beams) and the modification of the two high-luminosity interaction regions, ATLAS and CMS. To achieve these increases, the energy of the beams at the point that they are injected into the (Super) LHC should also be increased to 1 TeV. This will require an upgrade of the full pre-injector system, the needed changes in the Super Proton Synchrotron being the most expensive. Currently the collaborative research effort of LHC Accelerator Research Program, LARP, is conducting research into how to achieve these goals.*[133]

8.7.8 Safety of particle collisions

Main article: Safety of high energy particle collision experiments

The experiments at the Large Hadron Collider sparked fears that the particle collisions might produce doomsday phenomena, involving the production of stable microscopic black holes or the creation of hypothetical particles called strangelets.*[134] Two CERN-commissioned safety reviews examined these concerns and concluded that the experiments at the LHC present no danger and that there is no reason for concern,*[135]*[136]*[137] a conclusion expressly endorsed by the American Physical Society.*[138]

The reports also noted that the physical conditions and collision events which exist in the LHC and similar experiments occur naturally and routinely in the universe without hazardous consequences,*[136] including ultra-high-energy cosmic rays observed to impact Earth with energies far higher than those in any man-made collider.

8.7.9 Popular culture

The Large Hadron Collider gained a considerable amount of attention from outside the scientific community and its progress is followed by most popular science media. The LHC has also inspired works of fiction including novels, TV series, video games and films.

The novel *Angels & Demons*, by Dan Brown, involves antimatter created at the LHC to be used in a weapon against the Vatican. In response, CERN published a "Fact or Fiction?" page discussing the accuracy of the book's portrayal of the LHC, CERN, and particle physics in general.*[139] The movie version of the book has footage filmed on-site at one of the experiments at the LHC; the director, Ron Howard, met with CERN experts in an effort to make the science in the story more accurate.*[140]

The novel *FlashForward*, by Robert J. Sawyer, involves the search for the Higgs boson at the LHC. CERN published a "Science and Fiction" page interviewing Sawyer and physicists about the book and the TV series based on it.*[141]

The short story "Breakthrough" by Tom Morris considers possible unexpected and alarming consequences of the LHC experiments. Available online

CERN employee Katherine McAlpine's "Large Hadron Rap" *[142] surpassed 7 million YouTube views.*[143]*[144] The band Les Horribles Cernettes was founded by women from CERN. The name was chosen so to have the same initials as the LHC.*[145]*[146]

National Geographic Channel's *World's Toughest Fixes*, Season 2 (2010), Episode 6 "Atom Smasher" features the replacement of the last superconducting magnet section in the repair of the supercollider after the 2008 quench incident. The episode includes actual footage from the repair facility to the inside of the supercollider, and explanations of the function, engineering, and purpose of the LHC.*[147]

The Large Hadron Collider was the focus of the 2012 student film Decay, with the movie being filmed on location in CERN's maintenance tunnels.*[148]

The feature documentary Particle Fever follows the experimental physicists at CERN who run the experiments, as well as the theoretical physicists who attempt to provide a conceptual framework for the LHC's results. It won the

Sheffield International Doc/Fest in 2013.

Onion News Network featured a parodied news story about the LHC titled "Bored Scientists Now Just Sticking Random Things Into Large Hadron Collider".[*][149]

8.7.10 See also

- Compact Linear Collider
- High Luminosity Large Hadron Collider
- International Linear Collider
- List of accelerators in particle physics
- Particle Fever
- Very Large Hadron Collider

8.7.11 References

[1] "The Large Hadron Collider". *cern.ch*.

[2] Highfield, Roger (16 September 2008). "Large Hadron Collider: Thirteen ways to change the world". *The Daily Telegraph* (London). Retrieved 2008-10-10.

[3] "CERN LHC sees high-energy success" (Press release). BBC News. 30 March 2010. Retrieved 2010-03-30.

[4] CERN Press Office (13 February 2012). "LHC to run at 4 TeV per beam in 2012". CERN.

[5] "LHC smashes energy record with test collisions". *BBC News*. Retrieved 28 August 2015.

[6] Jonathan Webb (5 April 2015). "Large Hadron collider restarts after pause". BBC. Retrieved 5 April 2015.

[7] O'Luanaigh, Cian. "Proton beams are back in the LHC". *CERN: Accelerating science*. CERN. Retrieved 24 April 2015.

[8] "Large Hadron Collider turns on 'data tap'". *BBC News*. Retrieved 28 August 2015.

[9] "Missing Higgs". CERN. 2008. Retrieved 2008-10-10.

[10] "Towards a superforce". CERN. 2008. Retrieved 2008-10-10.

[11] "What is the Worldwide LHC Computing Grid?". CERN. January 2011. Retrieved 2012-01-11.

[12] "Welcome". CERN. January 2011. Retrieved 2012-01-11.

[13] "Hunt for Higgs boson hits key decision point - Technology & science - Science - NBC News". *msnbc.com*.

[14] Worldwide LHC Computing Grid main page 14 November 2012: "[A] global collaboration of more than 170 computing centres in 36 countries ... to store, distribute and analyse the ~25 Petabytes (25 million Gigabytes) of data annually generated by the Large Hadron Collider"

[15] What is the Worldwide LHC Computing Grid? (Public 'About' page) 14 November 2012: "Currently WLCG is made up of more than 170 computing centers in 36 countries...The WLCG is now the world's largest computing grid"

[16] G. F. Giudice, *A Zeptospace Odyssey: A Journey into the Physics of the LHC*, Oxford University Press, Oxford 2010, ISBN 978-0-19-958191-7.

[17] Brian Greene (11 September 2008). "The Origins of the Universe: A Crash Course". *The New York Times*. Retrieved 2009-04-17.

[18] "... in the public presentations of the aspiration of particle physics we hear too often that the goal of the LHC or a linear collider is to check off the last missing particle of the Standard Model, this year's *Holy Grail* of particle physics, the Higgs boson. *The truth is much less boring than that!* What we're trying to accomplish is much more exciting, and asking what the world would have been like without the Higgs mechanism is a way of getting at that excitement." – Chris Quigg (2005). "Nature's Greatest Puzzles". *Econf C:I*. 040802 (1). arXiv:hep-ph/0502070. Bibcode:2005hep.ph....2070Q.

[19] "Why the LHC". CERN. 2008. Retrieved 2009-09-28.

[20] "Zeroing in on the elusive Higgs boson". US Department of Energy. March 2001. Retrieved 2008-12-11.

[21] "Accordingly, in common with many of my colleagues, I think it highly likely that both the Higgs boson and other new phenomena will be found with the LHC." ..."This mass threshold means, among other things, that something new – either a Higgs boson or other novel phenomena – is to be found when the LHC turns the thought experiment into a real one."Chris Quigg (February 2008). "The coming revolutions in particle physics". *Scientific American*. pp. 38–45. Retrieved 2009-09-28.

[22] Shaaban Khalil (2003). "Search for supersymmetry at LHC". *Contemporary Physics* **44** (3): 193–201. Bibcode:2003ConPh..44..193K. doi:10.1080/0010751031000077378.

[23] Alexander Belyaev (2009). "Supersymmetry status and phenomenology at the Large Hadron Collider". *Pramana* **72** (1): 143–160. Bibcode:2009Prama..72..143B. doi:10.1007/s12043-009-0012-0.

[24] Anil Ananthaswamy (11 November 2009). "In SUSY we trust: What the LHC is really looking for". *New Scientist*.

[25] Lisa Randall (2002). "Extra Dimensions and Warped Geometries" (PDF). *Science* **296** (5572): 1422–1427. Bibcode:2002Sci...296.1422R. doi:10.1126/science.1072567. PMID 12029124.

8.7. LARGE HADRON COLLIDER

[26] Panagiota Kanti (2009). "Black Holes at the LHC". *Lecture Notes in Physics*. Lecture Notes in Physics **769**: 387–423. arXiv:0802.2218. doi:10.1007/978-3-540-88460-6_10. ISBN 978-3-540-88459-0.

[27] CERN (18 July 2012). "Heavy ions and quark-gluon plasma".

[28] "LHC experiments bring new insight into primordial universe" (Press release). CERN. November 26, 2010. Retrieved December 2, 2010.

[29] G. Aad *et al.* (2010). *Phys. Rev. Lett.* 105 252303.

[30] "What is LHCb" (PDF). *CERN FAQ*. CERN Communication Group. January 2008. p. 44. Retrieved 2010-04-02.

[31] Joel Achenbach (March 2008). "The God Particle". *National Geographic Magazine*. Retrieved 2008-02-25.

[32] "The Z factory". CERN. 2008. Retrieved 2009-04-17.

[33] Henley, E. M.; Ellis, S. D., eds. (2013). *100 Years of Subatomic Physics*. World Scientific. ISBN 978-981-4425-80-3.

[34] Dr. Stephen Myers (4 October 2013). "The Large Hadron Collider 2008-2013". *International Journal of Modern Physics A* **28** (25): 1330035-1–1330035-65. Bibcode:2013IJMPA..2830035M. doi:10.1142/S0217751X13300354.

[35] "LHC: How Fast do These Protons Go?". *yogiblog*. Retrieved 2008-10-29.

[36] "LHC commissioning with beam". CERN. Retrieved 2009-04-17.

[37] "Operational Experience of the ATLAS High Level Trigger with Single-Beam and Cosmic Rays" (PDF). Retrieved 2010-10-29.

[38] Jörg Wenninger (November 2007). "Operational challenges of the LHC" (PowerPoint). p. 53. Retrieved 2009-04-17.

[39] "Ions for LHC (I-LHC) Project". CERN. 1 November 2007. Retrieved 2009-04-17.

[40] Paul Rincon (10 September 2008). "'Big Bang' experiment starts well". BBC News. Retrieved 2009-04-17.

[41] "Worldwide LHC Computing Grid". CERN. 2008. Retrieved 2 October 2011.

[42] "grille de production : les petits pc du lhc". Cite-sciences.fr. Retrieved 2011-05-22.

[43] "Worldwide LHC Computing Grid". *Official public website*. CERN. Retrieved 2 October 2011.

[44] "LHC@home". *berkeley.edu*.

[45] Craig Lloyd (18 Dec 2012). "First LHC proton run ends in success, new milestone". Retrieved 26 Dec 2014.

[46] "First beam in the LHC – Accelerating science" (Press release). CERN Press Office. 10 September 2008. Retrieved 2008-10-09.

[47] Paul Rincon (23 September 2008). "Collider halted until next year". BBC News. Retrieved 2008-10-09.

[48] "Large Hadron Collider – Purdue Particle Physics". Physics.purdue.edu. Retrieved 2012-07-05.

[49] Large Hadron Collider.

[50] "The LHC is back" (Press release). CERN Press Office. 20 November 2009. Retrieved 2009-11-20.

[51] "Two circulating beams bring first collisions in the LHC" (Press release). CERN Press Office. 23 November 2009. Retrieved 2009-11-23.

[52] Amina Khan (31 March 2010). "Large Hadron Collider rewards scientists watching at Caltech". *Los Angeles Times*. Retrieved 2010-04-02.

[53] M. Hogenboom (24 July 2013). "Ultra-rare decay confirmed in LHC". BBC. Retrieved 2013-08-18.

[54] "Challenges in accelerator physics". CERN. 14 January 1999. Retrieved 2009-09-28.

[55] John Poole (2004). "Beam Parameters and Definitions" (PDF). Missing or empty |title= (help)

[56] "CERN – The Large Hadron Collider". Public.web.cern.ch. Retrieved 2010-08-28.

[57] Agence Science-Presse (7 December 2009). "LHC: Un (très) petit Big Bang" (in French). Lien Multimédia. Retrieved 2010-10-29. Google translation

[58] "How much does it cost?". CERN. 2007. Retrieved 2009-09-28.

[59] Luciano Maiani (16 October 2001). "LHC Cost Review to Completion". CERN. Retrieved 2001-01-15.

[60] Toni Feder (2001). "CERN Grapples with LHC Cost Hike". *Physics Today* **54** (12): 21. Bibcode:2001PhT....54l..21F. doi:10.1063/1.1445534.

[61] "Bursting magnets may delay CERN collider project". Reuters. 5 April 2007. Retrieved 2009-09-28.

[62] Paul Rincon (23 September 2008). "Collider halted until next year". BBC News. Retrieved 2009-09-28.

[63] Robert Aymar (26 October 2005). "Message from the Director-General" (Press release). CERN Press Office. Retrieved 2013-06-12.

[64] "Fermilab 'Dumbfounded' by fiasco that broke magnet". Photonics.com. 4 April 2007. Archived from the original on 2008-06-16. Retrieved 2009-09-28.

[65] "Fermilab update on inner triplet magnets at LHC: Magnet repairs underway at CERN" (Press release). CERN Press Office. 1 June 2007. Retrieved 2009-09-28.

[66] "Updates on LHC inner triplet failure". *Fermilab Today*. Fermilab. 28 September 2007. Retrieved 2009-09-28.

[67] Paul Rincon (23 September 2008). "Collider halted until next year". BBC News. Retrieved 2009-09-29.

[68] "LHC to restart in 2009" (Press release). CERN Press Office. 5 December 2008. Retrieved 2008-12-08.

[69] Dennis Overbye (5 December 2008). "After repairs, summer start-up planned for collider". *New York Times*. Retrieved 2008-12-08.

[70] "News on the LHC". CERN. 16 July 2009. Retrieved 2009-09-28.

[71] "Restarting the LHC: Why 13 Tev?". *cern.ch*. Retrieved 28 August 2015.

[72] "First LHC magnets prepped for restart". *symmetry magazine*. Retrieved 28 August 2015.

[73] "First beam in the LHC – Accelerating science" (Press release). CERN Press Office. 10 September 2008. Retrieved 2008-09-10.

[74] Mark Henderson (10 September 2008). "Scientists cheer as protons complete first circuit of Large Hadron Collider". *Times Online* (London). Retrieved 2008-10-06.

[75] "Interim Summary Report on the Analysis of the 19 September 2008 Incident at the LHC" (PDF). CERN. 15 October 2008. EDMS 973073. Retrieved 2009-09-28.

[76] "Incident in LHC sector 3–4" (Press release). CERN Press Office. 20 September 2008. Retrieved 2009-09-28.

[77] "CERN releases analysis of LHC incident" (Press release). CERN Press Office. 16 October 2008. Retrieved 2009-09-28.

[78] "Final LHC magnet goes underground" (Press release). CERN Press Office. 30 April 2009. Retrieved 2009-08-04.

[79] L. Rossi (2010). "Superconductivity: its role, its success and its setbacks in the Large Hadron Collider of CERN" (PDF). *Superconductor Science and Technology* **23** (3): 034001. Bibcode:2010SuScT..23c4001R. doi:10.1088/0953-2048/23/3/034001.

[80] "CERN announces start-up date for LHC" (Press release). CERN Press Office. 7 August 2008.

[81] "CERN management confirms new LHC restart schedule" (Press release). CERN Press Office. 9 February 2009. Retrieved 2009-02-10.

[82] "CERN inaugurates the LHC" (Press release). CERN Press Office. 21 October 2008. Retrieved 2008-10-21.

[83] "LHC sets new world record" (Press release). CERN. 30 November 2009. Retrieved 2010-03-02.

[84] "Big Bang Machine sets collision record". *The Hindu*. Associated Press. 30 March 2010.

[85] "CERN completes transition to lead-ion running at the LHC" (Press release). CERN. 8 November 2010. Retrieved 2010-11-08.

[86] "The Latest from the LHC : Last period of proton running for 2010. – CERN Bulletin". Cdsweb.cern.ch. 1 November 2010. Retrieved 2011-08-17.

[87] CERN Press Office (17 December 2012). "The first LHC protons run ends with new milestone". CERN.

[88] "Long Shutdown 1: Exciting times ahead". *cern.ch*. Retrieved 28 August 2015.

[89] "CERN". *cern.ch*. Retrieved 28 August 2015.

[90] "One LHC sector up to full energy". *cern.ch*. Retrieved 28 August 2015.

[91] O'Luanaigh, Cian. "First successful beam at record energy of 6.5 TeV". *CERN: Accelerating Science*. CERN. Retrieved 24 April 2015.

[92] Record breaking collision at 13TeV, CERN Press release

[93] "http://www.sciencedaily.com/releases/2015/06/150603181744.htm?utm_source=feedburner&utm_medium=feed&utm_campaign=Feed%25253A+sciencedaily%25252Fmost_popular+%252528Most+Popular+News+-+ScienceDaily%252529". *www.sciencedaily.com*. Retrieved 2015-06-04.

[94] "The LHC is back" (Press release). CERN. 20 November 2009. Retrieved 2010-03-02.

[95] First Science Produced at LHC 2009-12-15

[96] "Large Hadron Collider to come back online after break". BBC News. 19 February 2010. Retrieved 2010-03-02.

[97] "LHC sees first stable-beam 3.5 TeV collisions of 2011". symmetry breaking. 13 March 2011. Retrieved 2011-03-15.

[98] CERN Press Office (22 April 2011). "LHC sets world record beam intensity". Press.web.cern.ch. Retrieved 2011-05-22.

[99] "Densest Matter Created in Big-Bang Machine". *nationalgeographic.com*.

[100] "LHC achieves 2011 data milestone". Press.web.cern.ch. 17 June 2011. Retrieved 2011-06-20.

[101] "One recorded inverse femtobarn".

[102] Jonathan Amos (22 December 2011). "LHC reports discovery of its first new particle". BBC News.

[103] "LHC physics data taking gets underway at new record collision energy of 8TeV". Press.web.cern.ch. 5 April 2012. Retrieved 2012-04-05.

[104] "New results indicate that new particle is a Higgs boson". CERN. 14 March 2013. Retrieved 14 March 2013.

[105] Ghosh, Pallab (12 Nov 2012). "Popular physics theory running out of hiding places". *BBC News*. Retrieved 14 November 2012.

[106] "The first LHC protons run ends with new milestone". CERN. 17 December 2012. Retrieved 10 March 2014.

[107] "First successful beam at record energy of 6.5 TeV". CERN. 10 April 2015. Retrieved 5 May 2015.

[108] P. Rincon (17 May 2010). "LHC particle search 'nearing', says physicist". BBC News.

[109] V. Khachatryan *et al.* (CMS collaboration) (2010). "Transverse momentum and pseudorapidity distributions of charged hadrons in pp collisions at $\sqrt{s}$ = 0.9 and 2.36 TeV". *Journal of High Energy Physics* **2010** (2): 1–35. arXiv:1002.0621. Bibcode:2010JHEP...02..041K. doi:10.1007/JHEP02(2010)041.

[110] V. Khachatryan *et al.* (CMS collaboration) (2011). "Search for Microscopic Black Hole Signatures at the Large Hadron Collider". *Physics Letters B* **697** (5): 434. arXiv:1012.3375. Bibcode:2011PhLB..697..434C. doi:10.1016/j.physletb.2011.02.032.

[111] V. Khachatryan *et al.* (CMS collaboration) (2011). "Search for Supersymmetry in pp Collisions at 7 TeV in Events with Jets and Missing Transverse Energy". *Physics Letters B* **698** (3): 196. arXiv:1101.1628. Bibcode:2011PhLB..698..196C. doi:10.1016/j.physletb.2011.03.021.

[112] G. Aad *et al.* (ATLAS collaboration) (2011). "Search for supersymmetry using final states with one lepton, jets, and missing transverse momentum with the ATLAS detector in $\sqrt{s}$ = 7 TeV pp". *Physical Review Letters* **106** (13): 131802. arXiv:1102.2357. Bibcode:2011PhRvL.106m1802A. doi:10.1103/PhysRevLett.106.131802.

[113] G. Aad *et al.* (ATLAS collaboration) (2011). "Search for squarks and gluinos using final states with jets and missing transverse momentum with the ATLAS detector in $\sqrt{s}$ = 7 TeV proton-proton collisions". *Physics Letters B* **701** (2): 186–203. arXiv:1102.5290. Bibcode:2011PhLB..701..186A. doi:10.1016/j.physletb.2011.05.061.

[114] Chalmers, M. Reality check at the LHC, physicsworld.com, Jan 18, 2011

[115] McAlpine, K. Will the LHC find supersymmetry?, physicsworld.com, Feb 22, 2011

[116] Geoff Brumfiel (2011). "Beautiful theory collides with smashing particle data". *Nature* **471** (7336): 13–14. Bibcode:2011Natur.471...13B. doi:10.1038/471013a.

[117] CERN Press Office (21 July 2011). "LHC experiments present their latest results at Europhysics Conference on High Energy Physics". Press.web.cern.ch. Retrieved 2011-09-01.

[118] CERN Press Office (22 August 2011). "LHC experiments present latest results at Mumbai conference". Press.web.cern.ch. Retrieved 2011-09-01.

[119] Pallab Ghosh (22 August 2011). "Higgs boson range narrows at European collider". BBC News.

[120] Pallab Ghosh (27 August 2011). "LHC results put supersymmetry theory 'on the spot'". BBC News.

[121] "LHCb experiment sees Standard Model physics". *Symmetry Breaking*. SLAC/Fermilab. 29 August 2011. Retrieved 2011-09-01.

[122] "ATLAS and CMS experiments present Higgs search status". CERN. 13 December 2011. Retrieved 2 January 2012.

[123] "CERN experiments observe particle consistent with long-sought Higgs boson". CERN. 4 July 2012. Retrieved 4 July 2012.

[124] First evidence for the decay $B_0{}^0 \to \mu^*{}^+-$, 8 Nov 2012, draft, LCHb collaboration.

[125] CMS collaboration (5 September 2013). "Measurement of the $B^0{}_s \to \mu+\mu-$ Branching Fraction and Search for $B^0 \to \mu+\mu-$ with the CMS Experiment". *Physical Review Letters*. arXiv:1307.5025. Bibcode:2013PhRvL.111j1804C. doi:10.1103/PhysRevLett.111.101804. Retrieved 23 December 2014.

[126] "Hints of New Physics Detected in the LHC?".

[127] New subatomic particles predicted by Canadians found at CERN, 19 November 2014

[128] , 19 November 2014

[129] R. Aaij et al. (LHCb collaboration) (2015). "Observation of J/ψp resonances consistent with pentaquark states in Λ0 b→J/ψK−
p decays". *Physical Review Letters* **115** (7). doi:10.1103/PhysRevLett.115.072001.

[130] "CERN's LHCb experiment reports observation of exotic pentaquark particles". cern.ch. Retrieved 28 August 2015.

[131] Rincon, Paul (1 July 2015). "Large Hadron Collider discovers new pentaquark particle". *BBC News*. Retrieved 2015-07-14.

[132] F. Ruggerio (29 September 2005). "LHC upgrade (accelerator)" (PDF). *8th ICFA Seminar*. Retrieved 2009-09-28.

[133] "DOE Review of LARP". Fermilab. 5–6 June 2007. Retrieved 2011-06-09.

[134] Alan Boyle (2 September 2008). "Courts weigh doomsday claims". *Cosmic Log*. MSNBC. Retrieved 2009-09-28.

[135] J.-P. Blaizot, J. Iliopoulos, J. Madsen, G.G. Ross, P. Sonderegger, H.-J. Specht (2003). "Study of Potentially Dangerous Events During Heavy-Ion Collisions at the LHC". CERN. Retrieved 2009-09-28.

[136] J. Ellis J, G. Giudice, M.L. Mangano, T. Tkachev, U. Wiedemann (LHC Safety Assessment Group) (5 September 2008). "Review of the Safety of LHC Collisions". *Journal of Physics G* **35** (11): 115004. arXiv:0806.3414. Bibcode:2008JPhG...35k5004E. doi:10.1088/0954-3899/35/11/115004.

[137] "The safety of the LHC". CERN. 2008. Retrieved 2009-09-28.

[138] Division of Particles & Fields. "Statement by the Executive Committee of the DPF on the Safety of Collisions at the Large Hadron Collider" (PDF). American Physical Society. Retrieved 2009-09-28.

[139] "Angels and Demons". CERN. 2011. Retrieved 2015-08-02.

[140] Ceri Perkins (2 June 2008). "ATLAS gets the Hollywood treatment". *ATLAS e-News*. Retrieved 2015-08-02.

[141] "FlashForward". CERN. September 2009. Retrieved 2009-10-03.

[142] Katherine McAlpine (28 July 2008). "Large Hadron Rap". YouTube. Retrieved 2011-05-08.

[143] Roger Highfield (6 September 2008). "Rap about world's largest science experiment becomes YouTube hit". *Daily Telegraph* (London). Retrieved 2009-09-28.

[144] Jennifer Bogo (1 August 2008). "Large Hadron Collider rap teaches particle physics in 4 minutes". *Popular Mechanics*. Retrieved 2009-09-28.

[145] Malcolm W Brown (29 December 1998). "Physicists Discover Another Unifying Force: Doo-Wop" (PDF). *New York Times*. Retrieved 2010-09-21.

[146] Heather McCabe (10 February 1999). "Grrl Geeks Rock Out" (PDF). *Wired News*. Retrieved 2010-09-21.

[147] "Atom Smashers". *World's Toughest Fixes*. Season 2. Episode 6. National Geographic Channel. Retrieved 15 June 2014.

[148] Boyle, Rebecca (2012-10-31). "Large Hadron Collider Unleashes Rampaging Zombies". Retrieved 22 November 2012.

[149] "Bored Scientists Now Just Sticking Random Things Into Large Hadron Collider".

8.7.12 External links

- Official website
- Overview of the LHC at CERN's public webpage
- CERN Courier magazine
- LHC Portal Web portal
- Lyndon Evans and Philip Bryant (eds) (2008). "LHC Machine". *Journal of Instrumentation* **3** (8): S08001. Bibcode:2008JInst...3S8001E. doi:10.1088/1748-0221/3/08/S08001. Full documentation for design and construction of the LHC and its six detectors (2008).

Video

- CERN, how LHC works on YouTube
- "Petabytes at the LHC". *Sixty Symbols*. Brady Haran for the University of Nottingham.
- Animation of LHC in collision production mode (June 2015)

News

- Eight Things To Know As The Large Hadron Collider Breaks Energy Records

Coordinates: 46°14′N 06°03′E / 46.233°N 6.050°E

8.8 SLAC National Accelerator Laboratory

"SLAC" redirects here. For other uses, see SLAC (disambiguation).

SLAC National Accelerator Laboratory, originally named **Stanford Linear Accelerator Center**,[*][2][*][3] is a United States Department of Energy National Laboratory operated by Stanford University under the programmatic direction of the U.S. Department of Energy Office of Science and located in Menlo Park, California.

The SLAC research program centers on experimental and theoretical research in elementary particle physics using electron beams and a broad program of research in atomic and solid-state physics, chemistry, biology, and medicine using synchrotron radiation.

8.8. SLAC NATIONAL ACCELERATOR LABORATORY

The entrance to SLAC in Menlo Park.

8.8.1 History

Founded in 1962 as the Stanford Linear Accelerator Center, the facility is located on 426 acres (1.72 square kilometers) of Stanford University-owned land on Sand Hill Road in Menlo Park, California—just west of the University's main campus. The main accelerator is 2 miles long—the longest linear accelerator in the world—and has been operational since 1966.

Nobel Prize

Research at SLAC has produced three Nobel Prizes in Physics:

- 1976: The charm quark—see J/ψ meson*[4]
- 1990: Quark structure inside protons and neutrons*[5]
- 1995: The tau lepton*[6]

SLAC's meeting facilities also provided a venue for the Homebrew Computer Club and other pioneers of the home computer revolution of the late 1970s and early 1980s.

In 1984 the laboratory was named an ASME National Historic Engineering Landmark and an IEEE Milestone.*[7]

SLAC developed and, in December 1991, began hosting the first World Wide Web server outside of Europe.*[8]

In the early-to-mid 1990s, the Stanford Linear Collider (SLC) investigated the properties of the Z boson using the Stanford Large Detector.

As of 2005, SLAC employs over 1,000 people, some 150 of which are physicists with doctorate degrees, and serves over 3,000 visiting researchers yearly, operating particle accelerators for high-energy physics and the Stanford Synchrotron Radiation Laboratory (SSRL) for synchrotron light radiation research, which was "indispensable" in the research leading to the 2006 Nobel Prize in Chemistry awarded to Stanford Professor Roger D. Kornberg.*[9]

In October 2008, the Department of Energy announced that the Center's name would be changed to SLAC National Accelerator Laboratory. The reasons given include a better representation of the new direction of the lab and the ability to trademark the laboratory's name. Stanford University had legally opposed the Department of Energy's attempt to trademark "Stanford Linear Accelerator Center".*[2]*[10]

In March 2009 it was announced that the SLAC National Accelerator Laboratory was to receive $68.3 Million in Recovery Act Funding to be disbursed by Department of Energy's Office of Science.*[11]

8.8.2 Components

SLAC 1.9 mile (3 kilometer) long Klystron Gallery above the beamline Accelerator

Accelerator

Part of the SLAC beamline

The main accelerator is an RF linear accelerator that can accelerate electrons and positrons up to 50 GeV. At 2.0 miles (about 3.2 kilometers) long, the accelerator is the longest linear accelerator in the world, and is claimed to be "the world's straightest object." *[12] The main accelerator is buried 30 feet (about 10 meters) below ground*[13] and passes underneath Interstate Highway 280. The aboveground klystron gallery atop the beamline is the longest building in the United States.

SLC pit and detector

Stanford Linear Collider

The Stanford Linear Collider was a linear accelerator that collided electrons and positrons at SLAC.*[14] The center of mass energy was about 90 GeV, equal to the mass of the Z boson, which the accelerator was designed to study. Grad student Barrett D. Milliken discovered the first Z event on 12 April 1989 while poring over the previous day's computer data from the Mark II detector.*[15] The bulk of the data was collected by the SLAC Large Detector, which came online in 1991. Although largely overshadowed by the Large Electron-Positron Collider at CERN, which began running in 1989, the highly polarized electron beam at SLC (close to 80%*[16]) made certain unique measurements possible, such as parity violation in Z Boson-b quark coupling.

Presently no beam enters the south and north arcs in the machine, which leads to the Final Focus, therefore this section is mothballed to run beam into the PEP2 section from the beam switchyard.

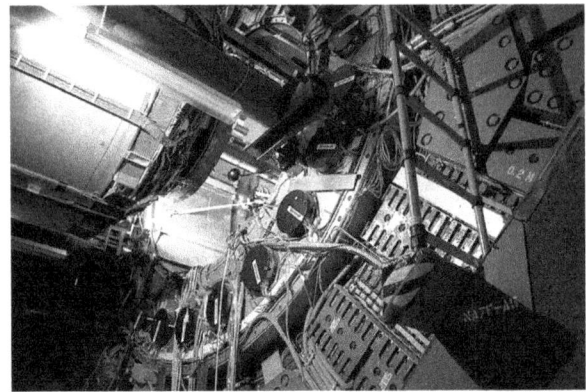

Inside view of the SLD

SLAC Large Detector

The SLAC Large Detector (SLD) was the main detector for the Stanford Linear Collider. It was designed primarily to detect Z bosons produced by the accelerator's electron-positron collisions. The SLD operated from 1992 to 1998.

PEP

PEP (Positron-Electron Project) began operation in 1980, with center-of-mass energies up to 29 GeV. At its apex, PEP had five large particle detectors in operation, as well as a sixth smaller detector. About 300 researchers made used of PEP. PEP stopped operating in 1990, and PEP-II began construction in 1994.*[17]

PEP-II

From 1999 to 2008, the main purpose of the linear accelerator was to inject electrons and positrons into the PEP-II accelerator, an electron-positron collider with a pair of storage rings 1.4 miles (2.2 km) in circumference. PEP-II was host to the BaBar experiment, one of the so-called B-Factory experiments studying charge-parity symmetry.

Stanford Synchrotron Radiation Lightsource

Main article: Stanford Synchrotron Radiation Lightsource

The Stanford Synchrotron Radiation Lightsource (SSRL) is a synchrotron light user facility located on the SLAC campus. Originally built for particle physics, it was used in experiments where the J/ψ meson was discovered. It is now used exclusively for materials science and biology experiments which take advantage of the high-intensity synchrotron radiation emitted by the stored electron beam to study the structure of molecules. In the early 1990s, an independent electron injector was built for this storage ring, allowing it to operate independently of the main linear accelerator.

Fermi Gamma-ray Space Telescope

Fermi Gamma-ray Space Telescope

Main article: Fermi Gamma-ray Space Telescope

SLAC plays a primary role in the mission and operation of the Fermi Gamma-ray Space Telescope, launched in August 2008. The principle scientific objectives of this mission are:

- To understand the mechanisms of particle acceleration in AGNs, pulsars, and SNRs.
- To resolve the gamma-ray sky: unidentified sources and diffuse emission.
- To determine the high-energy behavior of gamma-ray bursts and transients.
- To probe dark matter and fundamental physics.

KIPAC

Main article: Kavli Institute for Particle Astrophysics and Cosmology

The Kavli Institute for Particle Astrophysics and Cosmology (KIPAC) is partially housed on the grounds of SLAC, in addition to its presence on the main Stanford campus.

PULSE

The Stanford PULSE Institute (PULSE) is a Stanford Independent Laboratory located in the Central Laboratory at SLAC. PULSE was created by Stanford in 2005 to help Stanford faculty and SLAC scientists develop ultrafast x-ray research at LCLS. PULSE research publications can be viewed here.

LCLS

The Linac Coherent Light Source (LCLS) is a free electron laser facility located at SLAC. The LCLS is partially a reconstruction of the last 1/3 of the original linear accelerator at SLAC, and can deliver extremely intense x-ray radiation for research in a number of areas. It achieved first lasing in April 2009.*[18]

Aerial photo of the Stanford Linear Accelerator Center, with detector complex at the right (east) side

The laser produces hard X-rays, 10^9 times the relative brightness of traditional synchrotron sources and is the most powerful x-ray source in the world. LCLS enables a variety of new experiments and provides enhancements for existing experimental methods. Often, x-rays are used to take "snapshots" of objects on the nearly atomic level before obliterating samples. The laser's wavelength, ranging from 200 to 2000 electron volts (eV)*[19] is similar to the width of an atom, providing extremely detailed images for objects previously unattainable.*[20] Additionally, the laser is capable of capturing images with a "shutter speed" measured in femtoseconds, or million-billionths of a second, necessary because the intensity of the beam is often high enough so that the sample explodes on the femtosecond timescale.*[21]

LCLS-II

The LCLS-II project is to provide a major upgrade to LCLS by adding two new X-ray laser beams. The new system will utilize the 500 metres (1,600 ft) of existing tunnel to add new superconducting accelerator at 4 GeV and two undulators. The advancement from the discoveries using this new capabilities may include new drugs, next-generation computers, and new materials.*[22]

FACET

In 2012, the first two thirds (~2 km) of the original SLAC LINAC were re-commissioned for a new user facility, the Facility for Advanced Accelerator Experimental Tests (FACET). This new facility is capable of delivering 23 GeV, 3 nC electron (and positron) beams with short bunch lengths and small spot sizes, ideal for beam-driven Plasma Acceleration studies. *[23]

NLCTA

The Next Linear Collider Test Accelerator (NLCTA) is a 60-120 MeV high-brightness electron beam linear accelerator used for experiments on advanced beam manipulation and acceleration techniques. It is located at SLAC's end station B. A list of relevant research publications can be viewed here.

8.8.3 Other discoveries

- SLAC has also been instrumental in the development of the klystron, a high-power microwave amplification tube.

- There is active research on plasma acceleration with recent successes such as the doubling of the energy of 42 GeV electrons in a meter-scale accelerator.

- There was a *Paleoparadoxia* found at the SLAC site, and its skeleton can be seen at a small museum there in the Breezeway.*[24]

- The SSRL facility was used to reveal hidden text in the Archimedes Palimpsest. X-rays from the synchrotron radiation lightsource caused the iron in the original ink to glow, allowing the researchers to photograph the original document that a Christian monk had scrubbed off.*[25]

8.8.4 See also

- Accelerator physics
- Beamline
- CERN
- Cyclotron
- Dipole magnet
- Electromagnetism
- List of particles
- List of United States college laboratories conducting basic defense research
- Particle beam
- Particle physics
- Quadrupole magnet
- Spallation Neutron Source
- Wolfgang Panofsky (1961–84, SLAC Director; Professor, Stanford University)

8.8.5 References

[1] Labs at a glance - SLAC http://science.energy.gov/laboratories/slac-national-accelerator-laboratory/

[2] "SLAC renamed to SLAC Natl. Accelerator Laboratory" . *The Stanford Daily*. 16 October 2008. Archived from the original on 2012-01-29. Retrieved 2008-10-16.

[3] "Stanford Linear Accelerator Center renamed SLAC National Accelerator Laboratory" (Press release). SLAC National Accelerator Laboratory. 15 October 2008. Archived from the original on 2011-07-20. Retrieved 2011-07-20.

[4] Nobel Prize in Physics 1976. Half prize awarded to Burton Richter.

[5] Nobel Prize in Physics 1990 Award split between Jerome I. Friedman, Henry W. Kendall, and Richard E. Taylor.

[6] Nobel Prize in Physics 1995 Half prize awarded to Martin L. Perl.

[7] "Milestones:Stanford Linear Accelerator Center, 1962". *IEEE Global History Network*. IEEE. Retrieved 3 August 2011.

[8] The Early World Wide Web at SLAC: Early Chronology and Documents

[9] "2006 Nobel Prize in Chemistry". *SLAC Virtual Visitor Center*. Stanford University. n.d. Archived from the original on 5 August 2011. Retrieved 19 March 2015.

[10] A New Name for SLAC

[11] 23, 2009 - SLAC National Accelerator Laboratory to Receive $68.3 Million in Recovery Act Funding

[12] Saracevic, Alan T. "Silicon Valley: It's where brains meet bucks." *San Francisco Chronicle* 23 October 2005. p J2. Accessed 2005-10-24.

[13] Neal, R. B. (1968). "Chap. 5". *The Stanford Two-Mile Accelerator* (PDF). New York, New York: W.A. Benjamin, Inc. p. 59. Retrieved 2010-09-17.

[14] Loew, G. A. (1984). "The SLAC Linear Collider and a few ideas on Future Linear Colliders" (PDF). *Proceedings of the 1984 Linear Accelerator Conference*.

[15] Rees, J. R. (1989). "The Stanford Linear Collider". *Scientific American 1989* **261**: 36–43. See also a colleague's logbook at http://www.symmetrymagazine.org/cms/?pid=1000294.

[16] Ken Baird, Measurements of A_{LR} and A_{lepton} from SLD http://hepweb.rl.ac.uk/ichep98/talks_1/talk101.pdf

[17] http://www.slac.stanford.edu/gen/grad/GradHandbook/slac.html

[18] Linac Coherent Light Source webpage

[19] "SOFT X-RAY MATERIALS SCIENCE (SXR) url=https://portal.slac.stanford.edu/sites/lcls_public/Instruments/SXR/Pages/Specifications.aspx accessdate=2015-03-22".

[20] Bostedt, C. et al. (2013). "Ultra-fast and ultra-intense x-ray sciences: First results from the Linac Coherent Light Source free-electron laser". *Journal of Physics B* **46** (16): 164003. Bibcode:2013JPhB...46p4003B. doi:10.1088/0953-4075/46/16/164003.

[21] Rachel Ehrenberg, ScienceNews.org

[22] "LCLS-II Upgrade to Enable Pioneering Research in Many Fields". *Cryogenic Society of America*. 8 July 2015. Retrieved 15 August 2015.

[23] FACET: SLAC's new user facility

[24] Stanford's SLAC Paleoparadoxia much thanks to Adele Panofsky, Dr. Panofsky's wife, for her reassembly of the bones of the Paleoparadoxia uncovered at SLAC.

[25] Bergmann, Uwe. "X-Ray Fluorescence Imaging of the Archimedes Palimpsest: A Technical Summary" (PDF). SLAC National Accelerator Laboratory. Retrieved 2009-10-04.

8.8.6 External links

- SLAC official webpage
    - SLAC Today, SLAC's online newspaper, published weekdays
    - *symmetry* magazine, SLAC's monthly particle physics magazine, with Fermilab

Coordinates: 37°24′53″N 122°13′18″W / 37.41472°N 122.22167°W

8.9 Tevatron

Coordinates: 41°49′55″N 88°15′06″W / 41.831904°N 88.251715°W

The **Tevatron** is a circular particle accelerator in the United States, at the Fermi National Accelerator Laboratory (also known as *Fermilab*), just east of Batavia, Illinois, and holds the title of the second highest energy particle collider in the world, after the Large Hadron Collider (LHC) near Geneva, Switzerland. The Tevatron was a synchrotron that accelerated protons and antiprotons in a 6.86 km, or 4.26 mi, ring to energies of up to 1 TeV, hence its name.[1] The Tevatron was completed in 1983 at a cost of $120 million and significant upgrade investments were made in 1983–2011.

The main achievement of the Tevatron was the discovery in 1995 of the top quark—the last fundamental fermion predicted by the standard model of particle physics. On July 2, 2012, scientists of the CDF and DØ collider experiment teams at Fermilab announced the findings from the analysis of around 500 trillion collisions produced from the Tevatron collider since 2001, and found that the existence of the suspected Higgs boson was highly likely with only a 1-in-550 chance that the signs were due to a statistical fluctuation. The findings were confirmed two days later as being correct with a likelihood of error less than 1 in a million by data from the LHC experiments.[2]

The Tevatron ceased operations on 30 September 2011,[3] due to budget cuts[4] and because of the completion of the LHC, which began operations in early 2010 and was far more powerful (planned energies were two 7 TeV beams at the LHC compared to 1 TeV at the Tevatron). The main ring of the Tevatron will probably be reused in future experiments, and its components may be transferred to other particle accelerators.[5]

8.9.1 History

December 1, 1968 saw the breaking of ground for the linear accelerator (linac). The construction of the Main Accelerator Enclosure began on October 3, 1969 when the first shovel of earth was turned by Robert R. Wilson, NAL's director. This would become the 6.4 km circumference Fermilab's Main Ring.[6]

The linac first 200 MeV beam started on December 1, 1970. The booster first 8 GeV beam was produced on May 20, 1971. On June 30, 1971, a proton beam was guided for the first time through the entire National Accelerator Laboratory accelerator system including the Main Ring. The beam was accelerated to only 7 Gev. Back then, the Booster Accelerator took 200 MeV protons from the Linac and "boosted" their energy to 8 billion electron volts. They were then injected into the Main Accelerator.[6]

On the same year before the completion of the Main Ring, Wilson testified to the Joint Committee on Atomic Energy on March 9, 1971 that it was feasible to achieve a higher energy by using superconducting magnets. He also suggested that it could be done by using the same tunnel as the main ring and the new magnets would be installed in the same locations to be operated in parallel to the existing magnets of the Main Ring. That was the starting point of the Tevatron project.[7] The Tevatron was in research and development phase between 1973 and 1979 while the acceleration at the Main Ring continued to be enhanced.[8]

A series of milestones saw acceleration rise to 20 GeV on January 22, 1972 to 53 GeV on February 4 and to 100 GeV on February 11. On March 1, 1972, the then NAL accelerator system accelerated for the first time a beam of protons to its design energy of 200 GeV. By the end of 1973, NAL's accelerator system operated routinely at 300 GeV.[6]

On 14 May 1976 Fermilab took its protons all the way to 500 GeV. This achievement provided the opportunity to introduce a new energy scale, the teraelectronvolt (TeV), equal to 1000 GeV. On 17 June of that year, the European Super Proton Synchrotron accelerator (SPS) had achieved an initial circulating proton beam (with no accelerating radio-frequency power) of only 400 GeV.[9]

The conventional magnet Main Ring was shut down in 1981 for installation of superconducting magnets underneath it. The Main Ring continued to serve as an injector for the Tevatron until the Main Injector was completed west of the Main Ring in 2000.[7] The 'Energy Doubler', as it was known then, produced its first accelerated beam—512 GeV —on July 3, 1983.[10]

Its initial energy of 800 GeV was achieved on February 16, 1984. On October 21, 1986 acceleration at the Tevatron was pushed to 900 GeV, providing a first proton–antiproton collision at 1.8 TeV on November 30, 1986.[11]

The *Main Injector*, which replaced the Main Ring,[12] was the most substantial addition, built over six years from 1993 at a cost of $290 million.[13] Tevatron collider Run II begun on March 1, 2001 after successful completion of that facility upgrade. From then, the beam had been capable of delivering an energy of 980 GeV.[12]

On July 16, 2004 the Tevatron achieved a new peak luminosity, breaking the record previously held by the old European Intersecting Storage Rings (ISR) at CERN. That very Fermilab record was doubled on September 9, 2006, then a bit more than tripled on March 17, 2008 and ultimately multiplied by a factor of 4 over the previous 2004 record on April 16, 2010 (up to 4×10^{32} cm^{-2} s^{-1}).[11]

The Tevatron ceased operations on 30 September 2011. By the end of 2011, the Large Hadron Collider (LHC) at CERN had achieved a luminosity almost ten times higher than Tevatron's (at 3.65×10^{33} cm^{-2} s^{-1}) and a beam energy of 3.5 TeV each (doing so since March 18, 2010), already ~3.6 times the capabilities of the Tevatron (at 0.98 TeV).

8.9.2 Mechanics

The acceleration occurs in a number of stages. The first stage is the 750 keV *Cockcroft-Walton* pre-accelerator, which ionizes hydrogen gas and accelerates the negative ions created using a positive voltage. The ions then pass into the 150 meter long linear accelerator (linac) which uses oscillating electrical fields to accelerate the ions to 400 MeV. The ions then pass through a carbon foil, to remove the electrons, and the charged protons then move into the *Booster*.[14]

The Booster is a small circular synchrotron, around which the protons pass up to 20,000 times to attain an energy of around 8 GeV. From the Booster the particles pass into the Main Injector, which was completed in 1999 to perform a number of tasks. It can accelerate protons up to 150 GeV; it can produce 120 GeV protons for antiproton creation; it can increase antiproton energy to 150 GeV and it can inject protons or antiprotons into the Tevatron. The antiprotons are created by the *Antiproton Source*. 120 GeV protons are collided with a nickel target producing a range of particles including antiprotons which can be collected and stored in the accumulator ring. The ring can then pass the antiprotons to the Main Injector.

The Tevatron can accelerate the particles from the Main Injector up to 980 GeV. The protons and antiprotons are accelerated in opposite directions, crossing paths in the CDF and DØ detectors to collide at 1.96 TeV. To hold the particles on track the Tevatron uses 774 niobium-titanium superconducting dipole magnets cooled in liquid helium producing 4.2 teslas. The field ramps over about 20 seconds as the particles are accelerated. Another 240 NbTi quadrupole magnets are used to focus the beam.[*][1]

The initial design luminosity of the Tevatron was 10^{30} cm[*]−2 s[*]−1, however the accelerator has following upgrades been able to deliver luminosities up to $4\times10^{*32}$ cm[*]−2 s[*]−1.[*][15]

On September 27, 1993 the cryogenic cooling system of the Tevatron Accelerator was named an International Historic Landmark by the American Society of Mechanical Engineers. The system, which provides cryogenic liquid helium to the Tevatron's superconducting magnets, was the largest low-temperature system in existence upon its completion in 1978. It keeps the coils of the magnets, which bend and focus the particle beam, in a superconducting state so that they consume only 1/3 of the power they would require at normal temperatures.[*][8]

8.9.3 Discoveries

The Tevatron confirmed the existence of several subatomic particles that were predicted by theoretical particle physics, or gave suggestions to their existence. In 1995, the CDF experiment and DØ experiment collaborations announced the discovery of the top quark, and by 2007 they measured its mass to a precision of nearly 1%. In 2006, the CDF collaboration reported the first measurement of B_s oscillations, and observation of two types of sigma baryons.[*][16] In 2007, the DØ and CDF collaborations reported direct observation of the "Cascade B" (Ξ−
b) Xi baryon.[*][17]

In September 2008, the DØ collaboration reported detection of the Ω−
b, a "double strange" Omega baryon with the measured mass significantly higher than the quark model prediction.[*][18][*][19] In May 2009 the CDF collaboration made public their results on search for Ω−
b based on analysis of data sample roughly four times larger than the one used by DØ experiment.[*][20] The mass measurements from the CDF experiment were 6054.4±6.8 MeV/c^2 and in excellent agreement with Standard Model predictions, and no signal has been observed at the previously reported value from the DØ experiment. The two inconsistent results from DØ and CDF differ by 111±18 MeV/c^2 or by 6.2 standard deviations. Due to excellent agreement between the mass measured by CDF and the theoretical expectation, it is a strong indication that the particle discovered by CDF is indeed the Ω−
b. It is anticipated that new data from LHC experiments will clarify the situation in the near future.

On July 2, 2012, two days before a scheduled announcement at the Large Hadron Collider (LHC), scientists at the Tevatron collider from the CDF and DØ collaborations announced their findings from the analysis of around 500 trillion collisions produced since 2001: They found that the existence of the Higgs boson was likely with a mass in the region of 115 to 135 GeV.[*][21][*][22] The statistical significance of the observed signs was 2.9 sigma, which meant that there is only a 1-in-550 chance that a signal of that magnitude would have occurred if no particle in fact existed with those properties. The final analysis of data from the Tevatron did however not settle the question of whether the Higgs particle exists.[*][2][*][23] Only when the scientists from the Large Hadron Collider announced the more precise LHC results on July 4, 2012, with a mass of 125.3 ± 0.4 GeV (CMS)[*][24] or 126 ± 0.4 GeV (ATLAS)[*][25] respectively, was there strong evidence through consistent measurements by the LHC and the Tevatron for the possible existence of a Higgs particle at that mass range.

8.9.4 Earthquake detection

Earthquakes, even if they were thousands of miles away, did cause strong enough movements in the magnets to negatively affect the beam quality and even disrupt it. Therefore tiltmeters were installed on Tevatron's magnets to monitor minute movements and to help identify the cause of problems quickly. The first known earthquake to disrupt the beam was the 2002 Denali earthquake, with another collider shutdown caused by a moderate local quake on June 28, 2004.[*][26] Since then, the minute seismic vibrations emanating from over 20 earthquakes were detected at the Tevatron without a shutdown, like the 2004 Indian Ocean earthquake, the 2005 Sumatra earthquake, New Zealand's 2007 Gisborne earthquake, the 2010 Haiti earthquake and the 2010 Chile earthquake.[*][27]

8.9.5 See also

- Bevatron

- Large Hadron Collider

- Superconducting Super Collider

- Zevatron

8.9.6 References

[1] R. R. Wilson (1978). "The Tevatron". Fermilab. FERMILAB-TM-0763.

[2] "Tevatron scientists announce their final results on the Higgs particle". Fermi National Accelerator Laboratory. July 2, 2012. Retrieved July 7, 2012.

[3] "'Nothing lasts forever at the edge of science': U.S. drops behind in race to find origins of the universe as huge particle collider is shut down". *Daily Mail*. 1 October 2011. Retrieved 7 October 2012.

[4] Mark Alpert (29 September 2011). "Future of Top U.S. Particle Physics Lab in Jeopardy". *Scientific American*. Retrieved 7 October 2012.

[5] Prachi Patel-Predd (4 September 2008). "What Happens to Particle Accelerators After They Are Shut Down?". *Scientific American*. Retrieved 7 October 2012.

[6] "Accelerator History—Main Ring". Fermilab History and Archives Project. Retrieved 7 October 2012.

[7] "Accelerator History—Main Ring transition to Energy Doubler/Saver". Fermilab History and Archives Project. Retrieved 7 October 2012.

[8] "The Fermilab Tevatron Cryogenic Cooling System". ASME. 1993. Retrieved 2015-08-12.

[9] "Super Proton Synchrotron marks its 25th birthday". *CERN courier*. 2 July 2011. Retrieved 7 October 2012.

[10] "1983—The Year the Tevatron Came to Life". *Fermi News* **26** (15). 2003.

[11] "Interactive timeline". Fermilab. Retrieved 7 October 2012.

[12] "Run II begins at the Tevatron". *CERN courier*. 30 April 2001. Retrieved 7 October 2012.

[13] "Main Injector and Recycler Ring History and Public Information". Fermilab Main Injector department. Retrieved 7 October 2012.

[14] "Accelerators —Fermilab's Chain of Accelerators". Fermilab. 15 January 2002. Retrieved 2 December 2009.

[15] The TeVatron Collider: A Thirty-Year Campaign

[16] "Experimenters at Fermilab discover exotic relatives of protons and neutrons". Fermilab. 2006-10-23. Retrieved 2006-10-23.

[17] "Back-to-Back b Baryons in Batavia". Fermilab. 2007-07-25. Retrieved 2007-07-25.

[18] "Fermilab physicists discover "doubly strange" particle". Fermilab. September 3, 2008. Retrieved 2008-09-04.

[19] V. M. Abazov *et al.* (DØ collaboration) (2008). "Observation of the doubly strange b baryon Ω_b^-". *Physical Review Letters* **101** (23): 231002. arXiv:0808.4142. Bibcode:2008PhRvL.101w2002A. doi:10.1103/PhysRevLett.101.232002.

[20] T. Aaltonen *et al.* (CDF Collaboration) (2009). "Observation of the Ω_b^- and Measurement of the Properties of the Ξ_b^- and Ω_b^-". *Physical Review D* **80**: 072003. arXiv:0905.3123. Bibcode:2009PhRvD..80g2003A. doi:10.1103/PhysRevD.80.072003.

[21] "Updated Combination of CDF and DØ's Searches for Standard Model Higgs Boson Production with up to 10.0 fb-1 of Data". Tevatron New Phenomena & Higgs Working Group. June 2012. Retrieved August 2, 2012.

[22] "Evidence for a particle produced in association with weak bosons and decaying to a bottom-antibottom quark pair in Higgs boson searches at the Tevatron". Tevatron New Phenomena & Higgs Working Group. July 2012. Retrieved August 2, 2012.

[23] Rebecca Boyle (July 2, 2012). "Tantalizing Signs of Higgs Boson Found By U.S. Tevatron Collider". Popular Science. Retrieved July 7, 2012.

[24] CMS collaboration (31 July 2012). "Observation of a new boson at a mass of 125 GeV with the CMS experiment at the LHC". arXiv:1207.7235.

[25] ATLAS collaboration (31 July 2012). "Observation of a New Particle in the Search for the Standard Model Higgs Boson with the ATLAS Detector at the LHC". arXiv:1207.7214.

[26] Was that a quake? Ask the Tevatron

[27] Tevatron Sees Haiti Earthquake

8.9.7 Further reading

- Valery Lebedev, Vladimir Shiltsev, ed. (2014). *Accelerator Physics at the Tevatron Collider*. Springer. ISBN 978-1-4939-0884-4.

8.9.8 External links

- Live Tevatron status
- The Hunt for the Higgs at Tevatron
- Technical details of the accelerators

Chapter 9

Text and image sources, contributors, and licenses

9.1 Text

- **Quark** *Source:* https://en.wikipedia.org/wiki/Quark?oldid=681925889 *Contributors:* AxelBoldt, Derek Ross, Vicki Rosenzweig, Mav, Bryan Derksen, The Anome, Gareth Owen, Andre Engels, PierreAbbat, Peterlin~enwiki, Ben-Zin~enwiki, Zoe, Heron, Montrealais, Hfastedge, Edward, Dante Alighieri, Ixfd64, CesarB, Card~enwiki, NuclearWinner, Looxix~enwiki, Ahoerstemeier, Elliot100, Docu, J-Wiki, Nanobug, Aarchiba, Julesd, Glenn, Schneelocke, Jengod, A5, Timwi, Dysprosia, DJ Clayworth, Phys, Ed g2s, Bevo, Olathe, MD87, Jni, Phil Boswell, Sjorford, Donarreiskoffer, Robbot, Sanders muc, Moncrief, Merovingian, PxT, Texture, Bkell, UtherSRG, Widsith, Ancheta Wis, Giftlite, ShaunMacPherson, Harp, Nunh-huh, Lupin, Herbee, Leflyman, Monedula, 0x6D667061, Xerxes314, Anville, Hoho~enwiki, Alison, Beardo, Moogle10000, Wronkiew, Jackol, Bobblewik, Bodhitha, Piotrus, Kaldari, Elroch, Icairns, Zfr, TonyW, Ukexpat, BrianWilloughby, Grunt, O'Dea, Jiy, Discospinster, Rich Farmbrough, Guanabot, T Long, Vsmith, Saintswithin, SocratesJedi, Mani1, Bender235, Lancer, RJHall, Mr. Billion, El C, Kwamikagami, Laurascudder, Susvolans, Triona, Axezz, Bobo192, Army1987, C S, Ziggurat, Rangelov, Matt McIrvin, Jojit fb, Nk, Pentalis, Obradovic Goran, Fwb22, Lysdexia, Benjonson, Alansohn, Gary, Gintautasm, Guy Harris, Keenan Pepper, MonkeyFoo, Lectonar, Mac Davis, Wdfarmer, Snowolf, Schapel, Knowledge Seeker, Evil Monkey, VivaEmilyDavies, CloudNine, Kusma, Kazvorpal, Kay Dekker, Crosbiesmith, Mogigoma, Linas, Mindmatrix, JarlaxleArtemis, ScottDavis, LOL, Wdyoung, Before My Ken, Tylerni7, Jwanders, Dataphiliac, AndriyK, Noetica, Wayward, Wisq, Palica, Marudubshinki, Calréfa Wéná, GSlicer, Graham87, Deltabeignet, Kbdank71, Yurik, Crzrussian, Rjwilmsi, Bremen, Marasama, SpNeo, Mike Peel, Bubba73, DoubleBlue, Matt Deres, Yamamoto Ichiro, Algebra, Dsnow75, RobertG, Nihiltres, Jeff02, RexNL, TeaDrinker, Chobot, DVdm, Jpacold, Gwernol, Elfguy, Roboto de Ajvol, YurikBot, Wavelength, Bambaiah, Sceptre, Hairy Dude, Jimp, Phantomsteve, TheDoober, Dobromila, JabberWok, CambridgeBayWeather, Chaos, Salsb, Wimt, Ugur Basak, NawlinWiki, Spike Wilbury, Bossrat, SCZenz, Randolf Richardson, Danlaycock, Tony1, DRosenbach, Robertbyrne, Dna-webmaster, WAS 4.250, Closedmouth, Pietdesomere, Heathhunnicutt, Kevin, Banus, RG2, Kamickalo, That Guy, From That Show!, Veinor, MacsBug, SmackBot, Aigarius, BBandHB, Incnis Mrsi, InverseHypercube, C.Fred, Bazza 7, Ikip, Anastrophe, Jrockley, Eskimbot, AnOddName, Jonathan Karlsson, Edgar181, Gilliam, Dauto, NickGarvey, Vvarkey, Bluebot, KaragouniS, Keegan, Dahn, Bigfun, Miquonranger03, OrangeDog, Silly rabbit, Metacomet, Tripledot, Nbarth, DHN-bot~enwiki, Sbharris, Colonies Chris, Hallenrm, Scwlong, Gsp8181, Can't sleep, clown will eat me, Mallorn, Jeff DLB, TKD, Addshore, Mqjjb30e, Cybercobra, Khukri, B jonas, Jdlambert, Lpgeffen, Nrcprm2026, Akriasas, Zadignose, Jóna Þórunn, Bdushaw, Beyazid, TriTertButoxy, SashatoBot, SciBrad, Doug Bell, Soap, Richard L. Peterson, John, Mgiganteus1, SpyMagician, Edconrad, Loadmaster, 2T, Waggers, SandyGeorgia, Ravi12346, Dbzfrk15146, Peyre, Newone, GDallimore, Happy-melon, Majora4, Chovain, Tawkerbot2, Cryptic C62, JForget, Vaughan Pratt, Hello789, ZICO, SUPRATIM DEY, Ruslik0, CuriousEric, Paulfriedman7, Logical2u, Myasuda, RoddyYoung, Typewritten, Cydebot, Abeg92, Mike Christie, Grahamec, Gogo Dodo, Jayen466, 879(CoDe), Michael C Price, Tawkerbot4, Ameliorate!, Akcarver, Gimmetrow, SallyScot, Casliber, Thijs!bot, Epbr123, NeoPhyteRep, LeBofSportif, Markus Pössel, Anupam, Sopranosmob781, Headbomb, Marek69, John254, KJBurns, MichaelMaggs, Escarbot, Eleuther, Ice Ardor, Aadal, AntiVandalBot, SmokeyTheCat, Tyco.skinner, Exteray, RobJ1981, Rsocol, Ke garne, Deflective, Husond, MER-C, CosineKitty, Andonic, East718, Pkoppenb, DanPMK, Magioladitis, WolfmanSF, Thasaidon, Bongwarrior, VoABot II, باسم, Inertiatic076, Kevinmon, Christoph Scholz~enwiki, Aka042, Giggy, Tanvirzaman, Johnbibby, Cyktsui, ArchStanton69, Ace42, Allstarecho, Shijualex, DerHexer, Elandra, Denis tarasov, MartinBot, Poeloq, Dorvaq, CommonsDelinker, HEL, J.delanoy, Nev1, Ops101ex, DrKiernan, Hgpot, Ferdyshenko, Jigesh, DJ1AM, Tarotcards, Coppertwig, TomasBat, Nikbuz, SJP, FJPB, Vainamainien, Tiggydong, Robprain, Sheliak, Cuzkatzimhut, Lights, X!, VolkovBot, CWii, ABF, John Darrow, Holme053, Nousernamesleft, Ryan032, GimmeBot, Davehi1, A4bot, Captain Courageous, Guillaume2303, Anonymous Dissident, Drestros power, Qxz, Anna Lincoln, Eldaran~enwiki, Leafyplant, Don4of4, PaulTanenbaum, Abdullais4u, Jbryancoop, Mbalelo, Gilisa, Eubulides, Chronitis, Seresin, Dustybunny, Insanity Incarnate, Upquark, Edge1212, Ollieho, AOEU Warrior, SieBot, Graham Beards, WereSpielChequers, Csmart287, Guguma5, Winchelsea, Jbmurray, Caltas, Vanished User 8a9b4725f8376, Keilana, Bentogoa, Aillema, RadicalOne, Arbor to SJ, Elcobbola, Physics one, Dhatfield, RSStockdale, Son of the right hand, Ngexpert5, Ngexpert6, Ngexpert7, Psycherevolt, Sean.hoyland, Myger-ardromance, Dabomb87, Nergaal, Muhends, Romit3, SallyForth123, Atif.t2, ClueBot, The Thing That Should Not Be, Wwheaton, Xeno malleus, Harland1, Piledhigheranddeeper, Maxtitan, DragonBot, Glopso, Choonkiat.lee, Himynameisdumb, Worth my salt, Arthur Quark, Estirabot, Brews ohare, Jotterbot, PhySusie, Brianboulton, Dekisugi, ANOMALY-117, Sallicio, Yomangan, Jtle515, Katanada, DumZiBoT, TimothyRias, XLinkBot, Vayalir, Oldnoah, Saintlucifer2008, Nathanwesley3, Dragonfiremage, Devilist666, Mancune2001, Jbeans, WikiDao, SkyLined,

Truthnlove, Airplaneman, Eklipse, Addbot, Eric Drexler, AVand, Some jerk on the Internet, Captain-tucker, Giants2008, Iceblock, Ronhjones, Quarksci, Mseanbrown, Looie496, LaaknorBot, Peti610botH, AgadaUrbanit, Tide rolls, Vicki breazeale, Gail, Extruder~enwiki, Abduallah mohammed, Dealer77, Luckas-bot, Yobot, Fraggle81, Cflm001, Legobot II, Amble, Mmxx, Superpenguin1984, Worm That Turned, The Vector Kid, Planlips, Fangfyre, TestEditBot, Azcolvin429, Vroo, Synchronism, Bility, Orion11M87, AnomieBOT, Xi rho, Rubinbot, Jim1138, Bookaneer, Yotcmdr, Crystal whacker, Sonic h, Materialscientist, Citation bot, Pitke, Vuerqex, Bci2, ArthurBot, LilHelpa, Xqbot, Jeffrey Mall, AbigailAbernathy, Srich32977, Alex2510, Almabot, Uscbino, Pmlineditor, RibotBOT, Shmomuffin, Gunjan verma81, Chotarocket, Ernsts, Renverse, A. di M., Weekendpartier, FrescoBot, Paine Ellsworth, DelphinidaeZeta, Steve Quinn, Citation bot 1, AstaBOTh15, Pinethicket, Jonesey95, Calmer Waters, Skyerise, Pmokeefe, Jschnur, Searsshoesales, Jrobbinz123, Lissajous, Turian, Lando Calrissian, Wotnow, Ansumang, Reaper Eternal, 564dude, Jackvancs, Bobotast, MINTOPOINT, TjBot, DexDor, Антон Глїнисты, Daggersteel10, Chiechiecheist, EmausBot, John of Reading, WikitanvirBot, Duskbrood, FergalG, Slightsmile, Barak90, Wikipelli, TheLemon1234, Manofgrass, Brazmyth, H3llBot, Stoneymufc29, GeorgeBarnick, Brandmeister, Ego White Tray, TYelliot, ClueBot NG, Gilderien, A520, Cheeseequalsyum, Timothy jordan, 123Hedgehog456, Maplelanefarm, 336, Helpful Pixie Bot, Jeffreyts11, 123456789malm, Bibcode Bot, BG19bot, Hurricanefan25, MusikAnimal, Davidiad, MosquitoBird11, Mydogpwnsall, MrBill3, Njavallil, Glacialfox, Walterpfeifer, Thebannana, CE9958, Marioedesouza, Mediran, Dexbot, Rishab021, Cjean42, Sriharsh1234, Sam boron100, Wankybanky, Wikitroll12345, RojoEsLardo, Jwratner1, NottNott, Saebre, JNrgbKLM, KheltonHeadley, AspaasBekkelund, HectorCabreraJr, Hazinho93, Quadrupedi, QuantumMatt101, Philipphilip0001, Monkbot, RiderDB, Egfraley, Tetra quark, Weed305, KasparBot and Anonymous: 705

- **Eightfold Way (physics)** *Source:* https://en.wikipedia.org/wiki/Eightfold_Way_(physics)?oldid=672321057 *Contributors:* Charles Matthews, Jitse Niesen, Giftlite, Marcika, Xerxes314, Jason Quinn, Laurascudder, Eritain, BD2412, Strait, YurikBot, Bambaiah, Jimp, RussBot, Spike Wilbury, SCZenz, Sbyrnes321, SmackBot, Hmains, Ephraim33, Bluebot, Radagast83, Dmh~enwiki, Eassin, Jaeger5432, Headbomb, Gökhan, Cuzkatzimhut, VolkovBot, Le Pied-bot~enwiki, DumZiBoT, Addbot, Lightbot, Luckas-bot, Amirobot, Citation bot, Xqbot, I dream of horses, Skyerise, Sijothankam, 777sms, EmausBot, John of Reading, Jjspinorfield1, Isocliff, Bibcode Bot, Ekips39, Studzinski.daniel, Monkbot and Anonymous: 12

- **Quark model** *Source:* https://en.wikipedia.org/wiki/Quark_model?oldid=681414296 *Contributors:* Rmhermen, Stevertigo, AugPi, Chrisjj, Donarreiskoffer, Rorro, Xerxes314, Doshell, Rich Farmbrough, Masudr, David Schaich, Laurascudder, Andrew Gray, Feezo, Linas, Wdyoung, Tabletop, Kbdank71, Marasama, Chekaz, Nigosh, Srleffler, Commander Nemet, Elfguy, Bambaiah, Hairy Dude, Jmauro2000, Salsb, Zwobot, Ilmari Karonen, Finell, SmackBot, C.Fred, Kevin Ryde, RProgrammer, Voyajer, VMS Mosaic, Radagast83, Eassin, Drinibot, ShelfSkewed, Michael C Price, Headbomb, OrenBochman, AntiVandalBot, Magioladitis, Ludvikus, Tarotcards, Coppertwig, Cuzkatzimhut, LokiClock, The Stickler, Muhends, ClueBot, Rotational, Glopso, PixelBot, SchreiberBike, Addbot, Luckas-bot, Yobot, Citation bot, ArthurBot, Zhividya, TechBot, Omnipaedista, Cekli829, Kwiki, Citation bot 1, DEm, WikitanvirBot, Hesana, Jjspinorfield1, Suslindisambiguator, AManWithNoPlan, Brandmeister, ClueBot NG, Bibcode Bot, WikiDenvah, Dexbot, Tony Mach, Arlene47, Truocled, Alakzi and Anonymous: 35

- **Up quark** *Source:* https://en.wikipedia.org/wiki/Up_quark?oldid=666803858 *Contributors:* Bryan Derksen, Alfio, Jni, Giftlite, Xerxes314, Kjoonlee, Bookandcoffee, CharlesC, Rjwilmsi, Mike Peel, Chobot, Hairy Dude, Rt66lt, Spike Wilbury, SCZenz, Poulpy, Eog1916, Bluebot, Tamfang, T-borg, Eric Saltsman, Hetar, Lottamiata, Laplace's Demon, Merryjman, CmdrObot, Myasuda, Raoul NK, Headbomb, JAnDbot, Abyssoft, I310342~enwiki, Idioma-bot, Sheliak, Wilmot1, VolkovBot, TXiKiBoT, Anonymous Dissident, Gekritzl, AlleborgoBot, SieBot, Muhends, Bobathon71, DragonBot, DumZiBoT, TimothyRias, Addbot, Eivindbot, LaaknorBot, ChenzwBot, Naidevinci, Ehrenkater, Lightbot, Luckas-bot, Citation bot, ArthurBot, Xqbot, DSisyphBot, Ditimchanly, Almabot, A. di M., Paine Ellsworth, Citation bot 1, Trappist the monk, TjBot, Ripchip Bot, EmausBot, WikitanvirBot, StringTheory11, Quondum, Helpful Pixie Bot, Bibcode Bot, BG19bot, P76837, Oznitecki, Alexzhang2, The Great Leon, Monkbot and Anonymous: 38

- **Down quark** *Source:* https://en.wikipedia.org/wiki/Down_quark?oldid=663284022 *Contributors:* Bryan Derksen, Alfio, Timwi, Jni, Herbee, Xerxes314, Rich Farmbrough, Kjoonlee, Rjwilmsi, Mike Peel, Chobot, YurikBot, Rt66lt, Acidsaturation, Spike Wilbury, SCZenz, Poulpy, Otto ter Haar, Skizzik, Bluebot, Tamfang, Llwang, Eric Saltsman, MTSbot~enwiki, Hetar, Laplace's Demon, Myasuda, Raoul NK, Thijs!bot, Headbomb, Davidhorman, JAnDbot, Abyssoft, MartinBot, Jvineberg, I310342~enwiki, Sheliak, VolkovBot, TXiKiBoT, Anonymous Dissident, SieBot, Ngexpert6, Muhends, Bobathon71, Lawrence Cohen, Daigaku2051, Auntof6, NuclearWarfare, TimothyRias, Addbot, Lightbot, Luckas-bot, Yobot, Citation bot, ArthurBot, Xqbot, DSisyphBot, Paine Ellsworth, Citation bot 1, Tim1357, Trappist the monk, EmausBot, ZéroBot, Quondum, Rezabot, Helpful Pixie Bot, Bibcode Bot, TheMan4000, 786b6364, Monkbot and Anonymous: 22

- **Strange quark** *Source:* https://en.wikipedia.org/wiki/Strange_quark?oldid=663285537 *Contributors:* Bryan Derksen, Alfio, Jni, Owain, Xerxes314, Soman, Kjoonlee, Kwamikagami, Rsholmes, Esb82, Neonumbers, Rjwilmsi, Mike Peel, Gurch, Erik4, Chobot, YurikBot, Jimp, Salsb, SCZenz, Poulpy, SmackBot, Bluebot, NCurse, Vina-iwbot~enwiki, Yevgeny Kats, Zzzzzzzzzzz, Laplace's Demon, MightyWarrior, Myasuda, Thijs!bot, Headbomb, Chillysnow, JAnDbot, Abyssoft, Bongwarrior, Albmont, McSly, I310342~enwiki, Pdcook, Sheliak, VolkovBot, SieBot, Muhends, Auntof6, Iohannes Animosus, TimothyRias, IngerAlHaosului, Addbot, ProbablyAmbiguous, Luckas-bot, Yobot, AnomieBOT, Citation bot, Sarah12sarah, Erik9bot, Thehelpfulbot, Paine Ellsworth, Rkr1991, Citation bot 1, Skyerise, Johann137, Trappist the monk, Puzl bustr, Agrasa, Wikiborg4711, EmausBot, Hhhippo, ZéroBot, Quondum, CocuBot, Helpful Pixie Bot, Bibcode Bot, Vkpd11, P76837, Matthew gib, Glaisher, RhinoMind and Anonymous: 38

- **Charm quark** *Source:* https://en.wikipedia.org/wiki/Charm_quark?oldid=663286006 *Contributors:* Bryan Derksen, Alfio, Bogdangiusca, Xerxes314, Bodhitha, Perey, Kjoonlee, Rjwilmsi, Mike Peel, Chobot, YurikBot, Bambaiah, Conscious, Salsb, SCZenz, Scottfisher, Poulpy, SmackBot, Delldot, Warhol13, Rezecib, Vina-iwbot~enwiki, Happy-melon, Laplace's Demon, CRGreathouse, Michael C Price, Thijs!bot, Headbomb, Nisselua, JAnDbot, Abyssoft, Uncle.wink, Bryanhiggs, HEL, I310342~enwiki, Qoou.Anonimu, Idioma-bot, Sheliak, Anonymous Dissident, Kumorifox, BeIsKr, AlleborgoBot, SieBot, Muhends, TimothyRias, Addbot, Mjamja, Lightbot, Luckas-bot, Yobot, Nallimbot, Citation bot, ArthurBot, Quebec99, Xqbot, DSisyphBot, GrouchoBot, RibotBOT, SassoBot, A. di M., Paine Ellsworth, Dogposter, D'ohBot, Citation bot 1, Citation bot 4, RedBot, MastiBot, Trappist the monk, EarthCom1000, Alph Bot, EmausBot, ZéroBot, Quondum, Anita5192, CocuBot, Rezabot, Helpful Pixie Bot, Bibcode Bot, Penguinstorm300, Hoppeduppeanut, Leowestland and Anonymous: 35

- **Bottom quark** *Source:* https://en.wikipedia.org/wiki/Bottom_quark?oldid=676070719 *Contributors:* Bryan Derksen, Xerxes314, Bodhitha, Icairns, Kjoonlee, Bobo192, Pinar, WadeSimMiser, Rjwilmsi, Mike Peel, Erkcan, FlaBot, Itinerant1, Chobot, YurikBot, Bambaiah, Jimp, Conscious, Ozabluda, SpuriousQ, Salsb, SCZenz, Lexicon, Poulpy, Physicsdavid, SmackBot, Hmains, Luís Felipe Braga, Laplace's Demon, CmdrObot, Outriggr, Niubrad, הספר, Thijs!bot, Headbomb, JAnDbot, Abyssoft, Pkoppenb, Dr. Morbius, I310342~enwiki, Joshmt, Idioma-bot, Sheliak, VolkovBot, Antixt, AlleborgoBot, BartekChom, Muhends, Auntof6, TimothyRias, Lockalbot, Addbot, Mr Sme, Luckas-bot, THEN

WHO WAS PHONE?, Citation bot, ArthurBot, Xqbot, GrouchoBot, StevenVerstoep, Thehelpfulbot, Paine Ellsworth, Citation bot 1, Jonesey95, Double sharp, TjBot, EmausBot, Barak90, TuHan-Bot, ZéroBot, StringTheory11, Quondum, Chris857, ChuispastonBot, Widr, Helpful Pixie Bot, Bibcode Bot, P76837, ChrisGualtieri, Ajd268, Mfb, Monkbot, Axel Azzopardi, Kenijr and Anonymous: 33

- **Top quark** *Source:* https://en.wikipedia.org/wiki/Top_quark?oldid=679102575 *Contributors:* Damian Yerrick, Bryan Derksen, HPA, Haryo, Bkell, Giftlite, Xerxes314, Edcolins, Bodhitha, David Schaich, Kjoonlee, Axl, Woohookitty, Rjwilmsi, Strait, Mike Peel, Vegaswikian, Wikiliki, Goudzovski, Chobot, YurikBot, Bambaiah, JabberWok, Gaius Cornelius, Salsb, Howcheng, SCZenz, Emijrp, Physicsdavid, SmackBot, Incnis Mrsi, ZerodEgo, Mr.Z-man, Jgwacker, Pulu, Stikonas, Mets501, Peyre, RekishiEJ, Banedon, הסרפד, Headbomb, Davidhorman, Oreo Priest, AntiVandalBot, JAnDbot, Abyssoft, Maliz, HEL, Fatka, I310342~enwiki, Idioma-bot, Sheliak, Biggus Dictus, TXiKiBoT, Reibot, Kachuak, Ptrslv72, SieBot, Hatster301, Muhends, ClueBot, Niceguyedc, Noca2plus, Choonkiat.lee, Brews ohare, Kakofonous, Jtle515, TimothyRias, Prostarplayer321, SkyLined, Cockatoot, Addbot, Mr0t1633, Mjamja, ChenzwBot, Ginosbot, Zorrobot, Luckas-bot, Naudefjbot~enwiki, Dreamer08, AnomieBOT, Icalanise, Citation bot, ArthurBot, LilHelpa, DSisyphBot, Unready, GrouchoBot, RibotBOT, Soandos, Paine Ellsworth, Citation bot 1, Jonesey95, Thinking of England, Nomis2k, Higgshunter, RjwilmsiBot, Mophoplz, EmausBot, John of Reading, WikitanvirBot, Barak90, StringTheory11, Peter.poier, Quondum, Samlever, Whoop whoop pull up, Reify-tech, Helpful Pixie Bot, Bibcode Bot, Glevum, Kephir, Mmitchell10, Quadrupedi, Monkbot, BrunoUbaldo and Anonymous: 64

- **Flavour (particle physics)** *Source:* https://en.wikipedia.org/wiki/Flavour_(particle_physics)?oldid=681888935 *Contributors:* Schewek, Michael Hardy, Nurg, Xerxes314, Varlaam, Andycjp, R. fiend, DragonflySixtyseven, CALR, STGM, Andrew Gray, Knowledge Seeker, Egg, Alai, Sylvain Mielot, Linas, Mindmatrix, SpNeo, Drrngrvy, YurikBot, Bambaiah, Hairy Dude, NTBot~enwiki, Bhny, Cossy, Długosz, SCZenz, Nick, Karl Andrews, SmackBot, Incnis Mrsi, Dauto, Doug Bell, Zero sharp, Ompty, BFD1, Ruslik0, Cydebot, Hydraton31, Xxanthippe, Michael C Price, Thijs!bot, Headbomb, FelixP~enwiki, Rompe, Hayesgm, Knotwork, CosineKitty, Robin S, Askielboe, Yonidebot, Choihei, I310342~enwiki, Thecinimod, VolkovBot, A4bot, Kresadlo, Maxim, Odellus, Ptrslv72, SieBot, VVVBot, The Stickler, Muhends, PixelBot, Jtle515, Count Truthstein, DumZiBoT, MystBot, SkyLined, Addbot, ZeroOmega, SpBot, Ehrenkater, HerculeBot, Luckas-bot, Ptbotgourou, Magog the Ogre, Icalanise, Omnipaedista, Citation bot 1, Xtermin8R645, B2NVB2, Jrobbinz123, 777sms, Bizzurp, EmausBot, VinculumMan, AvocatoBot, Drift chambers, Skynden, Isambard Kingdom and Anonymous: 42

- **Isospin** *Source:* https://en.wikipedia.org/wiki/Isospin?oldid=682122579 *Contributors:* Stone, Giftlite, Xerxes314, Michael Devore, RScheiber, Jason Quinn, AmarChandra, Lumidek, Perey, Rich Farmbrough, Hidaspal, V79, Cmdrjameson, RJFJR, Linas, Robert K S, Jwanders, TPickup, Ddn2, FreplySpang, Rjwilmsi, Strait, Mike Peel, Margosbot~enwiki, Goudzovski, M7bot, Bambaiah, Bhny, Archelon, Welsh, Thiseye, SmackBot, Incnis Mrsi, Sue Anne, Colonies Chris, Sawran~enwiki, KI, Iridescent, Cydebot, Michael C Price, My Flatley, Zalgo, Thijs!bot, Headbomb, Knotwork, JAnDbot, Madmarigold, Avicennasis, Lilac Soul, KIAaze, Tarotcards, Fylwind, VolkovBot, Quilbert, Anonymous Dissident, Antixt, OlekG, PaddyLeahy, SieBot, Likebox, OsamaBinLogin, Uzdzislaw, Albambot, Addbot, Luckas-bot, Citation bot, ArthurBot, Bozzochet, Obersachsebot, Glenmark, Br77rino, J04n, Ernsts, RedAcer, Citation bot 1, Minivip, FoxBot, WikitanvirBot, Helpful Pixie Bot, Bibcode Bot, BG19bot, Jamisonsloan, Monkbot, Kfitzell29, GioComitini and Anonymous: 45

- **Strangeness** *Source:* https://en.wikipedia.org/wiki/Strangeness?oldid=674819737 *Contributors:* Xavic69, Ahoerstemeier, Timwi, Herbee, Xerxes314, JeffBobFrank, RScheiber, Icairns, Xeroc, Mike Rosoft, Jkl, Jag123, LostLeviathan, Fred Condo, Mel Etitis, Tevatron~enwiki, Eyu100, Donotresus, FlaBot, Who, Fresheneesz, Srleffler, Roboto de Ajvol, Bambaiah, Hairy Dude, Conscious, Shawn81, Kyorosuke, SCZenz, 99 Willys on Wheels on the wall, 99 Willys on Wheels..., SmackBot, Stepa, JSpudeman, Complexica, Richard L. Peterson, ZICO, ShelfSkewed, Cydebot, Dchristle, Mbell, Headbomb, AntiVandalBot, NE2, The sage, I310342~enwiki, Pernogr~enwiki, Anonymous Dissident, Pamputt, Riwnodennyk, Callie.hoon, SilvonenBot, Addbot, Mr0t1633, Zorrobot, Citation bot, Wnme, Ernsts, A. di M., Qwarx, Yutsi, Johann137, Turian, Alarichus, Dinamik-bot, EmausBot, Vacation9, ⊠⊠⊠, Furkhaocean, JamesMoose, Ibnbaja and Anonymous: 33

- **Charm (quantum number)** *Source:* https://en.wikipedia.org/wiki/Charm_(quantum_number)?oldid=669956667 *Contributors:* Xavic69, Giftlite, RScheiber, Pol098, Pdelong, Bambaiah, SmackBot, Tim Q. Wells, Cydebot, Thijs!bot, Headbomb, TXiKiBoT, Count Truthstein, ArthurBot, Ulm, Ernsts, Erik9bot, Carlog3, EmausBot, CocuBot, Ibnbaja, JuhoSchultz and Anonymous: 3

- **Bottomness** *Source:* https://en.wikipedia.org/wiki/Bottomness?oldid=673493654 *Contributors:* Xavic69, Fredrik, RScheiber, Strait, Goudzovski, Bambaiah, D Monack, SmackBot, Tim Q. Wells, Cydebot, Headbomb, VolkovBot, Wikiwide, TXiKiBoT, Count Truthstein, Luckas-bot, AnomieBOT, Xqbot, Ernsts, Wetman88, Alph Bot, EmausBot, Rezabot and Anonymous: 3

- **Topness** *Source:* https://en.wikipedia.org/wiki/Topness?oldid=673493535 *Contributors:* Xavic69, Fredrik, RScheiber, Mako098765, Strait, Goudzovski, Bambaiah, SmackBot, Chris the speller, Tim Q. Wells, Cydebot, Headbomb, JAnDbot, Fatka, VolkovBot, Wikiwide, Count Truthstein, Addbot, Luckas-bot, ArthurBot, Ernsts, Carlog3, Craig Pemberton, DrilBot, TobeBot, Dinamik-bot, Ttsush, EmausBot, CocuBot and Anonymous: 4

- **Baryon number** *Source:* https://en.wikipedia.org/wiki/Baryon_number?oldid=675888853 *Contributors:* Andre Engels, Stevertigo, Delirium, Phys, Sanders muc, Securiger, Herbee, Xerxes314, Dratman, RScheiber, Jason Quinn, Discospinster, Pjacobi, Brim, Guy Harris, H2g2bob, Linas, Ted BJ, Isnow, Ddn2, BD2412, Raymond Hill, Bubba73, Margosbot~enwiki, Fresheneesz, Cannywizard, PointedEars, Roboto de Ajvol, YurikBot, Bambaiah, Tom Lougheed, V1adis1av, QFT, Doug Bell, Dan Gluck, Cydebot, Thijs!bot, Headbomb, Barakitty, Richard n, JAnDbot, CosineKitty, Bbi5291, Siryendor, FaTTshady74, STBotD, TXiKiBoT, A4bot, Venny85, SieBot, Muhends, Erodium, Addbot, WikiDreamer Bot, Legobot, Luckas-bot, Amirobot, ArthurBot, XZeroBot, Ernsts, MastiBot, Ttsush, Sahimrobot, Ernest3.141 and Anonymous: 17

- **Color charge** *Source:* https://en.wikipedia.org/wiki/Color_charge?oldid=643420544 *Contributors:* CYD, Daniel Mahu, Looxix~enwiki, Theresa knott, Glenn, AugPi, Sethmahoney, Coren, The Anomebot, DJ Clayworth, Phys, Robbot, Wereon, M-Falcon, Herbee, Xerxes314, Dratman, Alison, JeffBobFrank, Jason Quinn, Eequor, RetiredUser2, 4pq1injbok, Lovelac7, MuDavid, Bender235, Jlcooke, Gauge, Ben Webber, El C, Euyyn, Wiki-Ed, Matt McIrvin, Fwb22, Njaard, Keenan Pepper, DrGaellon, Velho, Linas, Cruccone, Southwest, BD2412, Crazynas, Jeff02, FrankTobia, Bambaiah, Gaius Cornelius, E2mb0t~enwiki, Finell, KasugaHuang, Felisse, SmackBot, Incnis Mrsi, Bluebot, Ahnood, UNV, Alphathon, VMS Mosaic, Daniel bg, Dcamp314, Michael Bednarek, Keisetsu, CRGreathouse, Joelholdsworth, ChrisKennedy, TicketMan, Keraunos, Headbomb, West Brom 4ever, Cultural Freedom, JAnDbot, Yill577, Shim'on, Melamed katz, KylieTastic, Liometopum, Aaron Rotenberg, SieBot, WereSpielChequers, RadicalOne, BartekChom, Beofluff, Muhends, Dstebbins, ClueBot, Drmies, Manishearth, Andrewjmcneil, RexxS, Truthnlove, Addbot, Download, CarsracBot, AgadaUrbanit, Mrmister99823, Luckas-bot, Yobot, Amirobot, Götz, Jim1138, Citation bot, ArthurBot, Obersachsebot, Xqbot, BenzolBot, TobeBot, Dinamik-bot, Scgtrp, Serketan, Hhhippo, AvicBot, JSquish, Maschen, Kate460, Helpful Pixie Bot, Super Duper Fixer Upper, BattyBot, Qashqaiilove, Cjean42, Graphium, Haezlgrace, Samproctor125 and Anonymous: 61

- **Color confinement** *Source:* https://en.wikipedia.org/wiki/Color_confinement?oldid=675749735 *Contributors:* Lorenzarius, Michael Hardy, Schneelocke, Timwi, Doradus, Phys, Secretlondon, Robbot, Ruakh, Isopropyl, Giftlite, Herbee, Xerxes314, Tagishsimon, Lumidek, MuDavid, Jag123, Cortonin, Jon.baldwin, Mpatel, Isnow, Ashmoo, Rjwilmsi, FlaBot, Pfctdayelise, Chobot, Mushin, Bambaiah, Hairy Dude, Ohwilleke, Salsb, E2mb0t~enwiki, Tetracube, WAS 4.250, Banus, SmackBot, Incnis Mrsi, Jjalexand, DHN-bot~enwiki, Sbharris, Colonies Chris, VMS Mosaic, Lambiam, Sinistrum, TaggedJC, Newone, Treue~enwiki, Thijs!bot, Headbomb, Cultural Freedom, Yill577, Natsirtguy, Melamed katz, A Nobody, Spoxjox, Anonymous Dissident, SieBot, FlowerFaerie087, Denisarona, Asher196, Manishearth, Dab240, Mustufailed, Texas Chainstore Manager, Addbot, Dudemanfellabra, Db1101, Eutactic, Luckas-bot, AnomieBOT, Materialscientist, Xqbot, D'ohBot, RedBot, TobeBot, Trappist the monk, DixonDBot, EmausBot, Peaceray, Maschen, StanS, Helpful Pixie Bot, Bibcode Bot, Orentago, BG19bot, Slinkblot, Jdellamalva, Tikki and Anonymous: 48

- **Electric charge** *Source:* https://en.wikipedia.org/wiki/Electric_charge?oldid=681550518 *Contributors:* AxelBoldt, Mav, Andre Engels, Roadrunner, Peterlin~enwiki, Heron, JohnOwens, Michael Hardy, Ixfd64, Delirium, Looxix~enwiki, Ellywa, Mdebets, Glenn, Rossami, Nikai, Andres, Raven in Orbit, Reddi, Omegatron, Gakrivas, Lumos3, Rogper~enwiki, Gentgeen, Robbot, Fredrik, Dukeofomnium, Wikibot, Fuelbottle, Wjbeaty, Giftlite, DavidCary, Herbee, Snowdog, Dratman, Valen~enwiki, RScheiber, Jason Quinn, Brockert, OldakQuill, Manuel Anastácio, LiDaobing, Karol Langner, Icairns, Iantresman, GNU, Vincom2, Discospinster, Guanabot, Jpk, Dbachmann, ZeroOne, Laurascudder, Bobo192, Rbj, Giraffedata, Kjkolb, Scentoni, Mdd, Alansohn, Atlant, ABCD, Velella, Wtshymanski, HenkvD, Mikeo, DV8 2XL, Gene Nygaard, HenryLi, Oleg Alexandrov, Nuno Tavares, Cimex, Rocastelo, StradivariusTV, Oliphaunt, BillC, Eleassar777, Cyberman, Palica, BD2412, Demonuk, Edison, SMC, Krash, Dougluce, FlaBot, Psyphen, Nivix, Alfred Centauri, Gurch, Kri, Gdrbot, Manscher, YurikBot, Bambaiah, Lucinos~enwiki, Stephenb, Manop, Pseudomonas, JDoorjam, TDogg310, Chichui, Kkmurray, Wknight94, Light current, Enormousdude, Johndburger, Tcsetattr, Pinikas, Reyk, Canley, Geoffrey.landis, JDspeeder1, GrinBot~enwiki, Mejor Los Indios, Sbyrnes321, Marquez~enwiki, Moeron, Vald, Thunderboltz, Dmitry sychov, HalfShadow, Gilliam, Oscarthecat, Andy M. Wang, Chris the speller, Lenko, DHN-bot~enwiki, Dual Freq, Hallenrm, Rrburke, The tooth, MichaelBillington, Hgilbert, Drphilharmonic, Daniel.Cardenas, Springnuts, Yevgeny Kats, Andrei Stroe, DJIndica, Naui~enwiki, Nmnogueira, SashatoBot, Richard L. Peterson, Slowmover, Cronholm144, Mgiganteus1, Bjankuloski06en~enwiki, Nonsuch, Ben Moore, RandomCritic, MarkSutton, Stikonas, Dicklyon, Levineps, Igoldste, Tawkerbot2, Chetvorno, JForget, CmdrObot, Kehrli, Jsd, Myasuda, Cydebot, Fl, Bvcrist, Meno25, Gogo Dodo, WISo, Christian75, Ssilvers, Thijs!bot, Epbr123, Barticus88, N5iln, Mojo Hand, Headbomb, Gerry Ashton, Escarbot, Aadal, AntiVandalBot, Seaphoto, Prolog, DarkAudit, Lyricmac, Tim Shuba, WikifingHelper, Asgrrr, JAnDbot, Acroterion, Bongwarrior, VoABot II, J2thawiki, Sstolper, Jjurik, Bubba hotep, User A1, DerHexer, InvertRect, Robin S, MartinBot, M. Bilal Shafiq, LedgendGamer, Pharaoh of the Wizards, Numbo3, Hans Dunkelberg, NightFalcon90909, Uncle Dick, Ginsengbomb, Katalaveno, DarkFalls, NewEnglandYankee, QuickClown, Juliancolton, ACBest, Treisijs, Lseixas, Jefferson Anderson, Sheliak, Philip Trueman, TXiKiBoT, The Original Wildbear, Ayan2289, Nickipedia 008, LuizBallotti, Monty845, Jpalpant, Biscuittin, Demmy100, SieBot, Gerakibot, Caltas, Gastin, Wing gundam, Msadaghd, JerrySteal, Jojalozzo, Oxymoron83, Faradayplank, Avnjay, Anchor Link Bot, Neo., Loren.wilton, ClueBot, The Thing That Should Not Be, Arakunem, Termine, Mild Bill Hiccup, Stephaninator, LeoFrank, Excirial, Kocher2006, Jusdafax, Brews ohare, Cenarium, Jotterbot, PhySusie, SchreiberBike, Wuzur, JDPhD, Versus22, Thinking Stone, Rror, Cernms, Truthnlove, Addbot, Some jerk on the Internet, CanadianLinuxUser, NjardarBot, LaaknorBot, Scottyferguson, LinkFA-Bot, Naidevinci, Ocwaldron, Tide rolls, Lightbot, JDSperling, Legobot, Luckas-bot, Yobot, CinchBug, Duping Man, AnomieBOT, DemocraticLuntz, Sertion, Jim1138, IRP, Pyrrhus16, Kingpin13, Bluerasberry, Materialscientist, Geek1337~enwiki, ImperatorExercitus, Xqbot, TheAMmollusc, Phazvmk, Addihockey10, Capricorn42, Nnivi, ProtectionTaggingBot, RibotBOT, Srr712, A. di M., Constructive editor, Frozenevolution, Ryryrules100, Jc3s5h, Drunauthorized, Mithrandir, Steve Quinn, Davidteng, Fast kartwheels, BenzolBot, DivineAlpha, AstaBOTh15, Pinethicket, I dream of horses, Jivee Blau, Calmer Waters, Tinton5, MastiBot, Serols, Meaghan, Lalrang2007, Logical Gentleman, FoxBot, TobeBot, SchreyP, Jonkerz, Ndkartik, Vrenator, Taytaylisious09, Ammodramus, Jamietw, DARTH SIDIOUS 2, Eshmate, Irfanyousufdar, EmausBot, John of Reading, GoingBatty, K6ka, Darkfight, Hhhippo, JSquish, Harddk, Stephen C Wells, Liam McM, Sonygal, L Kensington, Donner60, Peter Karlsen, Sven Manguard, Planetscared, ClueBot NG, Jack Greenmaven, Cking1414, Ihwood, Ulflund, CocuBot, MelbourneStar, O.Koslowski, Brickmack, AvocatoBot, Rm1271, Altaïr, F=q(E+v^B), Snow Blizzard, Brad7777, Bhaskarandpm, Eduardofeld, GoShow, Dexbot, JoshyyP, Brandonsmacgregor, Reeceyboii, Frosty, Reatlas, I am One of Many, Eyesnore, Tentinator, Germeten, Nablacdy, Spyglasses, Freddyboi69, 20M030810, SpecialPiggy, Marizperoj, Peterfreed, Rigid hexagon, Jiteshkumar727464, Dyeith, Podayeruma, Oleaster, Layfi, BlueDecker, GeneralizationsAreBad, Pritam kumar Barik, KasparBot, Ramprakashsfc and Anonymous: 427

- **Hypercharge** *Source:* https://en.wikipedia.org/wiki/Hypercharge?oldid=681928054 *Contributors:* Xavic69, Schneelocke, Charles Matthews, Phys, Robbot, Filemon, Giftlite, JeffBobFrank, Mike Rosoft, Pjacobi, Jag123, Xayma, Jörg Knappen~enwiki, Xauw, Roboto de Ajvol, YurikBot, Wavelength, Bambaiah, Hairy Dude, Gaius Cornelius, Paul D. Anderson, KnightRider~enwiki, SmackBot, Tom Lougheed, Dauto, V1adis1av, Radagast83, CRGreathouse, CmdrObot, Michael C Price, Headbomb, Zylorian, Alphachimpbot, JAnDbot, Robin S, HEL, Jnnnnn, STBotD, Anonymous Dissident, Venny85, Antixt, COBot, Zenofox, SkyLined, Addbot, Luckas-bot, Amirobot, Götz, Citation bot, Svix, Wiles stunlalt, MastiBot, DixonDBot, EmausBot, Preon, Halfb1t, ChrisGualtieri, Lesser Cartographies, SJ Defender and Anonymous: 20

- **Spin (physics)** *Source:* https://en.wikipedia.org/wiki/Spin_(physics)?oldid=682215602 *Contributors:* AxelBoldt, CYD, The Anome, Larry_Sanger, Andre Engels, XJaM, David spector, Stevertigo, Xavic69, Michael Hardy, Tim Starling, Dominus, Cyp, Stevenj, Glenn, AugPi, Rossami, Nikai, Andres, Med, Mxn, Charles Matthews, Timwi, Kbk, 4lex, Reina riemann, E23~enwiki, Phys, Wtrmute, Bevo, Elwoz, Robbot, Gandalf61, Blainster, DHN, Hadal, Papadopc, Jheise, Anthony, Diberri, Xanzzibar, Giftlite, Smjg, Lethe, Lupin, MathKnight, Xerxes314, Average Earthman, AlistairMcMillan, Ato, Andycjp, Gzuckier, Beland, Karol Langner, Spiralhighway, Elroch, B.d.mills, Tsemii, Frau Holle, Mike Rosoft, Igorivanov~enwiki, FT2, MuDavid, Paul August, Pt, Susvolans, Army1987, Wood Thrush, SpeedyGonsales, Physicistjedi, Obradovic Goran, Neonumbers, Keenan Pepper, Count Iblis, Egg, Linas, Palica, Torquil~enwiki, Ashmoo, Grammarbot, Zoz, Rjwilmsi, Zbxgscqf, Drrngrvy, FlaBot, Mathbot, TheMidnighters, Itinerant1, Ewlyahoocom, Gurch, Fresheneesz, Srleffler, Kri, Chobot, DVdm, YurikBot, Bambaiah, Hairy Dude, JabberWok, Rsrikanth05, NawlinWiki, Buster79, Hwasungmars, Kkmurray, Werdna, Djdaedalus, Simen, Netrapt, Mpjohans, KSevcik, GrinBot~enwiki, Joshronsen, Bo Jacoby, Sbyrnes321, DVD R W, Shanesan, KasugaHuang, That Guy, From That Show!, SmackBot, Unyoyega, Bluebot, Complexica, DHN-bot~enwiki, Sergio.ballestrero, V1adis1av, QFT, Voyajer, Terryeo, Ryanluck, Radagast83, Jgrahamc, MichaelBillington, Richard001, DMacks, Daniel.Cardenas, Bidabadi~enwiki, Sadi Carnot, Bdushaw, Andrei Stroe, Tesseran, SashatoBot, Leo C Stein, Vanished user 9i39j3, UberCryxic, Jonas Ferry, Vgy7ujm, Loodog, Jaganath, Terry Bollinger, Wierdw123, Inquisitus, Beefyt, Jc37, Dreftymac, Newone, RokasT~enwiki, Jaksmata, Aepryus, JRSpriggs, Joostvandeputte~enwiki, CRGreathouse, David s graff, Ahmes, Jason-Hise, Eric Le Bigot, Bmk, Myasuda, Mct mht, FilipeS, Cydebot, A876, Thijs!bot, Barticus88, Headbomb, Brichcja, Davidhorman, Oreo Priest, Widefox, Orionus, Tlabshier, Accordionman, Astavats, JAnDbot, Em3ryguy, MER-C, Igodard, PhilKnight, .anacondabot, Sangak, Magioladitis,

Swpb, LorenzoB, Monurkar~enwiki, TechnoFaye, Brilliand, R'n'B, CommonsDelinker, Victor Blacus, J.delanoy, Numbo3, Sackm, Maurice Carbonaro, Klatkinson, Cmichael, Uberdude85, Craklyn, CardinalDan, VolkovBot, Error9312, JohnBlackburne, Bolzano~enwiki, TXiKiBoT, Hqb, Anonymous Dissident, Costela, BotKung, Kganjam, Petergans, Kbrose, SieBot, BotMultichill, The way, the truth, and the light, RadicalOne, Jasondet, Paolo.dL, R J Sutherland, Lightmouse, JackSchmidt, Martarius, ClueBot, JonnybrotherJr, Warbler271, Mild Bill Hiccup, David Trochos, Outerrealm, Sbian, Peachypoh, SchreiberBike, Ant59, Crowsnest, XLinkBot, Addbot, Mathieu Perrin, Narayansg, Imeriki al-Shimoni, Sriharsha.karnati, Numbo3-bot, Tide rolls, Lightbot, Luckas-bot, Yobot, Nallimbot, AnomieBOT, Lendtuffz, Citation bot, Nepahwin, ArthurBot, Obersachsebot, Xqbot, Sionus, WandringMinstrel, Francine Rogers, Pradameinhoff, Tom1936, Ernsts, A. di M., NoldorinElf, Daleang, Baz.77.243.99.32, LucienBOT, Paine Ellsworth, Tobby72, Freddy78, Craig Pemberton, C.Bluck, Jondn, Pokyrek, Citation bot 1, I dream of horses, Adlerbot, Casimir9999, Kallikanzarid, Jkforde, Michael9422, Miracle Pen, Sgravn, 8af4bf06611c, Garuh knight, EmausBot, Beatnik8983, GoingBatty, JustinTime55, Zhenyok 1, Atomicann, JSquish, ZéroBot, Harddk, Neh0000, Quondum, Jacksccsi, Maschen, Zueignung, Rasinj, Eg-T2g, ClueBot NG, Paolo328, Gilderien, Frietjes, Widr, PhiMAP, Helpful Pixie Bot, Bibcode Bot, BG19bot, PUECH P.-F., Mark Arsten, Yudem, F=q(E+v^B), Blaspie55, Halfb1t, Robertwilliams2011, Dexbot, Foreverascone, ScitDei, Mark viking, Pedantchemist, W. P. Uzer, Francois-Pier, Mathphysman, Aidan Clark, Brotter121, KasparBot and Anonymous: 230

- **Weak hypercharge** *Source:* https://en.wikipedia.org/wiki/Weak_hypercharge?oldid=679239006 *Contributors:* Xavic69, Pjacobi, MeltBanana, David Schaich, BD2412, Chobot, Roboto de Ajvol, YurikBot, Bambaiah, Paul D. Anderson, Incnis Mrsi, Dauto, Michael C Price, Headbomb, Andre.holzner, HEL, Anonymous Dissident, Pamputt, Antixt, Tvine, MystBot, Addbot, Luckas-bot, Yobot, Götz, Ernsts, Slightsmile, QuantumSquirrel, ResidentAnthropologist, Helpful Pixie Bot, Bambi12~enwiki, EzPz4 and Anonymous: 20

- **Weak isospin** *Source:* https://en.wikipedia.org/wiki/Weak_isospin?oldid=679239808 *Contributors:* Xavic69, Charles Matthews, Giftlite, RScheiber, Hidaspal, Ian Pitchford, Chobot, Roboto de Ajvol, YurikBot, Bambaiah, Jimp, RussBot, Paul D. Anderson, Jheriko, Bbabba, Cydebot, Michael C Price, Headbomb, Igodard, Tokei-so, Andre.holzner, Pamputt, AlleborgoBot, Muhends, L.smithfield, Tvine, Addbot, Icalanise, ArthurBot, Ernsts, Puzl bustr, John of Reading and Anonymous: 17

- **B − L** *Source:* https://en.wikipedia.org/wiki/B_%E2%88%92_L?oldid=621966745 *Contributors:* Phys, Herbee, Xerxes314, Jag123, Starwed, Mike Peel, Bgwhite, Roboto de Ajvol, RussBot, Conscious, Gaius Cornelius, SmackBot, QFT, Doug Bell, Cydebot, Michael C Price, Headbomb, Maliz, Andre.holzner, Pamputt, ClueBot, Auntof6, MystBot, SkyLined, Addbot, Mjamja, Luckas-bot, Yobot, Ptbotgourou, Ernsts, A. di M., Erik9bot and Anonymous: 7

- **X (charge)** *Source:* https://en.wikipedia.org/wiki/X_(charge)?oldid=605281425 *Contributors:* XJaM, BD2412, SmackBot, Michael C Price, Headbomb, Addbot, Xqbot, Ernsts, A. di M. and Anonymous: 1

- **Hadron** *Source:* https://en.wikipedia.org/wiki/Hadron?oldid=681027720 *Contributors:* Bryan Derksen, Manning Bartlett, Peterlin~enwiki, Edward, Erik Zachte, ESnyder2, Fruge~enwiki, TakuyaMurata, Darkwind, Glenn, Nikai, Ehn, Olya, Phys, Bevo, Topbanana, BenRG, Twang, Donarreiskoffer, Korath, Wjhonson, Merovingian, Ojigiri~enwiki, Sunray, JesseW, Xanzzibar, Giftlite, Xerxes314, Dratman, Physicist, Mikro2nd, LiDaobing, Pthompson, Icairns, Jimaginator, Mike Rosoft, Vsmith, Goochelaar, Sunborn, Livajo, El C, Kwamikagami, Shanes, Fwb22, Jumbuck, Cookiemobsta, Velella, Rebroad, Vuo, Kusma, DV8 2XL, Linas, GrouchyDan, Palica, Marudubshinki, Kbdank71, Mana Excalibur, Kinu, Strait, FlaBot, RexNL, Goudzovski, FrankTobia, YurikBot, Radishes, Bambaiah, Hydrargyrum, Salsb, NawlinWiki, Wiki alf, SCZenz, Davemck, Bota47, Scriber~enwiki, Modify, Katieh5584, Eog1916, SmackBot, McGeddon, Gilliam, Benjaminevans82, Dingar, Persian Poet Gal, Telempe, DHN-bot~enwiki, Audriusa, Acepectif, Kokot.kokotisko, JorisvS, JarahE, BranStark, SJCrew, Eratticus, Chrumps, Jtuggle, Q43, Epbr123, Wikid77, Headbomb, Escarbot, Deflective, Gcm, NE2, Trapezoidal, Naval Scene, KEKPΩΨ, NeverWorker, Wwmbes, Alexllew, Lvwarren, Jebus0, DariusU, Khalid Mahmood, Adriaan, Rustyfence, Ron2, Leyo, J.delanoy, Maurice Carbonaro, JVersteeg, Rod57, Way2Smart22, Hugh Hudson, Y2H, Ansans, Bobxii, Chris Longley, Useight, Dylan bossart, VolkovBot, TXiKiBoT, Kinkydarkbird, Anonymous Dissident, Don4of4, Wordsmith, LeaveSleaves, Antixt, Enviroboy, Insanity Incarnate, Nibios, AlleborgoBot, SieBot, Yintan, LeadSongDog, RadicalOne, Paolo.dL, OKBot, JohnSawyer, Lazarus1907, Pinkadelica, Danthewhale, Martarius, ClueBot, Amaamaddq, Authoritative Physicist, Wwheaton, Rotational, DragonBot, Sciencedude9998, Tuchomator, El planeto, Kaiba, Thingg, Koshoid, Aitias, Apparition11, Rishi.bedi, TimothyRias, InternetMeme, Jbeans, MystBot, Sgpsaros, Tayste, Addbot, Pkkphysicist, Ehrenkater, Lightbot, Luckas-bot, Yobot, Nallimbot, Dagus2000, Fangfyre, LOLx9000, Thisaccountwillbebanned, Citation bot, Xqbot, Drilnoth, Br77rino, Wikiedit33, Ajahnjohn, Omnipaedista, RibotBOT, Mashmeister, Tjbright2, My cat's breath smells like catfood, Haeinous, Citation bot 1, Javert, Gil987, I dream of horses, Jonesey95, Rameshngbot, Thinking of England, Alarichus, SkyMachine, FoxBot, Johnshnappay, Антон Глинисты, Teravolt, Racerx11, Naznin farhah, Tommy2010, Rafabaez, Wikipelli, ZéroBot, StringTheory11, Hadron12, Donner60, Bobogoobo, Petrb, ClueBot NG, Gareth Griffith-Jones, Bibcode Bot, BG19bot, Dwightboone, Njavallil, Walterpfeifer, Pfeiferwalter, ChrisGualtieri, Ugog Nizdast, Lithelimbs, RoKo89, Michikohundred, KasparBot, Wwilliam726 and Anonymous: 169

- **Baryon** *Source:* https://en.wikipedia.org/wiki/Baryon?oldid=681412960 *Contributors:* AxelBoldt, Tobias Hoevekamp, Bryan Derksen, BenZin~enwiki, Heron, Tim Starling, Alan Peakall, Paul A, Salsa Shark, Glenn, Mxn, Charles Matthews, The Anomebot, ElusiveByte, Phys, Bevo, Traroth, Donarreiskoffer, Robbot, Korath, Kristof vt, Merovingian, Ojigiri~enwiki, Sunray, Wikibot, Giftlite, DocWatson42, ShaunMacPherson, Herbee, Xerxes314, Dratman, DÅ,ugosz, Kaldari, OwenBlacker, Icairns, JohnArmagh, Rich Farmbrough, Guanabot, Mani1, E2m, Tompw, El C, Bobo192, I9Q79oL78KiL0QTFHgyc, Giraffedata, Physicistjedi, Jumbuck, Gary, ABCD, Oleg Alexandrov, Woohookitty, Tevatron~enwiki, BD2412, Kbdank71, Nightscream, Ae77, MZMcBride, Chekaz, R.e.b., Erkcan, Maxim Razin, Oo64eva, Chobot, Roboto de Ajvol, YurikBot, Bambaiah, Jimp, Salsb, Ergzay, DragonHawk, SCZenz, E2mb0t~enwiki, Bota47, Simen, Sbyrnes321, Lainagier, Timotheus Canens, Bluebot, Colonies Chris, Kingdon, Shadow1, Bigmantonyd, Drphilharmonic, Kseferovic, Wierdw123, Physicsdog, Torrazzo, Verdy p, Michael C Price, Thijs!bot, Headbomb, Hcobb, Orionus, QuiteUnusual, Spartaz, Plantsurfer, Amateria1121, Diamond2, Swpb, BatteryIncluded, Hveziris, Saxophlute, Gwern, Ben MacDui, R'n'B, Ash, Tgeairn, Maurice Carbonaro, STBotD, VolkovBot, GimmeBot, NoiseEHC, Tearmeapart, BotKung, BrianADesmond, Antixt, AlleborgoBot, Lou427, SieBot, VVVBot, Gerakibot, LeadSongDog, Keilana, Paolo.dL, Doctorfluffy, TrufflesTheLamb, OKBot, Hamiltondaniel, TubularWorld, ClueBot, Artichoker, ChandlerMapBot, CalumH93, Addbot, LaaknotBot, CarsracBot, Jonhstone12, Legobot, Luckas-bot, Bugbrain 04, AnomieBOT, JackieBot, Materialscientist, Citation bot, ArthurBot, Xqbot, Omnipaedista, SassoBot, Spellage, WaysToEscape, FrescoBot, Citation bot 1, FoxBot, Noommos, EmausBot, John of Reading, JSquish, ZéroBot, StringTheory11, Stibu, Ethaniel, Markinvancouver, ClueBot NG, Koornti, Kasirbot, Rezabot, Bibcode Bot, Atomician, Zedshort, Marioedesouza, ChrisGualtieri, WorldWideJuan, CoolHandLouis, Monkbot, KasparBot and Anonymous: 106

- **List of baryons** *Source:* https://en.wikipedia.org/wiki/List_of_baryons?oldid=675785999 *Contributors:* Cherkash, GPHemsley, Donarreiskoffer, Giftlite, Rich Farmbrough, ZeroOne, Tompw, Keenan Pepper, Oleg Alexandrov, Woohookitty, Astrowob, Kbdank71, Strait, Mike Peel,

R.e.b., Arctic.gnome, YurikBot, Jonrock, Jimp, Cryptic, SCZenz, Gadget850, F15mos, Banus, GrinBot~enwiki, Sbyrnes321, EvanJPW, SmackBot, Tom Lougheed, Oceanh, Jiminy pop, JorisvS, UncleDouggie, Happy-melon, JRSpriggs, Neelix, Cydebot, WillowW, Mike Christie, Abtract, Wikid77, Headbomb, Magioladitis, Mollwollfumble, Randyfurlong, Leyo, Gogobera, Beatnik Party, MichaelSchoenitzer, GimmeBot, Anonymous Dissident, Sascha.baumeister~enwiki, Antixt, PaddyLeahy, Richard Ye, Wing gundam, Thisisnotatest, Lightmouse, Dabomb87, Muhends, NuclearWarfare, SkyLined, Addbot, DOI bot, Mjamja, Debresser, Vectorboson, SamatBot, Luckas-bot, Yobot, Materialscientist, Citation bot, Carlog3, Citation bot 1, Thinking of England, EmausBot, Markinvancouver, Rmashhadi, Bibcode Bot, BG19bot, P76837, Tony Mach, YiFeiBot, Monkbot and Anonymous: 37

- **Meson** *Source:* https://en.wikipedia.org/wiki/Meson?oldid=672777660 *Contributors:* AxelBoldt, Bryan Derksen, Josh Grosse, PierreAbbat, Ben-Zin~enwiki, Xavic69, TakuyaMurata, Fwappler, Ahoerstemeier, Ping, Phys, Bcorr, Jeffq, Donarreiskoffer, Robbot, Fredrik, Sanders muc, Merovingian, Rursus, Ojigiri~enwiki, Davidl9999, DocWatson42, Harp, Marcika, Xerxes314, Niteowlneils, Eequor, Physicist, Eroica, Icairns, Sam Hocevar, Lehi, Rich Farmbrough, Pjacobi, Tjic, Robotje, Nicke Lilltroll~enwiki, Pearle, Jumbuck, Jérôme, Bucephalus, Falcorian, Palica, Tevatron~enwiki, Mandarax, Kbdank71, Strait, Titoxd, FlaBot, Jeremygbyrne, Chobot, YurikBot, Wavelength, Bambaiah, Phmer, Jimp, Ozabluda, JabberWok, Salsb, Leutha, Długosz, SCZenz, Ravedave, Gadget850, Antiduh, Tetracube, SmackBot, Melchoir, Eskimbot, Chris the speller, DHN-bot~enwiki, Sbharris, Kevinpurcell, Mesons, DMacks, Jashank, JorisvS, Mgiganteus1, Geologyguy, Ryulong, JarahE, Myasuda, ChrisKennedy, Michael C Price, Thijs!bot, Headbomb, Escarbot, Orionus, Spartaz, Gökhan, Deflective, Magioladitis, Swpb, Khalid Mahmood, Tercer, Kostisl, Hans Dunkelberg, Tarotcards, Xiahou, JeffreyRMiles, VolkovBot, Prizrak, TXiKiBoT, Muro de Aguas, Martin451, LeaveSleaves, Antixt, SieBot, Majeston, Gerakibot, Graf Von Crayola, Humanityisthedisease, Mimihitam, Fratrep, OKBot, ClueBot, Terrorist96, Diagramma Della Verita, Brews ohare, Neville35, RMFan1, WikHead, Stephen Poppitt, Addbot, Gtakanis, Chzz, Debresser, CosmiCarl, AgadaUrbanit, Dickdock, Magog the Ogre, AnomieBOT, StratoWiki, Altruism2010, Citation bot, ArthurBot, Xqbot, Omnipaedista, WaysToEscape, FrescoBot, Paine Ellsworth, Ironboy11, Steve Quinn, 000ojjo000, Yehoshua2, Citation bot 1, Wdcf, Thinking of England, Puzl bustr, Ale And Quail, Discovery4, Mean as custard, Dkzico007, John of Reading, WikitanvirBot, GoingBatty, Hanretty, ZéroBot, StringTheory11, Markinvancouver, ClueBot NG, Christian.kolen, Wallace Kneeland, Helpful Pixie Bot, Bibcode Bot, Glevum, DerekWinters, Mark viking, Justin567Hicks, Prokaryotes, SJ Defender, Monkbot, KasparBot and Anonymous: 87

- **List of mesons** *Source:* https://en.wikipedia.org/wiki/List_of_mesons?oldid=679943368 *Contributors:* Cherkash, Donarreiskoffer, Giftlite, Xerxes314, Michael Devore, Eequor, Rich Farmbrough, ZeroOne, Tompw, Physicistjedi, Pearle, Keenan Pepper, Zyqqh, TenOfAllTrades, Woohookitty, Ch'marr, Kbdank71, JVz, Strait, Nihiltres, Agerom, RussBot, David McCormick, SCZenz, Gadget850, Sbyrnes321, That Guy, From That Show!, SmackBot, JorisvS, Happy-melon, Charles Baynham, Chrumps, Usgnus, Cydebot, Christian75, Coccoinomane, Headbomb, Stannered, JAnDbot, Magioladitis, Mollwollfumble, Gwern, Leyo, Potatoswatter, VolkovBot, Antixt, Ocsenave, Muhends, Mikaey, SkyLined, Addbot, Yobot, Kan8eDie, 4th-otaku, Rubinbot, Citation bot, ArthurBot, Xqbot, Ulm, Carlog3, W-C, Yehoshua2, Citation bot 1, Thinking of England, John of Reading, StringTheory11, Markinvancouver, Helpful Pixie Bot, Bibcode Bot, Maysens, Srebre, YiFeiBot, Monkbot and Anonymous: 16

- **Exotic hadron** *Source:* https://en.wikipedia.org/wiki/Exotic_hadron?oldid=675789447 *Contributors:* AxelBoldt, Glenn, Xerxes314, Rich Farmbrough, YUL89YYZ, Keenan Pepper, Count Iblis, April Arcus, Linas, Bambaiah, Wogsland, V1adis1av, Acjohnson55, Happy-melon, JRSpriggs, Postmodern Beatnik, ZICO, Thijs!bot, Whatever1111, Headbomb, Stannered, Leyo, VolkovBot, Antixt, Muhends, Curtis95112, MystBot, SkyLined, Addbot, Tide rolls, Luckas-bot, Yobot, Carlog3, Tarsilia, Acather96, Carbosi, ZéroBot, StringTheory11, Ethaniel, ChuispastonBot, Comicboy1996, Trompedo and Anonymous: 16

- **Tetraquark** *Source:* https://en.wikipedia.org/wiki/Tetraquark?oldid=676856701 *Contributors:* Bryan Derksen, Phys, Phil Boswell, Merovingian, Herbee, Varlaam, Physicist, Setokaiba, Icairns, SeaDour, Drbogdan, Rjwilmsi, Bubba73, Nhussein, Bambaiah, Todd Vierling, Antiduh, 2over0, Teply, SmackBot, V1adis1av, Wiki me, Yevgeny Kats, Newone, Headbomb, Sobreira, Hcobb, Ron2, VolkovBot, Antixt, Muhends, Psyden, Addbot, Luckas-bot, AnomieBOT, Citation bot, ArthurBot, Omnipaedista, Citation bot 1, Jonesey95, Jakeukalane, Meier99, WikitanvirBot, Peaceray, JSquish, ZéroBot, ChuispastonBot, Polosa, Bibcode Bot, Pedro.bicudo, Mesonic Interference, Tony Mach, Faizan, Bicyclegeek, Monkbot and Anonymous: 29

- **Pentaquark** *Source:* https://en.wikipedia.org/wiki/Pentaquark?oldid=682556411 *Contributors:* Bryan Derksen, The Anome, Bdesham, EddEdmondson, J'raxis, Glenn, Wfeidt, Schneelocke, Joquarky, Tpbradbury, Phys, Bevo, Phil Boswell, Robbot, Sanders muc, Merovingian, Herbee, Codepoet, Xerxes314, DÅ‚ugosz, Elektron, Icairns, Metahacker, Kenny TM~~enwiki, Pjacobi, Bender235, Kaganer, Fwb22, Mac Davis, Lokedhs, Dismas, Mpatel, GregorB, Mekong Bluesman, Ashmoo, Kbdank71, Marasama, Bubba73, Bambaiah, Wester, Phmer, 4C~enwiki, Salsb, Welsh, Simen, Arthur Rubin, Smurrayinchester, Physicsdavid, Bluebot, Hgrosser, George Ho, V1adis1av, Doug Bell, JorisvS, Thijs!bot, Headbomb, Sobreira, Davidhorman, Oreo Priest, Widefox, Ericoides, BatteryIncluded, Hweimer, Chiswick Chap, 83d40m, VolkovBot, FourteenDays, Lamro, Antixt, Spinningspark, Nubiatech, JackSchmidt, Denisarona, Muhends, Curtis95112, BobKawanaka, Muro Bot, Facts707, Addbot, Masur, LaaknorBot, Lightbot, PV=nRT, Yobot, Dreamer08, AnomieBOT, Materialscientist, Citation bot, ArthurBot, Ace111, Citation bot 2, Citation bot 1, Redrose64, Thinking of England, Winner 42, ZéroBot, Mandula, Brandmeister, Chris857, Ad Orientem, ClueBot NG, Ben morphett, Bibcode Bot, Pedro.bicudo, BG19bot, Enervation, Khazar2, AutisticCatnip, Jamietwells, KevinLiu, Qwerty123uiop, 22merlin, Belle, Oiyarbepsy, DN-boards1, GeneralizationsAreBad, Lollipop, Pentaquarksuperfan, Aashish Khadka, Vcydx and Anonymous: 67

- **Antiparticle** *Source:* https://en.wikipedia.org/wiki/Antiparticle?oldid=680600374 *Contributors:* AxelBoldt, CYD, Mav, Bryan Derksen, Andre Engels, Josh Grosse, Stevertigo, Mrwojo, Patrick, RTC, Paddu, CesarB, Nikai, Nikola Smolenski, Charles Matthews, The Anomebot, Wik, Omegatron, Bevo, Altenmann, Merovingian, Intangir, Wikibot, Martinwguy, Giftlite, Bogdanb, Harp, BenFrantzDale, Herbee, Spencer195, Fleminra, Jason Quinn, Zeimusu, Mako098765, Karol Langner, Mike Rosoft, Helohe, Rich Farmbrough, Guanabot, Pjacobi, Guanabot2, Mr. Billion, Joanjoc~enwiki, Kghose, Cmdrjameson, Giraffedata, Matt McIrvin, HasharBot~enwiki, Pediddle, Deror avi, Woohookitty, Mindmatrix, Wdyoung, GregorB, SeventyThree, Justin Ormont, Palica, Marudubshinki, Tevatron~enwiki, Rjwilmsi, Ae77, MZMcBride, KaiMartin, FlaBot, Krackpipe, Commander Nemet, Roboto de Ajvol, YurikBot, Borgx, Bambaiah, Zhaladshar, Spike Wilbury, Bota47, Terbospeed, Mkossick, Tim314, タチコマ robot, SmackBot, FocalPoint, Alsandro, Srnec, Dauto, Octahedron80, Drphilharmonic, Marcus Brute, Vinaiwbot~enwiki, Jake-helliwell, Grumpyyoungman01, Newone, Mellery, Van helsing, Tim1988, Myasuda, Gogo Dodo, Goldencako, Thijs!bot, Headbomb, Tyco.skinner, JAnDbot, Steveprutz, Ferritecore, Jpod2, Singularity, Dbiel, TomasBat, Eternalmatt, Joshmt, DorganBot, Cuckooman4, VolkovBot, TXiKiBoT, Red Act, Anonymous Dissident, AlleborgoBot, SieBot, Likebox, RadicalOne, Flyer22, KoenDelaere, Thomega, RW Marloe, BrightRoundCircle, Davidmosen, Jacob.jose, Anyeverybody, ClueBot, Diagramma Della Verita, Alexbot, Eeekster, Rishi.bedi, SilvonenBot, NellieBly, Lilaspastia, SkyLined, AkhtaBot, CarsracBot, Lightbot, Legobot, Luckas-bot, Yobot, Planlips, Csmallw, AnomieBOT,

Citation bot, Vuerqex, ArthurBot, Xqbot, Omnipaedista, RibotBOT, Muhwang, EmausBot, John of Reading, L Kensington, Benazhack, ClueBot NG, Geekingreen, Mesoderm, Bibcode Bot, B wik, Mark Arsten, Rm1271, Penguinstorm300, Robotsheepboy, YFdyh-bot, 77Mike77, संजीव कुमार, Dert567, Monkbot, Nazo!nin, KasparBot and Anonymous: 100

- **Beta decay** *Source:* https://en.wikipedia.org/wiki/Beta_decay?oldid=682280135 *Contributors:* AxelBoldt, Chenyu, Trelvis, Mav, Peterlin~enwiki, Ellywa, Andrewa, Cyan, Mxn, Robertb-dc, Shizhao, Pstudier, Jusjih, Twang, Donarreiskoffer, Robbot, Enochlau, Giftlite, Donvinzk, Harp, Herbee, Xerxes314, Radius, Mdob, Antandrus, Icairns, Ukexpat, Jørgen Friis Bak, Discospinster, Guanabot, Vsmith, Roo72, Gianluigi, Joanjoc~enwiki, Neilrieck, Bobo192, Army1987, Drw25, Nk, Haham hanuka, AjAldous, Wtmitchell, Saga City, Falcorian, Flying fish, Eleassar777, Gimboid13, Graham87, Rjwilmsi, FlaBot, Ground Zero, Itinerant1, Goudzovski, Chobot, AllyD, Bgwhite, Roboto de Ajvol, YurikBot, JWB, Jimp, JabberWok, Romanc19s, Spike Wilbury, Johantheghost, Reyk, Geoffrey.landis, JLaTondre, MacsBug, SmackBot, Incnis Mrsi, Ixtli, Gilliam, Skizzik, Dauto, Chris the speller, Octahedron80, Yurigerhard, Sbharris, Audriusa, Tsca.bot, Vladis1av, Decltype, Akulkis, Hgilbert, Polonium, Adj08, JorisvS, Steipe, Mets501, Tuttt, Happy-melon, Civil Engineer III, Timrem, CRGreathouse, JohnCD, Rwflammang, Joelholdsworth, WeggeBot, Kanags, HPaul, Neil9999, Barticus88, Headbomb, Escarbot, AntiVandalBot, Edokter, LibLord, Salgueiro~enwiki, MSBOT, .anacondabot, VoABot II, SHCarter, Pixel ;-), Geekmansworld, Kevinmon, Johnbibby, Dirac66, Edward321, Geboy, Andre.holzner, Catmoongirl, Happyfacesrock, It Is Me Here, Rominandreu, Mcat2, KylieTastic, DorganBot, Y2H, Sheliak, Club house, Milesisgreat, VolkovBot, Hqb, JhsBot, BotKung, Pishogue, FMasic, Cnilep, Tresiden, PlanetStar, Jasondet, Paolo.dL, Arjen Dijksman, Sean.hoyland, Rjc34, Muhends, Sidhu ghanta, Loren.wilton, ClueBot, R000t, Maxtitan, EhJJ, Mikaey, SoxBot III, Directormq, RP459, SkyLined, Addbot, Yakiv Gluck, DOI bot, Man utd suger, Ayrenz, Mdnahas, Zorrobot, Spacy73, Skippy le Grand Gourou, आशीष भटनागर, Luckas-bot, Yobot, Fraggle81, TaBOTzerem, THEN WHO WAS PHONE?, Azylber, Kulmalukko, AnomieBOT, SamuraiBot, Citation bot, Xqbot, FrescoBot, Tekmeme, Jonesey95, Achim1999, Minivip, ApusChin, Double sharp, Bj norge, Andrea105, John of Reading, Acather96, Dewritech, GoingBatty, JSquish, ∆, Coasterlover1994, Illinikiwi, ClueBot NG, Movses-bot, Widr, Meea, MerlIwBot, Bibcode Bot, BG19bot, ElphiBot, Onewhohelps, Cadiomals, Ragnarstroberg, Glevum, Currb, CeraBot, Idenshi, Goyala1, Stigmatella aurantiaca, ChrisGualtieri, Pvoytas, Monkbot, Raytuzio, Hyperclassic, Scipsycho, KasparBot and Anonymous: 174

- **Cabibbo–Kobayashi–Maskawa matrix** *Source:* https://en.wikipedia.org/wiki/Cabibbo%E2%80%93Kobayashi%E2%80%93Maskawa_matrix?oldid=663541514 *Contributors:* Michael Hardy, Charles Matthews, DJ Clayworth, BenRG, MathMartin, Giftlite, Xerxes314, Rich Farmbrough, Hidaspal, RJHall, Army1987, Cmdrjameson, Jag123, Physicistjedi, Mennato, RJFJR, AndyBuckley, Oleg Alexandrov, Linas, Ruziklan, Rjwilmsi, Mathbot, Itinerant1, Goudzovski, YurikBot, Bambaiah, Jimp, JabberWok, Pseudomonas, Black Falcon, RG2, Teply, GrinBot~enwiki, SmackBot, Jim62sch, Stepa, Hmains, Dauto, QFT, Ligulembot, Lambiam, Sipos0, CmdrObot, Markus Pössel, ايل جونغ كيم اناا...الليله!, Headbomb, I310342~enwiki, Drgnrave, Barraki, Cuzkatzimhut, VolkovBot, Anonymous Dissident, Pamputt, J-ishikawa~enwiki, Jim E. Black, Quasirandom, PipepBot, NuclearWarfare, DumZiBoT, TimothyRias, Addbot, Mjamja, Lightbot, PV=nRT, Luckas-bot, Yobot, AnomieBOT, Citation bot, TechBot, A. di M., FrescoBot, Citation bot 1, E-Soter, Bphyswiki, Brandmeister, Ebehn, Isocliff, PhysicsAboveAll, Widr, Fiddleyhead, Bibcode Bot, Nikos Papadakis, Sushant.2811, Monkbot and Anonymous: 39

- **Color–flavor locking** *Source:* https://en.wikipedia.org/wiki/Color%E2%80%93flavor_locking?oldid=664029913 *Contributors:* Incnis Mrsi, RayAYang, Lambiam, Dark Formal, JorisvS, Headbomb, Sean.hoyland, JL-Bot, Mild Bill Hiccup, UnCatBot, Legobot, Yobot, Kav2k, Citation bot, Steve Quinn, Bibcode Bot, Fittold27 and Anonymous: 3

- **Constituent quark** *Source:* https://en.wikipedia.org/wiki/Constituent_quark?oldid=661674231 *Contributors:* Bryan Derksen, Linas, Feydey, Ramin Shemirani~enwiki, Jimp, SmackBot, Hmains, Headbomb, Forgetting curve, Outlook, Blackjack4124, VolkovBot, Muhends, Addbot, Mjamja, Erik9bot, BenzolBot, Puzl bustr, Jesse V., Aesir.le, Egregie and Anonymous: 4

- **Current quark** *Source:* https://en.wikipedia.org/wiki/Current_quark?oldid=623024787 *Contributors:* Bryan Derksen, Ignis~enwiki, Linas, Ramin Shemirani~enwiki, SmackBot, Bluebot, Cryptic C62, Keraunos, Headbomb, Vanished user ty12kl89jq10, Muhends, Jovianeye, Addbot, Yobot, AnomieBOT, Franco3450, 老陳, Skr15081997 and Anonymous: 1

- **CP violation** *Source:* https://en.wikipedia.org/wiki/CP_violation?oldid=678895835 *Contributors:* Roadrunner, Stevertigo, Michael Hardy, Albertplanck, TakuyaMurata, Angela, Julesd, Netsnipe, Palfrey, Raven in Orbit, Coren, Charles Matthews, The Anomebot, Phys, Donarreiskoffer, Pigsonthewing, COGDEN, Ruakh, Giftlite, Jmnbpt, Harp, Xerxes314, Gracefool, ConradPino, HorsePunchKid, Mako098765, WhiteDragon, Karol Langner, Pmanderson, Lumidek, Rich Farmbrough, Hidaspal, Bobo192, Davidruben, Elipongo, Foobaz, I9Q79oL78KiL0QTFHgyc, M0rph, MPerel, Pearle, Sligocki, Evil Monkey, Dirac1933, BlastOButter42, Kusma, Kay Dekker, Oleg Alexandrov, Linas, Nopherox, Marudubshinki, Strait, Tawker, Ligulem, Mathbot, Lmatt, Goudzovski, Chobot, Gdrbot, Bhny, Limulus, JabberWok, NawlinWiki, Grafen, Crasshopper, Tonywalton, Square87~enwiki, Fram, ArielGold, GrinBot~enwiki, MacsBug, SmackBot, HalfShadow, PeterSymonds, Dauto, Chris the speller, Tigerhawkvok, Can't sleep, clown will eat me, QFT, Voyajer, Wen D House, Pwjb, Ligulembot, Drunken Pirate, GTFleming, Erwin, Ryulong, Dan Gluck, Lottamiata, Qqs83, IRevLinas, CRGreathouse, CmdrObot, Wafulz, Vyznev Xnebara, Friendofthehose, Vanished user vjhsduheuiui4t5hjri, Simon Brady, DumbBOT, Gimmetrow, Raoul NK, Mbell, Cosmi, Applecore91, Headbomb, WilliamH, Insane99, Dinagling, Leevclarke, Txomin, Igodard, Kaonslau~enwiki, Thasaidon, Parsecboy, Kevinmon, Homunq, Tonyfaull, DerHexer, Dr. Morbius, MartinBot, Rettetast, Warrickball, Felixbecker2, The dark lord trombonator, Extransit, I310342~enwiki, Larryisgood, Rich Janis, Jackfork, Venny85, Rknasc, PaddyLeahy, SieBot, Nintendostar, Wing gundam, Scasa~enwiki, LonelyMarble, ClueBot, Likebreakfe, Rotational, Chimesmonster, Yakrami, NuclearWarfare, DumZiBoT, Saeed.Veradi, MystBot, Airplaneman, Addbot, Cxz111, Landon1980, Leszek Jańczuk, Debresser, Tide rolls, Lightbot, Zorrobot, Micko.hjort~enwiki, LuK3, Legobot, Luckas-bot, Yobot, Amirobot, AnomieBOT, Killiondude, Citation bot, Quebec99, Xqbot, Spidern, False vacuum, FrescoBot, Paine Ellsworth, Citation bot 1, Sunandclouds, Elockid, Mutinus, Thinking of England, Sanomi, The Perfection, Dizanl, Ofercomay, John of Reading, Bphyswiki, Zerkroz, Fæ, Arbnos, Ebrambot, AManWithNoPlan, Kweckzilber, AfroScientist, Maschen, QuantumSquirrel, ClueBot NG, Jj1236, Snotbot, Ghartshaw, Fascismsucks555, Keithtacokeithsta, Sndfnsdfsdffdd, Togtfo, Honestguy55543, Anonymous5555, Helpful Pixie Bot, Bibcode Bot, Petermahlzahn, MeanMotherJr, ChrisGualtieri, Ranze, Andyhowlett, Mtdevans, Nigellwh, Spyglasses, Prokaryotes, Justinvasel, Monkbot, Soham92 and Anonymous: 105

- **Fermion** *Source:* https://en.wikipedia.org/wiki/Fermion?oldid=682673767 *Contributors:* AxelBoldt, Chenyu, Derek Ross, CYD, Mav, Bryan Derksen, The Anome, Ben-Zin~enwiki, Alan Peakall, Dominus, Dcljr, Looxix~enwiki, Glenn, Nikai, Andres, Wikiborg, David Latapie, Phys, Bevo, Stormie, Olathe, Donarreiskoffer, Robbot, Merovingian, Rorro, Wikibot, HaeB, Giftlite, Fropuff, Xerxes314, Vivektewary, JoJan, Karol Langner, Tothebarricades.tk, Icairns, Hidaspal, Vsmith, Laurascudder, Lysdexia, Ashlux, Graham87, Magister Mathematicae, Kbdank71, Syndicate, Strait, Protez, Drrngrvy, FlaBot, Srleffler, Chobot, YurikBot, RobotE, Jimp, Bhny, Captaindan, SpuriousQ, Salsb, Lomn, Enormousdude, CharlesHBennett, Federalist51, Tom Lougheed, Unyoyega, Jrockley, MK8, BabuBhatt, Complexica, Zachorious, Shalom Yechiel,

QFT, Garry Denke, Daniel.Cardenas, SashatoBot, Flipperinu, Dan Gluck, LearningKnight, Happy-melon, Paulfriedman7, Cydebot, Meno25, Zalgo, Thijs!bot, Mbell, Headbomb, Nick Number, Orionus, Shlomi Hillel, CosineKitty, NE2, Mwarren us, ZPM, Vanished user ty12kl89jq10, Joshua Davis, R'n'B, Tensegrity, Rod57, Dgiraffes, Alpvax, VolkovBot, TXiKiBoT, Red Act, Anonymous Dissident, Abdullais4u, בל יכול, Tanhueiming, Antixt, Haiviet~enwiki, EmxBot, Kbrose, SieBot, Likebox, Jojalozzo, Dhatfield, Oxymoron83, TubularWorld, ClueBot, Seervoitek, Rodhullandemu, Jorisverbiest, Feebas factor, ChandlerMapBot, Nilradical, Wikeepedian, Stephen Poppitt, Addbot, Vectorboson, Luckas-bot, Yobot, Planlips, Dickdock, AnomieBOT, Icalanise, Materialscientist, Xqbot, Br77rino, Balaonair, 老陳, Paine Ellsworth, Blackoutjack, Kikeku, Rameshngbot, Tom.Reding, RedBot, Alarichus, Michael9422, Silicon-28, TjBot, EmausBot, WikitanvirBot, Quazar121, Solomonfromfinland, JSquish, Fimin, Quondum, AManWithNoPlan, EdoBot, ClueBot NG, PBot1, EthanChant, Bibcode Bot, BG19bot, Petermahlzahn, KingKhan85, ChrisGualtieri, BoethiusUK, DerekWinters, Tentinator, JNrgbKLM, Mohit rajpal, KasparBot, Jiswin1992 and Anonymous: 120

- **Gell-Mann–Nishijima formula** *Source:* https://en.wikipedia.org/wiki/Gell-Mann%E2%80%93Nishijima_formula?oldid=585830388 *Contributors:* Wi, Bkell, Giftlite, Pcd72, Woohookitty, Tone, Hwasungmars, Reyk, Itserpol, Michael C Price, Thijs!bot, Headbomb, Maliz, STBot, HEL, AlleborgoBot, DragonBot, PixelBot, RandomTool2, Addbot, Vectorboson, Luckas-bot, Yobot, Citation bot, Citation bot 1, WikitanvirBot, Snicol1401, Bibcode Bot, ChrisGualtieri and Anonymous: 10

- **Generation (particle physics)** *Source:* https://en.wikipedia.org/wiki/Generation_(particle_physics)?oldid=672806813 *Contributors:* Selket, Arkuat, Aleron235, Giftlite, Harp, Xerxes314, Keith Edkins, Discospinster, FT2, Army1987, John Vandenberg, PaulHanson, DannyWilde, Fosnez, Vossman, Mushin, Conscious, Bhny, FFLaguna, SmackBot, InverseHypercube, Anastrophe, Dauto, Jmnbatista, Titus III, Aeluwas, SchmittM, Vyznev Xnebara, Q43, Thijs!bot, Headbomb, Nick Number, JAnDbot, Kborland, Choihei, TomasBat, Idioma-bot, VolkovBot, Nxavar, SieBot, Muhends, TimothyRias, MystBot, Addbot, Luckas-bot, HieronymousCrowley, Icalanise, Citation bot, Xqbot, LucienBOT, JIROT, Akesich, Afteread, Bookalign, Barak90, ZéroBot, ChuispastonBot, Bibcode Bot, Blaspie55, Prokaryotes, Isambard Kingdom and Anonymous: 24

- **Gluon** *Source:* https://en.wikipedia.org/wiki/Gluon?oldid=681546689 *Contributors:* AxelBoldt, CYD, Bryan Derksen, Gdarin, TakuyaMurata, Card~enwiki, Looxix~enwiki, Ellywa, Ahoerstemeier, Med, Schneelocke, Phys, Phil Boswell, Donarreiskoffer, Fredrik, Merovingian, Hadal, Giftlite, Herbee, Xerxes314, Eequor, Darrien, Keith Edkins, RetiredUser2, Icairns, Mike Rosoft, AlexChurchill, HedgeHog, Kenny TM~~enwiki, David Schaich, Ioliver, Mashford, El C, Kwamikagami, Ardric47, Obradovic Goran, Alansohn, Guy Harris, Dachannien, Ricky81682, Batmanand, Velella, Kazvorpal, April Arcus, Forteblast, Mpatel, Palica, BD2412, Kbdank71, Rjwilmsi, Macumba, Strait, Mike Peel, Bubba73, Klortho, FlaBot, Srleffler, Chobot, YurikBot, Wavelength, Bambaiah, Hairy Dude, Jimp, JabberWok, Zelmerszoetrop, Salsb, SCZenz, Randolf Richardson, Ravedave, Danlaycock, Bota47, LeonardoRob0t, Anclation~enwiki, Physicsdavid, Erudy, GrinBot~enwiki, Kgf0, SmackBot, Melchoir, Cessator, Benjaminevans82, Abtal, MK8, Colonies Chris, Can't sleep, clown will eat me, Decltype, Qcdmaestro, Edconrad, Darkpoison99, FredrickS, Omsharan, Pegasusbot, Gregbard, ProfessorPaul, Thijs!bot, Headbomb, Rriegs, Oreo Priest, AntiVandalBot, Shambolic Entity, Deflective, Mujokan, Yill577, Happycool, Mother.earth, Martynas Patasius, WiiWillieWiki, HEL, Hans Dunkelberg, Gombang, Inwind, Sheliak, Jonthaler, VolkovBot, TXiKiBoT, Davehi1, Kriak, Anonymous Dissident, Imasleepviking, AlleborgoBot, EJF, SieBot, Steven Crossin, OKBot, ClueBot, Wwheaton, Qsaw, Nucularphysicist, Ottava Rima, Gordon Ecker, Rhododendrites, Brews ohare, Cacadril, RexxS, JKeck, Against the current, SkyLined, Addbot, DOI bot, Lightbot, Skippy le Grand Gourou, Luckas-bot, Planlips, AnomieBOT, Jim1138, JackieBot, Citation bot, Bci2, ArthurBot, Xqbot, Neil95, Triclops200, Omnipaedista, TorKr, 老陳, Paine Ellsworth, Ivoras, Citation bot 1, Pekayer11, Rameshngbot, PNG, RjwilmsiBot, TjBot, Lilcal89012, EmausBot, Socob, JSquish, StringTheory11, Quondum, TyA, Maschen, RolteVolte, ClueBot NG, Timothy jordan, Maplelanefarm, Bibcode Bot, BG19bot, Gravitoweak, Cadiomals, Tropcho, Fraulein451, DrHjmHam, Rhlozier, D.shinkaruk, Yaara dildaara, BronzeRatio, Monkbot, Yikkayaya, KasparBot and Anonymous: 142

- **Hadronization** *Source:* https://en.wikipedia.org/wiki/Hadronization?oldid=665417782 *Contributors:* Alexwatson, Charles Matthews, Phys, Sam Hocevar, Fwb22, RJFJR, TenOfAllTrades, Isaac Rabinovitch, Ohwilleke, SCZenz, Zwobot, Whobot, SmackBot, Incnis Mrsi, Colonies Chris, Chrumps, A876, Headbomb, Pichote, Sir Link, Moogwrench, Sothisislife101, VVVBot, Addbot, Luckas-bot, Yobot, Citation bot, Double sharp, Naznin farhah, Chaudière, Raktimabir, Jj1236, Bibcode Bot, Prokaryotes and Anonymous: 16

- **Lepton** *Source:* https://en.wikipedia.org/wiki/Lepton?oldid=679886360 *Contributors:* Bryan Derksen, Andre Engels, PierreAbbat, BenZin~enwiki, Heron, Xavic69, Fruge~enwiki, Fwappler, Ahoerstemeier, Julesd, Glenn, Mxn, A5, Wikiborg, Dysprosia, Radiojon, Imc, Morwen, Fibonacci, Bcorr, Phil Boswell, Donarreiskoffer, Robbot, Merovingian, Wikibot, Giftlite, Smjg, DocWatson42, Harp, Herbee, Xerxes314, Sysin, Knutux, LiDaobing, LucasVB, ClockworkLunch, RetiredUser2, Icairns, Mike Rosoft, Chris j wood, Martinl~enwiki, Smalljim, Giraffedata, Jumbuck, RobPlatt, Neonumbers, Ahruman, Computerjoe, Simon M, Woohookitty, Mindmatrix, Rjwilmsi, Strait, Erkcan, FlaBot, DannyWilde, Mastorrent, Celebere, Peterl, YurikBot, Bambaiah, Jimp, Salsb, Spike Wilbury, Jaxl, SCZenz, DeadEyeArrow, Tetracube, Smoggyrob, Dmuth, Jaysbro, Sbyrnes321, That Guy, From That Show!, SmackBot, Bazza 7, KocjoBot~enwiki, Jrockley, Mom2jandk, Cool3, Hmains, Complexica, DHN-bot~enwiki, Mesons, Yevgeny Kats, TriTertButoxy, SashatoBot, Ouzo~enwiki, Happy-melon, Kurtan~enwiki, Myasuda, Cydebot, Meno25, Photocopier, Michael C Price, Casliber, Thijs!bot, Headbomb, Newton2, Mentifisto, Autotheist, Steveprutz, NeverWorker, NicoSan, MartinBot, Arjun01, HEL, J.delanoy, Numbo3, Gombang, Num1dgen, Ceoyoyo, VolkovBot, Macedonian, Mocirne, TXiKiBoT, Anonymous Dissident, Abdullais4u, Antixt, Jhb110, Thanatos666, AlleborgoBot, SieBot, ToePeu.bot, RadicalOne, Ngexpert7, Jacob.jose, Hamiltondaniel, TubularWorld, Muhends, ClueBot, ICAPTCHA, UniQue tree, Snigbrook, Fyyer, IceUnshattered, Cmj91uk, LieAfterLie, Manu-ve Pro Ski, TimothyRias, Addbot, Betterusername, AgadaUrbanit, Ehrenkater, OlEnglish, Zorrobot, Andy2308, Legobot, Luckas-bot, Ptbotgourou, Maxim Sabalyauskas, Planlips, JackieBot, Icalanise, Citation bot, أحمد.غامدي.24, ArthurBot, Almabot, Omnipaedista, Alexeymorgunov, 老陳, Tormine, MathFacts, Citation bot 1, MastiBot, Earthandmoon, EmausBot, John of Reading, Az29, Galaktiker, StringTheory11, Quondum, Surajt88, I hate whitespace, ClueBot NG, Scimath Genius, Braincricket, Widr, Helpful Pixie Bot, Bibcode Bot, Tyler6360534, Katagun5, Melenc, DerekWinters, Prasanna4s, Machosquirrel, Devinhorn, KasparBot and Anonymous: 149

- **Leptoquark** *Source:* https://en.wikipedia.org/wiki/Leptoquark?oldid=674819966 *Contributors:* Bcorr, Herbee, Xerxes314, Pjacobi, David Schaich, Bender235, Erkcan, Timboe, Hairy Dude, Conscious, PJTraill, Colonies Chris, Mesons, Linus M., Headbomb, Sanitycult, Calwiki, Waltoncats, Addbot, Mjamja, Wireader, Citation bot, Ernsts, Citation bot 1, Slightsmile, ResidentAnthropologist, Bibcode Bot and Anonymous: 9

- **J/ψ meson** *Source:* https://en.wikipedia.org/wiki/J/psi_meson?oldid=678932131 *Contributors:* Kowloonese, Cyp, Andrewa, Schneelocke, Doradus, Thue, Jeffq, Donarreiskoffer, Sanders muc, Sverdrup, JB82, Giftlite, Eequor, Plutor, DragonflySixtyseven, Sam Hocevar, Eb.hoop, Grutter, Kjoonlee, Kwamikagami, Menscher, Tripodics, Palica, Kbdank71, Rjwilmsi, Strait, Erkcan, Bgwhite, YurikBot, Bambaiah, Jimp, Salsb,

SCZenz, Ravedave, Poulpy, Physicsdavid, Wizofaus, SmackBot, Vald, Hmains, Colonies Chris, OrphanBot, JorisvS, Dicklyon, Adriferr, Norm mit, Kurtan~enwiki, Headbomb, WinBot, Certain, JAnDbot, Matthew Fennell, Igodard, Jondaman21, HEL, MITBeaverRocks, STBotD, Sheliak, VolkovBot, TimAdye, Senemmar, Tomaxer, SieBot, Nihil novi, TubularWorld, Gabor Denes, SkyLined, Addbot, LaaknorBot, Lightbot, Luckas-bot, Yobot, Citation bot, ArthurBot, TheAMmollusc, GrouchoBot, Carlog3, Paine Ellsworth, Xiglofre, Johann137, Trappist the monk, EmausBot, StringTheory11, SporkBot, Timetraveler3.14, Bibcode Bot, Hmainsbot1, Bill2239, Trwood, Monkbot, Grand'mere Eugene and Anonymous: 37

- **Parton (particle physics)** *Source:* https://en.wikipedia.org/wiki/Parton_(particle_physics)?oldid=679736336 *Contributors:* The Anome, Mako098765, Ylai, Euyyn, Rjwilmsi, Strait, RE, Erkcan, Chobot, Jimp, RDBury, Bluebot, Sprocketonline, TriTertButoxy, Haus, Harej bot, Rotiro, Thijs!bot, Headbomb, D.H, Escarbot, HEL, Natsirtguy, LordAnubisBOT, VolkovBot, Neparis, Humanino, AnonyScientist, Crocodilecoup, Cmcphys, Addbot, Mjamja, SpBot, OlEnglish, Luckas-bot, AnomieBOT, Citation bot, Bci2, RedBot, MondalorBot, EmausBot, Ornithikos, ClueBot NG, Helpful Pixie Bot, Bibcode Bot, Richiebful, Keroniddron and Anonymous: 31

- **Preon** *Source:* https://en.wikipedia.org/wiki/Preon?oldid=646794027 *Contributors:* Maury Markowitz, Heron, Ewen, Edward, Kickaha~enwiki, Dcljr, Timwi, David Latapie, Chrisjj, BenRG, Altenmann, Merovingian, Xanzzibar, David Gerard, Giftlite, Graeme Bartlett, Herbee, Monedula, Semorrison, Mboverload, Eequor, Icairns, Urhixidur, Rich Farmbrough, Pjacobi, Drhex, John Vandenberg, Jag123, Calton, Alai, Uncle G, GregorB, BD2412, Ketiltrout, Rjwilmsi, Fragglet, Mathrick, Smithbrenon, CJLL Wright, Chobot, ScottAlanHill, Jpkotta, YurikBot, Ugha, Bambaiah, Phmer, Ohwilleke, Merick, Gcapp1959, Dialectric, Buster79, Trovatore, Długosz, Closedmouth, Iellwood, Paul D. Anderson, Lserni, SmackBot, RockMaestro, Bayardo, Stepa, GwydionM, Kmarinas86, DocKrin, JesseStone, Trekphiler, Vladislav, Fatla00, SilverStar, Jaganath, Md2perpe, Will314159, Friendly Neighbour, CmdrObot, Doc W, AlphaNumeric, LactoseTI, Mglg, Keraunos, Headbomb, Thadius856, Joe Schmedley, Ph.eyes, GurchBot, Yill577, Randyfurlong, Dr. Morbius, Lexivore, Experiential, ChauriCh, Tanaats, OliverHarris, VolkovBot, Fences and windows, Calwiki, Anonymous Dissident, Thrawn562, Dirkbb, Synthebot, Antixt, Pegasus1965, AHMartin, PlanetStar, Work permit, WereSpielChequers, 1ForTheMoney, SkyLined, Addbot, Lightbot, OlEnglish, Zorrobot, The Bushranger, Luckas-bot, Yobot, EnochBethany, AnomieBOT, Icalanise, Materialscientist, Citation bot, Jsharpminor, GrouchoBot, FrescoBot, Goodbye Galaxy, Citation bot 1, MastiBot, Bj norge, RjwilmsiBot, Ofercomay, Detogain, WikitanvirBot, Slightsmile, The Mysterious El Willstro, Hhhippo, Wyvern Rex., Suslindisambiguator, Gilderien, Preon, Helpful Pixie Bot, Bibcode Bot, BG19bot, Marioedesouza, Andyhowlett, I am One of Many, FrigidNinja, Draconnis caput, Delbert7 and Anonymous: 101

- **QCD matter** *Source:* https://en.wikipedia.org/wiki/QCD_matter?oldid=650185710 *Contributors:* Ixfd64, Xerxes314, Niteowlneils, WhiteDragon, The Land, FT2, Army1987, Mac Davis, Vuo, Kazvorpal, Ceyockey, Davidfstr, GregorB, SeventyThree, Marudubshinki, Leapfrog314, Nanite, Goudzovski, YurikBot, Wavelength, Bambaiah, Hairy Dude, Archelon, Salsb, SCZenz, Dhollm, Nekura, SmackBot, Mcneile, Colonies Chris, Dark Formal, Mgiganteus1, Xionbox, Dan Gluck, CmdrObot, Headbomb, "", Beasticles, Igodard, WolfmanSF, HiB2Bornot2B, Maurice Carbonaro, LokiClock, Fences and windows, KP-Adhikari, Jmath666, SieBot, Martarius, SteelSoul, Brews ohare, Ordovico, TimothyRias, IngerAlHaosului, Addbot, Rbwolf, OlEnglish, Luckas-bot, Yobot, Citation bot, Eumolpo, Mnmngb, Tom.Reding, 23790AD, Johann137, Trappist the monk, Naviguessor, Gerasime, Suslindisambiguator, Helpful Pixie Bot, Curb Chain, Bibcode Bot, Anwarwaseem, Mogism, RhinoMind and Anonymous: 35

- **Quark epoch** *Source:* https://en.wikipedia.org/wiki/Quark_epoch?oldid=677763687 *Contributors:* Gandalf61, Rpyle731, Army1987, Eclipsed, Colonies Chris, Michael C Price, Headbomb, Peter Gulutzan, Z10x, Escarbot, TXiKiBoT, Muhends, Addbot, Luckas-bot, Xqbot, FrescoBot, Dogaru Florin, Tom.Reding, 13XIII, Slightsmile, ResidentAnthropologist, Helpful Pixie Bot, Jwratner1, Tetra quark and Anonymous: 3

- **Quark star** *Source:* https://en.wikipedia.org/wiki/Quark_star?oldid=681326105 *Contributors:* Bryan Derksen, Oliver Pereira, Skysmith, Alfio, Glenn, Robbot, Rursus, Giftlite, Jyril, Curps, Peter Ellis, Rrw, Icairns, Djyang~enwiki, Adashiel, Mike Rosoft, Mormegil, Rich Farmbrough, AdamSolomon, DaveGorman, I9Q79oL78KiL0QTFHgyc, Cherlin, Mac Davis, BRW, Vuo, Pauli133, Daniel Newby, Mazca, Pol098, Eyreland, Marasama, Maxim Razin, Patrick1982, Algri, Smithbrenon, YurikBot, Spacepotato, Bambaiah, Jengelh, Salsb, Smkolins, Th1rt3en, The Yeti, Abramul, MacsBug, SmackBot, Edgar181, Kmarinas86, Colonies Chris, Ron g, Mrwuggs, Downwards, Pwjb, ChowRiit, Doug Bell, Titus III, Dark Formal, Zzzzzzzzzzz, OS2Warp, Asymptote, Thijs!bot, Headbomb, "", Mashiah Davidson, WolfmanSF, Witch-King, Morris729, Tarotcards, Dorftrottel, Idioma-bot, TXiKiBoT, Someguy1221, Mzmadmike, Lamro, The Mad Genius, SieBot, Triwbe, Hamiltondaniel, Gorkymalorki, Ktr101, Alexbot, Coinmanj, Millionsandbillions, MilesAgain, DumZiBoT, Addbot, SpellingBot, IcelandicTundra, Peti610botH, OlEnglish, Potekhin, Yobot, Amirobot, Trinitrix, Baxxterr, Sailorleo, Robert Treat, AnomieBOT, Citation bot, Daners, Gap9551, Moxy, Bobicmon, Perlscrypt, Tom.Reding, Nacen, IVAN3MAN, Puzl bustr, Chungers1, 564dude, RjwilmsiBot, GoingBatty, Metaxy, ZéroBot, StringTheory11, Ebrambot, Suslindisambiguator, L1A1 FAL, David C Bailey, Shaneanthonypitts, Alex Nico, Qubuntu, Bibcode Bot, BG19bot, Yukterez, Mustaphffgha, A.E.U.M., BattyBot, DarafshBot, Tentinator, Ferreira932, RhinoMind, PSR150958, GrizzlyPear, Erikprantare and Anonymous: 71

- **Quantum chromodynamics** *Source:* https://en.wikipedia.org/wiki/Quantum_chromodynamics?oldid=678061083 *Contributors:* AxelBoldt, CYD, Zundark, Youandme, Ewen, Stevertigo, Michael Hardy, Ahoerstemeier, Whkoh, Emperorbma, Jitse Niesen, Phys, Robbot, Fredrik, Ojigiri~enwiki, Seth Ilys, Alan Liefting, Giftlite, JamesMLane, Monedula, Xerxes314, JeffBobFrank, Jason Quinn, Elroch, Icairns, Sam Hocevar, Lumidek, Sctfn, Eep[2], David Schaich, JonL, Goplat, AdamSolomon, Pt, El C, CDN99, Robotje, Slicky, Physicistjedi, Azn king28, Fwb22, Guy Harris, Ricky81682, TenOfAllTrades, Skyring, Kusma, Alai, Mpatel, Betsythedevine, Mendaliv, VermillionBird, Rjwilmsi, Coemgenus, FlaBot, Thenewdeal87, Adoniscik, Algebraist, YurikBot, Wavelength, Bambaiah, Hairy Dude, Moto Perpetuo, Ohwilleke, JabberWok, Kirill Lokshin, Spike Wilbury, BlackAndy, Thiseye, CecilWard, Voidxor, Zzuuzz, Banus, Finell, SmackBot, Henriok, Vald, ProveIt, GaeusOctavius, Chris the speller, Bluebot, TimBentley, Complexica, Colonies Chris, Modest Genius, Berland, Grover cleveland, Garry Denke, TriTertButoxy, DJIndica, Jaganath, RoboDick~enwiki, NNemec, Slakr, Ryulong, Tawkerbot2, Memetics, Capefeather, Runningonbrains, Cydebot, DavidMcCabe, Headbomb, WVhybrid, Noclevername, Escarbot, Salgueiro~enwiki, Shambolic Entity, Andonic, Hut 8.5, Pkoppenb, .anacondabot, Robomojo, Corvidaecorvus, Maliz, Connor Behan, TechnoFaye, R'n'B, HEL, DrKiernan, Acalamari, Shomroni, Lseixas, Skullfunk, GrahamHardy, Idioma-bot, Sheliak, Cuzkatzimhut, VolkovBot, TXiKiBoT, Calwiki, Rei-bot, Saibod, KP-Adhikari, Ptrslv72, SieBot, Dawn Bard, Likebox, Anchor Link Bot, ClueBot, WDavis1911, Pechmerle, PixelBot, Brews ohare, Chrisarnesen, XLinkBot, SilvonenBot, SkyLined, Truthnlove, Addbot, DOI bot, AnnaFrance, SpBot, Lightbot, Zorrobot, Legobot, Luckas-bot, Yobot, Tamtamar, Nallimbot, Citation bot, LilHelpa, Info21, Chrisfox8, Pra1998, Petros000, FrescoBot, Ecuqkindler, Timmeken, Ganondolf, Meier99, Heurisko, Earthandmoon, Tarsilia, McSaks, Autumnalmonk, EmausBot, Mnkyman, Wikipelli, Brazmyth, Quondum, Aschwole, Rcsprinter123, Maschen, Fwilczek, RolteVolte, Neduard, QuantumSquirrel, Teaktl17, ClueBot NG, Helpful Pixie Bot, Bibcode Bot, Dalit Llama, BG19bot, PhnomPencil, Vkpd11, Snow Blizzard, Cjean42, Joeinwiki, Trompedo, KasparBot and Anonymous: 137

- **Quarkonium** *Source:* https://en.wikipedia.org/wiki/Quarkonium?oldid=677791042 *Contributors:* AxelBoldt, Xavic69, Rursus, Mako098765, Icairns, Mike Rosoft, Anthony Appleyard, Kbdank71, Rjwilmsi, MZMcBride, MatthewMastracci, FlaBot, Goudzovski, YurikBot, Bambaiah, Hairy Dude, Salsb, Długosz, SCZenz, Physicsdavid, Reedy, Colonies Chris, Teki D, Akriasas, Kurtan~enwiki, Thijs!bot, Headbomb, Jimbobl, Magioladitis, Choihei, Reedy Bot, Tarotcards, Joshmt, Muro de Aguas, Antixt, Neparis, SieBot, BartekChom, Bobathon71, MystBot, Addbot, Amirobot, Citation bot, MIRROR, Metrictensor, Ricolai, Citation bot 1, Jonesey95, Johann137, 564dude, RjwilmsiBot, EmausBot, SanchoOoPansa, ZéroBot, Chris857, Bibcode Bot, Khazar2, Illia Connell, Giu8888, JJ1976, Dbjergaard and Anonymous: 19

- **Quark–lepton complementarity** *Source:* https://en.wikipedia.org/wiki/Quark%E2%80%93lepton_complementarity?oldid=667466538 *Contributors:* Chuunen Baka, RJFJR, Stephenb, SmackBot, BryanG, JorisvS, Cydebot, BetacommandBot, Koeplinger, Headbomb, Mentifisto, Yahel Guhan, STBotT, Ergo leu, Paulfharrison, BartekChom, TubularWorld, Yobot, Omnipaedista, Citation bot 1, Bibcode Bot, BattyBot and Anonymous: 8

- **Standard Model** *Source:* https://en.wikipedia.org/wiki/Standard_Model?oldid=679087631 *Contributors:* AxelBoldt, Derek Ross, CYD, Bryan Derksen, The Anome, Ed Poor, Andre Engels, Roadrunner, David spector, Isis~enwiki, Youandme, Ram-Man, Stevertigo, Edward, Patrick, Boud, Michael Hardy, SebastianHelm, Looxix~enwiki, Julesd, Glenn, AugPi, Mxn, Raven in Orbit, Reddi, Phr, Tpbradbury, Populus, Haoherb428, Phys, Floydian, Bevo, Pierre Boreal, AnonMoos, BenRG, Jeffq, Dmytro, Drxenocide, Robbot, Nurg, Securiger, Texture, Roscoe x, Fuelbottle, Mattflaschen, Tobias Bergemann, Alan Liefting, Ancheta Wis, Giftlite, Dbenbenn, Harp, Herbee, Monedula, LeYaYa, Xerxes314, Dratman, Alison, JeffBobFrank, Dmmaus, Pharotic, Brockert, Bodhitha, Andycjp, Sonjaaa, HorsePunchKid, APH, Icairns, AmarChandra, Gscshoyru, Kate, Arivero, FT2, Rama, Vsmith, David Schaich, Xezbeth, D-Notice, Dfan, Bender235, Pt, El C, Laurascudder, Shanes, Drhex, Fogger~enwiki, Brim, Rbj, Jeodesic, Jumbuck, Alansohn, Gary, ChristopherWillis, Guy Harris, Axl, Sligocki, Kocio, Stillnotelf, Alinor, Wtmitchell, Egg, TenOfAllTrades, H2g2bob, Killing Vector, Linas, Mindmatrix, Benbest, Dodiad, Mpatel, Faethon, TPickup, Faethon34, Palica, Dysepsion, Faethon36, Qwertyca, Drbogdan, Rjwilmsi, Zbxgscqf, Macumba, Strangethingintheland, Dstudent, R.e.b., Bubba73, Drrngrvy, Agasicles, FlaBot, Naraht, Agasides, DannyWilde, Dave1g, Itinerant1, Gparker, Jrtayloriv, Goudzovski, Chobot, Bgwhite, FrankTobia, YurikBot, Bambaiah, Ohwilleke, VoxMoose, Bhny, JabberWok, Bovineone, Krbabu, SCZenz, JulesH, Davemck, Lomn, E2mb0t~enwiki, Dna-webmaster, Jrf, Dv82matt, Tetracube, Hirak 99, Arthur Rubin, Netrapt, JLaTondre, Caco de vidro, RG2, GrinBot~enwiki, That Guy, From That Show!, Hal peridol, SmackBot, YellowMonkey, Tom Lougheed, Melchoir, Bazza 7, KocjoBot~enwiki, Jagged 85, Thunderboltz, Setanta747 (locked), Skizzik, Dauto, Chris the speller, Bluebot, TimBentley, Sirex98, Silly rabbit, Complexica, Metacomet, DHN-bot~enwiki, MovGP0, QFT, Kittybrewster, Addshore, Jmnbatista, Cybercobra, Jgwacker, BullRangifer, Soarhead77, Daniel.Cardenas, Yevgeny Kats, Byelf2007, TriTertButoxy, Craig Bolon, Ajnosek, Ekjon Lok, Bjankuloski06, Tarcieri, Waggers, JarahE, Michaelbusch, Lottamiata, Newone, Twas Now, IanOfNorwich, Srain, Patrickwooldridge, J Milburn, Mosaffa, Gatortpk, Vessels42, Geremia, Van helsing, Harrigan, Phatom87, Cydebot, David edwards, Verdy p, Michael C Price, Xantharius, Crum375, JamesAM, Thijs!bot, Epbr123, Headbomb, Phy1729, Stannered, Tariqhada, Seaphoto, Orionus, Voyaging, Gnixon, Jbaranao, Jrw@pobox.com, Len Raymond, Narssarssuaq, Bakken, CattleGirl, Davidoaf, Vanished user ty12kl89jq10, Lvwarren, Taborgate, Leyo, HEL, J.delanoy, Hans Dunkelberg, Stephanwehner, Wbellido, Aoosten, Jacksonwalters, The Transliterator, DadaNeem, Student7, Joshmt, WJBscribe, Jozwolf, Hexane2000, BernardZ, Awren, Sheliak, Physicist brazuca, Schucker, Goop Goop, Fences and windows, Dextrose, Mcewan, Swamy g, TXiKiBoT, Sharikkamur, Thrawn562, Voorlandt, Escalona, Setreset, PDFbot, Pleroma, UnitedStatesian, Piyush Sriva, Kacser, Billinghurst, Francis Flinch, Moose-32, Ptrslv72, David Barnard, SieBot, ShiftFn, Robdunst, Jim E. Black, SheepNotGoats, Gerakibot, Nozzer42, Mr swordfish, Wing gundam, Bamkin, Likebox, Arthur Smart, HungarianBarbarian, Commutator, KathrynLybarger, Iomesus, C0nanPayne, Crazz bug 5, ClueBot, Superwj5, Wwheaton, Garyzx, SuperHamster, Elsweyn, Maldmac, DragonBot, Djr32, Diagramma Della Verita, Nymf, Eeekster, Brews ohare, NuclearWarfare, PhySusie, Ordovico, Mastertek, DumZiBoT, BodhisattvaBot, Guarracino, Mitch Ames, Truthnlove, Stephen Poppitt, Tayste, Addbot, Deepmath, Eric Drexler, DWHalliday, Mjamja, Leszek Jańczuk, NjardarBot, Mwoldin, Bassbonerocks, Barak Sh, AgadaUrbanit, Lightbot, Smeagol 17, Abjiklam, Ve744, Luckas-bot, Yobot, Orion11M87, AnomieBOT, JackieBot, Icalanise, Citation bot, ArthurBot, Northryde, LilHelpa, Xqbot, Sionus, Professor J Lawrence, Tomwsulcer, Edsegal, GrouchoBot, Trongphu, QMarion II, Ernsts, A. di M., Bytbox, FrescoBot, Paine Ellsworth, Aliotra, Steve Quinn, Citation bot 1, Rameshngbot, MJ94, RedBot, MastiBot, Aknochel, Sijothankam, Puzl bustr, Beta Orionis, Physics therapist, Bj norge, Innotata, Jesse V., RjwilmsiBot, Mathewsyriac, Afteread, EmausBot, Bookalign, WikitanvirBot, Wilhelm-physiker, Bdijkstra, DerNeedle, Kenmint, Dbraize, Tanner Swett, HeptishHotik, مہمنشیں بہار, Suslindisambiguator, Quondum, Webbeh, UniversumExNihilo, Vanished user fijw983kjaslkekfhj45, Maschen, RockMagnetist, Stormymountain, Ζeta ζ, Whoop whoop pull up, Isocliff, ClueBot NG, Smtchahal, Snotbot, Tonypak, O.Koslowski, CharleyQuinton, Dsperlich, Theopolisme, ZakMarksbury, Helpful Pixie Bot, Bibcode Bot, BG19bot, Tirebiter78, AvocatoBot, Lukys~enwiki, Stapletongrey, Ownedroad9, Chip123456, ChrisGualtieri, Khazar2, Billyfesh399, Rhlozier, JYBot, Dexbot, Doom636, Rongended, Cerabot~enwiki, CuriousMind01, Cjean42, Jayanta mallick, Joeinwiki, Kowtje, JPaestpreornJeolhlna, Eyesnore, Euan Richard, Nigstomper, Particle physicist, Prokaryotes, Jernahthern, Ginsuloft, Dimension10, JNrgbKLM, Krabaey, 1codesterS, FelixRosch, Monkbot, Delbert7, BradNorton1979, Lathamboyle, Tetra quark, KasparBot, Buckbill10 and Anonymous: 357

- **Strong interaction** *Source:* https://en.wikipedia.org/wiki/Strong_interaction?oldid=681830498 *Contributors:* AxelBoldt, Sodium, Bryan Derksen, RK, Andre Engels, Danny, Peterlin~enwiki, Heron, Xavic69, Tim Starling, EddEdmondson, Gdarin, Ixfd64, Wintran, Salsa Shark, AugPi, Timwi, Bemoeial, Wikiborg, Fuzheado, ElusiveByte, Populus, Phys, Omegatron, Bevo, PuzzletChung, Robbot, Mayooranathan, Henrygb, Giftlite, DocWatson42, Sj, Harp, Monedula, Xerxes314, Remy B, Jason Quinn, Utcursch, Zfr, AmarChandra, Sam Hocevar, Nicobn~enwiki, Jørgen Friis Bak, Discospinster, FT2, Vsmith, ArnoldReinhold, Trekie8472, Roybb95~enwiki, David Schaich, Gianluigi, Marknewlyn~enwiki, Drhex, CDN99, Army1987, Matt McIrvin, Danski14, VivaEmilyDavies, TenOfAllTrades, Vuo, DV8 2XL, Kazvorpal, Oleg Alexandrov, Superstring, Spettro9, BillC, Eras-mus, Tevatron~enwiki, Mendaliv, Zbxgscqf, Strait, Loudenvier, Donotresus, Yamamoto Ichiro, Siv0r, Lmatt, Srleffler, Chobot, DVdm, Wavelength, Borgx, Bambaiah, Phmer, Jimp, Supasheep, Limulus, Salsb, RazorICE, Expensivehat, Dhollm, Lobwedge, Superiority, Tetracube, Willtron, GrinBot~enwiki, Asterion, Finell, AndrewWTaylor, SmackBot, RDBury, Tom Lougheed, Trojo~enwiki, Jrockley, Dauto, Chris the speller, Jjalexand, Complexica, Sbharris, Colonies Chris, Richard001, TTE, Spiritia, Titus III, FrozenMan, Newone, Happy-melon, Conrad.Irwin, Rowellcf, Chrisahn, Cydebot, Danny Bierek, Mtpaley, Wannabe Runny, Irigi, Headbomb, Niduzzi, KP Botany, Tlabshier, Hanzoro5, Dougher, Steelpillow, JAnDbot, Supertheman, Roleplayer, Magioladitis, Kopovoi, Vssun, Hoverfish, Khalid Mahmood, Pan Dan, Robin S, Vortimer, Sujaybhu, Natsirtguy, Peter Chastain, Cpiral, Tygrrr, Treisijs, Alpvax, VolkovBot, JohnBlackburne, TXiKiBoT, Marskuzz, Muro de Aguas, Qxz, Sintaku, SieBot, Gerakibot, Escape Artist Swyer, Proton666, ObfuscatePenguin, ClueBot, GorillaWarfare, Jackey0105, DragonBot, Jefflayman, Nownownow, Cenarium, Razorflame, Zahnrad, Silvercromagnon, InternetMeme, Rreagan007, WikHead, Drogs630, Addbot, Guoguo12, Omega Squad, ThisIsMyWikipediaName, Seratna, CarsracBot, Purple Emu, CosmiCarl, AgadaUrbanit, Tide rolls, Lightbot, Luckas-bot, Timeroot, Donthedev, Rifter0x0000, Umnum, AnomieBOT, VanishedUser sdu9aya9fasdsopa, Orange Knight of

9.1. TEXT

Passion, Piano non troppo, Citation bot, Obersachsebot, Xqbot, DSisyphBot, Barelistido, Almabot, GrouchoBot, RibotBOT, SassoBot, Mnmngb, CES1596, Gummer85, Citation bot 1, Boulaur, RedBot, Jauhienij, Surf5270, ElPeste, Slon02, EmausBot, John of Reading, Mnkyman, JSquish, Cogiati, Bamyers99, Rexprimoris, Donner60, ChuispastonBot, ClueBot NG, Jj1236, Helpful Pixie Bot, Bolatbek, ElphiBot, J.wong.wiki, Glevum, Zedtwitz, Zedshort, Nishantkumar19, Kisokj, YFdyh-bot, Andyhowlett, Reatlas, CsDix, EvergreenFir, Aurelianjh, Jwratner1, Diggerh, Kshitizarora2993, Tetra quark, KasparBot and Anonymous: 171

- **Weak interaction** *Source:* https://en.wikipedia.org/wiki/Weak_interaction?oldid=679571094 *Contributors:* AxelBoldt, Chenyu, Sodium, Bryan Derksen, Tarquin, AstroNomer~enwiki, Andre Engels, XJaM, Heron, JohnOwens, Gdarin, Delirium, Andrewa, Andres, Emperorbma, Timwi, Fibonacci, Phys, Phil Boswell, Lowellian, Mayooranathan, Tobias Bergemann, Giftlite, Sj, Herbee, Xerxes314, Jcobb, Mckaysalisbury, Munkee, Toby Woodwark, Bbbl67, Icairns, AmarChandra, Lumidek, Jørgen Friis Bak, Discospinster, ArnoldReinhold, Roybb95~enwiki, Gianluigi, Joanjoc~enwiki, Shanes, AJP, AtomicDragon, Danski14, Alansohn, Arthena, Axl, SidneySM, Hwefhasvs, DV8 2XL, Nightstallion, Kazvorpal, Linas, StradivariusTV, Benbest, Bbatsell, Palica, Tevatron~enwiki, Graham87, BD2412, Ketiltrout, Rjwilmsi, Strait, Erkcan, The wub, FlaBot, Naraht, Itinerant1, Srleffler, Chobot, Krishnavedala, YurikBot, Borgx, Bambaiah, Hairy Dude, Jimp, Sillybilly, Conscious, Epolk, JabberWok, Gaius Cornelius, Shaddack, SCZenz, Irishguy, Shimei, Willtron, RG2, Phr en, That Guy, From That Show!, Luk, SmackBot, David Kernow, Tom Lougheed, WookieInHeat, Dauto, Chris the speller, Philosopher, Moshe Constantine Hassan Al-Silverburg, Complexica, DHN-bot~enwiki, Zirconscot, BIL, Wen D House, "alyosha", Maxwahrhaftig, Akriasas, Vina-iwbot~enwiki, Bdushaw, TTE, SashatoBot, Fontenello, Herr apa, Condem, Tony Fox, MottyGlix, JRSpriggs, Heartofgoldfish, Calmargulis, Green caterpillar, Joelholdsworth, Cydebot, Michael C Price, Mtpaley, Thijs!bot, ChKa, Kichwa Tembo, Headbomb, Hcobb, Icep, Escarbot, AntiVandalBot, Jimeree, Steelpillow, JAnDbot, Magioladitis, Swpb, مباس, Wormcast, DAGwyn, Giggy, Khalid Mahmood, Gah4, Tarotcards, 2help, Lighted Match, DorganBot, Halmstad, Idiomabot, VolkovBot, Jcuadros, Hilarious Bookbinder, TXiKiBoT, Rei-bot, CaptinJohn, Awl, Shenanegins, BotKung, Wingedsubmariner, Antixt, Xxxlilbritxxx, Ptrslv72, Monty845, AlleborgoBot, SieBot, Paolo.dL, Skyentist, Ptr123, ClueBot, Bondchic007, SuperHamster, Erudecorp, Rotational, Jackey0105, Alexbot, Cenarium, Zomno, Zahnrad, He6kd, TimothyRias, InternetMeme, Timo Metzemakers, Stephen Poppitt, Addbot, Some jerk on the Internet, Markdman, ChenzwBot, Ehrenkater, Tide rolls, Luckas-bot, Yobot, Les boys, Kilom691, THEN WHO WAS PHONE?, Rifter0x0000, Duping Man, Dickdock, Magog the Ogre, AnomieBOT, Materialscientist, Citation bot, Quebec99, Kreigiron, Xqbot, Drilnoth, BurntSynapse, GrouchoBot, Omnipaedista, RibotBOT, Workanode, Jaz1305, Mnmngb, Dave3457, FrescoBot, Charles.walker, LucienBOT, Ionutzmovie, Grandiose, Pinethicket, Boulaur, Rameshngbot, RedBot, 23790AD, Tea with toast, Jauhienij, FoxBot, Earthandmoon, RjwilmsiBot, Itamarhason, Newty23125, EmausBot, WikitanvirBot, GA bot, GoingBatty, Splibubay, StringTheory11, Braswiki, Git2010, Wayne Slam, Jsayre64, Maschen, ChuispastonBot, ClueBot NG, VinculumMan, Physics is all gnomes, Fjpyanez, Mouse20080706, Helpful Pixie Bot, Geo7777, Bibcode Bot, Junaid2754, Bolatbek, Phbarnacle, Neutral current, Glevum, Idenshi, Marioedesouza, Dexbot, Spray787, Reatlas, CsDix, Jamesmcmahon0, Ihatedirac2k13, Kharkiv07, Jwratner1, YimmyYohnson, Monkbot, BalderdashVonDrivel, ASCarretero, Malerisch, Lachlan Newland, Tetra quark, KasparBot and Anonymous: 155

- **List of particles** *Source:* https://en.wikipedia.org/wiki/List_of_particles?oldid=682746251 *Contributors:* AxelBoldt, Danny, Rmhermen, Stevertigo, Bdesham, Ahoerstemeier, Stan Shebs, Docu, Salsa Shark, Nikai, Evercat, Schneelocke, Charles Matthews, Jitse Niesen, CBDunkerson, Bevo, Raul654, Donarreiskoffer, Robbot, Sanders muc, Merovingian, Pengo, Giftlite, Herbee, Xerxes314, Dratman, Jeremy Henty, Alensha, Bodhitha, Physicist, Hayne, Quadell, RetiredUser2, Mysidia, Icairns, Asbestos, D6, Urvabara, Discospinster, Rich Farmbrough, FT2, Qutezuce, ArnoldReinhold, Neko-chan, El C, Laurascudder, Susvolans, EmilJ, Physicistjedi, Minghong, Gbrandt, Eddideigel, Axl, Mac Davis, David Ko, Radical Mallard, RJFJR, Count Iblis, Dirac1933, TenOfAllTrades, LFaraone, Oleg Alexandrov, Linas, JarlaxleArtemis, Duncan.france, GregorB, Cedrus-Libani, Karam.Anthony.K, Palica, Rjwilmsi, Zbxgscqf, JLM~enwiki, Strait, Ems57fcva, Krash, Dan Guan, DannyWilde, Lmatt, Goudzovski, Chobot, YurikBot, Bambaiah, Vuvar1, Madkayaker, Hydrargyrum, Presscorr, Chaos, Salsb, Tavilis, SCZenz, Lexicon, TUSHANT JHA, Dna-webmaster, Tomvds, Poulpy, Cstmoore, TLSuda, NeilN, MacsBug, Tom Lougheed, McGeddon, Bazza 7, WookieInHeat, Derdeib, Yamaguchi 先生, Betacommand, Bluebot, Master of Puppets, DHN-bot~enwiki, Raistuumum, Juancnuno, Kittybrewster, Acepectif, Ligulembot, TriTertButoxy, ArglebargleIV, Khazar, John, FrozenMan, JorisvS, 041744, Dr Greg, Slakr, Mets501, Scorpion0422, Cbuckley, Iridescent, TwistOfCain, Happy-melon, JRSpriggs, Flickboy, Van helsing, Lithium6, Neelix, Rotiro, Cydebot, Quibik, Christian75, Omicronpersei8, Thijs!bot, Qwyrxian, TauLibrus, Headbomb, Inner Earth, 49, Guptasuneet, Scottmsg, WinBot, Elmoosecapitan, Tyco.skinner, AubreyEllenShomo, Arch dude, Johnman239, Mwarren us, TheEditrix2, CalamusFortis, MartinBot, Sadisticsuburbanite, Bissinger, Anaxial, CommonsDelinker, Maurice Carbonaro, Zojj, OliverHarris, Joshmt, Adanadhel, Lseixas, Graphite Elbow, VolkovBot, Jmrowland, Quilbert, Anonymous Dissident, Dstary, Escalona, JPMasseo, Figureskatingfan, Inx272, Meters, Antixt, Hamish a e fowler, GoddersUK, Bluetryst, SieBot, Ishvara7, WereSpielChequers, Audrius u, VovanA, Paolo.dL, RSStockdale, Anchor Link Bot, StewartMH, Explicit, ClueBot, Unbuttered Parsnip, Nolimitownass, DragonBot, Atomic7732, TimothyRias, SkyLined, Addbot, DOI bot, Jojhutton, Favonian, LinkFA-Bot, OlEnglish, Teles, Legobot, Luckas-bot, Yobot, Dov Henis, Azcolvin429, AnomieBOT, Götz, Icalanise, Flewis, Materialscientist, OllieFury, Vuerqex, ArthurBot, Vulcan Hephaestus, Blennow, Reality006, Coretheapple, Jcimorra, RibotBOT, Ernsts, A. di M., Axelfoley12, Zosterops, FrescoBot, Paine Ellsworth, Citation bot 1, JIK1975, Tom.Reding, Diffequa, WikitanvirBot, Racerx11, 112358sam, Aegnor.erar, Hops Splurt, HESUPERMAN, Hhhippo, AvicBot, JSquish, StringTheory11, Waperkins, Bamyers99, Suslindisambiguator, L Kensington, DennisIsMe, RockMagnetist, ClueBot NG, Snotbot, Primergrey, Vio45lin, Widr, MsFionnuala, Oklahoma3477, Bibcode Bot, CityOfSilver, Cap'n G, BML0309, Dan653, Twocount, Penguinstorm300, Dexbot, LightandDark2000, Ohiggy, TwoTwoHello, Andyhowlett, Printersmoke, Orion 2013, ARUNEEK, Seino van Breugel, AspaasBekkelund, TheMagikCow, Vyom27, ParkersComments, Selva Ganapathy and Anonymous: 290

- **Timeline of particle discoveries** *Source:* https://en.wikipedia.org/wiki/Timeline_of_particle_discoveries?oldid=679101637 *Contributors:* Rmhermen, Tempshill, Harp, Xerxes314, Bodhitha, Perey, Discospinster, Cmdrjameson, JohnAlbertRigali, Crosbiesmith, Rjwilmsi, Strait, Bubba73, Goudzovski, David H Braun (1964), Yamara, SCZenz, Tony1, GrinBot~enwiki, Attilios, SmackBot, Onionmon, Dl2000, JeffW, Lottamiata, Newone, Jhlawr, Headbomb, D.H, Pkoppenb, JNW, Joshmt, Chronitis, SieBot, A. Carty, BartekChom, Muhends, Wprlh, Addbot, DOI bot, MizzoulaB, Luckas-bot, Icalanise, Pepo13, Citation bot, Gogiva, Xqbot, Omnipaedista, Ulm, Carlog3, Citation bot 1, Wbm1058, Bibcode Bot, Dexbot, Trinitresque, Makecat-bot, JanJaeken, ElŞahin, Revolution1221, Monkbot and Anonymous: 24

- **James Bjorken** *Source:* https://en.wikipedia.org/wiki/James_Bjorken?oldid=665810432 *Contributors:* Timrollpickering, Giftlite, Xerxes314, Mako098765, Ukexpat, Klemen Kocjancic, David Schaich, Snowolf, Etacar11, Fram, SmackBot, GaeusOctavius, Pokerpoodle, Skinnyweed, John, Wtwilson3, Sir Nicholas de Mimsy-Porpington, CapitalR, Jarszick, Cydebot, Gonzo fan2007, Thijs!bot, Headbomb, Waacstats, Cuzkatzimhut, Mozak50, Qwfp, Kbdankbot, Addbot, Tncnv, Luckas-bot, Yobot, Lildyson314, Das Kollektiv, Citation bot, Omnipaedista, Sophus Bie, Citation bot 1, Foobarnix, Badger M., RjwilmsiBot, Suslindisambiguator, Bibcode Bot, Brad7777, PCbot, VIAFbot, Gnoldog, Jonarnold1985, KasparBot and Anonymous: 15

- **Nicola Cabibbo** *Source:* https://en.wikipedia.org/wiki/Nicola_Cabibbo?oldid=678181121 *Contributors:* Stone, Timrollpickering, Herbee, CALR, Army1987, ADM, Reaverdrop, Rjwilmsi, Chobot, RussBot, Cinik, Attilios, Mitchan, Stepa, Pitam, Hmains, Ephraim33, Ser Amantio di Nicolao, Jetman, Cydebot, Synergy, Thijs!bot, Headbomb, Escarbot, .anacondabot, Magioladitis, Bongomatic, Cuzkatzimhut, VolkovBot, Naohiro19 revertvandal, Duncan.Hull, Phe-bot, Arjen Dijksman, Sean.hoyland, RS1900, Lord Horatio Nelson, Djr32, Alexbot, Flavus, RogDel, CrackerJack7891, Deineka, Addbot, Arzewski, SpBot, PV=nRT, Zorrobot, Yobot, Piano non troppo, Citation bot, DirlBot, Xqbot, Howard McCay, CES1596, Imbalzanog, DrilBot, Foobarnix, SchreyP, The grey side, RjwilmsiBot, Molimaging, Suslindisambiguator, JeanneMish, ClueBot NG, 林佳臻, Bibcode Bot, Stephenwanjau, Brad7777, BattyBot, Ymytm, Dexbot, VIAFbot, Raoul Bertorello, Monkbot, Jonarnold1985, Miguel Sapienza and Anonymous: 28

- **Richard Feynman** *Source:* https://en.wikipedia.org/wiki/Richard_Feynman?oldid=682548616 *Contributors:* Tobias Hoevekamp, Magnus Manske, CYD, Mav, MarXidad, Gareth Owen, Ed Poor, RK, Andre Engels, XJaM, MadSurgeon, Little_guru, Peterlin~enwiki, Ben-Zin~enwiki, Ajdecon, Mswake, Heron, Zippy, KYSoh, Olivier, Stevertigo, Svenbor, RTC, Infrogmation, JohnOwens, Michael Hardy, Zocky, Isomorphic, Norm, Gabbe, Gaurav, Zanimum, Dcljr, IZAK, Skysmith, Paul A, Minesweeper, Looxix~enwiki, Mkweise, Mortene, Ahoerstemeier, DavidWBrooks, Samuelsen, Angela, LittleDan, Jacquez, Julesd, Ffx, Sugarfish, Ciphergoth, Nikai, Susurrus, Cimon Avaro, Jiang, EdH, Lukobe, Jod, Skyfaller, Smack, Etaoin, Hike395, Agtx, Redjar, Timwi, Incandenza, Ww, Doradus, Zoicon5, Tpbradbury, Maximus Rex, Furrykef, Grendelkhan, Saltine, Whcernan, Taxman, Phys, Thue, Bevo, Fvw, Raul654, Johnleemk, David.Monniaux, Mjmcb1, Lumos3, Nufy8, Robbot, Paranoid, Tlockney, Pigsonthewing, Fredrik, Ray Radlein, Zandperl, Chris 73, Matt me, Sanders muc, Goethean, Psychonaut, Postdlf, MathMartin, Dmadeo, Clngre, Goofyheadedpunk, Blainster, Timrollpickering, DHN, Rebrane, Intangir, SC, Saforrest, Wereon, MOiRe, Ianml, Pps, Lupo, Mattflaschen, David Gerard, Enochlau, Mor~enwiki, Hexii, Ancheta Wis, Connelly, Giftlite, DocWatson42, Jyril, Paisa, Jonth, Wolfkeeper, Whtknt, Tom harrison, Boojit, Fastfission, HangingCurve, Peruvianllama, No Guru, Anville, Curps, Henry Flower, Gamaliel, Nsh, Cantus, Daniel Brockman, Wikisux, Siroxo, Blizzarex, Eequor, Taak, Christofurio, Darrien, Rjyanco, Mckaysalisbury, Mmm~enwiki, Lawrennd, K7jeb, LordSimonofShropshire, Pgan002, Weasel75, Alexf, Metlin, SarekOfVulcan, LiDaobing, Abu badali, Danko Georgiev, Quadell, Ran, Fangz, Lightst, Antandrus, Mako098765, Robert Brockway, MisfitToys, PDH, Rdsmith4, Xtreambar, Anythingyouwant, DragonflySixtyseven, Johnflux, Reagle, Tomruen, Joyous!, ArthurDenture, Vsb, Benzh~enwiki, Klemen Kocjancic, 朝彦, Hax0rw4ng, Safety Cap, Thorwald, Gazpacho, Smiller933, Lucidish, Ornil, D6, TheBlueWizard, Jayjg, CALR, Shipmaster, JimJast, Discospinster, Marsvin, Rich Farmbrough, Guanabot, ObsessiveMathsFreak, Liwanshan, FT2, Cacycle, Pjacobi, Rama, Vsmith, Pie4all88, Andrewferrier, Ponder, Altmany, Paul August, Bender235, Djordjes, Janderk, Kjoonlee, Brian0918, Pmcm, CanisRufus, Livajo, Chvsanchez, Julius.kusuma, Kwamikagami, Cafzal, Chairboy, Shanes, Susvolans, RoyBoy, Palm dogg, Gershwinrb, Bobo192, Thisuser, Army1987, Stesmo, Duk, Viriditas, MITalum, L33tminion, I9Q79oL78KiL0QTFHgyc, La goutte de pluie, Matt McIrvin, Jojit fb, Nk, Darwinek, Alphax, Physicistjedi, Wikinaut, Rje, Lilymaiden, Hooperbloob, Jumbuck, Schissel, Rernst, Alansohn, ChristopherWillis, Keenan Pepper, Walkerma, Stack, Theodore Kloba, A Kit, DreamGuy, Brinkost, Samohyl Jan, Wtmitchell, Dhartung, Bbsrock, Cburnett, Evil Monkey, Arenhaus, Dirac1933, CorwinLe CLOGIC, Computerjoe, GabrielF, Reaverdrop, Alai, Eric Herboso, DSatz, Nick Mks, Ceyockey, Tintin1107, Oleg Alexandrov, Ashujo, Crosbiesmith, Feezo, Richard Arthur Norton (1958-), Jasonm, Woohookitty, Justinlebar, LOL, Ekem, Robert K S, JFG, Kosher Fan, Ruud Koot, Psiphim6, DavidMendoza, Lkjhgfdsa, Orz, Trödel, Mpatel, GregorB, CharlesC, Wayward, HiFiGuy, Emerson7, LexCorp, Jaia, RichardWeiss, Graham87, BD2412, Kbdank71, Melesse, Lord.lucan, Ryan Norton, Altman, Rjwilmsi, Nightscream, Koavf, Zbxgscqf, SeeFood, War, Binary, HonoluluMan, Amire80, Quiddity, Dennis Estenson II, JHMM13, Bruce1ee, Salix alba, MZMcBride, SpNeo, Chris Purves, Stilgar135, HappyCamper, Sohmc, Niffe, Dudegalea, Ghepeu, Keimzelle, Williamborg, GregAsche, Kasparov, Rangek, Titoxd, FlaBot, Moskvax, Ian Pitchford, SchuminWeb, RobertG, Doc glasgow, Mathbot, Nihiltres, RachelBrown, Wars, DevastatorIIC, Fosnez, DannyDaWriter, Kapitolini, Srleffler, Kri, DJsunkid, Valentinian, Danielfong, Chobot, Jaraalbe, DVdm, Mhking, Bgwhite, Manscher, Dj Capricorn, Shervinafshar, YurikBot, RobotE, Vecter, MJustice, RussBot, FrenchIsAwesome, Runningmouse, Bhny, Splash, Bhumburg, WLGades, Azucar~enwiki, Gaius Cornelius, Philopedia, Anomalocaris, Hawkeye7, Badagnani, Rjensen, AndyHedges, JocK, Cleared as filed, Equilibrial, Brandon, Krakatoa, PhilipO, Scs, Tony1, Gmosaki, EEMIV, Vlad, Dr. Ebola, Bota47, IceWeasel, DRosenbach, Mholland, FF2010, Zargulon, Ario, MCB, Ninly, Shimei, TheMadBaron, The Fish, Theda, Rb82, WHRupp, Eezbub, Petri Krohn, Tevildo, GraemeL, CWenger, LeonardoRob0t, Back ache, QmunkE, VodkaJazz, Rearden9, Whobot, T. Anthony, Altsarc, Curpsbot-unicodify, Garion96, Kaicarver, MagneticFlux, David Biddulph, Banus, Rwwww, Jentizzle, Vulturell, Sardanaphalus, Veinor, A13ean, A bit iffy, SmackBot, YellowMonkey, Fireworks, TheBilly, Liwyatan, Roger Hui, 1dragon, JohnRussell, InverseHypercube, CRKingston, C.Fred, Batman Jr., AndyZ, HeWhoE, Prateek.agrawal, ScaldingHotSoup, Zyxw, Gabrielleitao, Timotheus Canens, UrsaFoot, GaeusOctavius, Dbnull, Jcrobert, Commander Keane bot, Eclecticerudite, Hmains, The Gnome, Amatulic, Fintler, RDBrown, Iain.dalton, Thumperward, Lollerskates, Happylobster, Malenien, Papa November, Stevage, Kevin Ryde, Bbq332, DHN-bot~enwiki, Sbharris, Danilsuits, Konstable, Oatmeal batman, AdamSmithee, Hgrosser, Modest Genius, Can't sleep, clown will eat me, Ww2censor, Bowin~enwiki, Malor, WestA, Ourhomeplanet, Cybercobra, Downwards, Gylgamesh, Marc-André Aßbrock, G716, Derek R Bullamore, Jbergquist, DMacks, Jaedglass, Salamurai, Steiger, Sadi Carnot, Ck lostsword, Bejnar, Pilotguy, Qmwne235, Blahm, SashatoBot, Eliyak, Michael Romanov, Harryboyles, John, Writtenonsand, Slowmover, Agencius, Loodog, Gobonobo, TauntingElf, Clare., Devan Miller, Tktktk, Edwy, Jodamn, Profjim, Terry Bollinger, Mgiganteus1, Mary Read, XinJeisan, IronGargoyle, Jollyroger, Alpha Omicron, Syrcatbot, Camilo Sanchez, Shimmera, WhiteHatLurker, Fishbowlbob, Wikidrone, Rglovejoy, Dicklyon, Hypnosifl, Shinryuu, Doczilla, Killua, Flipperinu, Lee Carre, Hu12, Dkwong323, Ramuman, BranStark, Iridescent, Clarityfiend, Skapur, Sander Säde, Benplowman, PetiteFadette, LethargicParasite, Cain47, Courcelles, Wikidude1, Hauberg, Codelieb, Doceddi, Gifuoh, RSido, Paulmlieberman, Altales Teriadem, CmdrObot, Mattbr, Van helsing, Thinkdunson, Marcelo Pinto, Sjoerd22, ShelfSkewed, Chicheley, Myasuda, Fletcher, Kd5npf, Phenylfairy, Shanoman, HalJor, Cydebot, Peripitus, MC10, Jack O'Lantern, Clayoquot, JFreeman, MWaller, A Softer Answer, David edwards, Teratornis, Lbertybell, Quartic, LaGrange, Daveross26, David from Downunder, Headbomb, Marek69, Tellyaddict, Cj67, Bunzil, Oneshotaccount, Heroeswithmetaphors, MichaelMaggs, Roponor, Escarbot, KrakatoaKatie, AntiVandalBot, RobotG, Luna Santin, Kitty Davis, Yomangani, Petrsw, Tjmayerinsf, Danger, Mikevans, LegitimateAndEvenCompelling, Bigjimr, JAnDbot, Deflective, Davewho2, MER-C, CosineKitty, Epeefleche, Avaya1, IanOsgood, Benzzene, Workaphobia, Krazykenny, Delius1967, Rothorpe, .anacondabot, WolfmanSF, Bdalevin, Xb2u7Zjzc32, JamesBWatson, Twsx, Hekerui, Cgingold, Diggernet, LorenzoB, BagelCarr, Duendeverde, Philg88, Tomgreeny, Apdevries, Gwern, Abebenjoe, Euneirophrenia, Felvalen, Peltio, CarlFeynman, Nichlok, Gatos, CommonsDelinker, PapalAuthoritah, Slash, J.delanoy, Trusilver, RSRScrooge, Nbauman, SureFire, Super Girl, Snowfalcon cu, Tikiwonto, Choihei, Vanished user 342562, Madzyzome, Hair Commodore, Ndokos, Ycdkwm, Janus Shadowsong, Ilikerps, Ipigott, Sunidesus, Floaterfluss, DadaNeem, Ferahgo the Assassin, 83d40m, BostonRed, MatthewBurton, Doctorpete, Student7, Plindenbaum, Kenneth M Burke, Scewing, Sheliak, Devraj5000, Grvs22, Cuzkatzimhut, Deor, FloydRTurbo, Johan1298~enwiki, Maile66, JamesBHunt, Aesopos, Philip Trueman, Martinevans123, Jimmyeatskids, Technopat, Yowhatsshakin, Dj thegreat, Rei-bot, Creonlevit, Mathwhiz 29, Anonymous Dissident, GcSwRhIc, Starnold, Zimbardo Cookie Experiment, Ast4, Wordsmith, Slysplace, Geometry guy, Duncan.Hull, Katimawan2005,

Gilisa, Madhero88, Girona7, Brian Huffman, Gabrielsleitao, Lamro, Analogica, CoolKid1993, Falcon8765, Enviroboy, Spinningspark, The Devil's Advocate, Ninian1, StevenJohnston, Oortcloudy, EmxBot, Schlammer, Cornswalled, Stimmj, JulesN, SieBot, ToePeu.bot, Paradoctor, Dawn Bard, Alex Middleton, Strombomboli, Djdan4961, Seijihyouronka, Utternutter, Dredea, Abhishikt, RucasHost, Likebox, Flyer22, The Sunshine Man, Physikalk, Masgatotkaca, PolarBot, Arbor to SJ, Razereas, Razeres, Feynman245, AWeishaupt, Afernand74, MadmanBot, Randomblue, Tradereddy, ImperialismGo, Extensive~enwiki, StewartMH, Myrvin, ImageRemovalBot, GorillaWarfare, WurmWoode, All Hallow's Wraith, QueenAdelaide, MikeVitale, RODERICKMOLASAR, Ct4ul4u, Nsk92, Pertin, Joao Xavier, Msgarrett, Switchcraft, Kannie, Masterpiece2000, DragonBot, Canis Lupus, Alexbot, John Nevard, Gulmammad, Jotterbot, Ngebendi, TYie34, Brianbjparker, CMW275, Cardinalem, Timshiels, Johnuniq, Darkicebot, XLinkBot, MessinaRagazza, Shadow600, TheStrongForce, WikHead, Good Olfactory, Kbdankbot, RandomTool2, Calvinash, Tayste, Addbot, Brumski, Mortense, Dgroseth, Neodop, Yobmod, Farquezy, Qediagrams, Download, Debresser, CUSENZA Mario, Dr. Universe, LinkFA-Bot, Numbo3-bot, Tide rolls, Steak, Legobot, Luckas-bot, Yobot, Pink!Teen, Bunnyhop11, Donfbreed, NLWASTI, Mmxx, Sms dpd, AnomieBOT, DoctorJoeE, Götz, Steamturn, Kingpin13, Materialscientist, FudicialPoint, Citation bot, Mcoogan75, Xqbot, Werthless5, Tripodian, Sionus, Ekwos, St.nerol, 5464536, Nasnema, Davshul, Etherealstill, Ramakv, Anna Frodesiak, Srich32977, Mrtraska, Omnipaedista, BulldogBeing, Dannykaye85, Siddharth9200, Amaury, Johnofjack, Canned Soul, Bigbird77, A. di M., Erik9, A.amitkumar, Swealtsje, Fingerz, LucienBOT, BlasingL4, Mxipp, Zenecan, Khakistani, PorkoltLover60, Fusion156, JohnnyGerms, Radiohead40540057, Citation bot 1, Anibar E, Plucas58, Skyerise, A8UDI, Fat&Happy, Olibroman, Primaler, Jaytip, DmitryVinokurov, Corinne68, Jmorykin, Sciencemeds, Jordgette, Jonkerz, Tova Hella, Thestgman, Pkos99, Reaper Eternal, Everyone Dies In the End, Canuckian89, Reach Out to the Truth, Minimac, Atreklin, TjBot, Beyond My Ken, Javaweb, Massieu, Ccrazymann, Androstachys, DASH-Bot, EmausBot, John of Reading, Sprout333, Bazookafox1, Jim Michael, Kkm010, Ida Shaw, Illegitimate Barrister, Col. Sweeto, WeijiBaike-Bianji, Finemann, KnownLoop, H3llBot, Suslindisambiguator, Baruta07, Piantanida31, Cf. Hay, RaqiwasSushi, OpenlibraryBot, YnnusOiramo, Grams46, Bill william compton, ChuispastonBot, AndyTheGrump, GP modernus, Kinkreet, Petrb, Kelainoss, Xanchester, ClueBot NG, Kkddkkdd, Nancyds, Sjjones85, Science&HiTechReviewer, Rulew, Clearlyfakeusername, Chriscook54321, HazelAB, Widr, Darrenwershler, Helpful Pixie Bot, MiraS21, Accedie, Bibcode Bot, Lowercase sigmabot, HydrusGemini, BG19bot, Secarrothers, ServiceAT, Imseakin, Blake Burba, CatPath, MusikAnimal, B87lar, Exercisephys, Robert the Devil, Teika kazura, Brad7777, Glacialfox, Scotstarvit, Fylbecatulous, BattyBot, Fedora96, The Wish List, Contributor1969104, Richardfeynmanslefttesticle, Ninmacer20, NATPsbs, Theo Buckley, Khazar2, Wiwi Samsul, Hvakshahtrah, Eoxenford, Themoother, Dexbot, Pganju, Inayity, Konbini, Mattermenot, Churn and change, Graphium, Jamesx12345, Augustus Leonhardus Cartesius, Kankerbakker, Corinne, Limit-theorem, Nicojonesgodel, Rfassbind, FlutteringCarp, Epicgenius, FreeCorrect-dot-com, Airborntroll, Wiki.correct.1, Jrgauthier, Tentinator, Pishwiki, Aqwsdezz, Fwood2, EewayneSmithe, DaltonBantz, Hitchjk, Monochrome Monitor, Spyglasses, Zenibus, CurlyLoop, Ginsuloft, Cockatoo123, Luxure, AndreSalles, Dide101, Lurgey, J.arnob, Lordbowtie, Lagoset, Monkbot, Samama191, HiYahhFriend, Redlink Ranger, RVadder, Srenmus, HouseOfChange, Gronk Oz, Tf2manu994, Garfield Garfield, InfoDataMonger, Heuh0, Parisstreatham, Mark Fall, Joseph2302, Jamesmoriartyandsherlock, Hellosirmynameisbob, Sleepy Geek, UncommonInevitable, Wright1234567890, Indigo1HalJordanThaalSinestroAtrocitusCarolFerrisAndMore, 3027D, NilubonT, HammerHead67, Kthibault, StrikerTex, PrinceMilliband, Brominator, The pack king, IrepresentBob, GrantJPeterson and Anonymous: 882

- **Murray Gell-Mann** *Source:* https://en.wikipedia.org/wiki/Murray_Gell-Mann?oldid=681196340 *Contributors:* The Cunctator, Mav, Seb, XJaM, Rsabbatini, Dante Alighieri, Mic, Looxix~enwiki, Ahoerstemeier, RodC, Phr, Maximus Rex, SHeumann, Bevo, Joseaperez, Drxenocide, Fredrik, Timrollpickering, Andrew Levine, Bkell, JesseW, Wikibot, Giftlite, Harp, Mintleaf~enwiki, Marcika, Xerxes314, Curps, Duncharris, Henryhartley, Beardo, Jason Quinn, Edrex, Neilc, Danko Georgiev, Antandrus, Beland, Joi, PDH, Balcer, Hyperneural, Tomruen, Lumidek, Hkpawn~enwiki, Tsemii, Klemen Kocjancic, Calwatch, Chris Howard, D6, Jayjg, Rich Farmbrough, Guanabot, Pjacobi, Aris Katsaris, Srbauer, Kwamikagami, Orlady, Tgeller, Bobo192, Mike Schwartz, Duk, Viriditas, MITalum, CoolGuy, Mdd, Ahruman, Sligocki, A Kit, Snowolf, Ksnow, Dirac1933, Papovik, 霧木諒二, Ghirlandajo, TheCoffee, April Arcus, Tintin1107, Etacar11, BillC, Mat813, Emerson7, Behack, Terryn3, Kbdank71, Angusmclellan, Koavf, Sorenr, Lockley, MZMcBride, SpNeo, Brighterorange, Amelio Vázquez, FlaBot, Wikiliki, Mujib, DannyWilde, Gurch, Srleffler, OpenToppedBus, Valentinian, Mstroeck, Chobot, YurikBot, TexasAndroid, Bambaiah, Dnik, RussBot, Hwasungmars, GhostInTheShell, Mccready, Irishguy, DYLAN LENNON~enwiki, E2mb0t~enwiki, Nick123, 2over0, Arthur Rubin, Curpsbot-unicodify, Garion96, Johnpseudo, MagneticFlux, Qero, Schizobullet, SmackBot, Davepape, Roger Hui, CRKingston, Zyxw, Eskimbot, Ogdred, Chris the speller, RDBrown, EncMstr, Hibernian, Kevin Ryde, TheLeopard, Josteinn, VikSol, Newport, Pgoerner, Khukri, G716, Andrei Stroe, John, Ergative rlt, NYCJosh, Syrcatbot, MarcAurel, ThePI, Norm mit, Shaneq, Albertod4, KyleGardiner, Ale jrb, Andrew Delong, Myasuda, MaxEnt, Cydebot, Island Dave, Hebrides, MWaller, JodyB, Thijs!bot, Headbomb, JustAGal, Bunzil, CharlotteWebb, Glj1952, Escarbot, AntiVandalBot, RobotG, Kramden4700, Jrw@pobox.com, Deflective, Matthew Fennell, Xeno, Connormah, VoABot II, Newyears, Duendeverde, Gwern, Greg Salter, Arjun01, CommonsDelinker, FANSTARbot, Bongomatic, SureFire, SuperGirl, Vanished user 342562, Stan J Klimas, SimulacrumDP, DadaNeem, Master shepherd, Jamesontai, Thismightbezach, Cuzkatzimhut, VolkovBot, Rvilbig, Aesopos, TXiKi-BoT, EvanCarroll, Anonymous Dissident, GcSwRhIc, Starnold, JhsBot, Broadbot, Gilisa, CoolKid1993, MCTales, Spinningspark, Alcmaeonid, Resurgent insurgent, Ponyo, SieBot, D2lraXBlZGlh, Nathan, WRK, Likebox, Belinrahs, Lightmouse, LarRan, RS1900, Tyfrazier, ClueBot, Binksternet, All Hallow's Wraith, Dave.bradi, MikeVitale, Joao Xavier, Masterpiece2000, Alexbot, Terra Xin, Ammar.sajdi, Cardinalem, RogDel, MessinaRagazza, SilvonenBot, Good Olfactory, Kbdankbot, RandomTool2, Addbot, Kamuichikap, Numbo3-bot, Luckas-bot, Yobot, NLWASTI, TheWindyCity, Joan kingston, KamikazeBot, AnomieBOT, Adeliine, Ulric1313, Citation bot, Happyhuman, Davshul, RibotBOT, Mnmngb, Preobrazhenskiy, Ironboy11, Haeinous, DivineAlpha, Skyerise, A8UDI, Badger M., ItsZippy, Vrenator, OKleavemealonealready, Updatehelper, EmausBot, ZéroBot, Lemeza Kosugi, Suslindisambiguator, Fabian Hassler, Fitzrovia calling, JeanneMish, Orange Suede Sofa, ClueBot NG, FSchlaghecken, LJosil, Helpful Pixie Bot, Aaaeditor, Bibcode Bot, Allecher, Contact '97, MrBill3, Chucktesta, Brad7777, Tutelary, JoshuSasori, Ninmacer20, VIAFbot, IBhanap, DavidPKendal, Aloneinthewild, Arabba, Bibliophilen, Rkyuen, Jonarnold1985, NilubonT, Jerodlycett, KasparBot, Csumstudent and Anonymous: 177

- **Sheldon Lee Glashow** *Source:* https://en.wikipedia.org/wiki/Sheldon_Lee_Glashow?oldid=675778251 *Contributors:* Amillar, Mic, Cyde, Jcajacob, Maximus Rex, Pigsonthewing, Timrollpickering, JerryFriedman, Aomarks, Ancheta Wis, Giftlite, Harp, Everyking, Curps, Macrakis, Tonymaric, Lumidek, Cvalente, D6, Nicobn~enwiki, Aris Katsaris, Bender235, AmosWolfe, A Kit, Snowolf, Ksnow, Dirac1933, RyanGerbil10, Etacar11, Carcharoth, Emerson7, Kbdank71, MZMcBride, Yamamoto Ichiro, FlaBot, Bdolicki, Srleffler, Valentinian, Chobot, Anetode, Diotti, Melanchthon, Zargulon, Reyk, LeonardoRob0t, GrinBot~enwiki, SmackBot, Roger Hui, Lestrade, KocjoBot~enwiki, Stepa, Delldot, Eskimbot, GaeusOctavius, Mrmewe, Kuningas, RDBrown, Nixeagle, Mhym, Threeafterthree, G716, Andrei Stroe, Thomaspaine, John, Dicklyon, Eastfrisian, HelloAnnyong, Tawkerbot2, Dlohcierekim, Albertod4, HennessyC, Drinibot, Summonmaster13, Cydebot, Ntsimp, MWaller, DumbBOT, Jmg38, Headbomb, RobotG, Gcm, MER-C, Txomin, .anacondabot, Enoent, Duendeverde, Tvoz, R'n'B, Commons-

Delinker, Seanhud, Amikake3, Rvilbig, Kyle the bot, Epson291, HowardFrampton, TXiKiBoT, Jimmyeatskids, Anonymous Dissident, GcSwRhIc, BeIsKr, Leafyplant, Antixt, EmxBot, MarkRCarterPhD, Ponyo, SieBot, Toddst1, PolarBot, Wuhwuzdat, RS1900, All Hallow's Wraith, Nsk92, Joao Xavier, Masterpiece2000, Djr32, Cardinalem, DumZiBoT, Kbdankbot, Addbot, Jacopo Werther, GargoyleBot, Numbo3-bot, Lightbot, زرش, Skippy le Grand Gourou, Luckas-bot, Yobot, NLWASTI, Joan kingston, Materialscientist, ArthurBot, Davshul, Mnmngb, Shadowjams, LucienBOT, LaurenRueda, Fat&Happy, Full-date unlinking bot, W E Hill, TobeBot, TheMesquito, Beyond My Ken, Afteread, Groundandskyhuge, WikitanvirBot, Jenks24, Suslindisambiguator, JeanneMish, QuantumSquirrel, Kelainoss, ClueBot NG, Usctommytrojan, Helpful Pixie Bot, Allecher, Brad7777, Ninmacer20, Alexandergrutendik, VIAFbot, Avampace, Bibliophilen, Bettina Maris, KasparBot, Csumstudent and Anonymous: 68

- **Haim Harari** Source: https://en.wikipedia.org/wiki/Haim_Harari?oldid=662061881 Contributors: Ijon, Timrollpickering, Gershom, Giftlite, RScheiber, El C, GregorB, Rjwilmsi, Angusmclellan, SmackBot, Royalguard11, Betacommand, Chris the speller, Bluebot, Shuki, CmdrObot, Cydebot, Thijs!bot, Headbomb, Nthep, Waacstats, Moritherapy, Aemely, RS1900, Nudve, Cpt.schoener, TimothyRias, RogDel, Addbot, דוד55, Yobot, AnomieBOT, Citation bot, Historicist, Davshul, FrescoBot, RjwilmsiBot, Barak90, Suslindisambiguator, Helpful Pixie Bot, Furor Teutonicus, Bibcode Bot, Brad7777, VIAFbot, Monochrome Monitor, Checkthe, KasparBot and Anonymous: 16

- **John Iliopoulos** Source: https://en.wikipedia.org/wiki/John_Iliopoulos?oldid=668001606 Contributors: Timrollpickering, Giftlite, David Schaich, Daranz, Woohookitty, Gurch, Stormbay, Ephraim33, Ex nihil, Jarszick, Cydebot, Headbomb, Scottcmu, Waacstats, AlleborgoBot, BotMultichill, Amalgam82, Sean.hoyland, Rhododendrites, Addbot, Lightbot, Luckas-bot, Yobot, Citation bot, Xqbot, Omnipaedista, Merongb10, DefaultsortBot, RjwilmsiBot, Nivekin, Bibcode Bot, Brad7777, Whynancy, JonathanHMDavis, KasparBot and Anonymous: 8

- **Makoto Kobayashi (physicist)** Source: https://en.wikipedia.org/wiki/Makoto_Kobayashi_(physicist)?oldid=646534063 Contributors: Yearofthedragon, XJaM, Gabbe, Anobo, Timrollpickering, Giftlite, Fukumoto, Robert Weemeyer, Piotrus, Ranma9617, Gene Nygaard, Canadian Paul, Gryffindor, Goudzovski, Prolineserver, Wogsland, Kintetsubuffalo, Seann, V1adis1av, Andrei Stroe, John, PetaRZ, Tfccheng, Cydebot, CieloEstrellado, Thijs!bot, Headbomb, Shikeishu, Yellowdesk, TRBlom, Gcm, Pkoppenb, Moralist, Dekimasu, Waacstats, Sodabottle, Aboutmovies, Idioma-bot, Hugo999, VolkovBot, TXiKiBoT, Abtinb, Broadbot, Pamputt, Enkyo2, SieBot, Arjen Dijksman, Lightmouse, Correogsk, Sean.hoyland, Florentino floro, CristianCantoro, Joao Xavier, Namazu-tron, Djr32, Alexbot, Liberal Saudi, Cardinalem, DumZiBoT, Chanakal, Nick84, Addbot, AkhtaBot, Hapsala, Zorrobot, Luckas-bot, Yobot, Amirobot, Nallimbot, Sumivec, Materialscientist, 1234eatmoore, Xqbot, Davshul, I am Me true, RibotBOT, Mxipp, Cocu, GGT, WikitanvirBot, Molimaging, Suslindisambiguator, Mjbmrbot, Helpful Pixie Bot, Bibcode Bot, Brad7777, VIAFbot and Anonymous: 16

- **Leon M. Lederman** Source: https://en.wikipedia.org/wiki/Leon_M._Lederman?oldid=678049221 Contributors: Kpjas, RjLesch, The Cunctator, XJaM, Rmhermen, SimonP, Rsabbatini, Mic, Maximus Rex, Bevo, Dunhamrc, Bearcat, Xanzzibar, Giftlite, Tom harrison, Fastfission, Everyking, Curps, Wmahan, Isidore, PDH, Klemen Kocjancic, D6, Rich Farmbrough, Guanabot, FT2, Aris Katsaris, CanisRufus, El C, PhilHibbs, Ctrl build, Darwinek, Jumbuck, Ksnow, Dirac1933, MIT Trekkie, Japanese Searobin, Siafu, Chochopk, Tabletop, Zzyzx11, Emerson7, Jack Cox, Rjwilmsi, Koavf, Lockley, Bubba73, The wub, Pbhayani, FlaBot, Srleffler, Valentinian, CJLL Wright, Chobot, TexasAndroid, NTBot~enwiki, UDScott, Krakatoa, ShaiM, GrinBot~enwiki, Finell, Vulturell, SmackBot, Chris the speller, Zapiens, Nixeagle, Threeafterthree, AFP~enwiki, Andrei Stroe, SashatoBot, Ser Amantio di Nicolao, JzG, John, Loodog, John Cumbers, Simkiott, Dicklyon, PetaRZ, Blinking Spirit, HennessyC, Tnystran, Drinibot, Evan7257, FlyingToaster, Keithh, Cydebot, Headbomb, RobotG, JAnDbot, Deflective, Gcm, MER-C, Blaine Steinert, Sspillers, Magioladitis, Waacstats, Drdavidhill, Cgingold, David Eppstein, Parunach, Dkriegls, Johnpacklambert, Maurice Carbonaro, Ginsengbomb, Michael.holland, Plindenbaum, DorganBot, VolkovBot, TXiKiBoT, Jimmyeatskids, Bcody80, Gilisa, AlleborgoBot, VivaLeet, Ponyo, SieBot, Alex Middleton, PolarBot, Wuhwuzdat, DeirdreB, Highland14, RS1900, ClueBot, Binksternet, All Hallow's Wraith, Pertin, Joao Xavier, Jelly Beanie, Cirt, Masterpiece2000, Another Believer, Cardinalem, Rishi.bedi, XLinkBot, Zwinglisjubilee, Airplaneman, Hm29168, Addbot, Ronhjones, EconoPhysicist, Ginosbot, Numbo3-bot, Kar1897, OlEnglish, زرش, Legobot, Luckas-bot, NLWASTI, AnomieBOT, ArthurBot, Xqbot, Davshul, J JMesserly, Omnipaedista, Mnmngb, Shadowjams, FrescoBot, Steve Quinn, LaurenRueda, TRBP, TobeBot, Americandemeter, RjwilmsiBot, EmausBot, John of Reading, HiMyNameIsFrancesca, Outreachscientist, Werieth, Kkm010, DonToto, Ὁ οἶστρος, A930913, JeanneMish, Jonke20, Roanotto, ClueBot NG, Asalrifai, BG19bot, Allecher, Ĉiuĵaŭde, Ninmacer20, Cyberbot II, ChrisGualtieri, VIAFbot, Bibliophilen, Pinocchio3000, Arosariorivera, PADDTWILL007, KasparBot, Csumstudent and Anonymous: 80

- **Luciano Maiani** Source: https://en.wikipedia.org/wiki/Luciano_Maiani?oldid=679216568 Contributors: Magnus Manske, Kaihsu, Timrollpickering, Klemen Kocjancic, David Schaich, Goudzovski, Jaraalbe, RussBot, Closedmouth, Lserni, SmackBot, Wogsland, Ph7five, Khukri, Cydebot, Headbomb, Okki, Waacstats, Hans Dunkelberg, TXiKiBoT, Hqb, Azukimonaka, PolarBot, Sean.hoyland, StewartMH, RS1900, DumZiBoT, Addbot, Lightbot, Yobot, Bbb23, Ironboy11, Merongb10, Badger M., RjwilmsiBot, Baroc, Primergrey, BG19bot, Brad7777, Dexbot, Bibliophilen, Nexusfirs, KasparBot, Csumstudent and Anonymous: 18

- **Toshihide Maskawa** Source: https://en.wikipedia.org/wiki/Toshihide_Maskawa?oldid=659961751 Contributors: XJaM, Rmhermen, Takuya-Murata, Anobo, BenRG, Timrollpickering, Giftlite, Aphaia, Piotrus, David Schaich, Gene Nygaard, Rjwilmsi, Koavf, Chobot, Carl Daniels, Prolineserver, Wogsland, Ephraim33, Andrei Stroe, John, PetaRZ, Tfccheng, Picaroon, Cydebot, CieloEstrellado, Thijs!bot, Headbomb, Yellowdesk, JAnDbot, Gcm, Dekimasu, Waacstats, Nightshadow28, Aboutmovies, Idioma-bot, VolkovBot, TXiKiBoT, Broadbot, Jarke, Pamputt, Enkyo2, SieBot, Oda Mari, Lightmouse, Calle Widmann, Sean.hoyland, Florentino floro, CristianCantoro, Joao Xavier, Rotational, Namazu-tron, Djr32, Alexbot, Tomeasy, Liberal Saudi, Cardinalem, DumZiBoT, Chanakal, SilvonenBot, Vegas949, Nick84, Addbot, Hapsala, Zorrobot, Doraemonplus, Luckas-bot, Yobot, Amirobot, Ciphers, Sumivec, Materialscientist, ArthurBot, LemonairePaides, Davshul, CES1596, Seibun, RedBot, TobeBot, RjwilmsiBot, WikitanvirBot, Suslindisambiguator, ChuispastonBot, ClueBot NG, Bibcode Bot, Brad7777, VIAFbot, KasparBot and Anonymous: 17

- **Yuval Ne'eman** Source: https://en.wikipedia.org/wiki/Yuval_Ne'eman?oldid=678049640 Contributors: Tobias Conradi, Charles Matthews, Zero0000, Gidonb, Xerxes314, Gamaliel, Woggly, Charm, Remuel, Haham hanuka, SHIMONSHA, Ynhockey, Jheald, Dirac1933, Drbreznjev, Japanese Searobin, Angr, Etacar11, Ruud Koot, Chochopk, Deltabeignet, Wikix, Koavf, Mikedelsol, The wub, Wikiliki, Valentinian, YurikBot, RussBot, Occuserpens, Anomalocaris, Joshdboz, Number 57, Moe Epsilon, GrinBot~enwiki, DVD R W, SmackBot, Bluebot, Shuki, Shamir1, Shmuliko, Yevgeny Kats, Ser Amantio di Nicolao, Goodnightmush, Wtg87, DGtal, Gilabrand, Cydebot, Przemek Jahr, Lugnuts, Faigl.ladislav, Headbomb, Epeefleche, Waacstats, Cgingold, Yonidebot, Derwig, Plindenbaum, DorganBot, Anonymous Dissident, Gilisa, AlleborgoBot, SieBot, RS1900, NuclearWarfare, DumZiBoT, Good Olfactory, Addbot, דוד55, Matěj Grabovský, Yobot, Davshul, Omnipaedista, MerlLinkBot, MegaSloth, RjwilmsiBot, EmausBot, Gilbertavalon, ZéroBot, Suslindisambiguator, ChuispastonBot, CocuBot, Ninmacer20, Jethro B, PHert, VIAFbot, Idanmezan, Monochrome Monitor, CyberXRef, Wwikix, KasparBot and Anonymous: 26

- **Kazuhiko Nishijima** *Source:* https://en.wikipedia.org/wiki/Kazuhiko_Nishijima?oldid=682649679 *Contributors:* Timrollpickering, Rich Farmbrough, SmackBot, Chris the speller, RayAYang, Cydebot, Michael C Price, Headbomb, Waacstats, Peachypoh, Addbot, Luckas-bot, Yobot, AnomieBOT, Materialscientist, Citation bot, NocturneNoir, Shubinator, A. di M., CES1596, Citation bot 1, RjwilmsiBot, JeanneMish, Noodleki, QuantumSquirrel, Helpful Pixie Bot, Bibcode Bot, Brad7777, EagerToddler39, VIAFbot, Monkbot, KasparBot and Anonymous: 5

- **Burton Richter** *Source:* https://en.wikipedia.org/wiki/Burton_Richter?oldid=660779906 *Contributors:* The Cunctator, Mic, El~enwiki, Maximus Rex, Pakaran, Pigsonthewing, Carnildo, Ancheta Wis, Giftlite, Everyking, ChicXulub, D6, Aris Katsaris, Bender235, Ht1848, CanisRufus, Alansohn, TheParanoidOne, Calton, Snowolf, Ksnow, RyanGerbil10, Japanese Searobin, Emerson7, Graham87, Rjwilmsi, MZMcBride, FlaBot, Srleffler, Valentinian, YurikBot, Azucar~enwiki, Takwish, Hawkeye7, Tearlach, Scottfisher, Zargulon, LeonardoRob0t, KocjoBot~enwiki, Wogsland, Betacommand, Nixeagle, Andrei Stroe, John, Dicklyon, QueensFinest86, Margoz, HennessyC, Drinibot, Ken Gallager, Cydebot, Thijs!bot, Headbomb, Z10x, Escarbot, RobotG, Gcm, .anacondabot, Waacstats, Greg Daniels, BrokenSphere, Plindenbaum, TXiKiBoT, Jimmyeatskids, McM.bot, Pachtova, SieBot, Nihil novi, All Hallow's Wraith, Joao Xavier, Masterpiece2000, DragonBot, Alexbot, BOTarate, Cardinalem, Ladsgroup, Trabelsiismail, Kbdankbot, Addbot, Jacopo Werther, Vsun, Ginosbot, Numbo3-bot, DubaiTerminator, Lightbot, Luckas-bot, Yobot, NLWASTI, ErwinsMoggie, AnomieBOT, Materialscientist, Dangshei, Davshul, Skyerise, RjwilmsiBot, John of Reading, Wikfr, Jonke20, ChuispastonBot, Movses-bot, BattyBot, YFdyh-bot, JYBot, Mogism, VIAFbot, Churn and change, Jeff kerrison, Pinocchio3000, Jonarnold1985, KasparBot and Anonymous: 28

- **Samuel C. C. Ting** *Source:* https://en.wikipedia.org/wiki/Samuel_C._C._Ting?oldid=673641908 *Contributors:* Alex.tan, Olivier, Menchi, Ahoerstemeier, Salsa Shark, Jiang, Kaihsu, Maximus Rex, Shizhao, Robbot, SchmuckyTheCat, DHN, HaeB, Ancheta Wis, Harp, Spencer195, Everyking, ChicXulub, Xmnemonic, PDH, D6, Srbauer, Kwamikagami, Physicistjedi, A Kit, Ksnow, RyanGerbil10, Japanese Searobin, WadeSimMiser, Chochopk, Clemmy, SDC, Emerson7, Jack Cox, Kbdank71, Rjwilmsi, MZMcBride, FlaBot, Ground Zero, Srleffler, Bgwhite, NTBot~enwiki, RussBot, DAJF, Jpbowen, Acmuser, Curpsbot-unicodify, Garion96, YellowMonkey, KocjoBot~enwiki, Luenlin, Nixeagle, Threeafterthree, Multivariable, Fuhghettaboutit, Khukri, Andrei Stroe, Ser Amantio di Nicolao, Takamaxa, John, Guat6, Verdant04, Rglovejoy, Dicklyon, RichardF, HongQiGong, Ageoflo, Margoz, Patrickwooldridge, Raysonho, Myasuda, Cydebot, David A. Victor, MWaller, Thijs!bot, Qwyrxian, Headbomb, Bunzil, Escarbot, RobotG, BenJWoodcroft, Harryzilber, Gcm, Txomin, Connormah, Alan-526, CommonsDelinker, Rrostrom, Nigholith, Osund, LordAnubisBOT, VolkovBot, Kyle the bot, TXiKiBoT, Jimmyeatskids, GcSwRhIc, Luuva, Resurgent insurgent, AHMartin, Ponyo, SieBot, Nihil novi, PolarBot, Arjen Dijksman, Toastforbrekkie, Muhends, Showa61, The Thing That Should Not Be, Joao Xavier, Meiguoren, Masterpiece2000, Alexbot, Eustress, Mlaffs, Cardinalem, Versus22, DumZiBoT, Kolyma, MystBot, Kbdankbot, Addbot, Jacopo Werther, Ginosbot, Numbo3-bot, Pracchia-78, Legobot, Luckas-bot, Yobot, Amirobot, Aldebaran66, NLWASTI, ErwinsMoggie, Nallimbot, Eumolpo, DSisyphBot, Davshul, Teamjenn, Chen Guangming, Mnmngb, Ironboy11, Ysyoon, Crispyslice, Full-date unlinking bot, Badger M., ZhBot, Dps04, Some Wiki Editor, El Mayimbe, DARTH SIDIOUS 2, AndyHe829, 68g, Nlzonnetje, Suslindisambiguator, JeanneMish, YaeHeart, ChuispastonBot, Woonhocho, ClueBot NG, Wasserla, QES girl, Bo543, Bilderbear, Sweet Elixir, The Elixir Of Life, Bibliophilen, Grand'mere Eugene, Icebox5210, Hiram Abiff, KasparBot, Csumstudent and Anonymous: 66

- **George Zweig** *Source:* https://en.wikipedia.org/wiki/George_Zweig?oldid=678347076 *Contributors:* XJaM, Andres, Lunchboxhero, Fredrik, Line, Klemen Kocjancic, Arminius, Rich Farmbrough, Vsmith, Simfish, Matt McIrvin, Lectonar, Tintin1107, Japanese Searobin, Woohookitty, Etacar11, Cbustapeck, Wikiliki, DannyWilde, Bgwhite, Roboto de Ajvol, Rsrikanth05, Bill-on-the-Hill, A314268, Zwobot, Carabinieri, CRKingston, Khukri, Wizardman, Clicketyclack, Ser Amantio di Nicolao, Dicklyon, Newone, Cydebot, Headbomb, CosineKitty, Hamsterlopithecus, Waacstats, JadeNB, Dorian Mode, Cuzkatzimhut, TXiKiBoT, Muhends, ClueBot, NuclearWarfare, Mlaffs, Mifter, MystBot, Addbot, GargoyleBot, Lightbot, Luckas-bot, AnomieBOT, JackieBot, Materialscientist, Citation bot, Xqbot, Gap9551, GrouchoBot, Omnipaedista, Ironboy11, Citation bot 1, DrilBot, ZTCRV, Badger M., RjwilmsiBot, EmausBot, ChuispastonBot, Helpful Pixie Bot, Bibcode Bot, BG19bot, Dexbot, Bibliophilen, Monochrome Monitor, Bluedudemi, Sleepy Geek and Anonymous: 29

- **Brookhaven National Laboratory** *Source:* https://en.wikipedia.org/wiki/Brookhaven_National_Laboratory?oldid=676931554 *Contributors:* Vicki Rosenzweig, RTC, Docu, Jengod, Robbot, Vespristiano, Dina, Harp, Fastfission, Xerxes314, Leonard G., Bobblewik, Mako098765, Dubaduba~enwiki, Ylai, Brian0918, Ylee, Orlady, Mdd, PaulHanson, Calton, Grenavitar, Dirac1933, DV8 2XL, Tripodics, Saperaud~enwiki, FlaBot, Naraht, Ground Zero, NekoDaemon, Zimbabweed, YurikBot, LittleSmall~enwiki, SCZenz, Scottfisher, Mike Selinker, Revengeofthynerd, MartinGugino, SmackBot, Verne Equinox, Gjs238, Colonies Chris, ButtonwoodTree, WestA, Pulu, Gump Stump, Ohconfucius, Drumpler, Dale101usa, Eastlaw, Americasroof, N2e, Cydebot, Gogo Dodo, WISo, Wejstheman, Thijs!bot, Headbomb, Kauczuk, Ph.eyes, Magioladitis, Jllm06, Mcginley1, Cliffie97, The Anomebot2, BatteryIncluded, Sesesq, CommonsDelinker, HEL, 72Dino, Vsosin, Greatestrowerever, Funandtrvl, Yaanch, Iracaz, Venny85, Pavelock, Cmcnicoll, Ketone16, Wfhuocrke, Lightmouse, TTload, Niceguyedc, Ktr101, Jonverve, Addbot, LaaknorBot, Lightbot, Legobot, Yobot, Ptbotgourou, ArthurBot, LilHelpa, Xqbot, D'ohBot, El Mayimbe, TjBot, Ripchip Bot, GeneralZod1108, EmausBot, ZéroBot, Molimaging, MALLUS, BrekekekexKoaxKoax, Frietjes, BG19bot, Protein Chemist, BattyBot, Mirafori, Iop3456, RaphaelQS, Heeren10, Trackteur, KasparBot and Anonymous: 59

- **Belle experiment** *Source:* https://en.wikipedia.org/wiki/Belle_experiment?oldid=603743287 *Contributors:* HaeB, Vegaswikian, The wub, DannyWilde, Goudzovski, Chobot, Conscious, SCZenz, Jess Riedel, Finell, MacsBug, Reedy, C.Fred, Wogsland, Dicklyon, Calvero JP, Headbomb, Andante1980, Pkoppenb, The Anomebot2, Sue H. Ping, Tagir, Alexbot, Qwfp, Addbot, Luckas-bot, Icalanise, Omnipaedista, RedBot, AManWithNoPlan, Kndimov and Anonymous: 20

- **BaBar experiment** *Source:* https://en.wikipedia.org/wiki/BaBar_experiment?oldid=678330312 *Contributors:* SimonP, Theresa knott, HaeB, Bobblewik, Mako098765, MuDavid, S.K., Physicistjedi, Alexander Sommer, Keenan Pepper, Apoc2400, SidP, Gene Nygaard, Linas, Wayward, Kanasyy, Jivecat, Erkcan, The wub, DannyWilde, Goudzovski, RussBot, Archelon, Square87~enwiki, Deville, Open2universe, Ishamael, Kgf0, A13ean, Reedy, Wogsland, Tigerhawkvok, OrphanBot, Wen D House, Mellery, ShelfSkewed, Myasuda, Alii h, Headbomb, Second Quantization, MECU, Magioladitis, Miaers, DD2K, LordAnubisBOT, Joshmt, Michael H 34, Dirc, Inductiveload, Capuchin, EmxBot, Tmhong, CultureDrone, Ktr101, Addbot, M.nelson, SpBot, Tassedethe, AnomieBOT, LilHelpa, Whyfjord, Metrictensor, Abisoffer, Standardfact, Popovvk, AndyHe829, Wikfr, Eothred, ChrisGualtieri, Dexbot, Rongended and Anonymous: 31

- **Collider Detector at Fermilab** *Source:* https://en.wikipedia.org/wiki/Collider_Detector_at_Fermilab?oldid=675608973 *Contributors:* Bodhitha, Falcorian, Commander Keane, Strait, FlaBot, YurikBot, JabberWok, SCZenz, Mehtala, Roy Fultun, SmackBot, Bdrell, SchmittM, Betaeleven, CuriousEric, Xaariz, Mbell, Headbomb, Z22, Athaenara, W3stfa11, AlleborgoBot, Srushe, LaurenMW, Addbot, Mortense, WFPM, Debresser, Lightbot, Luckas-bot, Yobot, Archon 2488, Galoubet, Xqbot, MIRROR, Mnmngb, AmyJAllen, Thinking of England, ChuispastonBot and Anonymous: 21

- **DØ experiment** *Source:* https://en.wikipedia.org/wiki/D0_experiment?oldid=680368255 *Contributors:* The Anome, Nurg, Pmanderson, Helohe, Rich Farmbrough, Smalljim, Ash211, Rjwilmsi, Ttwaring, FlaBot, Heiko, SCZenz, SmackBot, Hydrogen Iodide, Betacommand, Bluebot, Burair, Fiziker, Jumping cheese, Skapur, Headbomb, D.H, Potatoswatter, Anna Lincoln, Grndfthrprdx, EmanWilm, Addbot, Mortense, Arbitrarily0, Luckas-bot, Yobot, Ptbotgourou, Götz, Xqbot, MIRROR, D'ohBot, Citation bot 1, DrilBot, Higgshunter, MisterDub, DnaX, EdoBot, Freddie711, Bibcode Bot, Visuall, TwoTwoHello, Monkbot, Nkurinsky and Anonymous: 23

- **Fermilab** *Source:* https://en.wikipedia.org/wiki/Fermilab?oldid=681315607 *Contributors:* Mav, Shsilver, Rmhermen, Minesweeper, Ahoerstemeier, Docu, Rob Hooft, Bevo, Robbot, Goethean, Gnomon Kelemen, Wikibot, Tom harrison, Fastfission, Xerxes314, Jason Quinn, Bodhitha, Andycjp, H Padleckas, Crawdad, Thorwald, N328KF, O'Dea, Adambondy, Paul August, Brian0918, Ylee, Arancaytar, Dralwik, PaulHanson, Andrew Gray, Clay1039, Ayeroxor, SidP, Freshraisin, Dirac1933, Linas, Riffsyphon1024, Jpers36, Acone, Jugger90, Tevatron~enwiki, Reisio, Saperaud~enwiki, Rjwilmsi, Jivecat, Strait, Vegaswikian, Erkcan, The wub, Ttwaring, Bob Wiyadabebe-Iytsaboi, Matt Deres, NekoDaemon, Chobot, Blando728, Kummi, YurikBot, RussBot, KevinCuddeback, Bovineone, SCZenz, Voidxor, Scottfisher, JasonAD, Besselfunctions, ThunderBird, Curpsbot-unicodify, Ybbor, GrinBot~enwiki, Timothyarnold85, Morgan wasko, SmackBot, Clockhappy, TestPilot, Stepa, Liaocyed, Frumpet, Nil Einne, HeartofaDog, Chris the speller, DroEsperanto, Dual Freq, Sct72, Backspace, Jumping cheese, Savidan, Kevlar67, Pulu, Ser Amantio di Nicolao, DavidGC, Soap, Mcshadypl, DavidBailey, Jaganath, Bucksburg, Simkiott, Rob Shanahan, PRRfan, IceHunter, CzarB, CapitalR, CmdrObot, Itomchandler, N2e, Ken Gallager, Studiousstud, Cydebot, Archange56, Thijs!bot, Ucanlookitup, Headbomb, Rjshade, EdJohnston, Neatpete86, AntiVandalBot, Yellowdesk, SamIAmNot, Ingolfson, Barek, Albany NY, Z22, Magioladitis, Jllm06, Swpb, MMD61764, The Anomebot2, MrWarMage, Speedracer0883, STBot, Emsox, CommonsDelinker, HEL, DandyDan2007, Maurice Carbonaro, Athaenara, Fastspinecho, Plasticup, DadaNeem, DorganBot, VIOLENTRULER, JeffreyRMiles, Funandtrvl, Master z0b, VolkovBot, Rodstur, TheQuandry, DavidBrahm, Mbehnkeil, TXiKiBoT, PKDASD, Vegeta206, Hughey, Hmwith, SieBot, Sonicology, WereSpielChequers, Jauerback, The Parsnip!, Yintan, Ketone16, Flyer22, Lightmouse, Anchor Link Bot, Martarius, Sfan00 IMG, ClueBot, Frmorrison, Blanchardb, DragonBot, Diagramma Della Verita, ResidueOfDesign, Beamjockey, Kakofonous, Joe N, Addbot, Aaronjhill, Mjamja, AkhtaBot, Zahd, Asippel89, FermilabUser, Write-out, Lightbot, Igor26, Luckas-bot, Yobot, Naudefjbot~enwiki, AnomieBOT, Archon 2488, AdjustShift, Citation bot, LouriePieterse, Multxfer, Shadowjams, Spellage, Steve Quinn, Jschnur, Kalmbach, Higgshunter, 777sms, DrCrisp, RjwilmsiBot, EmausBot, RA0808, Solomonfromfinland, Gayshark, H3llBot, Hotzemoerkerk, ClueBot NG, Markthoms1, Frietjes, Helpful Pixie Bot, Tirebiter78, Sunshine Warrior04, MeanTuring, BonifaceFR, BattyBot, Mogism, Kennethaw88, Ramendoctor, Nikevcowsky, KasparBot and Anonymous: 136

- **Large Hadron Collider** *Source:* https://en.wikipedia.org/wiki/Large_Hadron_Collider?oldid=682276784 *Contributors:* Bryan Derksen, The Anome, Andre Engels, XJaM, Deb, William Avery, Moravice, Stevertigo, Frecklefoot, Edward, Michael Hardy, Booyabazooka, Dante Alighieri, Eliasen, Gabbe, Ixfd64, TakuyaMurata, Arpingstone, Egil, Ahoerstemeier, Msablic, Julesd, Salsa Shark, Jll, Evercat, Oak~enwiki, Ehn, Timwi, Doradus, Tpbradbury, Furrykef, Buridan, Thomasedavis, Samsara, Bevo, Rls, David.Monniaux, BenRG, Frazzydee, Freerk~enwiki, Phil Boswell, Gromlakh, Branddobbe, Robbot, Sdedeo, Fredrik, Jredmond, Sanders muc, Nurg, Noplasma, Romanm, Naddy, Postdlf, Sverdrup, Rursus, Texture, Davodd, Trevor Johns, Alan Liefting, David Gerard, Connelly, Centrx, Giftlite, DocWatson42, Brouhaha, Harp, Herbee, Semorrison, Pharotic, Gregb, PenguiN42, SoWhy, Alberto da Calvairate~enwiki, Lode~enwiki, Beland, OverlordQ, Mako098765, ShakataGaNai, Fred Stober, Redroach, Jokestress, Kuralyov, Pmanderson, Lumidek, Jh51681, Thorwald, Mike Rosoft, DanielCD, Discospinster, Rich Farmbrough, Sladen, FT2, Franjesus, Rama, Vsmith, Florian Blaschke, Wk muriithi, YUL89YYZ, David Schaich, Alistair1978, Ylai, Bender235, Rubicon, ESkog, Felixhudson, RJHall, Lycurgus, Oldsoul, Laurascudder, Art LaPella, RoyBoy, Simfish, Adambro, Bastique, Bobo192, Army1987, Smalljim, Dreish, Davidruben, Elipongo, Kjkolb, Physicistjedi, Daf, RichardNeill, Gbrandt, Rsholmes, Alansohn, Rio~enwiki, Arthena, CyberSkull, Keenan Pepper, Linmhall, Axl, SlimVirgin, Scarecroe, Sligocki, Kocio, Batmanand, Stack, Arobic, Wdfarmer, Hu, Radical Mallard, Wtmitchell, Schapel, Velella, Ronark, Ssbarker, Mr flea, Suruena, Harej, Amorymeltzer, RainbowOfLight, Dirac1933, H2g2bob, Bsadowski1, Gortu, Gene Nygaard, Afowler, LukeSurl, Blaxthos, Ceyockey, AndyBuckley, Mahanga, Quirkie, Keaton, Linas, Mindmatrix, Camw, BillC, Robert K S, Pol098, RoToR~enwiki, Duncan.france, Mpatel, Miss Madeline, Adhalanay, MFH, Jdiemer, GregorB, Osric, Kralizec!, Jacj, Burfdl, Starwed, Dbutler1986, RichardWeiss, Ashmoo, Graham87, Kane5187, Rjwilmsi, Tim!, Mr.Unknown, Strait, Eyu100, XP1, Amire80, Josiah Rowe, Leeyc0, Seneka~enwiki, Zaak, Grahn, Jmcc150, MZMcBride, Mike Peel, SimonMenashy, Cakedamber, Bubba73, The wub, DoubleBlue, Reinis, Dirtygreek, Yamamoto Ichiro, Drrngrvy, FlaBot, Margosbot~enwiki, Ysangkok, Nihiltres, Trekkie4christ, Elliot Lipeles, RexNL, Sp00n, Baryn, Goudzovski, SVTCobra, Snailwalker, Chobot, Scoops, Theo Pardilla, Visor, DVdm, Simesa, Tone, Wjfox2005, Zimbabweed, Amaurea, YurikBot, Bambaiah, Tommyt, Phmer, Tznkai, Pip2andahalf, Geljamin, ZZ9pluralZalpha, Markhoney, Arado, Bhny, Ozabluda, JabberWok, Hydrargyrum, Stephenb, CambridgeBayWeather, Rsrikanth05, Oni Lukos, GeeJo, Shanel, NawlinWiki, Juhanson, Janke, Długosz, Mmenal, Taco325i, Toba~enwiki, Thiseye, SCZenz, Uni4dfx, Esthurin, Nucleusboy, Anetode, Nephron, Brandon, Ravedave, Ndavies2, PhilipO, Davemck, Voidxor, WeirdEars, Tony1, Zythe, Falcon9x5, Scottfisher, SColombo, Wangi, Cstaffa, Nlu, Wknight94, Rwxrwxrwx, Sandstein, Zargulon, Ageekgal, Nikkimaria, Closedmouth, Arthur Rubin, Adilch, Josh3580, Chaleur, Al Farnsworth, Gppande, RotoSequence, JoanneB, Chrishmt0423, Caco de vidro, Physicsdavid, RG2, Michael Farris, NeilN, Nekura, CIreland, SaveTheWhales, Per piotrr Edman, Wolf1728, SmackBot, Elonka, 1dragon, Ququ, Tarret, Mrcoolbp, Hydrogen Iodide, Bigbluefish, Pavlovič, Istvan, Mjspe1, Aetheling1125, Rrius, Felix Dance, Kilo-Lima, Jagged 85, QuantumShadow, Setanta747 (locked), Piksi, WookieInHeat, Hiflyer, ZerodEgo, Canthusus, Rachel Pearce, HalfShadow, Gilliam, Slaniel, Jushi, Skizzik, Dauto, Andy M. Wang, Viraz, PJTraill, Palad1n, Basejumper123~enwiki, Persian Poet Gal, PeterMcCready, Djinn65, Jprg1966, Thumperward, Miquonranger03, MalafayaBot, Hibernian, The Rogue Penguin, Cbetan, DHN-bot~enwiki, Colonies Chris, MaxSem, Cmanser, Philc 0780, Zsinj, Rogermw, Tsca.bot, Kavehmz, Kelvin Case, Jacob Poon, Neo256, Snowmanradio, Lantrix, Nunocordeiro, Krsont, Kittybrewster, Backspace, Jmnbatista, Teoryn, Wen D House, Khukri, Acepectif, Decltype, Earthsky, Hurricane Floyd, Lamikae~enwiki, Dreadstar, A.R., PsychoJosh, Esb, Wybot, DMacks, Stew560, ShadowUltra, Zonk43, Ihatetoregister, Salamurai, Ifrit, Ged UK, Ohconfucius, Lambiam, ArglebargleIV, Arnoutf, Swatjester, RBPierce, Ninjagecko, Dark Formal, Freewol, SilkTork, Adj08, Kylehung, Rara bb, DMurphy, Minna Sora no Shita, Cariniabean, Bjankuloski06en~enwiki, Bella Swan, Wiseoldbum, Phancy Physicist, Mr. Quertee, KyleMac, BillFlis, Rainwarrior, Joeylawn, Beetstra, Muadd, Griffles, Bigboy101011, Artman40, Ryulong, EdC~enwiki, Pseudoanonymous, Autonova, Dl2000, Cristi.falcas, Lee Carre, JarahE, Pejman47, Inferiorjon, Rygel, BranStark, ISD, Iridescent, Colonel Warden, Joseph Solis in Australia, Hurricanefloyd, Newone, Kludger, Wleizero, StephenBuxton, Cyclades, GDallimore, Cbrown1023, Benplowman, Blehfu, Courcelles, Chovain, PeterS32, Darklombax, Owen214, Eastlaw, SkyWalker, J Milburn, Friendly Neighbour, Unsuspected, Tanthalas39, TunaSushi, Deon, Wafulz, Zarex, Dycedarg, Chrumps, Olaf Davis, JohnCD, Runningonbrains, Scrivener72, Mix Bouda-Lycaon, Juhachi, Timothylord, FlyingToaster, Shandris, Karenjc, TheAdventMaster, Fletcher, Ebenonce, Rotiro, Jefchip, Jac16888, Kanags, Reywas92, Perfect Proposal, Kslotte, Meno25, DMeyering, Islander, Tnicol, Michael C Price, AndersFeder, DumbBOT, RotaryAce, Gimmetrow, Eintragung ins Nichts, Huddahbuddah, Epbr123, Pajz, Ewhite2, Paragon12321, Peter johnson4, Mojo Hand, Headbomb, Dtgriscom, Marek69, Dalahäst, A3RO, Electron9, James086, Wildthing61476, Uiteoi, Stoshmaster, Siawase, Leon7, Grayshi, Rtomas, MichaelMaggs, DewiMorgan, Baclough, MassKnowledgeLearner, GruBBy, Shirt58, Bm gub, TheBlueFox, Voortle, Daelian~enwiki, Tyco.skinner, Pwhitwor, 2bornot2b, Isilanes, Farosdaughter, Yellowdesk, Storkk, LttS, Gökhan, Jonterry4, Darrenhusted, Res2216firestar, DOSGuy, Omeganian, Smitty Mcgee, Seddon, Eurobas, Dricherby, Pkoppenb, Acroterion, DataMatrix, Addw, Connormah, Pedro, VoABot II, Mrund, Smarts53, Swpb, UnusedAccount, AMK1211, CTF83!, Hypergeek14, Nineko, Aka042, Tourettes1993, FrF, Catgut, Septuagent, Cgingold, BatteryIncluded, Japo, Error792, Mwvandersteen, Ponty Pirate, Chris G, Purslane, Richard1990, Fluteflute, Balcerzak, Ztobor, FisherQueen, BetBot~enwiki, Andre.holzner, Mpwheatley, Red Sunset, Mtbaldyred, NReitzel, ScorpO, TechnoFaye, Biguana, Verdatum, KTo288, Leyo, Wbrice83186, Lilac Soul, Fatpratmatt, HEL, Paul Suhler, Beehive101, Ssolbergj, J.delanoy, Pclover, Pharaoh of the Wizards, DavidB601, Victorgmartins, Trusilver, AstroHurricane001, Tuduser, Hom sepanta, Gaming4JC, Hans Dunkelberg, Uncle Dick, Maurice Carbonaro, Ginsengbomb, 4johnny, Nmajmani, Darth Mike, Smite-Meister, Mariekshan, Heggied, Tokyogirl79, Rod57, Arronax50, McSly, Austin512, Nocarrier813, Silas S. Brown, (jarbarf), Rossenglish, Plasticup, NewEnglandYankee, BreakerLOLZ, Bodhikun, Trilobitealive, Ohms law, Vanished user 47736712, ThinkBlue, Kae1is, Olegwiki, Pundit, Joshua Issac, Juliancolton, Entropy, STBotD, Lifeboatpres, Burzmali, Jamesofur, Gwen Gale, EarthRise33, DorganBot, Jtankers, Dpr101188, Pdcook, Zomglolwtfzor, Tkgd2007, Rpeh, CardinalDan, Idioma-bot, Sheliak, Kevin Mason Barnard, Alexgenaud, Priceman86, Iconoclast09, VolkovBot, Managerpants, Phasma Felis, Mackmgg, Siriusvector, Jeff G., Nburden, Michelle Roberts, Quentonamos, Mcewan, Funkysapien~enwiki, Philip Trueman, TXiKiBoT, Joopercoopers, Zidonuke, Sroc, Jeffakolb, Malinaccier, Red Act, Hqb, Andrius.v, NPrice, Nxavar, Justinb52, Anonymous Dissident, Atelerix, Omweb, Gjtorikian, Dormskirk, Someguy1221, ChuChingadas, CaptinJohn, Vanir-sama, Seraphim, Batista619, Cerebellum, Dtely, Ctmt, JhsBot, Broadbot, Praveen pillay, THC Loadee, Abdullais4u, Jackfork, LeaveSleaves, Cgwaldman, MacFodder, Henryodell, Spinal232, FunkDemon, WebScientist, Reddawnz, Pleriche, SheffieldSteel, RandomXYZb, SwordSmurf, Staka, Geepster, Synthebot, Chronitis, Falcon8765, MCTales, Alaniaris, ShandraShazam, Insanity Incarnate, Ptrslv72, Truthanado, Northfox, Shanmugammpl, Vitalikk, Finnrind, DJ JS9, Deconstructhis, Red, TimProof, Fanatix, SieBot, TJRC, Meltonkt, WereSpielChequers, Shawnlandden, Rob.bastholm, Yuefairchild, RJaguar3, The way, the truth, and the light, LeadSongDog, Chihuong bk~enwiki, Soler97, Keilana, Govtrust, Interchange88, Quest for Truth, Flyer22, Theevildoctorodewulfe, Arbor to SJ, Retbutler92, WannabeAmatureHistorian, Prestonmag, Mimihitam, CaelumArisen, Oxymoron83, Tefalstar, Paintman, BartekChom, Pac72, Lightmouse, Crisis, Techman224, Greatrobo76, Robfrost, Profgregory, OKBot, Maelgwnbot, LonelyMarble, Jóhann Heiðar Árnason, Torchwoodwho, Kurtilein, Casmiky, Chrisrus, Maxime.Debosschere, Capitalismojo, Veldin963, Sheps999, Sockies, Mygerardromance, Hamiltondaniel, WikiLaurent, Yhkhoo, Superbeecat, Florentino floro, Pinkadelica, Tmckeage, Babakathy, Randy Kryn, Jtyard, Scaler1112, TSRL, ImageRemovalBot, Llywelyn2000, Faithlessthewonderboy, Lethesl, Gourra, Atif.t2, Loren.wilton, Martarius, Beeblebrox, Darsie from german wiki pedia, Phyte, ClueBot, GorillaWarfare, Boodlesthecat, GrandDrake, Mhworth, The Thing That Should Not Be, Mriya, IceUnshattered, Techdawg667, General Epitaph, Wwheaton, Ndenison, Frmorrison, Dux is me, DanielDeibler, Boing! said Zebedee, Xertoz, Estevoaei, QuantumAmyrillis, TarzanASG, Piledhigheranddeeper, Kyurkewicz, Phenylalanine, Ooogly, Final Philosopher, Mspraveen, Polaroids4x5, Maccy69, Somno, Ktr101, Excirial, AssegaiAli, Alexbot, Hj FUN, Ottre, Eeekster, Haseth, Vanisheduser12345, GreenGourd, Vivio Testarossa, Sun Creator, Thomasmackat41, NuclearWarfare, Ice Cold Beer, Cenarium, Jotterbot, PhySusie, 842U, Wond3rbread1991, Snacks, Shivarudra, Shaunofthefuzz7, Lenary, Cellodont, Ottawa4ever, GlobetrottingAussie, Netanel h, Buckethed, Thingg, Colmsherry, Brickwall04, MAGZine, Spinoff, Geckon, Versus22, Apparition11, Wnt, Manu-ve Pro Ski, DumZiBoT, SlaterDeterminant, Wikidokman, Wonderflash1111, Homocion, Cvet~enwiki, Ohmyyes, InternetMeme, 03md, XLinkBot, Spitfire, PSimeon, DragonFury, Oldnoah, Stickee, RickWagnerPhD, Duncan, 02millers, Little Mountain 5, Mitch Ames, NellieBly, Pokesausage, Jlcoving, PL290, Alexius08, Dacool7, TravisAF, Nukes4Tots, Gouryella, JCDenton2052, ISGTW, HexaChord, Stephen Poppitt, Tayste, Stefinho360, Prowikipedians, Pyfan, JBsupreme, Master michael 90, Willking1979, Some jerk on the Internet, DOI bot, Guoguo12, TheNeutroniumAlchemist, Phlegm Rooster, Pedant75, AkhtaBot, GuyanaMan, Totakeke423, Ronhjones, Polishhill, Lets Enjoy Life, Gflashwnox, PlumCrumbleAndCustard, Frikandel~enwiki, CanadianLinuxUser, AcademyAD, Gmeyerowitz, Ajeetkumar81, MrOllie, Mentisock, Download, MrVanBot, Lihaas, AndersBot, Chzz, Debresser, Deamon138, Rodeo90, Barak Sh, Andefs, Chem-MTFC, Anonyfuss, Kisbesbot, Mkhan-95, AgadaUrbanit, 84user, Ehrenkater, Wikbot, F Notebook, Tide rolls, Verbal, Lightbot, ScAvenger, AvalonTreman, QuadrivialMind, SPat, Zorrobot, MuZemike, Trotter, Alamgir, Beatsbox, Num43, Jim, Ch1n4m4n, Narutolovehinata5, Everyme, Kyro, Clay Juicer, Luckas-bot, TheSuave, AzureFury, Kartano, Flukas, Ptbotgourou, Fraggle81, Legobot II, Donfbreed, II MusLiM HyBRiD II, Evans1982, Antimatter Dilbert, Anypodetos, Buddy431, Bugnot, Jimbob16314, Terryblack, THMRK1, Lukealanjohnson, Paaulinho, Csmallw, Princeb11, Magog the Ogre, Libriantichi, Orion11M87, AnomieBOT, Xasdas, HisNameIsChris, DoctorJoeE, Archon 2488, Autumn Fall, Rubinbot, Bsimmons666, PianoDan, Message From Xenu, Jim1138, Pyrrhus16, AloysiusLiliusBot, Teamnumberawesome, Piano non troppo, Contribut, Bmoc2012tms, Fahadsadah, Apau98, NORD74, Nacre 10, Adejam, Hedgehawk, Jacob2718, Citation bot, OllieFury, Eumolpo, Iloveham, ArthurBot, LovesMacs, Mavrisa, The Firewall, Dracoblaze4, Advertiseg, Zykure, Redwodka, Xqbot, Biologicithician, Capricorn42, AbhishekSinghRana, Newzebras, Jeffrey Mall, Mononomic, Stsang, Anna Frodesiak, Srich32977, Pikematerson, PimRijkee, GrouchoBot, Omnipaedista, Chickenweed, Hechser, Mark Schierbecker, MadGeographer, EmilyUndead, Seeleschneider, Ace111, Mgambentok, EuroWikiWorld, Der Falke, Locobot, JediMaster362, Natural Cut, Shadowjams, Spellage, WaysToEscape, Douchedoom, A. di M., Chriss789, SD5, Dailycare, FrescoBot, God=nocioni, Paine Ellsworth, PennsylvaniaGov, W Nowicki, Jeffrey Solimine, StaticVision, Recognizance, Thorenn, Bergdohle, Jamesooders, DivineAlpha, Cannolis, HamburgerRadio, Citation bot 1, PointOfPresence, Ntse, Camronwest, Biker Biker, Shiftyalex1, Pinethicket, I dream of horses, Schmitzhugen, HRoestBot, Edderso, Safetynut, Martinvl, RedBot, CamB42424, SpaceFlight89, Zestofalemon, Spaluch1, Tcnuk, Realtruth.co.nr, Shanmugamp7, Aknochel, December21st2012Freak, 343GuiltySpark343, IVAN3MAN, Cnwilliams, Lesdo234, Calle Cool, Fl4ian, ToBeBot, Puzl bustr, Wotnow, Beta Orionis, Fama Clamosa, Onanysunday, Hickorybark, Lotje, Dinamik-bot, TBloemink, Sammy00193, Muhammedpbuh, Jeffrd10, Excellt127, Tbhotch, Minimac, Swedernish, Remnar, Cathardic, Itain'tsobad, Mean as custard, Ztbbq, Sfsupro, RjwilmsiBot, TjBot, MMS2013, Alph Bot, Markos Strofyllas, Deagle AP, EmausBot, Black Shadow, Orphan Wiki, Mickeyhill, Ever388, Docjudith, Trickett rocks, Belismakr, Mrgalaxy01, Sumsum2010, RA0808, Jmv2009, Gimmetoo, Themorrissey, Slightsmile, Sandeepsuri, Wikipelli, K6ka, Piggyspider123, Hhhippo, ZéroBot, John Cline, A2soup, Shuipzv3, QWERTYMASTR, Ticklemefugly, 99chromehead, Emily Jensen, ElationAviation, Louprado, Hazard-SJ, Arbnos, Hashiq, Ltlighter, FreddoT, Christina Silverman, Timetraveler3.14, Bob drobbs, HelloDenyo, 123smellmyfeet, Brandmeister, Coasterlover1994, Nanouniverse, Tomásdearg92, Donner60, Aldnonymous, GermanJoe, Resonant.Interval, Citedegg, Philippe BINANT, GrayFullbuster, Alderepas, Kingexaldraw, ClueBot NG, Zucchinidreams, Ex Everest, Michaelmas1957, Tbonemalone123, Btcc11, Loliamnot13, HinduPundit, Chrisfex, Theimmaculatechemist, Kmchanw, Delusion23, Parcly Taxel, O.Koslowski, ScottSteiner, Anonymous5555, Widr, Kaileeslight, Wisconsinbadger, PooRadley, Bigbullhoodboy, Jbackroyd, Gustavoanaya, Hotswapster, Mophedd, Jemmalouisemay, Strike Eagle, Titodutta, Bibcode Bot, Trunks ishida, SidKemp, BG19bot, Mimzy2011, Eothred, Lhshammo, MusikAnimal, Wikiviks, Uk554, Umarfarooq111, Guy.shrimpton, Mmovchin, Knowsnothing613, Mannasoumya, Willknowsalmosteverything, Seniorlimpio, NeoTheChosenOne,

NotinREALITY, Eio, SkittleJuice, Tictac66, Achowat, Anbu121, BattyBot, Ant314159265, Mdann52, KyleRyanToth, Ytic nam, Rcw258, ChrisGualtieri, Abcadi, CrunchySkies, Mineville, GetTheShift, Rues~enwiki, EuroCarGT, KrazyKelle, 786b6346, Rhlozier, Blueprinteditor, Dexbot, Kulpreet33, 331dot, Deranged anna, Sidsandyy, Garuda0001, Athul av, Michael Anon, Febinmathew, EauOo, Hamid26747, Tony Mach, Frosty, Nilaykumar07, Jo-Jo Eumerus, Flatfatmat, OnlyShadab, Corn cheese, Reatlas, DaPanda44, Lyxkg007, Rfassbind, ETHJILA, संजीव कुमार, Abcdefghijklmnopqrstuvwxyz12345678, Glasstop, FiredanceThroughTheNight, Noctave, Evano1van, Giraffosaurus, Revolution1221, RaphaelQS, CensoredScribe, User-name929, Boone jenner, Zenibus, Prokaryotes, Neilroy1998, Giu8888, Mandruss, LaGeneralitat, StraightOuttaBrisbane, RoflCopter404, Inessa Alaverdyan, TaiSakuma, Mfb, MyNameIsn'tElvis, Jedipowerz01, Potassium 40, CaesarsPalaceDude, Tighef, WPratiwi, AsalKadal, AslanEntropy, Parabolooidal, Tlmpmt, Vorkel insignia, Swagit420, Vieque, Akro7, Rohan.benia, Medical physicist, HMSLavender, Calmyourfarm, PotatoNinja, Chryst Laxus, Wendy Sax, Onkar kasture2000, Kuber Kanade, Sonicwave32, Shantanu28Editor, Tetra quark, The Messenger of Hor-pen-abu, Infinite0694, Ellipapa, KasparBot, Jerryg480, Bhavya velani, Bunnie Saini, AMERIXANPSYCHO, DefinitelyNotBae and Anonymous: 1419

- **SLAC National Accelerator Laboratory** *Source:* https://en.wikipedia.org/wiki/SLAC_National_Accelerator_Laboratory?oldid=680114102 *Contributors:* Bryan Derksen, Edward, Minesweeper, Andrewa, Peter Kaminski, Mxn, Jengod, Nv8200pa, Ed g2s, Finlay McWalter, Sanders muc, Voyager640, DocWatson42, Dinomite, Fastfission, Peruvianllama, Leonard G., Xinoph, Wronkiew, Bobblewik, Neilc, Geni, CryptoDerk, Beland, Chrisn4255, Icairns, Jewbacca, Klemen Kocjancic, Squash, ChrisRuvolo, Rich Farmbrough, Art LaPella, Spoon!, Janna Isabot, Mike Schwartz, Dreish, Viriditas, Jag123, Rainer Bielefeld~enwiki, SPUI, Natelipkowitz, Tom Yates, Msh210, PaulHanson, Aranae, Wtshymanski, RJFJR, Dirac1933, Linas, RHaworth, Justinlebar, David Levy, Saperaud~enwiki, Erkcan, DannyWilde, NekoDaemon, Chobot, YurikBot, Darkstar949, JabberWok, Hydrargyrum, Rsrikanth05, SCZenz, Dionea, Scottfisher, TimK MSI, Karit~enwiki, Arthur Rubin, Brianlucas, ArielGold, Curpsbot-unicodify, Physicsdavid, GrinBot~enwiki, Kgf0, SmackBot, Hydrogen Iodide, Stepa, Miquonranger03, RadWorkerII, Colonies Chris, Annoyedgrunt, Cybercobra, PetesGuide, Akriasas, Pulu, Evil genius, Disavian, Iridescent, Astrobayes, Jerry-va, DanHickstein, CmdrObot, Raysonho, Old Guard, Johnlogic, Myasuda, Cydebot, Mblumber, JFreeman, Thijs!bot, Epbr123, Wikid77, Headbomb, WillMak050389, Thadius856, Qwerty Binary, Z22, Magioladitis, The Anomebot2, Wormcast, Hallonsten, JaGa, Jvimal, Racepacket, J.delanoy, Athaenara, BrokenSphere, Aboutmovies, Belovedfreak, Minesweeper.007, Joshmt, S, Philip Trueman, The Original Wildbear, Anonymous Dissident, SieBot, Umrguy42, Oxymoron83, Kqxr, Stepheng3, DumZiBoT, PSimeon, J Hazard, Addbot, Download, زرشک, User0529, Luckas-bot, Yobot, Nallimbot, Neutrinoless, Orion11M87, AnomieBOT, Archon 2488, Galoubet, İncelemeelemani, Citation bot, Cnwilliams, S4wilson, Iluvkitties991, DrCrisp, John of Reading, Look2See1, Profrocshae, Andyfreeberg, Helpful Pixie Bot, Bibcode Bot, PearlSt82, Edgitar86, Comfr, Darvii, Sonĝanto, This Name is Ironic, Phbuck, Maderthaner, Monkbot, MistyEye, Jtinsman, KasparBot and Anonymous: 56

- **Tevatron** *Source:* https://en.wikipedia.org/wiki/Tevatron?oldid=675865538 *Contributors:* Heron, Ahoerstemeier, Conti, Phoebe, Twang, Beefman, Bobblewik, Bodhitha, Beland, Pmanderson, Sam Hocevar, Lumidek, Deglr6328, FT2, David Schaich, Edgarde, E2m, Nrbelex, Andrewpmk, Saga City, Egg, Gene Nygaard, Itinerant, Linas, Rjwilmsi, Zbxgscqf, Strait, Bubba73, FlaBot, RobyWayne, Goudzovski, Hawkeye7, SCZenz, Cstaffa, F15mos, KasugaHuang, SmackBot, The Rogue Penguin, Burair, Glloq, Kittybrewster, Backspace, Ambix, Bigturtle, Pmbarros, Adrigon, Salamurai, Rsimmonds01, Oo7akbnd, Xornofxorn, Valoem, SchmittM, Ruslik0, **mech**, WISo, LeeSawyer, Headbomb, MikeLynch, Mwarren us, Z22, Jllm06, Stefansoldner, Evbassboy13, Rod57, Funandtrvl, VolkovBot, KP-Adhikari, Moose-32, PeterBFZ, Skipnicholson, Phe-bot, Mimihitam, Juvarra~enwiki, Vanished user qkqknjitkcse45u3, Martarius, ClueBot, Alexbot, Muro Bot, Joe N, Jtle515, XLinkBot, Fiskbil, Addbot, Lightbot, Dxia, Naudefjbot~enwiki, Amirobot, Dwayne, Citation bot, Xqbot, Orduna, Mnmngb, Rainald62, Dailycare, FrescoBot, Paine Ellsworth, Pknkly, Citation bot 1, DrilBot, Pinethicket, Im4evrsmrt, MondalorBot, Tea with toast, Wikicont97531, DixonDBot, Mary at CERN, Xor4200, RjwilmsiBot, EmausBot, Racerx11, Hhhippo, Arbnos, Bibcode Bot, M0rphzone, Rajathsbhat, BattyBot, Tony Mach, Mg393, Giraffosaurus, Go14smoke, ThaLibster, Rog7er and Anonymous: 82

9.2 Images

- **File:3gluon.png** *Source:* https://upload.wikimedia.org/wikipedia/commons/a/a8/3gluon.png *License:* Public domain *Contributors:* Transferred from en.wikipedia by SreeBot *Original artist:* Bambaiah at en.wikipedia

- **File:8foldway.png** *Source:* https://upload.wikimedia.org/wikipedia/commons/9/9f/8foldway.png *License:* CC-BY-SA-3.0 *Contributors:* Transferred from en.wikipedia to Commons. *Original artist:* Bambaiah

- **File:Aerial_View_of_Brookhaven_National_Laboratory.jpg** *Source:* https://upload.wikimedia.org/wikipedia/commons/7/75/Aerial_View_of_Brookhaven_National_Laboratory.jpg *License:* Public domain *Contributors:* Brookhaven *Original artist:* ENERGY.GOV

- **File:Albert_Einstein_Head.jpg** *Source:* https://upload.wikimedia.org/wikipedia/commons/d/d3/Albert_Einstein_Head.jpg *License:* Public domain *Contributors:* This image is available from the United States Library of Congress's Prints and Photographs division under the digital ID cph.3b46036.
This tag does not indicate the copyright status of the attached work. A normal copyright tag is still required. See Commons:Licensing for more information. *Original artist:* Photograph by Oren Jack Turner, Princeton, N.J.

- **File:Ambox_current_red.svg** *Source:* https://upload.wikimedia.org/wikipedia/commons/9/98/Ambox_current_red.svg *License:* CC0 *Contributors:* self-made, inspired by Gnome globe current event.svg, using Information icon3.svg and Earth clip art.svg *Original artist:* Vipersnake151, penubag, Tkgd2007 (clock)

- **File:Ambox_important.svg** *Source:* https://upload.wikimedia.org/wikipedia/commons/b/b4/Ambox_important.svg *License:* Public domain *Contributors:* Own work, based off of Image:Ambox scales.svg *Original artist:* Dsmurat (talk · contribs)

- **File:Asymmetricwave2.png** *Source:* https://upload.wikimedia.org/wikipedia/commons/0/0d/Asymmetricwave2.png *License:* CC BY 3.0 *Contributors:* Own work *Original artist:* TimothyRias

- **File:Baryon-decuplet-small.svg** *Source:* https://upload.wikimedia.org/wikipedia/commons/7/78/Baryon-decuplet-small.svg *License:* Public domain *Contributors:* Own work *Original artist:* Trassiorf

- **File:Baryon-octet-small.svg** *Source:* https://upload.wikimedia.org/wikipedia/commons/b/b5/Baryon-octet-small.svg *License:* Public domain *Contributors:* Own work *Original artist:* Trassiorf
- **File:Baryon_Supermultiplet_using_four-quark_models_and_half_spin.png** *Source:* https://upload.wikimedia.org/wikipedia/commons/3/39/Baryon_Supermultiplet_using_four-quark_models_and_half_spin.png *License:* CC BY-SA 3.0 *Contributors:* made by LaTeX/TikZ, based on David Griffiths "Introduction to Elementary Particles" *Original artist:* Studzinski.daniel
- **File:Baryon_decuplet.png** *Source:* https://upload.wikimedia.org/wikipedia/commons/e/ec/Baryon_decuplet.png *License:* CC-BY-SA-3.0 *Contributors:* Own work *Original artist:* Laurascudder
- **File:Baryon_decuplet.svg** *Source:* https://upload.wikimedia.org/wikipedia/commons/f/f6/Baryon_decuplet.svg *License:* Public domain *Contributors:* Own work (Original text: *self-made*) *Original artist:* Wierdw123 at English Wikipedia
- **File:Baryon_decuplet_w_mass.png** *Source:* https://upload.wikimedia.org/wikipedia/en/c/c1/Baryon_decuplet_w_mass.png *License:* PD *Contributors:*

self-made

Original artist:

Venny85 (talk)
- **File:Baryon_octet.png** *Source:* https://upload.wikimedia.org/wikipedia/commons/4/4b/Baryon_octet.png *License:* CC-BY-SA-3.0 *Contributors:* Own work *Original artist:* Laurascudder
- **File:Baryon_octet_w_mass.png** *Source:* https://upload.wikimedia.org/wikipedia/en/9/98/Baryon_octet_w_mass.png *License:* PD *Contributors:*

self-made

Original artist:

Venny85 (talk)
- **File:Bcoulomb.png** *Source:* https://upload.wikimedia.org/wikipedia/commons/0/04/Bcoulomb.png *License:* Public domain *Contributors:* http://en.wikipedia.org/wiki/Image:Bcoulomb.png *Original artist:* ?
- **File:Beta-minus_Decay.svg** *Source:* https://upload.wikimedia.org/wikipedia/commons/a/aa/Beta-minus_Decay.svg *License:* Public domain *Contributors:* This vector image was created with Inkscape. *Original artist:* Inductiveload
- **File:Beta_Negative_Decay.svg** *Source:* https://upload.wikimedia.org/wikipedia/commons/8/89/Beta_Negative_Decay.svg *License:* Public domain *Contributors:* This vector image was created with Inkscape. *Original artist:* Joel Holdsworth (Joelholdsworth)
- **File:Beta_spectrum_of_RaE.jpg** *Source:* https://upload.wikimedia.org/wikipedia/commons/e/e6/Beta_spectrum_of_RaE.jpg *License:* CC BY-SA 4.0 *Contributors:* Own work *Original artist:* HPaul
- **File:BosonFusion-Higgs.svg** *Source:* https://upload.wikimedia.org/wikipedia/commons/7/78/BosonFusion-Higgs.svg *License:* CC-BY-SA-3.0 *Contributors:*
- BosonFusion-Higgs.png *Original artist:* BosonFusion-Higgs.png: User:Harp 12:43, 28 March 2007
- **File:Brookhaven_National_Laboratory_logo.svg** *Source:* https://upload.wikimedia.org/wikipedia/en/4/45/Brookhaven_National_Laboratory_logo.svg *License:* Fair use *Contributors:*

Battelle 2008 Annual Report *Original artist:* ?
- **File:Burton_Richter_NSF_crop.jpg** *Source:* https://upload.wikimedia.org/wikipedia/commons/4/41/Burton_Richter_NSF_crop.jpg *License:* Public domain *Contributors:* This file was derived from Burton Richter NSF.jpg: <a href='//commons.wikimedia.org/wiki/File:Burton_Richter_NSF.jpg' class='image'><img alt='Burton Richter NSF.jpg' src='https://upload.wikimedia.org/wikipedia/commons/thumb/5/55/Burton_Richter_NSF.jpg/50px-Burton_Richter_NSF.jpg' width='50' height='36' srcset='https://upload.wikimedia.org/wikipedia/commons/thumb/5/55/Burton_Richter_NSF.jpg/75px-Burton_Richter_NSF.jpg 1.5x, https://upload.wikimedia.org/wikipedia/commons/thumb/5/55/Burton_Richter_NSF.jpg/100px-Burton_Richter_NSF.jpg 2x' data-file-width='360' data-file-height='260' /></a> *Original artist:* Burton_Richter_NSF.jpg: NSF
- **File:CERN_LHC_Tunnel1.jpg** *Source:* https://upload.wikimedia.org/wikipedia/commons/f/fc/CERN_LHC_Tunnel1.jpg *License:* CC BY-SA 3.0 *Contributors:* Own work *Original artist:* Julian Herzog (website)
- **File:CFN-600px.jpg** *Source:* https://upload.wikimedia.org/wikipedia/en/e/ea/CFN-600px.jpg *License:* PD *Contributors:*

Courtesy of Brookhaven National Laboratory

Original artist:

Brookhaven National Laboratory
- **File:COBE_cmb_fluctuations.png** *Source:* https://upload.wikimedia.org/wikipedia/commons/a/a3/COBE_cmb_fluctuations.png *License:* Public domain *Contributors:*
- http://lambda.gsfc.nasa.gov/product/cobe/dmr_image.cfm *Original artist:* The COBE datasets were developed by the NASA Goddard Space Flight Center under the guidance of the COBE Science Working Group.
- **File:CTEQ6_parton_distribution_functions.png** *Source:* https://upload.wikimedia.org/wikipedia/commons/0/0d/CTEQ6_parton_distribution_functions.png *License:* Public domain *Contributors:* Originally from en.wikipedia; description page is (was) here *Original artist:* User Ylai on en.wikipedia
- **File:Cabibbo_angle.svg** *Source:* https://upload.wikimedia.org/wikipedia/commons/5/50/Cabibbo_angle.svg *License:* Public domain *Contributors:* Own work *Original artist:* Headbomb

- **File:Challenger_explosion.jpg** *Source:* https://upload.wikimedia.org/wikipedia/commons/9/9f/Challenger_explosion.jpg *License:* Public domain *Contributors:* http://grin.hq.nasa.gov/ABSTRACTS/GPN-2004-00012.html *Original artist:* Kennedy Space Center

- **File:Charmed-dia-w.png** *Source:* https://upload.wikimedia.org/wikipedia/en/6/6d/Charmed-dia-w.png *License:* Fair use *Contributors:* http://www.bnl.gov/bnlweb/history/charmed.asp *Original artist:* ?

- **File:Coat_of_arms_of_Israel.svg** *Source:* https://upload.wikimedia.org/wikipedia/commons/8/8f/Emblem_of_Israel.svg *License:* Public domain *Contributors:* symbol created in 1948. *Original artist:* Original design by Max and Gabriel Shamir; Tonyjeff, based on national symbol.

- **File:Collider_Detector_at_Fermilab.jpg** *Source:* https://upload.wikimedia.org/wikipedia/commons/d/db/Collider_Detector_at_Fermilab.jpg *License:* CC-BY-SA-3.0 *Contributors:*

- Transferred from en.wikipedia by SreeBot *Original artist:* Bodhitha at en.wikipedia

- **File:Collider_Detector_at_Fermilab_(CDF)_silicon_vertex_detector.JPG** *Source:* https://upload.wikimedia.org/wikipedia/commons/1/15/Collider_Detector_at_Fermilab_%28CDF%29_silicon_vertex_detector.JPG *License:* CC BY-SA 4.0 *Contributors:* Own work *Original artist:* Z22

- **File:Collider_Detector_at_Fermilab_(CDF)_silicon_vertex_detector_cross_section.JPG** *Source:* https://upload.wikimedia.org/wikipedia/commons/0/09/Collider_Detector_at_Fermilab_%28CDF%29_silicon_vertex_detector_cross_section.JPG *License:* CC BY-SA 4.0 *Contributors:* Own work *Original artist:* Z22

- **File:Commons-logo.svg** *Source:* https://upload.wikimedia.org/wikipedia/en/4/4a/Commons-logo.svg *License:* ? *Contributors:* ? *Original artist:* ?

- **File:Copyright-problem.svg** *Source:* https://upload.wikimedia.org/wikipedia/en/c/cf/Copyright-problem.svg *License:* PD *Contributors:* ? *Original artist:* ?

- **File:Crab_Nebula.jpg** *Source:* https://upload.wikimedia.org/wikipedia/commons/0/00/Crab_Nebula.jpg *License:* Public domain *Contributors:* HubbleSite: gallery, release. *Original artist:* NASA, ESA, J. Hester and A. Loll (Arizona State University)

- **File:DZero.jpg** *Source:* https://upload.wikimedia.org/wikipedia/en/0/04/DZero.jpg *License:* PD *Contributors:* ? *Original artist:* ?

- **File:DZero_Control.jpg** *Source:* https://upload.wikimedia.org/wikipedia/commons/e/ef/DZero_Control.jpg *License:* Public domain *Contributors:* ? *Original artist:* ?

- **File:Edit-clear.svg** *Source:* https://upload.wikimedia.org/wikipedia/en/f/f2/Edit-clear.svg *License:* Public domain *Contributors:* The *Tango! Desktop Project*. *Original artist:*

    The people from the Tango! project. And according to the meta-data in the file, specifically: "Andreas Nilsson, and Jakub Steiner (although minimally)."

- **File:Eg1.png** *Source:* https://upload.wikimedia.org/wikipedia/en/b/be/Eg1.png *License:* PD *Contributors:*

    self-made

    Original artist:

    Venny85 (talk)

- **File:Eg2.png** *Source:* https://upload.wikimedia.org/wikipedia/en/3/3e/Eg2.png *License:* PD *Contributors:*

    self-made

    Original artist:

    Venny85 (talk)

- **File:Eg3.png** *Source:* https://upload.wikimedia.org/wikipedia/en/1/15/Eg3.png *License:* PD *Contributors:*

    self-made

    Original artist:

    Venny85 (talk)

- **File:Eg4.png** *Source:* https://upload.wikimedia.org/wikipedia/en/6/65/Eg4.png *License:* PD *Contributors:*

    self-made

    Original artist:

    Venny85 (talk)

- **File:Electric_field_point_lines_equipotentials.svg** *Source:* https://upload.wikimedia.org/wikipedia/commons/9/96/Electric_field_point_lines_equipotentials.svg *License:* Public domain *Contributors:* Own work *Original artist:* Sjlegg

- **File:Electron.svg** *Source:* https://upload.wikimedia.org/wikipedia/commons/8/8c/Electron.svg *License:* CC BY 3.0 *Contributors:* File:Standard Model of Elementary Particles.svg *Original artist:* user:MissMJ

- **File:Electron_neutrino.svg** *Source:* https://upload.wikimedia.org/wikipedia/commons/f/f4/Electron_neutrino.svg *License:* CC BY 3.0 *Contributors:* File:Standard Model of Elementary Particles.svg *Original artist:* user:MissMJ

- **File:Elementary_particle_interactions_in_the_Standard_Model.png** *Source:* https://upload.wikimedia.org/wikipedia/commons/a/a7/Elementary_particle_interactions_in_the_Standard_Model.png *License:* CC0 *Contributors:* Own work *Original artist:* Eric Drexler

- **File:Enrico_Fermi_1943-49.jpg** *Source:* https://upload.wikimedia.org/wikipedia/commons/d/d4/Enrico_Fermi_1943-49.jpg *License:* Public domain *Contributors:* This media is available in the holdings of the National Archives and Records Administration, cataloged under the ARC Identifier (National Archives Identifier) **558578**. *Original artist:* Department of Energy. Office of Public Affairs

9.2. IMAGES

- **File:Fermilab.jpg** *Source:* https://upload.wikimedia.org/wikipedia/commons/3/3f/Fermilab.jpg *License:* Public domain *Contributors:* [1] from [2]
  Original artist: Fermilab, Reidar Hahn

- **File:Fermilab_WilsonHall.JPG** *Source:* https://upload.wikimedia.org/wikipedia/commons/4/40/Fermilab_WilsonHall.JPG *License:* CC BY 2.5 *Contributors:* Transferred from en.wikipedia; Transfer was stated to be made by User:Jacopo Werther. *Original artist:* Original uploader was Besselfunctions at en.wikipedia

- **File:Fermilab_satellite.gif** *Source:* https://upload.wikimedia.org/wikipedia/commons/b/b4/Fermilab_satellite.gif *License:* Public domain *Contributors:* U.S. Geological Survey *Original artist:* U.S. Geological Survey

- **File:Fermilablogo.PNG** *Source:* https://upload.wikimedia.org/wikipedia/en/d/df/Fermilablogo.PNG *License:* Fair use *Contributors:* http://www.fnal.gov/ *Original artist:* ?

- **File:Feymanlibrary.JPG** *Source:* https://upload.wikimedia.org/wikipedia/commons/9/92/Feymanlibrary.JPG *License:* CC BY-SA 3.0 *Contributors:* Own work (Original text: self-made) *Original artist:* <a href='//en.wikipedia.org/wiki/User:DRosenbach' class='extiw' title='en:User:DRosenbach'>DRosenbach</a> *(<a href='//en.wikipedia.org/wiki/User_talk:DRosenbach' class='extiw' title='en:User talk:DRosenbach'>Talk</a> | <a href='//en.wikipedia.org/wiki/Special:Contributions/DRosenbach' class='extiw' title='en:Special:Contributions/DRosenbach'>Contribs</a>)

- **File:FeynmanAMNH.jpg** *Source:* https://upload.wikimedia.org/wikipedia/en/f/f1/FeynmanAMNH.jpg *License:* CC-BY-SA-3.0 *Contributors:*

  I (<a href='//en.wikipedia.org/wiki/User:DRosenbach' title='User:DRosenbach'>DRosenbach</a> *(<a href='//en.wikipedia.org/wiki/User_talk:DRosenbach' title='User talk:DRosenbach'>Talk</a>

  Original artist:

  <a href='//en.wikipedia.org/wiki/User:DRosenbach' title='User:DRosenbach'>DRosenbach</a> *(<a href='//en.wikipedia.org/wiki/User_talk:DRosenbach' title='User talk:DRosenbach'>Talk</a>

- **File:Feynman_Diagram_Y-3g.PNG** *Source:* https://upload.wikimedia.org/wikipedia/commons/1/1c/Feynman_Diagram_Y-3g.PNG *License:* CC BY-SA 3.0 *Contributors:* Own work *Original artist:* DrHjmHam

- **File:Feynman_and_Oppenheimer_at_Los_Alamos.jpg** *Source:* https://upload.wikimedia.org/wikipedia/commons/8/81/Feynman_and_Oppenheimer_at_Los_Alamos.jpg *License:* Public domain *Contributors:* ? *Original artist:* ?

- **File:Feynmann_Diagram_Gluon_Radiation.svg** *Source:* https://upload.wikimedia.org/wikipedia/commons/1/1f/Feynmann_Diagram_Gluon_Radiation.svg *License:* CC BY 2.5 *Contributors:* Non-Derived SVG of Radiate_gluon.png, originally the work of SilverStar at Feynmann-diagram-gluon-radiation.svg, updated by joelholdsworth. *Original artist:* Joel Holdsworth (Joelholdsworth)

- **File:Flag_of_Greece.svg** *Source:* https://upload.wikimedia.org/wikipedia/commons/5/5c/Flag_of_Greece.svg *License:* Public domain *Contributors:* own code *Original artist:* (of code) cs:User:-xfi- (talk)

- **File:Flag_of_Israel.svg** *Source:* https://upload.wikimedia.org/wikipedia/commons/d/d4/Flag_of_Israel.svg *License:* Public domain *Contributors:* http://www.mfa.gov.il/MFA/History/Modern%20History/Israel%20at%2050/The%20Flag%20and%20the%20Emblem *Original artist:* "The Provisional Council of State Proclamation of the Flag of the State of Israel" of 25 Tishrei 5709 (28 October 1948) provides the official specification for the design of the Israeli flag.

- **File:Flag_of_the_United_States.svg** *Source:* https://upload.wikimedia.org/wikipedia/en/a/a4/Flag_of_the_United_States.svg *License:* PD *Contributors:* ? *Original artist:* ?

- **File:Flag_of_the_United_States_Department_of_Energy.svg** *Source:* https://upload.wikimedia.org/wikipedia/commons/4/42/Flag_of_the_United_States_Department_of_Energy.svg *License:* Public domain *Contributors:* This vector image includes elements that have been taken or adapted from this: <a href='//commons.wikimedia.org/wiki/File:US-DeptOfEnergy-Seal.svg' class='image'><img alt='US-DeptOfEnergy-Seal.svg' src='https://upload.wikimedia.org/wikipedia/commons/thumb/b/bf/US-DeptOfEnergy-Seal.svg/20px-US-DeptOfEnergy-Seal.svg.png' width='20' height='20' srcset='https://upload.wikimedia.org/wikipedia/commons/thumb/b/bf/US-DeptOfEnergy-Seal.svg/30px-US-DeptOfEnergy-Seal.svg.png 1.5x, https://upload.wikimedia.org/wikipedia/commons/thumb/b/bf/US-DeptOfEnergy-Seal.svg/40px-US-DeptOfEnergy-Seal.svg.png 2x' data-file-width='600' data-file-height='600' /></a> US-DeptOfEnergy-Seal.svg. *Original artist:* Fry1989

- **File:Fluxtube_meson.png** *Source:* https://upload.wikimedia.org/wikipedia/commons/9/93/Fluxtube_meson.png *License:* Public domain *Contributors:* http://inspirehep.net/record/840296 , arXiv:0912.3181 [hep-lat] and M. Cardoso et al., *Lattice QCD computation of the colour fields for the static hybrid quark-gluon-antiquark system, and microscopic study of the Casimir scaling*, Physical Review D, **81** (2010)) *Original artist:* RolteVolte (<a href='//commons.wikimedia.org/wiki/User_talk:RolteVolte' title='User talk:RolteVolte'>talk</a>) 16:42, 23 October 2011 (UTC)

- **File:Folder_Hexagonal_Icon.svg** *Source:* https://upload.wikimedia.org/wikipedia/en/4/48/Folder_Hexagonal_Icon.svg *License:* Cc-by-sa-3.0 *Contributors:* ? *Original artist:* ?

- **File:GLAST_on_the_payload_attach_fitting.jpg** *Source:* https://upload.wikimedia.org/wikipedia/commons/4/4b/GLAST_on_the_payload_attach_fitting.jpg *License:* Public domain *Contributors:* http://mediaarchive.ksc.nasa.gov/detail.cfm?mediaid=36076 *Original artist:* Photo credit: NASA/Kim Shiflett

- **File:Gluon_coupling.svg** *Source:* https://upload.wikimedia.org/wikipedia/commons/1/13/Gluon_coupling.svg *License:* CC BY-SA 3.0 *Contributors:* Own work *Original artist:* Aurélian John-Herpin

- **File:Gluon_tube-color_confinement_animation.gif** *Source:* https://upload.wikimedia.org/wikipedia/commons/6/64/Gluon_tube-color_confinement_animation.gif *License:* CC BY-SA 3.0 *Contributors:* Own work *Original artist:* Manishearth

- **File:Gnome-preferences-desktop-accessibility2.svg** *Source:* https://upload.wikimedia.org/wikipedia/commons/4/4c/Gnome-preferences-desktop-accessibility2.svg *License:* CC BY-SA 3.0 *Contributors:* HTTP / FTP *Original artist:* GNOME icon artists

- **File:Hadron_colors.svg** *Source:* https://upload.wikimedia.org/wikipedia/commons/e/e4/Hadron_colors.svg *License:* CC BY-SA 3.0 *Contributors:*

- Hadron_colors.png *Original artist:* Hadron_colors.png: Army1987

- **File:Haim_harari.jpg** *Source:* https://upload.wikimedia.org/wikipedia/commons/6/62/Haim_harari.jpg *License:* CC BY-SA 3.0 *Contributors:* אתר פרס א.מ.ה *Original artist:* פרס א.מ.ה

- **File:He1523a.jpg** *Source:* https://upload.wikimedia.org/wikipedia/commons/5/5f/He1523a.jpg *License:* CC BY 4.0 *Contributors:* http://www.solstation.com/x-objects/he1523.htm *Original artist:* ESO, European Southern Observatory

- **File:Heinkel_He_111_during_the_Battle_of_Britain.jpg** *Source:* https://upload.wikimedia.org/wikipedia/commons/8/82/Heinkel_He_111_during_the_Battle_of_Britain.jpg *License:* Public domain *Contributors:* This is photograph MH6547 from the collections of the Imperial War Museums (collection no. 4700-05) *Original artist:* Unknown

- **File:Helium_atom_QM.svg** *Source:* https://upload.wikimedia.org/wikipedia/commons/2/23/Helium_atom_QM.svg *License:* CC-BY-SA-3.0 *Contributors:* Own work *Original artist:* User:Yzmo

- **File:Hydrogen300.png** *Source:* https://upload.wikimedia.org/wikipedia/commons/a/ad/Hydrogen300.png *License:* Public domain *Contributors:* Transferred from en.wikipedia; transferred to Commons by User:OverlordQ using CommonsHelper. *Original artist:* PoorLeno (talk) Original uploader was PoorLeno at en.wikipedia

- **File:Ilc_9yr_moll4096.png** *Source:* https://upload.wikimedia.org/wikipedia/commons/3/3c/Ilc_9yr_moll4096.png *License:* Public domain *Contributors:* http://map.gsfc.nasa.gov/media/121238/ilc_9yr_moll4096.png *Original artist:* NASA / WMAP Science Team

- **File:Interior_of_Fermi_Lab_Wilson_Hall.JPG** *Source:* https://upload.wikimedia.org/wikipedia/commons/d/d2/Interior_of_Fermi_Lab_Wilson_Hall.JPG *License:* CC BY-SA 3.0 *Contributors:* Own work *Original artist:* H Padleckas

- **File:J-psi_p_pentaquark_mass_spectrum.svg** *Source:* https://upload.wikimedia.org/wikipedia/commons/4/44/J-psi_p_pentaquark_mass_spectrum.svg *License:* CC BY 4.0 *Contributors:* Figure 3b in <a data-x-rel='nofollow' class='external text' href='http://arxiv.org/pdf/1507.03414v1.pdf'>"Observation of J/ψp resonances consistent with pentaquark states in $\Lambda^0_b \rightarrow J/\psi K^*$-p decays"</a> (arXiv:1507.03414, LHCb collaboration *Original artist:* CERN on behalf of the LHCb collaboration,

- **File:Jpsi-fit-mass.gif** *Source:* https://upload.wikimedia.org/wikipedia/commons/4/4d/Jpsi-fit-mass.gif *License:* Public domain *Contributors:* Fermilab *Original artist:* CDF Collaboration

- **File:Kaon-box-diagram-alt.svg** *Source:* https://upload.wikimedia.org/wikipedia/commons/7/7f/Kaon-box-diagram-alt.svg *License:* CC-BY-SA-3.0 *Contributors:* ? *Original artist:* ?

- **File:Kaon-box-diagram.svg** *Source:* https://upload.wikimedia.org/wikipedia/commons/8/8e/Kaon-box-diagram.svg *License:* CC-BY-SA-3.0 *Contributors:* ? *Original artist:* ?

- **File:Kaon-decay.png** *Source:* https://upload.wikimedia.org/wikipedia/commons/7/75/Kaon-decay.png *License:* CC-BY-SA-3.0 *Contributors:* Own work *Original artist:* User JabberWok on en.wikipedia

- **File:Kazuhiko_Nishijima.jpg** *Source:* https://upload.wikimedia.org/wikipedia/en/c/ca/Kazuhiko_Nishijima.jpg *License:* Fair use *Contributors:*
[1] *Original artist:* ?

- **File:Kepler-solar-system-2.gif** *Source:* https://upload.wikimedia.org/wikipedia/commons/1/1d/Kepler-solar-system-2.gif *License:* Public domain *Contributors:* ? *Original artist:* ?

- **File:Kkbar_had.svg** *Source:* https://upload.wikimedia.org/wikipedia/commons/f/ff/Kkbar_had.svg *License:* CC-BY-SA-3.0 *Contributors:* Own work based on w:File:Kkbar had.png *Original artist:* Bamse

- **File:LHC.svg** *Source:* https://upload.wikimedia.org/wikipedia/commons/7/74/LHC.svg *License:* CC BY-SA 2.5 *Contributors:* ? *Original artist:* ?

- **File:LHC_quadrupole_magnets.jpg** *Source:* https://upload.wikimedia.org/wikipedia/commons/0/06/LHC_quadrupole_magnets.jpg *License:* CC BY 2.0 *Contributors:* Flickr *Original artist:* gamsiz

- **File:Large_Hadron_Collider_dipole_magnets_IMG_0955.jpg** *Source:* https://upload.wikimedia.org/wikipedia/commons/e/ea/Large_Hadron_Collider_dipole_magnets_IMG_0955.jpg *License:* CC BY-SA 2.0 *Contributors:* Flickr *Original artist:* alpinethread

- **File:Lepton-interaction-vertex-eeg.svg** *Source:* https://upload.wikimedia.org/wikipedia/commons/b/b5/Lepton-interaction-vertex-eeg.svg *License:* CC BY 3.0 *Contributors:* Own work *Original artist:* TimothyRias

- **File:Lepton_isodoublets_fixed.png** *Source:* https://upload.wikimedia.org/wikipedia/en/9/93/Lepton_isodoublets_fixed.png *License:* CC-BY-SA-3.0 *Contributors:*
I (HEL (talk)) created this work entirely by myself. *Original artist:*
HEL (talk)

- **File:Location_Large_Hadron_Collider.PNG** *Source:* https://upload.wikimedia.org/wikipedia/commons/0/06/Location_Large_Hadron_Collider.PNG *License:* CC BY-SA 2.0 *Contributors:* en:OpenStreetMap: *Original artist:* diverse contributors; mashup by User:Zykure

- **File:Luciano_Maiani_1996.jpg** *Source:* https://upload.wikimedia.org/wikipedia/commons/2/23/Luciano_Maiani_1996.jpg *License:* CC BY-SA 3.0 *Contributors:* CERN Document Server *Original artist:* Laurent Guiraud

9.2. IMAGES

- **File:Map_of_Illinois_highlighting_DuPage_County.svg** *Source:* https://upload.wikimedia.org/wikipedia/commons/b/be/Map_of_Illinois_highlighting_DuPage_County.svg *License:* Public domain *Contributors:* The maps use data from nationalatlas.gov, specifically countyp020.tar.gz on the Raw Data Download page. The maps also use state outline data from statesp020.tar.gz. The Florida maps use hydrogm020.tar.gz to display Lake Okeechobee. *Original artist:* David Benbennick
- **File:Map_of_Illinois_highlighting_Kane_County.svg** *Source:* https://upload.wikimedia.org/wikipedia/commons/d/dc/Map_of_Illinois_highlighting_Kane_County.svg *License:* Public domain *Contributors:* The maps use data from nationalatlas.gov, specifically countyp020.tar.gz on the Raw Data Download page. The maps also use state outline data from statesp020.tar.gz. The Florida maps use hydrogm020.tar.gz to display Lake Okeechobee. *Original artist:* David Benbennick
- **File:Mergefrom.svg** *Source:* https://upload.wikimedia.org/wikipedia/commons/0/0f/Mergefrom.svg *License:* Public domain *Contributors:* ? *Original artist:* ?
- **File:Meson-Baryon-molecule-generic.svg** *Source:* https://upload.wikimedia.org/wikipedia/commons/0/0d/Meson-Baryon-molecule-generic.svg *License:* CC BY-SA 4.0 *Contributors:* Own work *Original artist:* Smurrayinchester
- **File:Meson.svg** *Source:* https://upload.wikimedia.org/wikipedia/commons/f/f9/Meson.svg *License:* Public domain *Contributors:* en:Image:Meson.gif *Original artist:* en:User:Wogsland, traced by User:Stannered
- **File:Meson_nonet_-_spin_0.svg** *Source:* https://upload.wikimedia.org/wikipedia/commons/c/c0/Meson_nonet_-_spin_0.svg *License:* Public domain *Contributors:* Image:Noneto mesônico de spin 0.png *Original artist:* User:E2m, User:Stannered
- **File:Meson_nonet_-_spin_1.svg** *Source:* https://upload.wikimedia.org/wikipedia/commons/1/13/Meson_nonet_-_spin_1.svg *License:* Public domain *Contributors:* Image:Noneto mesônico de spin 1.png *Original artist:* User:E2m, User:Stannered
- **File:Meson_octet.png** *Source:* https://upload.wikimedia.org/wikipedia/commons/3/3a/Meson_octet.png *License:* CC-BY-SA-3.0 *Contributors:* Own work *Original artist:* Laurascudder
- **File:Muon-Electron-Decay.svg** *Source:* https://upload.wikimedia.org/wikipedia/en/6/6f/Muon-Electron-Decay.svg *License:* Cc-by-sa-3.0 *Contributors:* ? *Original artist:* ?
- **File:Muon.svg** *Source:* https://upload.wikimedia.org/wikipedia/commons/d/d2/Muon.svg *License:* CC BY 3.0 *Contributors:* File:Standard Model of Elementary Particles.svg *Original artist:* user:MissMJ
- **File:Muon_g-2_building_at_Fermilab.jpg** *Source:* https://upload.wikimedia.org/wikipedia/commons/3/3f/Muon_g-2_building_at_Fermilab.jpg *License:* CC BY-SA 4.0 *Contributors:* Own work *Original artist:* Z22
- **File:Muon_neutrino.svg** *Source:* https://upload.wikimedia.org/wikipedia/commons/d/d0/Muon_neutrino.svg *License:* CC BY 3.0 *Contributors:* File:Standard Model of Elementary Particles.svg *Original artist:* user:MissMJ
- **File:Murray_Gell-Mann.jpg** *Source:* https://upload.wikimedia.org/wikipedia/commons/8/87/Murray_Gell-Mann.jpg *License:* CC BY 2.0 *Contributors:* http://flickr.com/photos/jurvetson/414368314/ *Original artist:* jurvetson of flickr.com
- **File:Murray_Gell-Mann_-_World_Economic_Forum_Annual_Meeting_2012.jpg** *Source:* https://upload.wikimedia.org/wikipedia/commons/2/2c/Murray_Gell-Mann_-_World_Economic_Forum_Annual_Meeting_2012.jpg *License:* CC BY-SA 2.0 *Contributors:* Flickr: Murray Gell-Mann - World Economic Forum Annual Meeting 2012 *Original artist:* World Economic Forum
- **File:Nagasakibomb.jpg** *Source:* https://upload.wikimedia.org/wikipedia/commons/e/e0/Nagasakibomb.jpg *License:* Public domain *Contributors:* http://www.archives.gov/research/military/ww2/photos/images/ww2-163.jpg National Archives image (208-N-43888) *Original artist:* Charles Levy from one of the B-29 Superfortresses used in the attack.
- **File:National_Synchrotron_Light_Source_II.jpg** *Source:* https://upload.wikimedia.org/wikipedia/commons/b/b3/National_Synchrotron_Light_Source_II.jpg *License:* CC BY-SA 3.0 *Contributors:* Photo taken during tour of Brookhaven National Laboratory. *Original artist:* Mcginley1
- **File:Ndslivechart.png** *Source:* https://upload.wikimedia.org/wikipedia/commons/b/b0/Ndslivechart.png *License:* Public domain *Contributors:* Own work *Original artist:* Minivip
- **File:Neutron_spin_dipole_field.jpg** *Source:* https://upload.wikimedia.org/wikipedia/commons/1/15/Neutron_spin_dipole_field.jpg *License:* CC BY-SA 4.0 *Contributors:* Own work *Original artist:* Bdushaw
- **File:Nobel_Prize.png** *Source:* https://upload.wikimedia.org/wikipedia/en/e/ed/Nobel_Prize.png *License:* ? *Contributors:*
Derivative of File:NobelPrize.JPG *Original artist:*
Photograph: JonathunderMedal: Erik Lindberg (1873-1966)
- **File:Noneto_mesônico_de_spin_0.png** *Source:* https://upload.wikimedia.org/wikipedia/commons/c/cd/Noneto_mes%C3%B4nico_de_spin_0.png *License:* Public domain *Contributors:* ? *Original artist:* ?
- **File:Noneto_mesônico_de_spin_1.png** *Source:* https://upload.wikimedia.org/wikipedia/commons/0/0a/Noneto_mes%C3%B4nico_de_spin_1.png *License:* Public domain *Contributors:* ? *Original artist:* ?
- **File:NuclearReaction.png** *Source:* https://upload.wikimedia.org/wikipedia/commons/7/7d/NuclearReaction.png *License:* CC BY-SA 3.0 *Contributors:* Own work *Original artist:* Michalsmid
- **File:Nuclear_Force_anim_smaller.gif** *Source:* https://upload.wikimedia.org/wikipedia/commons/3/35/Nuclear_Force_anim_smaller.gif *License:* CC BY-SA 3.0 *Contributors:* Own work *Original artist:* Manishearth
- **File:Nuvola_apps_edu_mathematics_blue-p.svg** *Source:* https://upload.wikimedia.org/wikipedia/commons/3/3e/Nuvola_apps_edu_mathematics_blue-p.svg *License:* GPL *Contributors:* Derivative work from Image:Nuvola apps edu mathematics.png and Image:Nuvola apps edu mathematics-p.svg *Original artist:* David Vignoni (original icon); Flamurai (SVG convertion); bayo (color)
- **File:Nuvola_apps_katomic.png** *Source:* https://upload.wikimedia.org/wikipedia/commons/7/73/Nuvola_apps_katomic.png *License:* LGPL *Contributors:* http://icon-king.com *Original artist:* David Vignoni / ICON KING

- File:Office-book.svg *Source:* https://upload.wikimedia.org/wikipedia/commons/a/a8/Office-book.svg *License:* Public domain *Contributors:* This and myself. *Original artist:* Chris Down/Tango project
- File:OiintLaTeX.svg *Source:* https://upload.wikimedia.org/wikipedia/commons/8/86/OiintLaTeX.svg *License:* CC0 *Contributors:* Own work *Original artist:* Maschen
- File:PQ_EB_ape_hyp_geom5_B_3D.jpg *Source:* https://upload.wikimedia.org/wikipedia/commons/8/85/PQ_EB_ape_hyp_geom5_B_3D.jpg *License:* CC BY-SA 4.0 *Contributors:* Own work *Original artist:* Pedro.bicudo
- File:PY_101_Energy.jpg *Source:* https://upload.wikimedia.org/wikipedia/commons/4/47/PY_101_Energy.jpg *License:* CC BY 3.0 *Contributors:* Own work *Original artist:* User:Isuckducks
- File:P_vip.svg *Source:* https://upload.wikimedia.org/wikipedia/en/6/69/P_vip.svg *License:* PD *Contributors:* ? *Original artist:* ?
- File:Particles_and_antiparticles.svg *Source:* https://upload.wikimedia.org/wikipedia/commons/c/cd/Particles_and_antiparticles.svg *License:* CC BY-SA 3.0 *Contributors:* This vector image was created with Inkscape. *Original artist:* Anynobody
- File:Parton_scattering.PNG *Source:* https://upload.wikimedia.org/wikipedia/commons/0/0d/Parton_scattering.PNG *License:* Public domain *Contributors:* Transferred from en.wikipedia
Original artist: Original uploader was AnonyScientist at en.wikipedia
- File:Pentaquark-Feynman.svg *Source:* https://upload.wikimedia.org/wikipedia/commons/3/31/Pentaquark-Feynman.svg *License:* CC BY 4.0 *Contributors:* Figure 1b in <a data-x-rel='nofollow' class='external text' href='http://arxiv.org/pdf/1507.03414v1.pdf'>"Observation of J/ψp resonances consistent with pentaquark states in $\Lambda^0_b \to J/\psi K^*$-p decays"</a> (arXiv:1507.03414, LHCb collaboration *Original artist:* CERN on behalf of the LHCb collaboration,
- File:Pentaquark-generic.svg *Source:* https://upload.wikimedia.org/wikipedia/commons/5/5b/Pentaquark-generic.svg *License:* CC BY-SA 4.0 *Contributors:* Own work *Original artist:* Headbomb
- File:Pentaquark.svg *Source:* https://upload.wikimedia.org/wikipedia/commons/c/cd/Pentaquark.svg *License:* CC0 *Contributors:* Own work *Original artist:* Smurrayinchester
- File:Phase_change_-_en.svg *Source:* https://upload.wikimedia.org/wikipedia/commons/0/0b/Phase_change_-_en.svg *License:* Public domain *Contributors:* Own work *Original artist:* F l a n k e r, penubag
- File:Photo_of_the_Week-_An_Incredible_Journey_--_Transporting_a_600-ton_Magnet_(9324124048).jpg *Source:* https://upload.wikimedia.org/wikipedia/commons/c/cc/Photo_of_the_Week-_An_Incredible_Journey_--_Transporting_a_600-ton_Magnet_%289324124048%29.jpg *License:* Public domain *Contributors:* Photo of the Week: An Incredible Journey -- Transporting a 600-ton Magnet *Original artist:* ENERGY.GOV
- File:PiPlus_muon_decay.svg *Source:* https://upload.wikimedia.org/wikipedia/commons/6/69/PiPlus_muon_decay.svg *License:* CC0 *Contributors:* Own work *Original artist:* Krishnavedala
- File:Portal-puzzle.svg *Source:* https://upload.wikimedia.org/wikipedia/commons/en/f/fd/Portal-puzzle.svg *License:* Public domain *Contributors:* ? *Original artist:* ?
- File:PositronDiscovery.jpg *Source:* https://upload.wikimedia.org/wikipedia/commons/6/69/PositronDiscovery.jpg *License:* Public domain *Contributors:* Anderson, Carl D. (1933). "The Positive Electron". *Physical Review* **43** (6): 491–494. DOI:10.1103/PhysRev.43.491. *Original artist:* Carl D. Anderson (1905–1991)
- File:Public.jpg *Source:* https://upload.wikimedia.org/wikipedia/en/e/e2/Public.jpg *License:* PD *Contributors:* ? *Original artist:* ?
- File:QCD.svg *Source:* https://upload.wikimedia.org/wikipedia/commons/2/2b/QCD.svg *License:* CC BY-SA 3.0 *Contributors:* Own work *Original artist:* Cjean42
- File:QCD_phase_diagram.png *Source:* https://upload.wikimedia.org/wikipedia/commons/8/8f/QCD_phase_diagram.png *License:* Public domain *Contributors:* Transferred from en.wikipedia; transferred to Commons by User:Quadell using CommonsHelper. *Original artist:* Original uploader was Dark Formal at en.wikipedia
- File:QCDphasediagram.svg *Source:* https://upload.wikimedia.org/wikipedia/commons/b/bc/QCDphasediagram.svg *License:* CC BY-SA 3.0 *Contributors:* Own work *Original artist:* TimothyRias
- File:Qcd_fields_field_(physics).svg *Source:* https://upload.wikimedia.org/wikipedia/commons/4/41/Qcd_fields_field_%28physics%29.svg *License:* CC0 *Contributors:* Own work *Original artist:* Maschen
- File:Quark_confinement.svg *Source:* https://upload.wikimedia.org/wikipedia/commons/6/6a/Quark_confinement.svg *License:* CC BY-SA 2.5 *Contributors:*
- Created by Lokal_Profil inspired by: *Original artist:* Lokal_Profil
- File:Quark_masses_as_balls.svg *Source:* https://upload.wikimedia.org/wikipedia/commons/b/b5/Quark_masses_as_balls.svg *License:* CC BY-SA 3.0 *Contributors:* Own work *Original artist:* Incnis Mrsi
- File:Quark_structure_proton.svg *Source:* https://upload.wikimedia.org/wikipedia/commons/9/92/Quark_structure_proton.svg *License:* CC BY-SA 2.5 *Contributors:* Own work *Original artist:* Arpad Horvath
- File:Quark_weak_interactions.svg *Source:* https://upload.wikimedia.org/wikipedia/commons/6/66/Quark_weak_interactions.svg *License:* Public domain *Contributors:* Derivative work, from public down work uploaded to en.wikipedia. original *Original artist:*
- Original work: [1]
- File:Queryensdf.jpg *Source:* https://upload.wikimedia.org/wikipedia/commons/5/5e/Queryensdf.jpg *License:* Public domain *Contributors:* Own work *Original artist:* Minivip

9.2. IMAGES

- **File:Question_book-new.svg** *Source:* https://upload.wikimedia.org/wikipedia/en/9/99/Question_book-new.svg *License:* Cc-by-sa-3.0 *Contributors:*
Created from scratch in Adobe Illustrator. Based on Image:Question book.png created by User:Equazcion *Original artist:* Tkgd2007

- **File:RaE1.jpg** *Source:* https://upload.wikimedia.org/wikipedia/commons/2/2b/RaE1.jpg *License:* CC BY-SA 4.0 *Contributors:* Own work *Original artist:* HPaul

- **File:Radioactive.svg** *Source:* https://upload.wikimedia.org/wikipedia/commons/b/b5/Radioactive.svg *License:* Public domain *Contributors:* Created by Cary Bass using Adobe Illustrator on January 19, 2006. *Original artist:* Cary Bass

- **File:RichardFeynman-PaineMansionWoods1984_copyrightTamikoThiel_bw.jpg** *Source:* https://upload.wikimedia.org/wikipedia/commons/1/1a/RichardFeynman-PaineMansionWoods1984_copyrightTamikoThiel_bw.jpg *License:* CC BY-SA 3.0 *Contributors:* OTRS communication from photographer *Original artist:* Copyright Tamiko Thiel 1984

- **File:Richard_Feynman_signature.svg** *Source:* https://upload.wikimedia.org/wikipedia/commons/5/5e/Richard_Feynman_signature.svg *License:* Public domain *Contributors:* Heritage Auction Gallery *Original artist:* Richard P. Feynman
Created in vector format by Scewing

- **File:Right_left_helicity.svg** *Source:* https://upload.wikimedia.org/wikipedia/commons/a/a9/Right_left_helicity.svg *License:* Public domain *Contributors:* en:Image:Right left helicity.jpg *Original artist:* en:User;HEL, User:Stannered

- **File:SLAC_Entrance.jpg** *Source:* https://upload.wikimedia.org/wikipedia/commons/7/7c/SLAC_Entrance.jpg *License:* CC BY 3.0 *Contributors:* Own work *Original artist:* Jvimal

- **File:SLAC_LogoSD.png** *Source:* https://upload.wikimedia.org/wikipedia/en/a/aa/SLAC_LogoSD.png *License:* Fair use *Contributors:* http://www-group.slac.stanford.edu/com/images/slac_logos_2012branding/SLAC_LogoSD.png *Original artist:* ?

- **File:SLAC_detector.jpg** *Source:* https://upload.wikimedia.org/wikipedia/commons/0/0a/SLAC_detector.jpg *License:* CC-BY-SA-3.0 *Contributors:* Own work *Original artist:* Justin Lebar

- **File:SLAC_long_view.jpg** *Source:* https://upload.wikimedia.org/wikipedia/commons/8/87/SLAC_long_view.jpg *License:* CC-BY-SA-3.0 *Contributors:* Own work *Original artist:* Justin Lebar

- **File:SLAC_pit_and_detector.jpg** *Source:* https://upload.wikimedia.org/wikipedia/commons/a/aa/SLAC_pit_and_detector.jpg *License:* CC-BY-SA-3.0 *Contributors:* Own work *Original artist:* Justin Lebar

- **File:SLAC_tunnel_2.jpg** *Source:* https://upload.wikimedia.org/wikipedia/commons/0/0d/SLAC_tunnel_2.jpg *License:* Public domain *Contributors:* http://energy.gov/articles/labchat-particle-accelerators-lasers-and-discovery-science-may-17-1pm-est *Original artist:* Brad Plummer

- **File:Satoshi_Ozaki_1991.jpg** *Source:* https://upload.wikimedia.org/wikipedia/commons/9/9b/Satoshi_Ozaki_1991.jpg *License:* Public domain *Contributors:* Satoshi Ozaki Named Senior Scientist Emeritus *Original artist:* Brookhaven National Laboratory

- **File:Science-symbol-2.svg** *Source:* https://upload.wikimedia.org/wikipedia/commons/7/75/Science-symbol-2.svg *License:* CC BY 3.0 *Contributors:* en:Image:Science-symbol2.png *Original artist:* en:User:AllyUnion, User:Stannered

- **File:Science.jpg** *Source:* https://upload.wikimedia.org/wikipedia/commons/5/54/Science.jpg *License:* Public domain *Contributors:* ? *Original artist:* ?

- **File:Sheldon_Glashow_at_Harvard_cropped.jpg** *Source:* https://upload.wikimedia.org/wikipedia/commons/c/c3/Sheldon_Glashow_at_Harvard_cropped.jpg *License:* Public domain *Contributors:* own work (Lumidek) *Original artist:* Luboš Motl - Lumidek

- **File:Speakerlink-new.svg** *Source:* https://upload.wikimedia.org/wikipedia/commons/3/3b/Speakerlink-new.svg *License:* CC0 *Contributors:* Own work *Original artist:* Kelvinsong

- **File:Spin_One-Half_(Slow).gif** *Source:* https://upload.wikimedia.org/wikipedia/commons/6/6e/Spin_One-Half_%28Slow%29.gif *License:* CC0 *Contributors:* Own work *Original artist:* JasonHise

- **File:Standard_Model_Feynman_Diagram_Vertices.png** *Source:* https://upload.wikimedia.org/wikipedia/commons/7/75/Standard_Model_Feynman_Diagram_Vertices.png *License:* CC BY-SA 3.0 *Contributors:* I made it in Adobe Illustrator *Original artist:* Garyzx

- **File:Standard_Model_of_Elementary_Particles.svg** *Source:* https://upload.wikimedia.org/wikipedia/commons/0/00/Standard_Model_of_Elementary_Particles.svg *License:* CC BY 3.0 *Contributors:* Own work by uploader, PBS NOVA [1], Fermilab, Office of Science, United States Department of Energy, Particle Data Group *Original artist:* MissMJ

- **File:Stanford-linear-accelerator-usgs-ortho-kaminski-5900.jpg** *Source:* https://upload.wikimedia.org/wikipedia/commons/8/8a/Stanford-linear-accelerator-usgs-ortho-kaminski-5900.jpg *License:* Public domain *Contributors:* United States Geological Survey *Original artist:* Peter Kaminski

- **File:Strong_force_charges.svg** *Source:* https://upload.wikimedia.org/wikipedia/commons/b/b6/Strong_force_charges.svg *License:* CC BY-SA 3.0 *Contributors:* Own work, Created from Garret Lisi's Elementary Particle Explorer *Original artist:* Cjean42

- **File:Stylised_Lithium_Atom.svg** *Source:* https://upload.wikimedia.org/wikipedia/commons/e/e1/Stylised_Lithium_Atom.svg *License:* CC-BY-SA-3.0 *Contributors:* based off of Image:Stylised Lithium Atom.png by Halfdan. *Original artist:* SVG by Indolences. Recoloring and ironing out some glitches done by Rainer Klute.

- **File:Symbol_book_class2.svg** *Source:* https://upload.wikimedia.org/wikipedia/commons/8/89/Symbol_book_class2.svg *License:* CC BY-SA 2.5 *Contributors:* Mad by Lokal_Profil by combining: *Original artist:* Lokal_Profil

- **File:TQ_EB_ape_hyp_r1_8_r2_14_Act_3D_Sim.jpg** *Source:* https://upload.wikimedia.org/wikipedia/commons/9/93/TQ_EB_ape_hyp_r1_8_r2_14_Act_3D_Sim.jpg *License:* CC BY-SA 4.0 *Contributors:* Own work *Original artist:* Pedro.bicudo

- **File:Table_isotopes_en.svg** *Source:* https://upload.wikimedia.org/wikipedia/commons/c/c4/Table_isotopes_en.svg *License:* CC BY-SA 3.0 *Contributors:*
- Table_isotopes.svg *Original artist:* Table_isotopes.svg: Napy1kenobi
- **File:Tau_lepton.svg** *Source:* https://upload.wikimedia.org/wikipedia/commons/f/f8/Tau_lepton.svg *License:* CC BY 3.0 *Contributors:* File:Standard Model of Elementary Particles.svg *Original artist:* user:MissMJ
- **File:Tau_neutrino.svg** *Source:* https://upload.wikimedia.org/wikipedia/commons/a/ac/Tau_neutrino.svg *License:* CC BY 3.0 *Contributors:* File:Standard Model of Elementary Particles.svg *Original artist:* user:MissMJ
- **File:Text_document_with_page_number_icon.svg** *Source:* https://upload.wikimedia.org/wikipedia/commons/3/3b/Text_document_with_page_number_icon.svg *License:* Public domain *Contributors:* Created by bdesham with Inkscape; based upon Text-x-generic.svg from the Tango project. *Original artist:* Benjamin D. Esham (bdesham)
- **File:Text_document_with_red_question_mark.svg** *Source:* https://upload.wikimedia.org/wikipedia/commons/a/a4/Text_document_with_red_question_mark.svg *License:* Public domain *Contributors:* Created by bdesham with Inkscape; based upon Text-x-generic.svg from the Tango project. *Original artist:* Benjamin D. Esham (bdesham)
- **File:The_2-in-1_structure_of_the_LHC_dipole_magnets.jpg** *Source:* https://upload.wikimedia.org/wikipedia/commons/b/b7/The_2-in-1_structure_of_the_LHC_dipole_magnets.jpg *License:* CC BY-SA 3.0 *Contributors:* http://www.worldscientific.com/worldscibooks/10.1142/8605 *Original artist:* E. M. Henley and S. D. Ellis
- **File:Top_antitop_quark_event.svg** *Source:* https://upload.wikimedia.org/wikipedia/commons/3/35/Top_antitop_quark_event.svg *License:* Public domain *Contributors:* Own work *Original artist:* Raeky
- **File:Two_AES_5-cells_cavities.jpg** *Source:* https://upload.wikimedia.org/wikipedia/commons/a/aa/Two_AES_5-cells_cavities.jpg *License:* CC BY 3.0 *Contributors:* http://accelconf.web.cern.ch/AccelConf/LINAC2014/papers/mopp052.pdf *Original artist:* M.H. Awida, M. Foley, I. Gonin, A. Grassellino, C. Grimm,T. Khabiboulline, A. Lunin, A. Rowe, V. Yakovlev, FNAL, Batavia, 60510, USA
- **File:US-DeptOfEnergy-Seal.svg** *Source:* https://upload.wikimedia.org/wikipedia/commons/b/bf/US-DeptOfEnergy-Seal.svg *License:* Public domain *Contributors:* Transferred from en.wikipedia to Commons. *Original artist:* The original uploader was K. Aainsqatsi at English Wikipedia
- **File:VFPt_charges_plus_minus_thumb.svg** *Source:* https://upload.wikimedia.org/wikipedia/commons/e/ed/VFPt_charges_plus_minus_thumb.svg *License:* CC BY-SA 3.0 *Contributors:* This plot was created with VectorFieldPlot *Original artist:* Geek3
- **File:VFPt_minus_thumb.svg** *Source:* https://upload.wikimedia.org/wikipedia/commons/d/d7/VFPt_minus_thumb.svg *License:* CC BY-SA 3.0 *Contributors:* This plot was created with VectorFieldPlot *Original artist:* Geek3
- **File:VFPt_plus_thumb.svg** *Source:* https://upload.wikimedia.org/wikipedia/commons/9/95/VFPt_plus_thumb.svg *License:* CC BY-SA 3.0 *Contributors:* This plot was created with VectorFieldPlot *Original artist:* Geek3
- **File:Vanguard_of_new_york_conscripts_arrive_at_Camp_Upton,_Long_Island._One_thousand-nine-hundred-forty_._._._-_NARA_-_533526.tif** *Source:* https://upload.wikimedia.org/wikipedia/commons/3/3e/Vanguard_of_new_york_conscripts_arrive_at_Camp_Upton%2C_Long_Island._One_thousand-nine-hundred-forty_._._._-_NARA_-_533526.tif *License:* Public domain *Contributors:* U.S. National Archives and Records Administration *Original artist:* Unknown or not provided
- **File:Vertex.png** *Source:* https://upload.wikimedia.org/wikipedia/en/d/db/Vertex.png *License:* Cc-by-sa-3.0 *Contributors:* ? *Original artist:* ?
- **File:Vertex_Detector.svg** *Source:* https://upload.wikimedia.org/wikipedia/commons/8/85/Vertex_Detector.svg *License:* Public domain *Contributors:* Own work *Original artist:* Inductiveload
- **File:View_inside_detector_at_the_CMS_cavern_LHC_CERN.jpg** *Source:* https://upload.wikimedia.org/wikipedia/commons/c/c7/View_inside_detector_at_the_CMS_cavern_LHC_CERN.jpg *License:* CC BY-SA 3.0 *Contributors:* Own work *Original artist:* Tighef
- **File:Weak_Decay_(flipped).svg** *Source:* https://upload.wikimedia.org/wikipedia/commons/4/4b/Weak_Decay_%28flipped%29.svg *License:* CC BY-SA 3.0 *Contributors:* Created in Inkscape based on :File:Weak decay diagram.svg *Original artist:* Niamh O'C
- **File:Wiki_letter_w_cropped.svg** *Source:* https://upload.wikimedia.org/wikipedia/commons/1/1c/Wiki_letter_w_cropped.svg *License:* CC-BY-SA-3.0 *Contributors:*
- Wiki_letter_w.svg *Original artist:* Wiki_letter_w.svg: Jarkko Piiroinen
- **File:Wikinews-logo.svg** *Source:* https://upload.wikimedia.org/wikipedia/commons/2/24/Wikinews-logo.svg *License:* CC BY-SA 3.0 *Contributors:* This is a cropped version of Image:Wikinews-logo-en.png. *Original artist:* Vectorized by Simon 01:05, 2 August 2006 (UTC) Updated by Time3000 17 April 2007 to use official Wikinews colours and appear correctly on dark backgrounds. Originally uploaded by Simon.
- **File:Wikiquote-logo.svg** *Source:* https://upload.wikimedia.org/wikipedia/commons/f/fa/Wikiquote-logo.svg *License:* Public domain *Contributors:* ? *Original artist:* ?
- **File:Wiktionary-logo-en.svg** *Source:* https://upload.wikimedia.org/wikipedia/commons/f/f8/Wiktionary-logo-en.svg *License:* Public domain *Contributors:* Vector version of Image:Wiktionary-logo-en.png. *Original artist:* Vectorized by Fvasconcellos (talk · contribs), based on original logo tossed together by Brion Vibber
- **File:Wilson_hall_fall_b.jpg** *Source:* https://upload.wikimedia.org/wikipedia/en/1/12/Wilson_hall_fall_b.jpg *License:* PD *Contributors:* ? *Original artist:* ?
- **File:Wis.jpg** *Source:* https://upload.wikimedia.org/wikipedia/commons/5/5e/Wis.jpg *License:* CC BY-SA 3.0 *Contributors:* Template: Weizmann Institute of Science *Original artist:* Weizmann Institute of Science
- **File:Wolfgang_Pauli_young.jpg** *Source:* https://upload.wikimedia.org/wikipedia/commons/4/43/Wolfgang_Pauli_young.jpg *License:* Public domain *Contributors:* ? *Original artist:* ?
- **File:École_de_Physique_des_Houches_(Les_Houches_Physics_School)_main_lecture_hall_1972.jpg** *Source:* https://upload.wikimedia.org/wikipedia/commons/a/ae/%C3%89cole_de_Physique_des_Houches_%28Les_Houches_Physics_School%29_main_lecture_hall_1972.jpg *License:* CC BY-SA 3.0 *Contributors:* Own work *Original artist:* A. T. Service

9.3 Content license

- Creative Commons Attribution-Share Alike 3.0

www.ingramcontent.com/pod-product-compliance
Lightning Source LLC
Chambersburg PA
CBHW081141180526
45170CB00006B/1881